MW00835403

ELECTRONIC PROPERTIES OF DIRAC AND WEYL SEMIMETALS

ELECTRONIC PROPERTIES OF DIRAC AND WEYL SEMIMETALS

EDUARD V GORBAR

*Taras Shevchenko National Kiev University, Ukraine &
Bogolyubov Institute for Theoretical Physics, Ukraine*

VLADIMIR A MIRANSKY

Western University, Canada

IGOR A SHOVKOVY

Arizona State University, USA

PAVLO O SUKHACHOV

*Western University, Canada &
Nordic Institute for Theoretical Physics, Sweden*

NEW JERSEY · LONDON · SINGAPORE · BEIJING · SHANGHAI · HONG KONG · TAIPEI · CHENNAI · TOKYO

Published by

World Scientific Publishing Co. Pte. Ltd.

5 Toh Tuck Link, Singapore 596224

USA office: 27 Warren Street, Suite 401-402, Hackensack, NJ 07601

UK office: 57 Shelton Street, Covent Garden, London WC2H 9HE

British Library Cataloguing-in-Publication Data
A catalogue record for this book is available from the British Library.

ELECTRONIC PROPERTIES OF DIRAC AND WEYL SEMIMETALS

ISBN 978-981-120-734-1 (hardcover)
ISBN 978-981-120-735-8 (ebook for institutions)
ISBN 978-981-120-736-5 (ebook for individuals)

For any available supplementary material, please visit
https://www.worldscientific.com/worldscibooks/10.1142/11475#t=suppl

Typeset by Stallion Press
Email: enquiries@stallionpress.com

Contents

Preface

Dirac and Weyl semimetals belong to a unique class of condensed matter materials, whose low-energy electron quasiparticles are described by relativistic-like Dirac and Weyl equations, respectively. The hallmark properties of the corresponding quasiparticles are their linear dispersion relations and well-defined chiralities. From the band-structural viewpoint, the valence and conduction bands of Dirac and Weyl semimetals touch at a number of points. The latter are known as the Dirac points and the Weyl nodes, respectively. While each Dirac point is doubly degenerate and consists of two overlapping Weyl nodes of opposite chiralities, the degeneracy is lifted in Weyl semimetals whose Weyl nodes are separated in energy and/or momentum.

Historically, a milestone in the search for the realization of relativistic-like dispersion relations in solids was the experimental discovery of graphene in 2004. In perfect agreement with theoretical predictions, graphene was confirmed to be a two-dimensional (2D) Dirac semimetal. Shortly afterwards, topological insulators were discovered experimentally. Unlike conventional insulators, the topological ones possess unusual gapless surface states with a relativistic-like dispersion relation that are protected by topology. The resulting surge of interest in topological phases of matter accelerated the search for novel materials, including those with three-dimensional (3D) relativistic-like quasiparticles. After about a decade of intensive theoretical and experimental investigations, it culminated in the discovery of 3D Dirac and Weyl semimetals.

Conceptually, Dirac and Weyl semimetals provide a rather direct synergetic link between high energy and condensed matter physics. Indeed, the relativistic-like energy spectrum in Dirac and Weyl materials opens the possibility of much simpler and more transparent realizations of many subtle high-energy phenomena by using tabletop experiments. Among the most fascinating examples are numerous observable properties of materials that result from the celebrated chiral anomaly. In addition to serving as alternative platforms for studying relativistic quantum effects, Dirac and Weyl semimetals also offer invaluable insights into the fundamental properties of matter on their own. In particular, they are characterized by nontrivial topological properties and demonstrate a rich variety of novel phenomena. They host topologically nontrivial surface Fermi arc states, have unconventional charge and heat transport, can realize strain-induced pseudoelectromagnetic (axial) gauge fields, as well as have new types of collective modes, unusual superconducting phases, etc.

The field of topological materials is still young and rapidly developing. Therefore, in this book, we concentrate primarily on the physical foundations of Weyl and Dirac semimetals and review only the main theoretical tools used in their studies. We also discuss only a limited range of phenomena, focusing predominantly on the electronic properties. At the same time, we leave out the discussion of numerous alternative realizations of Dirac and Weyl spectra in condensed matter systems such as the A-phase of superfluid ^{3}He, acoustic systems, optical lattices of cold atoms, high-T_c superconductors, etc. Also, we only briefly mention more complex forms of Dirac and Weyl semimetals such as multi-Weyl and type-II Weyl materials, which have been discovered most recently. Indeed, while some of these newest materials already demonstrate intriguing properties, many theoretical issues regarding the underlying physics remain to be understood.

It is worth noting that the properties of Weyl and Dirac systems have already been summarized in numerous reviews. For example, general overviews of Weyl and Dirac semimetals as well as other systems with relativistic-like dispersion relations are given in [1–5]. The transport properties are discussed in [6–16]. Many details regarding the *ab initio* calculations can be found in [17]. Reviews of experimental studies of Dirac and Weyl semimetals are presented in [18–23]. The nontrivial topological properties are considered, mostly from a mathematical viewpoint, in [24, 25]. There are also reviews like [26] that discuss a wider range of quantum field theoretical phenomena in external magnetic fields, including some applications to Dirac and Weyl semimetals. Finally, magnetic properties of Dirac and Weyl semimetals and strongly correlated quantum phenomena are described in [27–29].

Considering the existence of many reviews, some of which are truly excellent, it is natural to ask why we undertook the effort to write this book. There are two main reasons for that. First of all, a book provides an ideal format for discussing various aspects of the Dirac and Weyl semimetal physics from a unified viewpoint, without the usual size and scope restrictions of thematic reviews. The other reason is, as often happens, personal. Indeed, because of a serendipitous chain of events, we were fortunate enough to watch and get involved in the development of this new branch of condensed matter physics almost from the very beginning. By using this opportunity, we would like to share our experience and fascination with Dirac and Weyl semimetals with the readers.

In this book, we focus on the electronic properties of 3D Dirac and Weyl semimetals. In order to provide a historical perspective, however, we find it instructive to start from a general introduction to the electronic properties of graphene, which is a 2D Dirac semimetal. It is much simpler from the viewpoint of topology and internal symmetries, as well as provides a good reference point for discussing 3D Dirac and Weyl semimetals.

The emphasis of our presentation is often placed on the topological properties and discrete symmetries, which play a profound role in the underlying physics of Dirac and Weyl semimetals. However, it should be noted that, in the end, our main goal is to explain the key ideas, identify the roots of the unusual electronic properties, and review possible physical applications. Also, while we try to use a solid mathematical description throughout the book, the mathematical rigor by itself is not the ultimate objective. For example, we often rely on effective low-energy models of Dirac and Weyl semimetals, which are invaluable for a better insight into their physics. To supplement that, however, we also overview *ab initio* methods, some empirical details, and material realizations of Dirac and Weyl semimetals. In order to avoid repetitions and stay within a reasonable size limit, we sometimes sacrifice technical details. In most cases, such omissions should not be detrimental for the clarity of explanations and the overall understanding of basic physics. We hope that, for the most part, the presentation of the material is self-contained and should be sufficient for grasping the essentials of the Dirac and Weyl semimetal physics.

In this book, we also pay a special attention to the origin and properties of the surface Fermi arc states. The existence of such states is one of the distinctive features of Weyl and certain Dirac semimetals. In fact, their observation often serves as a primary experimental indicator of these topological materials. Not only the Fermi arc states have a special topologically protected status, but also rather unusual properties. One might argue that such a combination holds a great potential for applications.

A lion's share of the book is dedicated to the transport properties of Dirac and Weyl semimetals. Indeed, these materials reveal qualitatively new and often unusual phenomena arising, at least in part, from the non-trivial topology and a relativistic-like dispersion relation of quasiparticles. We discuss the use of several different methods in the studies of transport properties of Dirac and Weyl semimetals. The list includes the Green's function approach, the semiclassical chiral kinetic theory, as well as the hydrodynamic description of electron fluid. From the viewpoint of the book structure, we find it convenient to separate the most interesting anomalous transport properties into a separate chapter.

A part of the book is also dedicated to a discussion of various types of collective excitations in the electron plasma of Dirac and Weyl semimetals. As we point out, some of them are truly unique and have rather unusual properties. Considering the general interest in the topic and the practical importance of superconductivity, we also discuss the possibility of super-conductivity in Weyl and Dirac semimetals and emphasize some of its topological aspects. Although many questions about the superconductivity in topological semimetals remain unanswered, some recent findings are summarized in the last chapter.

We hope that this book will be useful for researchers interested in Dirac and Weyl materials, as well as for graduate and advanced undergraduate students studying condensed matter physics. By taking into account that the topic of Weyl and Dirac semimetals brings together many elements of fundamental physics from several subfields, including high energy and plasma physics, we hope that this book will be also a useful resource for a growing number of researchers involved in cross-disciplinary studies.

Acknowledgments

Most of the material in this book we learned by doing our own research, reading research articles in the field, and discussing specific topics with our colleagues around the world. It is truly impossible to mention everybody who shaped our understanding and triggered our insights. However, we are particularly grateful to our friends and collaborators Alexander Balatsky, Oleksandr Gamayun, Valery Gusynin, Vladimir Juričić, Vadim Loktev, Dmytro Oriekhov, Habib Rostami, Denys Rybalka, Gordon Semenoff, Sergei Sharapov, Oleksandr Sobol, Stanislav Vilchinskii, and Oleksandr Yakimenko from whom we learned a lot. We are also indebted to numerous colleagues around the world for various discussions, exchange of ideas, and communications over the years. A partial list of such people includes Maxim Chernodub, Alberto Cortijo, Matthias Geilhufe, Dmitri Kharzeev, Karl Landsteiner, Yuri Shtanov, Dam Son, Piotr Surowka, María Vozmediano, Mikhail Zubkov, Alexander Zyuzin, and Vladimir Zyuzin.

Chapter 1

Introduction

A journey of a thousand miles begins with a single step.

Laozi

The notion of quasiparticles is one of the most important concepts in condensed matter physics. In fact, the very diversity of this branch of physics is pivoted on the possibility of different energy dispersions and other characteristics of quasiparticles. Although, microscopically, the electrons in every solid-state material are described by nothing else but the standard nonrelativistic Schrödinger Hamiltonian with the electron–electron and electron–ion interactions, the dispersion relations of the electron Bloch states are incredibly diverse. In particular, they can have even a seemingly exotic relativistic-like form.

In this chapter, we provide a historical overview of the search for Dirac and Weyl semimetals. In our exposition, we purposefully pay special attention to numerous interconnections between high energy and condensed matter physics. It is worth noting, however, that this is achieved very naturally because many concepts used in the description of Dirac and Weyl semimetals were originally born in the realm of relativistic particle physics.

As is well known, Paul Dirac derived his eponymous equation in 1928 in order to describe relativistic electrons [30]. Over the years, the Dirac equation achieved the status of one of the most fundamental equations in physics. Not only it plays an essential role in the formulation of quantum electrodynamics as originally intended, but also it became indispensable in building the Standard Model of elementary particles.

The starting point in Dirac's derivation was the requirement that the relativistic quantum mechanical equation with the first-order time-derivative should be invariant under the Lorentz symmetry. As it turned out, this is possible only if the equation has a matrix form, which is in a drastic contrast to the nonrelativistic Schrödinger equation. As is well known now, such a matrix structure is a necessary attribute for the description of the intrinsic angular momentum (or the spin) of electrons. In its canonical form, the Dirac equation is written in terms of the so-called γ-matrices, which are 4×4 anticommuting complex matrices that generate the Clifford algebra $C\ell_{1,3}(R)$. The matrix structure of the equation also implies that the electron wave function is a four-component bispinor.

Soon after the formulation of the Dirac equation, Hermann Weyl proposed [31] a simpler equation that describes massless spin-1/2 particles. The Weyl equation is formulated in terms of the 2×2 Pauli matrices, which furnish a minimal framework for describing the spin of particles. Strictly speaking, there exist two inequivalent formulations of the Weyl equation describing massless spin-1/2 particles, namely those with the left- and right-handed chiralities. Here it might be appropriate to remind that, for massless three-dimensional fermions, the chirality coincides, up to a sign, with the helicity.[a]

By leaving out a more rigorous mathematical definition of the chirality for now, it is important to emphasize that the Weyl equation is at odds with the parity symmetry. This becomes immediately clear after noting that Weyl fermions of opposite chiralities transform into each other under the parity transformation. In contrast, the Dirac equation is invariant under parity. This is easy to understand after reviewing the more complicated structure of Dirac fermions, which can be represented as composites made of two opposite-chirality Weyl fermions. The degree of mixing of the Weyl components inside a Dirac fermion is determined by the value of the mass. As a result, the corresponding Weyl fermions of opposite chiralities decouple only in the limit of vanishing mass.

Historically, the Weyl equation was believed to be suitable for the description of neutrinos, which, for a long time, were presumed to be massless particles of a single chirality. However, the discovery of neutrino oscillations [32] unequivocally established that they are massive. This ruled out the possibility of the interpretation of neutrinos as massless Weyl fermions.[b]

---

[a]By definition, the helicity is the sign of the projection of particle's spin on the direction of its momentum. For particles with positive energies, the chirality is the same as their helicity. For antiparticles, which formally have negative energies, the chirality is opposite to their helicity.

[b]The nature of neutrino masses is still under debate. They could be either of the Dirac or Majorana type.

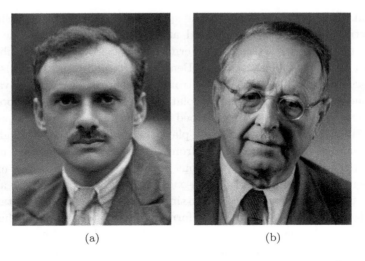

(a)   (b)

Fig. 1.1   (a) Paul A. M. Dirac (1902–1984) and (b) Hermann K. H. Weyl (1885–1955).

So, in the end, one must conclude that Weyl fermions found almost no utility as observable particles.

It is quite interesting that a mere decade after the discovery of the Dirac and Weyl equations, their manifestations also started to appear in the context of condensed matter physics. It was first realized by Conyers Herring [33] that a relativistic-like linear energy dispersion for electron quasiparticles in solids can appear in the vicinity of isolated points in the Brillouin zone, where neighboring energy bands touch. However, at that time, no direct connection to the Dirac or Weyl equation was made.

Later, the interest to materials, in which electron quasiparticles have linear dispersion relations in the vicinity of touching points of the conduction and valence bands, was revived by Aleksei Abrikosov and Sergey Beneslavskii [34]. Their study was motivated by the experimental data for gray tin [35] and mercury telluride (HgTe) [36].[c] Abrikosov and Beneslavskii showed that crystalline symmetries admit the possibility of a point contact of two energy bands and studied the cases of linear and quadratic energy spectra. They found that the Coulomb interaction produces only small (logarithmic) corrections to a linear low-energy dispersion

---

[c]Note that the first two-dimensional topological insulator was, in fact, experimentally discovered in HgTe/CdTe quantum wells almost 40 years later [37, 38].

relation. This is in contrast to the quadratic case, where the effective interaction is strongly renormalized at low energies and the concept of one-particle spectrum (Landau's Fermi-liquid) becomes inapplicable. In modern language, Abrikosov and Beneslavskii studied nothing else but Dirac semimetals, whose unique electronic structure comes from the valence and conduction bands touching at isolated Dirac points. The corresponding low-energy quasiparticles are described by a relativistic-like Dirac equation, where the Fermi velocity effectively plays the role of the speed of light.

In parallel and independently from the studies in condensed matter physics, the interest to chiral fermions in three-dimensional lattice models has been also developing in the high-energy physics community. To a large extent, this was driven by the need to simulate the effects of the spontaneous chiral symmetry breaking in lattice formulations of quantum field theories.

The chiral quantum anomaly was discovered by Stephen Adler, John Bell, and Roman Jackiw in 1969 [39, 40], who tried to resolve the acute discrepancy of about three orders of magnitude between the theoretical predictions for the rate of the neutral pion decay into a pair of photons, $\pi^0 \rightarrow \gamma\gamma$ in the current algebra framework and the experimental data. While calculating the relevant Feynman diagrams, Adler, Bell, and Jackiw found that some symmetries of the classical action could not remain good symmetries in the corresponding quantum field theory. From a technical viewpoint, such a chiral anomaly emerged from the inability to regularize the triangle diagrams for the axial-vector vertex in a way consistent with both chiral and gauge symmetries [41–45]. Without violating the electric charge conservation, the only consistent resolution of the dilemma was to sacrifice the conservation of chiral charge. The alternative, i.e., the breaking of gauge symmetry, which is directly connected to the conservation of the electric charge, would be unacceptable from the viewpoint of both theory and experiment. Surprisingly, the seemingly esoteric chiral anomaly happened to have a profound effect on the neutral pion decay into a pair of photons, leading to a prediction that agreed well with the experiment.

The lattice formulation of quantum field theory dates back to the original work of Kenneth Wilson in 1974 [46]. The main motivation behind his proposal was twofold: (i) to use the discretization as a mathematically consistent and physically transparent regularization of ultraviolet divergences in quantum field theory and (ii) to provide a convenient framework for numerical simulations, which were gaining power and popularity in all branches of science at that time, in order to study the nontrivial dynamics of strongly coupled gauge theories. Regarding the last point, it might be appropriate to mention that, even nowadays, the lattice methods remain among very few available tools to perform nonperturbative calculations in strongly coupled gauge field theories [47] such as quantum chromodynamics (QCD) [48, 49].

Since lattice field theories are ultraviolet finite by construction, one might question whether the chiral quantum anomaly could be realized in such a framework. As it turns out, simulating massless fermions on a lattice is indeed an extremely challenging task. Their naive implementation is plagued with the notorious "doubling problem". The latter refers to the fact that, in addition to the original massless fermion states, there are extra massless states in the spectrum, which come from different corners of the Brillouin zone and have no analogs in the continuum limit. Not only such doubler fermions give rise to additional unphysical particles, but also modify profoundly the implications of the chiral anomaly.

One of the solutions to avoid the unwanted doublers in the spectrum is to add the so-called Wilson term in the lattice action. It has the form of a discretized second derivative with respect to spatial coordinates and, therefore, produces a momentum-dependent mass term. Its special form ensures that the nature of the massless physical particles is not affected, but the doubler fermions become heavy and decouple from the infrared dynamics. Unfortunately, this solution comes at the cost of explicit breakdown of the chiral symmetry in a lattice model. Nonetheless, it is interesting to note that the theory has the correct continuum limit and even the exact anomaly is reproduced [50]. In practice, however, the limiting procedure requires some nontrivial fine tuning of bare parameters in the model.

Another important step in the understanding of lattice field theories with massless fermions was made by Holger Nielsen and Masao Ninomiya, who proved the famous "no-go" theorem about the fermion doubling and the lattice implementation of chirality in a series of seminal papers in 1981 [51–53]. The theorem states that any Hermitian, local, and translation-invariant fermion action that is defined on a regular lattice in a space–time with an even number of dimensions contains an equal number of right- and left-handed fermions. Needless to say, this finding plays an important role in modern state-of-the-art lattice simulations. Also, while it might not have been fully appreciated at the time, this theorem has direct implications for such condensed matter materials as Weyl semimetals. In particular, it explains why their energy spectra always contain an even number of band touching points with linear dispersion relations, which are known as the Weyl nodes. Although the nodes could be separated in momentum and/or energy, they must always come in pairs of opposite chirality.

In the case of electrically charged massless fermions, the anomalous non-conservation (i.e., production) of chiral charge can be triggered by parallel electric and magnetic fields. In this context, the chiral anomaly is closely connected with the topologically protected lowest Landau level of massless fermions [54]. In order to see this, one has to recall that the direction of the longitudinal velocity (i.e., the component along the direction of the background field) is opposite for the left- and right-handed quasiparticles in the lowest Landau level. Therefore, even an infinitesimally small external electric field causes counter-propagating spectral flows of states with opposite chirality (from negative to positive energies and vice versa). Overall, this fermion number conserving process is accompanied by the production of a chiral charge. The corresponding low-energy process is sometimes referred to as the "infrared face" of the chiral anomaly [41]. This is in contrast to the "ultraviolet face", which shows up, for example, in the calculation of the triangle diagrams mentioned earlier, where the ultraviolet divergencies have to be regularized.

The production of chiral charge is not the only infrared consequence of the chiral anomaly. The latter can also have profound effects on the transport properties in solids. In particular, as was first argued in [55], it could lead to a negative longitudinal magnetoresistivity in the zero-gap semiconductors whose quasiparticles are described by relativistic-like Dirac or Weyl equations.[d] The underlying physics and the role of the anomaly will be discussed in detail in Sec. 6.2. It might be appropriate to mention here that there are numerous experimental confirmations of negative magnetoresistivity in both Dirac [56–61] and Weyl [62–67] semimetals. On its own, this is an amazing fact considering the subtle origin of the quantum anomaly, which was first found as an oddity associated with the regularization of relativistic quantum field theories. Now, it emerges in the unusual charge transport in solids that can be measured with such conventional tools as voltmeter and ammeter.

Many aspects of electronic properties of Dirac and Weyl semimetals are rooted in the topology of their band structure. In general, the role

---

[d]Note that the resistivity is always positive and the term "negative magnetoresistivity" is a misnomer which, however, became conventional. It simply means that the resistivity decreases with magnetic field.

of topology in high energy and condensed matter physics has been discussed since the early days of quantum mechanics. Topology was used in theoretical attempts in high energy physics to introduce hypothetical magnetic monopoles into electrodynamics, to model baryons using the skyrmion configurations in the nonlinear sigma model, to get an insight into the nonperturbative dynamics of non-Abelian gauge theories, to explain the Aharonov–Bohm effect, and much more. It plays an equally important role in condensed matter phenomena, ranging from the quantum Hall effect and optical absorption spectra near the Van Hove singularities to the description of superconducting vortices in type-II superconductors and various types of domain walls.

One of the important steps in advancing topological ideas, which turned out to be of particular importance for Dirac and Weyl semimetals, was made by Michael Berry in 1984 [68]. He developed and popularized the concepts of the geometric phase and the geometric curvature, which are now often referred to as the Berry phase and the Berry curvature, respectively. In connection to Weyl semimetals, these concepts play a vital role since the individual Weyl nodes are monopoles of the Berry curvature. Furthermore, the flux of the Berry curvature through the sphere in momentum space surrounding an individual node is a topological invariant, whose value is proportional to the topological charge of the node. The charge is called the Chern number and its sign is directly related to the chirality. In application to Weyl semimetals, the Nielsen–Ninomiya theorem implies that the total topological charge in the system vanishes. This is not surprising since the flux of the Berry curvature through the whole Brillouin zone must be zero in view of the compact nature of the zone. As will be discussed in Sec. 3.2, the overall properties of Dirac and Weyl semimetals are affected not only by topology, but also by certain crystalline symmetries and the fundamental discrete symmetries such as time reversal and parity inversion. In many cases, such symmetries are crucial for the protection of the band-touching points.

One of the fundamental consequences of a nontrivial topology of the Bloch states in solids is the emergent bulk-boundary correspondence. In simple terms, the latter implies that there must exist spatially localized edge states at an interface between any pair of materials (or phases of matter) with different topological characteristics. The energies of such states should lie in the band gap of the two materials in contact. For the first time, such a correspondence appeared implicitly in the famous (1+1)-dimensional quantum field theoretic model of Roman Jackiw and Claudio Rebbi [69].

It was found that fermions in a background of a topologically nontrivial kink (or antikink) scalar-field configuration have localized zero-energy solutions. Later it was found [70, 71] that a similar phenomenon also takes place in the integer quantum Hall effect, where the current-carrying edge states are related to the Chern numbers of the Landau level wave functions. Despite the conceptual understanding of the importance of topology for electron states in solids, for a long time there was little experimental progress in discovering solid-state materials with topologically nontrivial band structures. The situation changed drastically in the last five to ten years.

It is worth noting that a few years after Nielsen and Ninomiya's work [55], suggesting that electron quasiparticles in zero-gap semiconductors can be described by the Weyl equation, Grigory Volovik argued that emergent Weyl fermions also appear in the A-phase of superfluid helium-3 (^{3}He-A) [72] (for a comprehensive overview of liquid helium physics, see the book by Volovik [73]). The Weyl quasiparticles in ^{3}He-A give rise to the same chiral anomaly as the electron quasiparticles in topological semimetals. Note, however, that the role of electric and magnetic fields in helium is played by effective synthetic gauge fields. Experimentally, the corresponding chiral anomalous properties were verified in the studies of vortex dynamics [74].

The monumental breakthrough in the electronic realization of the Dirac equation in condensed matter setting was made in 2004 when graphene was discovered by the group lead by Andre Geim and Konstantin Novoselov [75]. Graphene is a two-dimensional (2D) Dirac semimetal, which was predicted theoretically [76, 77] almost 20 years before its experimental discovery. The existence of such a 2D crystal seemed to be at odds with the universally accepted argument of Lev Landau and Rudolf Peierls [78, 79]. Formally, their studies showed that any 2D crystal should be destroyed by long-range thermal fluctuations [80]. After the discovery of graphene, however, it was realized that there are several loopholes that allow for the existence of graphene. The formal argument assumes infinite (or very large) samples, while real flakes of graphene may not be large enough for the most dangerous long-range fluctuations to develop. Indeed, the lower is the temperature, the longer should be the wavelengths of fluctuations that destroy the 2D crystal ordering. Additionally, graphene is often made on 3D substrates that further stabilize the fragile planar system. Nevertheless, one might also argue that the signs of the Landau–Peierls fluctuations are indirectly seen in graphene. Indeed, suspended graphene samples are never truly flat but rippled. As will be discussed in Chapter 2, the Dirac nature of graphene quasiparticles gives rise to an unconventional quantum Hall effect, as well as very high electrical and thermal conductivities [81]. Interestingly, graphene also reveals a number of relativistic phenomena such as the Klein paradox [82], the Zitterbewegung ("trembling motion") [83], the dynamics of supercritical charge ("atomic collapse") [84, 85], the Schwinger effect [86], and the dynamical generation of a Dirac mass in a magnetic field (magnetic catalysis) [26]. While these are relativistic phenomena that have been first predicted in the context of high energy physics, some of them were experimentally observed only in graphene so far.

The rise of graphene galvanized the condensed matter community in the search for materials with unusual relativistic-like energy dispersions. A few years later this search came to fruition when 2D topological insulators were discovered in experiment [37, 38]. This, in turn, further boosted the enormous interest in condensed matter topological systems. On the theoretical side, a general theorem relating the bulk topological number to the chiral edge states in 2D systems was formulated [87]. Then it was quickly realized that similar topological arguments should also apply to 3D materials. In particular, it was theoretically predicted [88–92] and then experimentally confirmed [93–95] that the surface states of 3D topological insulators are topologically nontrivial.[e] The low-energy dynamics of these states is described by the 2D Dirac equation. (For reviews on topological insulators, see [97–99], as well as books [100, 101].)

Despite having an energy gap between the valence and conduction bands, the ground state of a topological insulator is topologically different from vacuum or any "trivial" insulator. In other words, this means that one cannot smoothly deform the Hamiltonian of a topological insulator into that of a trivial one. From a mathematical viewpoint, all insulators can be classified according to their $\mathbb{Z}_2$ topological invariant, whose two eigenvalues correspond to either "trivial" or "topological" class. Furthermore, by using the topological arguments, one finds that the energy gap must necessarily close at an interface between the materials with different topological characteristics. In essence, this is the underlying reason for the bulk-boundary correspondence that explains the topologically protected gapless edge states.

---

[e]It is worth noting that, historically, the surface states with linear energy spectrum at the interface of two semiconductors with mutually inverted gaps were first described by Boris Volkov and Oleg Pankratov [96]. In essence, such a system is a topological insulator, whose topological properties were not recognized at that time, however.

In the context of Weyl semimetals, the nontrivial topology of Weyl nodes is responsible for the existence of special surface states known as the Fermi arcs [102]. While such states are spatially localized near the surface of the material, in reciprocal (momentum) space they form open intervals that connect the Fermi surfaces of the bulk Weyl fermions of opposite chiralities. A simple qualitative argument supporting the existence of the Fermi arcs follows from considering heuristically a Weyl semimetal as a stack of 2D topological insulators. Taken together, their edge states form the Fermi arcs. As was emphasized by Duncan Haldane [103], the Fermi arcs provide "plumbing" between the otherwise disconnected chiral pockets of the Fermi surface, which is essential for maintaining the same value of the electric chemical potential for the whole system. In view of their topological origin, the Fermi arc states are stable in the absence of disorder and have effectively a one-dimensional chiral energy dispersion. However, unlike the surface states of topological insulators, the stability of the Fermi arcs cannot be unambiguously inferred from the surface–bulk correspondence alone when disorder is present. The underlying reasons, as will be discussed in Sec. 5.5, are connected with the gapless energy spectrum of the bulk states.

Typically, quasiparticles in Dirac and Weyl semimetals have a very high mobility. This explains their excellent transport properties, including high electrical and thermoelectric conductivities, which were known and studied for a long time even before the relativistic-like character and nontrivial topology of their electron quasiparticles were revealed. Historically, it was the combination of a careful theoretical analysis and the power of novel experimental techniques that helped to confirm the unusual electronic properties of Dirac and Weyl semimetals. Ten years after the discovery of graphene, sodium bismuthide (Na_3Bi), which is a member of a larger family A_3X (where $A = $ Na, K, Bi and $X = $ As, Sb, Bi), and cadmium arsenide (Ca_3As_2) were the first materials experimentally confirmed as 3D Dirac semimetals in 2014 [104–109]. This was in agreement with the earlier theoretical predictions in [110, 111]. Similarly, shortly after the theoretical predictions in 2015 [112–118], Weyl semimetals were found experimentally in transition metal monopnictides, which include TaAs, TaP, NbAs, and NbP [116, 117, 119–126]. Later other, more complicated types of Dirac and Weyl semimetals were proposed. The corresponding list includes the type-II Weyl materials [20, 127, 128] with the naive Lorentz invariance explicitly broken by a large tilt of the Weyl cones, the multi-Weyl semimetals [129] with the topological charge of the nodes greater than one, and the multi-fold Dirac and Weyl materials [130–135] with many bands crossing at the same point. It was also realized that the Dirac and Weyl equations are quite ubiquitous and could appear in a much broader range of condensed matter systems, including the superfluid phases of liquid ^{3}He [73] mentioned earlier, superconductors [2, 3], optical lattices [5], phonon crystals [5], etc.

As is clearly demonstrated by the history of Dirac and Weyl semimetals, the intense exchange of ideas between high energy and condensed matter physics is very fruitful. We are confident that such an exchange is beneficial to both communities and could result in the appearance of new promising fields of research. Needless to say that the study of Dirac and Weyl semimetals is definitely not a singular example of such a mutually beneficial exchange. To support this claim, it suffices to recall several most representative examples from a relatively recent history of physics. The method of Green's functions, which was first developed in quantum electrodynamics, found numerous applications in many-body theory [136–139]. The idea of spontaneous symmetry breaking from the Bardeen–Cooper–Schrieffer theory of superconductivity [140] evolved into one of the most fundamental concepts in elementary particle physics that culminated in the experimental discovery of the Higgs boson in 2012. Another prominent example is Wilson's exact renormalization group [141, 142], which has roots in Kadanoff's block spin transformation [143]. Eventually, the corresponding chain of ideas led to the modern view of the Standard Model of elementary particles as an effective theory whose range of validity extends only up to a finite energy scale.

Most recently, the realization of the chiral anomaly in Dirac and Weyl semimetals contributed a lot to the understanding of its many facets in many-body systems connected, in particular, with the anomalous transport [6, 7, 11, 12]. For example, the nondissipative chiral magnetic [144–146], chiral separation [144, 147, 148], and chiral vortical [149–152] effects, which were originally proposed in a high-energy physics context in the 1980s, turned out to be important in solids too. Such effects are expected to have observable signatures also in heavy-ion collisions [26, 153–156]. In cosmology, they might be relevant for the generation and evolution of primordial magnetic fields in the early Universe [157, 158]. This fruitful exchange of ideas between seemingly different areas of physics not only allows one to predict novel, insightful, and sometimes very exotic physics phenomena, but also emphasizes the cohesiveness and unity of physics as a research discipline. It is natural to expect that the cross-disciplinary enrichment of ideas will continue and many new phenomena will be discovered as a result of such a collaboration in the future.

Before finalizing this chapter, let us briefly mention potential future applications of Dirac and Weyl semimetals. As will be discussed in Chapters 6 and 7, topological semimetals exhibit exceptional electronic transport properties related to their high charge carrier mobility[f] and unusual response to a magnetic field. Potentially, Dirac and Weyl semimetals could realize a new generation of electronic and spintronic devices with low power consumption [160]. Chirality of Weyl quasiparticles might even allow for "chiral electronic" devices [161]. Additionally, their gapless energy spectra and a fast relaxation time of excited carriers [162–164] make them promising for ultra-fast detectors of infrared light and low-energy particles. Finally, Weyl semimetals with topologically protected Fermi arcs might be useful in chemical catalysis [165], which takes place on the surface of a material. This is not surprising since the Fermi arcs, unlike the surface states in usual materials, are less sensitive to defects and disorder. Hopefully, other practical applications of Dirac and Weyl semimetals will appear in future.

---

[f]Charge carrier mobility in Dirac and Weyl semimetals could reach 10^7 cm^2/(V · s) at 5 K (see also Appendix D), which is about an order of magnitude higher than mobility in a free-standing graphene [159]. In both graphene as well as Dirac and Weyl semimetals, high mobility arises from a back-scattering suppression connected with a relativistic-like energy spectrum of electron quasiparticles.

Chapter 2

Graphene: Dirac Semimetal in Two Dimensions

> *I determined to endeavour to open up to him some*
> *glimpses of the truth, that is to say of the nature of things*
> *in Flatland.*
>
> Edwin Abbot

Although this book is devoted to the electronic properties of three-dimensional (3D) Dirac and Weyl semimetals, we find it logical to begin with the discussion of electronic properties of graphene. Indeed, its quasi-particles are governed by two-dimensional (2D) Dirac equation and demonstrate a variety of interesting properties. This makes graphene a perfect starting point.

2.1 Discovery and general properties

Graphene is a 2D crystal made of carbon atoms that are held in a hexagonal (honeycomb) lattice by very strong covalent bonds. It can also be viewed as an isolated single atomic layer of graphite. There also exist other types of graphene with a few layers of carbon atoms. In particular, bilayer and trilayer graphene are most actively studied. Since they have a nonlinear energy spectrum, such materials will not be discussed in this book.

In view of the general thermodynamic arguments against the existence of 2D crystals, which were discussed in Chapter 1, it is fair to say that graphene was discovered rather unexpectedly by Andre Geim and Konstantin Novoselov in 2004 [75]. The researchers applied the mechanical exfoliation technique, also known as the "scotch tape method", to a crystal of highly oriented pyrolytic graphite in order to peel off very thin layers of the bulk material. By a fortunate stroke of serendipity, they happened to place the resulting samples on top of an oxidized silicon wafer with the thickness of about 300 nm. Such a substrate, as it turned out, made the observation of graphene flakes possible even with a regular optical microscope. Among numerous flakes of various thickness, they found samples of monolayer graphene [166]. As it turned out in the end, sufficiently small flakes of graphene are stable on their own and large samples could be easily stabilized by substrates.

Graphene has a band structure of a gapless semimetal with the vanishing density of states at the Fermi surface. It is characterized by a high mobility of its quasiparticles making it a good electric and heat conductor. In addition, the density of charge carriers and their type (electron or hole) can be easily controlled by changing the gate electric potential. In a background magnetic field, graphene also reveals the unconventional (half-integer) quantum Hall effect that can be observed even at room temperature [167, 168].

Because of the vanishing band gap, monolayer graphene can have only a limited range of applications in modern semiconductor industry. Nevertheless, it can be used to improve the characteristics of traditional integrated circuits [169, 170], as well as to create new types of heterostructures with desired electronic properties. In particular, graphene is a promising platform for various terahertz applications [171, 172] including a highly sought-after terahertz transistor.

In terms of optical properties, graphene absorbs only about 2.3% of white light [173] and, therefore, is nearly transparent to the human eye. This makes graphene a promising material for creating transparent and flexible conductors for touchscreens, wearable electronics, medical biosensors, etc. On the other hand, the absorption level of 2.3% is quite substantial for 2D materials and may be useful in designing graphene-based solar cells. Also, when placed in a moderately strong perpendicular magnetic field of a few teslas, monolayer graphene becomes optically active. Incredibly, it can rotate the plane of polarization of the far-infrared light by several degrees [174]. This is a giant Faraday rotation for a single atomic layer crystal, which can potentially be used in creating ultrathin but efficient infrared magneto-optical devices.

The mechanical properties of graphene are also quite remarkable. The atomically thin graphene has the mass density of only about $0.76 \, \mathrm{mg/m^2}$, but holds the record of the strongest material ever tested [175, 176]. It is also very flexible and stretchable. Amazingly, graphene does not loose its structural integrity even when stretched by as much as 20% [177, 178]. Although it is only one-atom thick, graphene poses an impenetrable barrier to many atoms and gas molecules. The notable exceptions from this rule are water molecules that can easily permeate through graphene [179]. This property is quite valuable as it could be utilized, for example, in water purification and desalination [180–182].[a]

From the viewpoint of potential applications, it is very beneficial that graphene is chemically inert and stable under normal conditions. However, graphene has highly reactive edges, which make it toxic for living organisms. The other nuisance feature is that graphene can be easily contaminated by atoms and molecules (or "adparticles") from environment that strongly affect its electronic properties. Fortunately, the corresponding adparticles can be removed by annealing graphene in vacuum. Such a drawback can, however, be turned into an advantage by using graphene as microsensors capable of detecting individual gas molecules [183].

The discovery of graphene stimulated the search and subsequent experimental discovery of other atomically thin 2D materials. The corresponding partial list includes numerous crystalline monolayers such as borophene (a layer of boron atoms), germanene, silicene, stanene, phosphorene, and bismuthene. In addition, several more complicated 2D materials made of different atoms, such as monolayers of hexagonal boron nitride (h-BN), molybdenum disulfide (MoS_2), and tungsten diselenide (WSe_2), were synthesized. Some of them are gapless semimetals and/or metals, others are insulators and/or semiconductors. Such a variety of atomically thin materials makes them suitable for a wide range of applications in 2D electronics.

Before concluding this introductory section, it might be useful to note that detailed discussions of graphene and its properties can be found in many existing reviews [184–192] and books [81, 193, 194] that have been published to date. In the rest of this chapter, we will not attempt to provide a comprehensive overview of all properties of graphene. Instead, we will focus on electronic properties. This will set the stage for the subsequent discussion of 3D Dirac and Weyl semimetals.

2.2 Emergent Dirac quasiparticles in graphene

From the theoretical viewpoint, one of the most interesting features of graphene is the relativistic-like nature of its low-energy quasiparticles, which are described by the massless Dirac equation. Needless to say that

---

[a]It is rumored that the two discoverers of graphene tested the filtration property by distilling alcohol with the help of a graphene membrane placed over a glass of watered-down vodka.

the quasiparticles in graphene are not truly relativistic because they move much slower than the speed of light. In fact, their Fermi velocity v_F is about 300 times smaller than the speed of light c. As will be shown below, the Dirac nature of quasiparticles is an emergent low-energy phenomenon in graphene. In addition, although the spinor structure of the quasiparticle wave function is identical with that of a massless relativistic particle, its effective spin (or more correctly pseudospin) degree of freedom has a different meaning.

It is quite remarkable to think about the quasiparticles in graphene as the first example of massless charged Dirac fermions in nature.[b] Of course, massless Dirac particles have long been utilized in high energy physics as a starting point in the description of elementary particles. The corresponding approach is quite valuable as it emphasizes the role of an approximate chiral symmetry in particle physics. While this symmetry is not exact in nature, it helps to understand systematically the mass hierarchy of hadrons and some details of their low-energy interactions. Nevertheless, truly massless Dirac fermions are absent in particle physics. As is well known now, even electrically neutral neutrinos have nonzero masses [32]. It is fairly ironic, therefore, that the first exactly massless charged Dirac fermions have been discovered in the realm of solid-state physics.

2.2.1   *Tight-binding and low-energy models*

The first theoretical study of the electronic band structure of graphene dates back to 1947, when Philip Wallace used the tight-binding model of graphene as the first approximation in the description of graphite [195]. This is indeed natural since the coupling between the adjacent layers of carbon atoms in the lattice of graphite is much weaker than the bonding of carbon atoms in the same layer. One of the main reasons for this is that the spacing between the layers (0.337 nm) is more than twice as large as the hexagonal spacing between the carbon atoms in the layers ($a \approx 0.142$ nm).

In his pioneer paper, Wallace had described almost completely the band structure of a single hexagonal layer including the location of the band touching points and the linear dispersion relation of quasiparticles in the vicinity of those points. However, the Dirac spinor structure of the

---

[b]Strictly speaking, the theory of exactly massless charged Dirac fermions is inconsistent in the far infrared in the absence of a low-energy cutoff. In graphene, the latter is always present, e.g., due to a nonzero chemical potential from electron–hole puddles, impurity doping, finite size effects, etc.

quasiparticle wave function was not discussed. It was eventually revealed in 1984 by Gordon Semenoff, David DiVincenzo, and Eugene Mele [76, 77].

The derivation of graphene's electronic band structure within the tight-binding model is simple and provides an insight into the origin of the Dirac nature of low-energy quasiparticles. Therefore, it is instructive to briefly overview it here. To start with, let us note that the hexagonal lattice of graphene can be viewed as a superposition of two triangular sublattices of types A and B shown in Fig. 2.1(a). In other words, there are two atoms per unit cell in graphene. Note that all nearest neighbors of the type-A atoms are atoms of type B and, vice versa, all nearest neighbors of the type-B atoms are atoms of type A.

The position vectors of carbon atoms in sublattice A could be parameterized as follows:

$$\mathbf{r_n} = n_1 \mathbf{a_1} + n_2 \mathbf{a_2}, \tag{2.1}$$

where the site label $\mathbf{n} = (n_1, n_2)$ is determined by a pair of integers n_1 and n_2, and

$$\mathbf{a_1} = \left(\frac{\sqrt{3}a}{2}, \frac{3a}{2} \right), \quad \mathbf{a_2} = \left(\frac{\sqrt{3}a}{2}, -\frac{3a}{2} \right) \tag{2.2}$$

are the primitive translation vectors. Similarly, the position vectors of carbon atoms in sublattice B are given by

$$\mathbf{r'_n} = \boldsymbol{\delta}_1 + n_1 \mathbf{a_1} + n_2 \mathbf{a_2}, \tag{2.3}$$

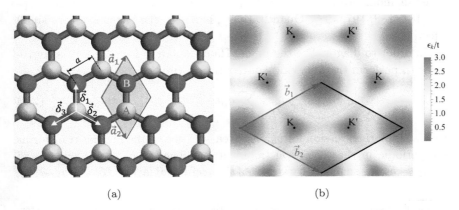

(a)   (b)

Fig. 2.1   (a) Schematic representation of graphene's lattice in coordinate space and (b) the corresponding reciprocal lattice in momentum space.

where $\boldsymbol{\delta}_1 = (\mathbf{a}_1 - \mathbf{a}_2)/3$ is the relative position of atom B with respect to atom A in the unit cell. The relative positions of the other two atoms of type B are given by $\boldsymbol{\delta}_2 = \boldsymbol{\delta}_1 + \mathbf{a}_2$ and $\boldsymbol{\delta}_3 = \boldsymbol{\delta}_1 - \mathbf{a}_1$, see Fig. 2.1(a). Explicitly, the corresponding relative positions are given by

$$\boldsymbol{\delta}_1 = (0, a), \quad \boldsymbol{\delta}_2 = \left(\frac{\sqrt{3}a}{2}, -\frac{a}{2}\right), \quad \boldsymbol{\delta}_3 = \left(-\frac{\sqrt{3}a}{2}, -\frac{a}{2}\right). \quad (2.4)$$

The simplest version of the tight-binding Hamiltonian for graphene can be written as follows:

$$H_0 = -t \sum_{\mathbf{n}} \sum_{i=1}^{3} a_{\mathbf{n}}^{\dagger} b_{\mathbf{n}+\boldsymbol{\delta}_i} + \text{h.c.}, \quad (2.5)$$

where $t \simeq 2.97\,\text{eV}$ [196] is the hopping parameter that models the interatomic matrix element between the sp^2 hybrid electron orbitals of the nearest-neighbor carbon atoms. The particle creation (annihilation) operators on the sublattices A and B are denoted by $a_{\mathbf{n}}^{\dagger}$ ($a_{\mathbf{n}}$) and $b_{\mathbf{n}}^{\dagger}$ ($b_{\mathbf{n}}$), respectively. This simple model, which includes only the most important interactions between the nearest atoms connected by the relative position vectors $\boldsymbol{\delta}_i$ ($i = 1, 2, 3$), see Fig. 2.1(a), is sufficient for illustrative purposes here. In a more refined analysis, however, it can be easily generalized to include next-to-nearest interactions and beyond. Such interactions lead to the breakdown of electron–hole symmetry. While the energy spectrum is isotropic at linear order, it becomes anisotropic at the quadratic order in momentum. The anisotropy has the form of the trigonal warping [187, 188, 197, 198]. Quantitatively, however, the latter is a very small effect.

In order to diagonalize the model Hamiltonian in Eq. (2.5), it is convenient to perform the following Fourier transform:

$$a_{\mathbf{k}} = \sum_{\mathbf{n}} a_{\mathbf{n}} e^{-i\mathbf{k}\cdot\mathbf{r}_{\mathbf{n}}}, \quad (2.6)$$

with a similar expression for $b_{\mathbf{n}}$. As is easy to verify, the Fourier transforms are periodic in the reciprocal (momentum) space,[c] i.e., $a_{\mathbf{k}+\boldsymbol{K}_l} = a_{\mathbf{k}}$ and $b_{\mathbf{k}+\boldsymbol{K}_l} = b_{\mathbf{k}}$ for an arbitrary vector of the reciprocal lattice $\boldsymbol{K}_l = l_1 \boldsymbol{b}_1 + l_2 \boldsymbol{b}_2$,

---

[c] In our discussions, we often use the terms of the momentum and the wave vector interchangeably. However, strictly speaking, they differ by a factor of the Planck constant, namely, $\boldsymbol{p} = \hbar \mathbf{k}$, where $\mathbf{k}$ is the wave vector and $\boldsymbol{p}$ is the momentum.

where l_1 and l_2 are integers and the basis vectors in the reciprocal space are given by

$$\boldsymbol{b}_1 = 2\pi \left(\frac{1}{\sqrt{3}a}, \frac{1}{3a} \right), \quad \boldsymbol{b}_2 = 2\pi \left(\frac{1}{\sqrt{3}a}, -\frac{1}{3a} \right). \tag{2.7}$$

It should be noted that the inverse Fourier transform is defined by

$$a_{\mathbf{n}} = \sqrt{A_{\text{cell}}} \int_{BZ} \frac{d^2k}{(2\pi)^2} a_{\mathbf{k}} e^{i\mathbf{k}\cdot\mathbf{r}_{\mathbf{n}}}, \tag{2.8}$$

where $A_{\text{cell}} = \sqrt{3}a^2/2$ is the area of the unit cell and the subscript BZ indicates that the integration runs over the Brillouin zone.

By making use of Eq. (2.8), the model Hamiltonian (2.5) can be rewritten in the following equivalent form:

$$H_0 = \int_{BZ} \frac{d^2k}{(2\pi)^2} (a_{\mathbf{k}}^\dagger, b_{\mathbf{k}}^\dagger) \begin{pmatrix} 0 & f_{\mathbf{k}} \\ f_{\mathbf{k}}^* & 0 \end{pmatrix} \begin{pmatrix} a_{\mathbf{k}} \\ b_{\mathbf{k}} \end{pmatrix}, \tag{2.9}$$

where

$$f_{\mathbf{k}} = -t \sum_{i=1}^{3} e^{i\mathbf{k}\cdot\boldsymbol{\delta}_i} = -t \left[e^{iak_y} + 2e^{-\frac{i}{2}ak_y} \cos\left(\frac{\sqrt{3}ak_x}{2} \right) \right]. \tag{2.10}$$

The eigenstates and eigenvalues of Hamiltonian (2.9) are given by

$$u_{+,\mathbf{k}} = \frac{1}{\sqrt{2}} \begin{pmatrix} 1 \\ f_{\mathbf{k}}^*/|f_{\mathbf{k}}| \end{pmatrix}, \quad u_{-,\mathbf{k}} = \frac{1}{\sqrt{2}} \begin{pmatrix} -f_{\mathbf{k}}/|f_{\mathbf{k}}| \\ 1 \end{pmatrix} \tag{2.11}$$

and

$$\epsilon_{\mathbf{k}} = \pm|f_{\mathbf{k}}|$$

$$= \pm t\sqrt{1 + 4\cos\left(\frac{\sqrt{3}ak_x}{2} \right) \left[\cos\left(\frac{\sqrt{3}ak_x}{2} \right) + \cos\left(\frac{3ak_y}{2} \right) \right]}, \tag{2.12}$$

respectively. The plus and minus signs in Eqs. (2.11) and (2.12) correspond to electrons in the conduction band and holes in the valence band, respectively. The corresponding dispersion relation is shown in Fig. 2.2. It is appropriate to mention in passing that this is exactly the same two-band structure that Wallace obtained already in 1947 [195].

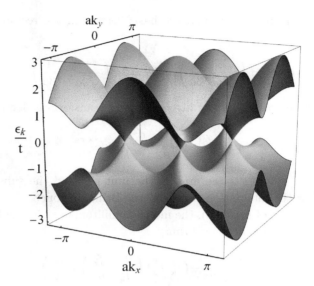

Fig. 2.2   Dispersion relation of quasiparticles given by Eq. (2.12) in a tight-binding model of graphene.

As will be discussed in more detail in Chapter 4, a direct method that can map the energy spectrum of the electron states in solids is the angle-resolved photoemission spectroscopy (ARPES). The corresponding experimental data in the case of graphene were obtained in [199–201] and agree very well with the theoretical results in Fig. 2.2.

As is easy to check, the electron and hole bands in Eq. (2.12) touch at the two *nonequivalent* Dirac points in the Brillouin zone: $\mathbf{k}_D^{\pm} = \pm(\mathbf{b}_1+\mathbf{b}_2)/3 = \pm 4\pi/(3\sqrt{3}a)\hat{\mathbf{x}}$, where the plus and minus signs correspond to K and K' points, respectively, see Fig. 2.1(b). The other four points are located at $\pm 2\pi/(3a)(\hat{\mathbf{x}}/\sqrt{3} + \hat{\mathbf{y}})$ and $\pm 2\pi/(3a)(\hat{\mathbf{x}}/\sqrt{3} - \hat{\mathbf{y}})$, where $\hat{\mathbf{x}}$ and $\hat{\mathbf{y}}$ are unit vectors in the x- and y-directions, respectively. In the vicinity of K and K' points (i.e., at $\mathbf{k} = \mathbf{k}_D^{\pm} + \tilde{\mathbf{k}}$, where $\tilde{\mathbf{k}}$ is small), the expression for $f_{\mathbf{k}}$ takes the following approximate form:

$$f_{\mathbf{k}_D^{\pm}+\tilde{\mathbf{k}}} \simeq \pm\hbar v_F(\tilde{k}_x \mp i\tilde{k}_y), \qquad (2.13)$$

where $v_F = 3a|t|/(2\hbar) \simeq 9.8 \times 10^5\,\text{m/s}$ is the Fermi velocity of low-energy quasiparticles. Note that, in the close vicinity of the K and K' points (also called the K and K' valleys), the energies of quasiparticles are given by the relativistic-like linear dispersion relation $\epsilon_{\mathbf{k}} \simeq \pm\hbar v_F|\tilde{\mathbf{k}}|$. Such a low-energy approximation provides a very good description for quasiparticles with energies up to about $|t|/3 \simeq 1\,\text{eV}$. At higher energies, deviations from the linear dispersion relation become nonnegligible (see also Fig. 2.2).

The phase space of the low-energy sector in graphene is represented by the two regions centered around K and K' points. Therefore, in order to formulate an effective low-energy description, it is convenient to introduce the following four-component wave function:

$$\Psi_{\mathbf{k}}^{\mathrm{T}} = (a_{K,\tilde{\mathbf{k}}}, b_{K,\tilde{\mathbf{k}}}, b_{K',\tilde{\mathbf{k}}}, a_{K',\tilde{\mathbf{k}}}), \tag{2.14}$$

which combines the relevant fields from both sublattices and both valleys. In the following, only the low-energy properties of graphene will be considered. It will be convenient, therefore, to omit the tilde in the notation of the wave vector, i.e., $\tilde{\mathbf{k}} \to \mathbf{k}$.

Note that the last two components in Eq. (2.14), associated with the sublattice degrees of freedom in the K' valley, were interchanged on purpose in the definition of the wave function $\Psi_{\mathbf{k}}$. As will be shown below, this order of components helps to reveal the Dirac nature of quasiparticles and, in turn, simplifies the analysis of the low-energy theory. Indeed, in terms of the four-component fermion field $\Psi_{\mathbf{k}}$, the low-energy Hamiltonian takes a particularly simple form, i.e.,

$$H_0 \simeq \hbar v_F \int \frac{d^2k}{(2\pi)^2} \Psi_{\mathbf{k}}^{\dagger} \begin{pmatrix} \mathbf{k} \cdot \boldsymbol{\tau} & 0 \\ 0 & -\mathbf{k} \cdot \boldsymbol{\tau} \end{pmatrix} \Psi_{\mathbf{k}}, \tag{2.15}$$

where $\boldsymbol{\tau} = (\tau_x, \tau_y)$ are the standard Pauli matrices acting in the space of two sublattices. As is easy to verify, the structure of Hamiltonian (2.15) is exactly the same as that of the 2D Dirac Hamiltonian for massless relativistic particles, provided the speed of light is replaced by v_F. This can be made even more evident by introducing the Dirac γ-matrices γ^0 and $\boldsymbol{\gamma}$ where

$$\gamma^0 = \begin{pmatrix} 0 & -\mathbb{1}_2 \\ -\mathbb{1}_2 & 0 \end{pmatrix}, \quad \boldsymbol{\gamma} = \begin{pmatrix} 0 & \boldsymbol{\tau} \\ -\boldsymbol{\tau} & 0 \end{pmatrix}, \tag{2.16}$$

and $\mathbb{1}_2$ is the 2×2 unit matrix.[d] Note that the three Dirac γ-matrices γ^μ with $\mu = 0, 1, 2$ generate a reducible 4×4 representation of the Clifford–Dirac algebra $Cl_{1,2}(\mathbb{C})$ in $2+1$ dimensions,[e] i.e., $\gamma^\mu\gamma^\nu + \gamma^\mu\gamma^\nu = 2g^{\mu\nu}$, where $g^{\mu\nu} = \text{diag}(1, -1, -1)$.

---

[d]We use the notations $\gamma^1 = \gamma_x = -\gamma_1$ and $\gamma^2 = \gamma_y = -\gamma_2$ interchangeably.
[e]Here we use the quantum field theory convention for the space–time dimension, i.e., $n+1$, where n is the number of spatial dimensions and 1 represents the time dimension.

By rewriting the Hamiltonian in coordinate space,[f] we obtain

$$H_0 \simeq v_F \int d^2r \bar{\Psi}(\mathbf{r})\boldsymbol{\gamma} \cdot (-i\hbar\boldsymbol{\nabla})\Psi(\mathbf{r}), \qquad (2.17)$$

where $\bar{\Psi}(\mathbf{r}) = \Psi^\dagger(\mathbf{r})\gamma^0$ is the conventional Dirac-conjugate bispinor. As we already mentioned, this explicit Dirac spinor structure of graphene quasiparticles was revealed in 1984 [76, 77]. It is instructive to remember that the "spinor" components of the field $\Psi_\mathbf{k}$ are not connected in any way with the actual spin of electrons. In fact, as is obvious from the derivation, they originate from the two sublattices (A and B) and two valley degrees of freedom (K and K'). For this reason, the corresponding "emergent spin" is called the *pseudospin*. As for the actual electron spin, until now it was ignored in the description. Its inclusion simply implies that the number of quasiparticle fields should be doubled by replacing $\Psi_\mathbf{k}$ with $\Psi_{\mathbf{k},s}$, where $s = \pm$ represents the spin-up ($s = +$) and spin-down ($s = -$) projections. Then the final momentum and coordinate space expressions for the Hamiltonian in Eqs. (2.15) and (2.17), respectively, should be modified by including additional sums over the spin index.

As we will discuss in details below, the relativistic-like nature of quasiparticles in graphene allows for many interesting observable phenomena. It also affects the internal symmetries and the topology of graphene, as well as leads to a wide range of unconventional transport properties.

2.2.2 *Relativistic-like effects in graphene*

In this subsection, the effects directly related to the quasirelativistic character of low-energy quasiparticles in graphene will be discussed. One of the most important features of such charge carriers is the momentum-independent velocity,[g] which is $v_F \approx c/300$ for graphene. In addition, the quasiparticles are characterized by a remarkably high mobility,[h] which

---

[f] Here we use the conventional Fourier transform $\Psi(\mathbf{r}) = \int d^2k/(2\pi)^2 \Psi_\mathbf{k} e^{i\mathbf{k}\cdot\mathbf{r}}$. Since the validity of the low-energy description breaks down at sufficiently large momenta, a suitable momentum cutoff should be introduced.

[g] Recall that, in usual metals, the velocity of the charge carriers is defined as $v = p/m^*$, where m^* is the effective mass and p is the momentum.

[h] The electron mobility, which is denoted by μ_e, is the coefficient that relates the electron drift velocity $\mathbf{v}_d$ to an applied electric field $\mathbf{E}$, i.e., $\mathbf{v}_d = \mu_e\mathbf{E}$. It also determines the partial contribution of the electron quasiparticles to the conductivity, i.e., $\sigma_e = en\mu_e$, where n is the number density of electrons.

is one of the underlying reason for the excellent transport properties of graphene.

Even at room temperature, the experimentally measured electron and hole mobilities in graphene are typically quite high. This has roots in the suppression of backscattering due to the linear energy spectrum and the pseudospin conservation (see also Sec. 2.3.1). Typical values of mobilities are of the order of $10^4 \, \text{cm}^2 \cdot \text{V}^{-1} \cdot \text{s}^{-1}$ [185] and could be much larger in clean samples.[i] Since the linear quasiparticle dispersion relation in graphene is valid in a wide range of energies (i.e., at $|\epsilon_{\mathbf{k}}| \lesssim 1 \, \text{eV}$), the mobility also remains high with increasing the density of charge carriers in doped samples [185] and depends only weakly on temperature. Such a high mobility is important for devices requiring both high conductivity and fast electron response.

The linear energy spectrum is not the only factor determining the properties of graphene's carriers. In particular, the Dirac spinor structure is responsible for several relativistic-like effects, including the Klein paradox, the Schwinger pair production, and the Zitterbewegung ("trembling motion"), which happen to affect the electron transport in a rather profound way [82, 83, 202–204].

Qualitatively, the Zitterbewegung [205] stems from the fact that any attempt to localize massless Dirac particles in coordinate space leads to a virtual creation of electron–hole pairs. Unfortunately, because of the large electron mass $m_e \approx 511 \, \text{keV}/c^2$, the experimental observation of the Zitterbewegung phenomenon in high energy physics is too difficult. However, as first suggested by Mikhail Katsnelson [83], the massless quasiparticles in graphene might be a perfect platform for the realization of Zitterbewegung in a condensed matter setup.

The Klein paradox [206] is related to the fact that ultrarelativistic particles can easily penetrate very high potential barriers exceeding the particle's rest energy. The massless quasiparticles in graphene provide a perfect setup for its realization [82, 207, 208] as was indeed observed experimentally in [209, 210]. It is worth noting that the Klein tunneling of massless Dirac quasiparticles is different from the conventional quantum tunneling. The potential barrier shifts the energy of the Dirac point. Therefore, when a positive energy quasiparticle crosses the barrier it turns into a negative energy state. The corresponding schematics is presented in Fig. 2.3(a). The

---

[i]Typical electron mobility of Si is about $10^3 \, \text{cm}^2 \cdot \text{V}^{-1} \cdot \text{s}^{-1}$.

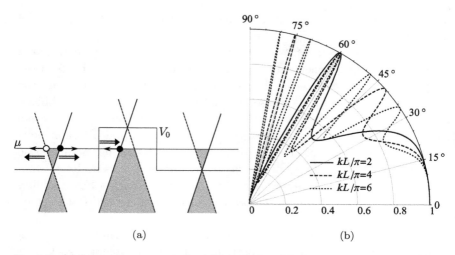

(a)　(b)

Fig. 2.3　(a) Schematic representation of the quasiparticle energy spectrum in graphene, where potential barrier V_0 is applied in the middle of the sample. While the double-lined arrows denote the momenta of quasiparticles, the single-lined ones correspond to the pseudospin. (b) The transmission coefficient T given in Eq. (2.18) for a few values of the wave vector k: $kL/\pi = 2$ (solid line), $kL/\pi = 4$ (dashed line), and $kL/\pi = 6$ (dotted line).

perfect tunneling can be understood in terms of the pseudospin conservation, which is indeed the case in the absence of short-range potentials that act differently on graphene's sublattices and allow for the intervalley scattering. As one can see from Fig. 2.3(a), the electron moving to the right always remains right-moving (i.e., with a positive group velocity as is shown by the thick arrows in Fig. 2.3(a)) because there are matching hole states with the same energy inside the barrier and the pseudospin is conserved. As will be discussed in Sec. 2.3.1, this is the origin of the absence of backscattering in graphene. For quasiparticles with a nonzero incident angle φ (measured from the normal to the barrier), the transmission coefficient T acquires the following simple form [82]:

$$T = \frac{\cos^2 \varphi}{1 - \sin^2 \varphi \cos^2 (k_\perp L)}, \tag{2.18}$$

where L is the width of the barrier and $k_\perp = k \cos \varphi$ is the component of the wave vector normal to it. The corresponding angular dependence is shown in Fig. 2.3(b). It is notable that the barrier remains transparent for $k_\perp = n\pi/L$, where $n = 0, \pm 1, \pm 2, \ldots$.

Since there is no exponential suppression, the Klein tunneling in graphene can be considered as an interband transition [208].[j] Combined together, the Klein paradox and the Zitterbewegung explain the minimal conductivity in graphene $\sigma_{\min} \propto e^2/(2\pi\hbar)$,[k] where the mean-free path of charge carriers in graphene cannot be less than de Broglie wavelength [83].

The same physics that is responsible for the Klein paradox could be also utilized for building the electronic Veselago lenses out of graphene. The conceptual idea behind such lenses goes back to the original paper by Viktor Veselago [214], who considered a possibility of the light refraction on a hypothetical material with a negative refractive index. In such a case, the group velocity of light changes sign at the boundary and leads to the focusing of the electron beam. This proposal was further supported by the theoretical work of John Pendry [215] and later confirmed experimentally in metamaterials [216–218].

In graphene, the Veselago lenses can be realized by p–n junctions, where positive energy quasiparticles on the n-doped side of the potential step transform into negative energy quasiparticles on the p-doped side. Then, as was argued in [219, 220], the resulting transmission of electron quasiparticles through the p–n junction indeed resembles the optical refraction in the Veselago lens and is schematically shown in Fig. 2.4 for the case of the

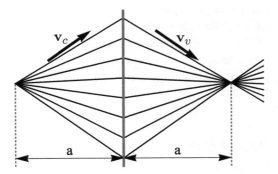

Fig. 2.4 Focusing of electrons by a symmetric p–n junction in graphene. The source is located at distance a to the left of the junction and the image is formed at distance a to the right. The velocities of electrons in the conduction and valence bands are denoted by $\mathbf{v}_c$ and $\mathbf{v}_v$.

---

[j]This can be also interpreted as the Schwinger effect.

[k]The experimental result [167] for the minimal conductivity $\sigma_{\min} \approx e^2/(2\pi\hbar)$ is roughly π times larger than that obtained theoretically [211, 212]. It appears that the minimum value is not universal, but depends on the concentration of defects [213].

vanishing temperature and equal carrier densities on both sides of the junction.[1] The experimental observation of an electronic negative refraction in graphene was reported in 2015 [221].

Finalizing this section, we note that the linear relativistic-like spectrum of quasiparticles is only one of many interesting properties of graphene. In the following sections, we will address its symmetries and topological properties.

2.3 Symmetries in graphene

The relativistic-like Dirac structure of the low-energy Hamiltonian in Eq. (2.17) suggests that the corresponding theory possesses an effective Lorentz symmetry in $2+1$ dimensions with v_F playing the role of the speed of light. Of course, this Lorentz symmetry is broken by interaction effects with electromagnetic fields, lattice defects, impurities, and phonons present in graphene. For example, the electromagnetic interaction is inconsistent with the effective Lorentz symmetry of quasiparticles in graphene because the gauge fields propagate with the speed of light, which is considerably larger than the Fermi velocity. In addition, electromagnetic fields spread into the whole 3D space, while the low-energy quasiparticles are confined to the 2D plane of graphene. Finally, the interaction of quasiparticles with lattice defects, impurities, and phonons implies the existence of a preferred frame of reference and, therefore, explicitly breaks the Lorentz symmetry.

In addition to the unusual space–time symmetry, graphene possesses a number of internal continuous and discrete symmetries that play a rather important role in shaping its physical properties. As will be briefly discussed in the subsections below, some of them are nearly exact, while others are explicitly broken.

2.3.1 *Chirality and chiral symmetry*

The emergence of the four-component Dirac bispinor in the low-energy description of graphene has profound symmetry consequences. In this connection, it is instructive to recall that such a bispinor provides a reducible representation of the Lorentz group in $2 + 1$ dimensions, which can be viewed as a direct sum of two irreducible representations. As a result, the Dirac Hamiltonian written in terms of the four-component bispinor enjoys

---

[1]If temperature is high and the densities of carriers are not fine-tuned to equal values on both sides of the barrier, a sharp image is transformed into a pair of caustics that coalesce in a cusp [219].

a rather high internal symmetry. In particular, it allows one to define the *chiral symmetry* and introduce a conserved quantum number of *chirality*.[m] As will be explained below, the latter is responsible for the suppression of quasiparticle backscattering in graphene.

Chirality, as a quantum number, can be formally identified with the eigenvalue of the following γ_5 matrix:

$$\gamma_5 = i\gamma^0 \gamma_x \gamma_y \gamma_z = \begin{pmatrix} \mathbb{1}_2 & 0 \\ 0 & -\mathbb{1}_2 \end{pmatrix}. \tag{2.19}$$

As is easy to verify, it anticommutes with the Dirac γ-matrices introduced in Eq. (2.16). Note that, in addition to γ_x and γ_y, there is also γ_z matrix, which has the same form as, e.g., γ_x in Eq. (2.16) but with $\tau_x \to \tau_z$. It does not appear in the low-energy Hamiltonian (2.17) and anticommutes with the rest of the Dirac γ-matrices (2.16). The ability to define such an independent 4×4 matrix is related to the existence of an additional internal symmetry in graphene.[n]

The chiral symmetry is the invariance of the Dirac Hamiltonian (2.15) under the chiral transformation $\Psi_{\mathbf{k}} \to e^{i\alpha\gamma_5}\Psi_{\mathbf{k}}$, where α is a real parameter. By taking into account that $(\gamma_5)^2 = \mathbb{1}_4$, it is trivial to find the eigenvalues of the chiral matrix $\zeta = \pm 1$. Since the first two components in the Dirac bispinor $\Psi_{\mathbf{k}}$ correspond to valley K and the other two represent valley K', chirality in graphene can be viewed as the valley degree of freedom.

In the absence of intervalley processes, chirality is conserved. Its conservation implies that the sublattice structure of the wave function is strongly correlated with the direction of the quasiparticle momentum $\mathbf{k}$. Indeed, this can be made explicit by using the Dirac equation for the wave function. For the quasiparticle solutions, one finds that the following identity holds: $\gamma_5 \Psi_{\mathbf{k}} = \pm(\hat{\mathbf{k}} \cdot \boldsymbol{\Sigma})\Psi_{\mathbf{k}}$, where $\hat{\mathbf{k}} = \mathbf{k}/|\mathbf{k}|$ is the unit vector in the direction of the momentum, $\boldsymbol{\Sigma} = \gamma_5\gamma^0\boldsymbol{\gamma} = \text{diag}(\boldsymbol{\tau}, \boldsymbol{\tau})$ is the "pseudospin" operator,[o] and the plus and minus signs correspond to quasiparticles with positive and negative energies, respectively.

---

[m] Strictly speaking, chirality is defined only in odd spatial dimensions. However, it is conventional in the literature to use this term in graphene too.

[n] Because of the high spin–valley symmetry that will be discussed in the following subsection, the notion of chirality is ambiguous for the Dirac Hamiltonian (2.15). One of the alternative definitions of chirality can be associated, for example, with the eigenvalues of the matrix $i\gamma_z$.

[o] The definition of the pseudospin matrix takes a more familiar form when written in components: $\Sigma_k = i\epsilon_{klm}\gamma_l\gamma_m/2$, where $k, l, m = x, y, z$ and ϵ_{klm} is the Levi–Civita tensor with $\epsilon_{xyz} = 1$. Because of the planar nature of graphene, however, the interpretation of $\boldsymbol{\Sigma}$ as a pseudospin should be used with caution.

From a physics viewpoint, the conservation of chirality and pseudospin has an important consequence. It is responsible for the suppression of backscattering of low-energy quasiparticles on disorder. Indeed, if the disorder potential is approximately constant on the scale of the interatomic distance, it is unlikely to change the pseudospin part of the wave function because that would require the mixing of the sublattice components A and B. In addition, the intervalley transfer is negligible for a typical scattering with a small momentum transfer. Then, by combining the conservation of the pseudospin and chirality (valley index), one finds that the backscattering is impossible [197, 222, 223]. With decreasing the range of the disorder potential, of course, the probability of backscattering will gradually increase. Largely, it is expected to be driven by the intervalley scatterings with a large momentum transfer. As will be discussed in Chapters 6 and 7, chirality plays an important role also in the transport properties of 3D Weyl and Dirac semimetals.

2.3.2   *Spin–valley symmetry*

Upon a careful inspection, one finds that the free Hamiltonian (2.15) has a rather high U(4) internal symmetry, associated with the two valleys and the electron spin degrees of freedom. Symmetries of this type are known in the high-energy physics nomenclature as flavor symmetries. The 16 generators of the U(4) spin–valley symmetry are [211]

$$\frac{1}{2}\sigma_\lambda \otimes \mathbb{1}_4, \quad \frac{1}{2}\sigma_\lambda \otimes i\gamma_z, \quad \frac{1}{2}\sigma_\lambda \otimes \gamma_5, \quad \frac{1}{2}\sigma_\lambda \otimes \gamma_z\gamma_5, \qquad (2.20)$$

where $\sigma_\lambda = (\mathbb{1}_2, \boldsymbol{\sigma})$ with $\lambda = 0, \ldots, 4$ are the Pauli matrices acting on the electron spin.

It is interesting to note that the U(4) symmetry is preserved even when the interaction with a background electromagnetic field is introduced through the minimal coupling prescription

$$H_0 \simeq v_F \sum_{s=\pm} \int d^2 r \bar{\Psi}_s(\mathbf{r}) \boldsymbol{\gamma} \cdot \left(-i\hbar\boldsymbol{\nabla} + \frac{e}{c}\mathbf{A} \right) \Psi_s(\mathbf{r}), \qquad (2.21)$$

where $\mathbf{A}$ is the electromagnetic vector potential and $-e$ ($e > 0$) is the electron charge. Similarly, the whole symmetry remains intact when the electric chemical potential μ is introduced via the term $-\mu\Psi_s^\dagger\Psi_s$ and the interaction with the electric scalar potential $\phi(\mathbf{r})$ is included. Note that, up to the replacement of μ with $e\phi(\mathbf{r})$, the interaction with the scalar potential is described by the same term in the Hamiltonian. The U(4) symmetry is also respected by the following electron–electron Coulomb interaction term:

$$H_C = \frac{1}{2} \sum_{s,s'=\pm} \int d^2r d^2r' \Psi_s^\dagger(\mathbf{r}) \Psi_s(\mathbf{r}) U_C(\mathbf{r}-\mathbf{r}') \Psi_{s'}^\dagger(\mathbf{r}') \Psi_{s'}(\mathbf{r}'), \qquad (2.22)$$

where $U_C(\mathbf{r})$ is the Coulomb potential that may include partial screening, but no spin dependence.

The U(4) spin–valley symmetry of the low-energy sector in graphene has several observational implications. Perhaps, one of the simplest of them is the height of the quantization steps of the Hall conductivity [167, 168]. As will be discussed in detail in Sec. 2.7, the corresponding value is given by the minimal quantization step $e^2/(2\pi\hbar)$ multiplied by a factor of 4. The latter is connected with the four-fold degeneracy of the quasiparticle Landau levels in graphene, which is a direct consequence of the U(4) symmetry.

The spin–valley symmetry can be broken by various interaction effects either explicitly or spontaneously. One of the simplest examples of an explicit breakdown is the Zeeman interaction with a background magnetic field B. It is described by the term $\mu_B B \Psi_s^\dagger \sigma_z \Psi_s$ in the low-energy Dirac Hamiltonian, where $\mu_B = |e|\hbar/(2m_e c)$ is the Bohr magneton and m_e is the electron mass. Here, for simplicity, it was assumed that the magnetic field points in the z-direction, which is perpendicular to the plane of graphene. As is easy to check, the Zeeman term breaks the U(4) symmetry down to $U_+(2) \times U_-(2)$, where $U_s(2)$ is the pseudospin symmetry at a fixed spin $s = \pm$. The eight generators of the remaining subgroup are

$$\mathcal{P}_s \otimes \mathbb{1}_4, \quad \mathcal{P}_s \otimes i\gamma_z, \quad \mathcal{P}_s \otimes \gamma_5, \quad \mathcal{P}_s \otimes \gamma_z\gamma_5, \qquad (2.23)$$

where $\mathcal{P}_s = (\mathbb{1}_2 + s\sigma_z)/2$ is the projector on the spin-up ($s = +$) or spin-down ($s = -$) quasiparticle states.

Let us recall that the simplest long-range Coulomb interaction in Eq. (2.22) does not break the $U(4)$ symmetry. However, there might exist other short-range interactions, which explicitly break the spin–valley symmetry in graphene [224]. Despite being small, they play an important role in determining the most stable ground state from a plethora of candidates with nearly degenerate free energies. In the presence of a background magnetic field, for example, such short-range interaction terms could be sufficiently strong to compete with the Zeeman term [225].

Another mechanism of an explicit symmetry breaking could be realized by placing graphene samples on substrates with commensurate lattices.

This can induce periodic out-of-plane lattice distortions and, therefore, provide a simple mechanism for lifting the on-site energy degeneracy between A and B sublattices [226]. Finally, under certain conditions, e.g., in the presence of a strong background magnetic field, the U(4) symmetry in graphene could be also spontaneously broken by the interaction effects. The dynamical symmetry breaking in a magnetic field is a universal phenomenon and is known as the magnetic catalysis [26, 227].

2.3.3　*Discrete symmetries*

In addition to the chiral and spin–valley symmetries, graphene possess a number of discrete symmetries that play an important role in determining its physical properties. The most notable of them are parity-inversion $\mathcal{P}$, time-reversal $\mathcal{T}$, and charge-conjugation $\mathcal{C}$ symmetries. In this subsection, their definitions and physics implications will be discussed. It is worth noting that these discrete symmetries are also very important in 3D Weyl and Dirac semimetals and will be discussed in detail in Sec. 3.2.

Let us start from the parity-inversion transformation $\mathcal{P}$. The corresponding transformation $\mathcal{P}$ involves the inversion of both spatial coordinates in the plane of graphene, $(x, y) \rightarrow (-x, -y)$, provided the sublattice sites A and B are simultaneously interchanged [228]. This leads to a notion of the parity in the low-energy theory of graphene that differs from the conventional notion used in $(2 + 1)$-dimensional QED [229]. Indeed, the latter requires that only one of the spatial coordinates is inverted because a naive inversion of both spatial coordinates is equivalent to a spatial rotation by π. When acting on the spinor fields in momentum space, the parity-inversion transformation in graphene is defined by

$$\mathcal{P}: \quad \Psi_s(\mathbf{k}) \rightarrow \hat{P}\Psi_s(\mathbf{k})\hat{P}^{-1} = \gamma^0 \Psi_s(-\mathbf{k}). \tag{2.24}$$

By taking into account the explicit form of γ^0 in Eq. (2.16), one can check that, as expected, such a transformation exchanges the A and B sublattices as well as the K and K' valleys.

As is clear from the underlying definition, one of the necessary conditions for the parity to be preserved in graphene is the equivalence of the electron properties on the A and B sublattices. Therefore, if for any reason (e.g., due to an induced potential from a substrate or a spontaneous Peierls-type instability) the electron energies on the sites of type A differ from those on the sites of type B, then the parity-inversion symmetry is broken. As was demonstrated already by Gordon Semenoff [76], the corresponding difference of the electron energies $\tilde{\Delta}_s$ induces a nonzero gap between the valence and conduction bands. In the low-energy Hamiltonian, this effect is captured by the Dirac mass term $\tilde{\Delta}_{z,s}\bar{\Psi}_s\gamma_z\Psi_s$,[P] which is parity odd.

In addition to the Dirac mass, there is another type of mass term that could potentially appear in the low-energy Hamiltonian of graphene [228]. It is the Haldane mass, which might be of special interest because it realizes the parity anomaly in $2+1$ dimensions [230] and could trigger the quantum Hall effect in the absence of an external magnetic field. In the Hamiltonian, it is captured by the parity-even term $\Delta_s\bar{\Psi}_s\gamma_z\gamma_5\Psi_s$, where Δ_s is the Haldane mass. It is worth noting that this term plays a qualitatively different role in 3D Weyl and Dirac semimetals. As will be discussed in the following chapter, instead of opening the gap, it shifts the Weyl nodes of opposite chiralities in momentum space. It is known as the chiral shift [231].

In the low-energy theory, the time-reversal transformation $\mathcal{T}$ acts on the Dirac field in momentum space as follows [228]:

$$\mathcal{T}: \quad \Psi_s(\mathbf{k}) \to \mathcal{T}\Psi_s(\mathbf{k})\mathcal{T}^{-1} = i\sigma_y \otimes \gamma_x\gamma_5\Psi_s(-\mathbf{k}). \quad (2.25)$$

As expected, such a transformation reverses the quasiparticle momenta and interchanges the spin-up and spin-down states. The latter, in particular, is achieved by adding the spin matrix $i\sigma_y$ in the definition. By making use of the explicit form of the γ-matrices, one can also verify that the time-reversal transformation interchanges the K and K' valleys, but not the sublattices. This is consistent with the fact that reversing the direction of time should not interchange the A and B sites in graphene. It should be noted that, unlike other symmetries, the time-reversal is an antiunitary transformation. Therefore, when acting on the matrix operators it should contain an additional complex conjugation operator.

Finally, the charge-conjugation transformation $\mathcal{C}$ is defined by [228]

$$\mathcal{C}: \quad \Psi_s(\mathbf{k}) \to \mathcal{C}\Psi_s(\mathbf{k})\mathcal{C}^{-1} = \gamma_x\bar{\Psi}_s^T(\mathbf{k}). \quad (2.26)$$

It is easy to check that this transformation interchanges the roles of electrons and holes, as well as the A and B sublattices.

In addition to the Dirac and Haldane masses $\tilde{\Delta}_{z,s}$ and Δ_s, it is possible to introduce two additional types of masses, $\tilde{\Delta}_s\bar{\Psi}_s\Psi_s$ and $i\tilde{\Delta}_{5,s}\bar{\Psi}_s\gamma_5\Psi_s$. The former does not break any of the three symmetries and corresponds

---

[P]Note that here, by definition, the Dirac mass $\tilde{\Delta}_{z,s}$ has the units of energy.

Table 2.1   Symmetry properties of a few low-energy operators in graphene with respect to the parity-inversion $\mathcal{P}$, time-reversal $\mathcal{T}$, and charge-conjugation $\mathcal{C}$ symmetries.

Name	Expression	$\mathcal{P}$	$\mathcal{T}$	$\mathcal{C}$
Electric charge density	$-e\bar{\Psi}_s\gamma^0\Psi_s$	$+$	$+$	$-$
Electric current density	$-ev_F\bar{\Psi}_s\boldsymbol{\gamma}\Psi_s$	$-$	$-$	$-$
Spin polarization	$\mu_s\Psi_s^\dagger\sigma_z\Psi_s$	$+$	$-$	$-$
$\mathcal{P}$-even Dirac mass	$\tilde{\Delta}_s\bar{\Psi}_s\Psi_s$	$+$	$+$	$+$
$\mathcal{P}$-odd Dirac mass	$\tilde{\Delta}_{z,s}\bar{\Psi}_s\gamma_z\Psi_s$	$-$	$+$	$+$
$\mathcal{C}\mathcal{P}$-odd Dirac mass	$\tilde{\Delta}_{5,s}\bar{\Psi}_s i\gamma_5\Psi_s$	$-$	$+$	$-$
Haldane mass	$\Delta_s\bar{\Psi}_s\gamma_z\gamma_5\Psi_s$	$+$	$-$	$+$

to the true Dirac mass used in high energy physics. It is straightforward to check that it intermixes valleys in the case of graphene. As for the gap $\tilde{\Delta}_{5,s}$, it preserves only $\mathcal{T}$ symmetry. The symmetry properties of the charge density $\bar{\Psi}_s\gamma^0\Psi_s$, the electric current density $\bar{\Psi}_s\boldsymbol{\gamma}\Psi_s$, the spin polarization $\Psi_s^\dagger\sigma_z\Psi_s$, the Dirac masses $\tilde{\Delta}_s\bar{\Psi}_s\Psi_s$, $\tilde{\Delta}_{z,s}\bar{\Psi}_s\Psi_s$ and $\tilde{\Delta}_{5,s}\bar{\Psi}_s i\gamma_5\Psi_s$, as well as the Haldane mass $\Delta_s\bar{\Psi}_s\gamma_z\gamma_5\Psi_s$ are summarized in Table 2.1.

In addition to rich symmetry properties, low-energy Hamiltonian in graphene has nontrivial topological properties. The latter are described in terms of the geometrical or, as it is commonly called, Berry phase and the Berry curvature, which will be discussed in the following section.

2.4   Berry phase and Berry curvature

The nontrivial topology of the quasiparticle Hamiltonian in graphene plays an important role in its transport and optical properties. For example, as will be discussed in Sec. 2.7, it affects the Hall conductivity leading to an anomalous quantization. Topology has an even bigger impact on 3D systems such as Weyl and certain Dirac semimetals, where it stabilizes the band touching points, protects the surfaces Fermi arc states, leads to unusual transport signatures, etc. The corresponding properties of 3D Dirac and Weyl semimetals will be discussed in detail in Chapter 3. Here, however, it is instructive to start by addressing first the topological properties of graphene.

The starting point is Hamiltonian (2.15) for a fixed spin, which is composed of two Hamiltonians for quasiparticles in the K and K' valleys given by $h_{\mathbf{k}}^{(K)} = \hbar v_F(\mathbf{k}\cdot\boldsymbol{\tau})$ and $h_{\mathbf{k}}^{(K')} = -\hbar v_F(\mathbf{k}\cdot\boldsymbol{\tau})$, respectively.

The eigenstates of the Hamiltonian for the K valley are (cf. with Eq. (2.11))

$$u_{+,\mathbf{k}} = \frac{1}{\sqrt{2}} \begin{pmatrix} 1 \\ e^{i\varphi_{\mathbf{k}}} \end{pmatrix}, \quad u_{-,\mathbf{k}} = \frac{1}{\sqrt{2}} \begin{pmatrix} -e^{-i\varphi_{\mathbf{k}}} \\ 1 \end{pmatrix}. \tag{2.27}$$

Here, the plus and minus signs correspond to the positive (electron) and negative (hole) energy states, respectively. By definition, $\varphi_{\mathbf{k}} = \arctan(k_y/k_x)$ is the azimuth angle of momentum $\mathbf{k}$. The eigenstates in the K' valley are obtained by interchanging $u_{+,\mathbf{k}} \leftrightarrow u_{-,\mathbf{k}}$.

The topological aspects of the electron-band structure in graphene can be understood by using the concept of the *Berry phase* [68].[q] In simple terms, the latter is a multiplicative phase factor $e^{i\gamma_n(\mathcal{C})}$ that the quasiparticle wave function acquires during an adiabatic evolution along a contour $\mathcal{C}$ in the phase space (the notions of the Berry phase, the Berry connection, and the Berry curvature will be discussed in more detail in Sec. 3.3). Since the wave functions in quantum mechanics are defined up to a phase factor, one might naively think that the Berry phase should be irrelevant for observable properties. This is not always the case, however. For example, a pair of quasiparticles evolving along different paths in the phase space may get nonequal Berry phases that will cause an interference and produce some distinctive observable effects. Also, a self-interference could arise for a single quasiparticle that gains a nontrivial Berry phase (modulo 2π) for a closed contour.

In the condensed matter setting, a nonzero Berry phase may appear for degenerate electron states [68]. For example, in graphene, the degeneracies occur at the apexes of the Dirac cones, where the valence and conductance bands touch. As we will see, the nontrivial topology in graphene is quantified by the nonvanishing Berry phase $\gamma_{\pm}(\mathcal{C}) = \mp\pi$, where the contour $\mathcal{C}$ encloses a single Dirac node.[r]

The Berry phase for quasiparticles in the nth band is formally defined by [68]

$$\gamma_n(\mathcal{C}) = -\oint_{\mathcal{C}} d\mathbf{k} \cdot \boldsymbol{\mathcal{A}}_{n,\mathbf{k}}, \tag{2.28}$$

where $\boldsymbol{\mathcal{A}}_{n,\mathbf{k}} = -iu_{n,\mathbf{k}}^{\dagger} \boldsymbol{\nabla}_{\mathbf{k}} u_{n,\mathbf{k}}$ is the *Berry connection* and $u_{n,\mathbf{k}}$ is the periodic part of the Bloch wave function in the nth band. In the model at hand,

---

[q]The Berry phase is also often called the geometric phase. For a representative collection of many original papers on the topic, see [232].
[r]The minus and plus signs correspond to the electron wave functions in the conductance and valence bands, respectively.

$n = \pm$ and $\mathcal{C}$ is a closed contour around the corresponding Dirac node. The Berry connection is not gauge invariant. Indeed, when the wave function is multiplied by a phase, i.e., $u_{n,\mathbf{k}} \to u'_{n,\mathbf{k}} = e^{i\alpha_{\mathbf{k}}} u_{n,\mathbf{k}}$, the Berry connection changes as follows: $\mathcal{A}_{n,\mathbf{k}} \to \mathcal{A}'_{n,\mathbf{k}} = \mathcal{A}_{n,\mathbf{k}} + \nabla_{\mathbf{k}} \alpha_{\mathbf{k}}$.

Because of the intrinsic (gauge) ambiguity in the definition of $\mathcal{A}_{n,\mathbf{k}}$, it is useful to introduce instead the gauge-invariant *Berry curvature*. In the 2D case, it is defined by

$$\Omega_{n,\mathbf{k}} = \frac{1}{\hbar} \epsilon_{ij} \partial_{k_i} (\mathcal{A}_{n,\mathbf{k}})_j, \tag{2.29}$$

where ϵ_{ij} is the 2D Levi-Civita tensor with $\epsilon_{xy} = 1$. As will be shown in Sec. 3.3, the Berry curvature in 3D Weyl and certain Dirac semimetal provides a very efficient means to identify the topological features of the electron dynamics. In the case of graphene, the use of the Berry curvature is helpful too, albeit a bit more subtle because of the reduced dimensionality of momentum space [233].

In order to simplify the analysis and avoid infrared ambiguities at intermediate stages of derivation, it is convenient to introduce a regularization parameter, e.g., the Haldane mass term $\Delta \tau_z$, in the low-energy valley Hamiltonian [233]. Then, the corresponding wave functions in the K valley are given by

$$u_{+,\mathbf{k}} = \begin{pmatrix} \cos(\theta_{\mathbf{k}}/2) \\ e^{i\varphi_{\mathbf{k}}} \sin(\theta_{\mathbf{k}}/2) \end{pmatrix}, \quad u_{-,\mathbf{k}} = \begin{pmatrix} -e^{-i\varphi_{\mathbf{k}}} \sin(\theta_{\mathbf{k}}/2) \\ \cos(\theta_{\mathbf{k}}/2) \end{pmatrix}, \tag{2.30}$$

where $\cos\theta_{\mathbf{k}} = \Delta/\sqrt{(\hbar v_F k)^2 + \Delta^2}$. It should be noted that the particular form of the wave functions (2.30), including also the phase factor, is very important for the correct Berry curvature and Berry phase (see also Sec. 3.3). The physical results for the gapless graphene will be then obtained by taking the limit $\Delta \to +0$, which agrees with the definition of the wave functions in Eq. (2.27). It is worth noting that a similar analysis, albeit with the Haldane mass Δ replaced by the z component of the wave vector k_z, also applies to Weyl fermions in 3D and will be discussed in Sec. 3.3. Note that the wave functions in the K' valley are obtained by interchanging $u_{+,\mathbf{k}} \leftrightarrow u_{-,\mathbf{k}}$ in Eq. (2.30).

By applying the Stokes theorem, the contour integral in Eq. (2.28) can be rewritten as a surface integral of the Berry curvature

$$\gamma_+(\mathcal{C}) = -\hbar \int_{S_{\mathcal{C}}} dS \, \Omega_{+,\mathbf{k}}. \tag{2.31}$$

Here $S_\mathcal{C}$ is the surface spanned over the contour $\mathcal{C}$ and the Berry curvature $\Omega_{\pm,\mathbf{k}}$ is given by

$$\Omega_{\pm,\mathbf{k}} = -\frac{i}{\hbar} \sum_{m,n=x,y} \epsilon_{mn} (\partial_{k_m} u^\dagger_{\pm,\mathbf{k}})(\partial_{k_n} u_{\pm,\mathbf{k}})$$

$$= \pm \frac{\Delta}{2\hbar[(\hbar v_F k)^2 + \Delta^2]^{3/2}}. \tag{2.32}$$

It is easy to check that the Berry curvature in the K' valley has the opposite sign. The regularizing role of the auxiliary mass Δ is clear now. Indeed, the Berry curvature at $\mathbf{k} = \mathbf{0}$ strongly depends on the order of limits. If Δ is set to zero first, then the Berry curvature (2.32) vanishes in view of $\lim_{\mathbf{k}\to\mathbf{0}} \lim_{\Delta\to+0} \Omega_{\pm,\mathbf{k}} = 0$. However, it diverges when the order of limits is interchanged, $\lim_{\Delta\to+0} \lim_{\mathbf{k}\to\mathbf{0}} \Omega_{\pm,\mathbf{k}} = \pm\infty$. As will be shown below, this result means that the Berry curvature is singular at $\mathbf{k} = \mathbf{0}$ in the gapless graphene.

The result of the integration in Eq. (2.31) should not depend on the specific choice of surface $S_\mathcal{C}$ spanned over the contour $\mathcal{C}$, provided only that such a surface does not contain the singular point $\mathbf{k} = \mathbf{0}$. For example, let us consider the northern hemisphere for the positive energy wave functions. Then the integral of $\Omega_{\pm,\mathbf{k}}$ is given by the product of the solid angle 2π and the value of the monopole charge $\mp1/2$, which produces $\gamma_\pm(\mathcal{C}) = \mp\pi$. Indeed, for any finite-radius contour around the Dirac point, we have

$$\gamma_+(\mathcal{C}) = -\lim_{\Delta\to+0} \int_0^{2\pi} d\varphi \int_0^{k_{\max}} \frac{\Delta k\, dk}{2[(\hbar v_F k)^2 + \Delta^2]^{3/2}} = -\pi. \tag{2.33}$$

This result together with Eq. (2.32) clearly shows that the Berry curvature in gapless graphene ($\Delta \to +0$) is given in terms of the δ-function, i.e., $\Omega_{\pm,\mathbf{k}} = \pm\pi\delta(k_x)\delta(k_y)$. Before finalizing this section, let us note that the Berry curvature in 3D Weyl semimetals is qualitatively different and is given by a monopole-like expression rather than the δ-function. This leads to richer manifestations of topology in quasiparticle properties.

Having established the symmetry and topology properties of graphene, we can now consider their implications in the electronic transport and discuss their signature effects activated by a magnetic field.

2.5 Electronic transport

Electrical conductivity is one of the most basic transport characteristics that determines the response of materials to an applied electric field.

Depending on several parameters, such as the size of the system, the concentration of charge carriers, the disorder potential, temperature, etc., various regimes of transport can be identified. For example, when the system is sufficiently small and the mean free path of charge carriers, which is determined primarily by the scattering on impurities and phonons, is large enough, a ballistic regime is realized. Another extreme is a hydrodynamic regime, which will be discussed in Sec. 2.5.2. It can be realized in ultra-pure materials at low temperatures when the rate of the electron–electron collisions is larger than the rate of collisions with impurities and phonons. If this is the case, the transport properties of electrons resemble those of a viscous fluid. Finally, the conventional Ohmic regime, where the charge carriers frequently scatter on disorder is also possible. Due to its great tunability, graphene offers a perfect platform for realizing all these types of transport.

It is instructive to explicitly show how the relativistic-like character of electron quasiparticles affects the charge transport in graphene. For illustrative purposes, we consider the simplest case without an external magnetic field. This is done in the next two subsections. The discussion of electric charge transport in graphene will provide a good starting point for the study of the transport properties of 3D Dirac and Weyl semimetals in Chapters 6 and 7.

2.5.1 *Linear response approach*

A convenient and reliable method for calculating the electrical conductivity is the standard Kubo linear response.[s] This approach will be also used for the investigation of transport properties of Weyl and Dirac semimetals in the subsequent chapters. In the linear response method, according to the Kubo formula, the optical (or, equivalently, the alternating current (AC)) electrical conductivity[t]

$$\sigma_{ij}(\Omega) = -\frac{i\hbar}{\Omega}\Pi_{ij}(\Omega + i0; \mathbf{0}) \qquad (2.34)$$

is defined through the retarded current–current correlator $\Pi_{ij}(\Omega + i0; \mathbf{0})$. The latter can be obtained from the corresponding imaginary time

---

[s]The Kubo linear response theory is discussed in many textbooks, e.g., [139, 234–236]. Therefore, only the key notions will be presented here.

[t]Note that conductivity (2.34) includes only the paramagnetic term but not the diamagnetic one. While the latter is exactly zero in the linearized approximation, it is important to reproduce the correct optical sum rule [228].

correlator

$$\Pi_{ij}(i\Omega_l; \mathbf{k}) = -\frac{1}{\hbar^2} \int_0^{\hbar/T} d\tau \int d^2 r \, e^{i\Omega_l \tau - i\mathbf{k}\cdot\mathbf{r}} \langle T_\tau J_i(\tau; \mathbf{r}) J_j(0; \mathbf{0}) \rangle \quad (2.35)$$

by using the analytical continuation $i\Omega_l \to \Omega + i0$. Here T is temperature in the energy units, $\Omega_l = 2l\pi T/\hbar$ is the bosonic Matsubara frequency, and $J_i(\tau; \mathbf{r})$ is the electric current density. In terms of the quasiparticle fields in graphene, the explicit expression for the latter reads

$$J_i(\tau; \mathbf{r}) = -ev_F \sum_s \bar{\Psi}_s(\tau; \mathbf{r}) \gamma^i \Psi_s(\tau; \mathbf{r}). \quad (2.36)$$

Note that unlike the conventional materials with the parabolic energy dispersion relation, the low-energy current operator in graphene (as well as in Dirac and Weyl semimetals) does not contain spatial derivatives.

To leading order in the coupling constant, the current–current correlator is represented by the one-loop diagram shown in Fig. 2.5(a). As will be also discussed in detail in Sec. 7.1, usually, the lowest-order approximation is not sufficient for the calculation of the electrical conductivity because it contains no information about the electron scattering on phonons and impurities. The latter processes are captured by the resummation of higher order diagrams shown in Figs. 2.5(b), 2.5(c), and 2.5(d), where the dashed lines represent either the phonon propagator or the impurity correlator.[u] Note, however, that the leading-order contribution might be sufficient for the optical conductivity at sufficiently high frequencies because the collisionless limit becomes a good approximation.

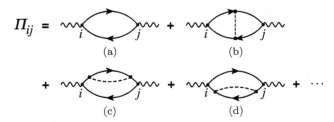

Fig. 2.5 Diagrammatic representation of several leading- and subleading-order contributions to the current–current correlator Π_{ij}. Solid lines denote the electron Green functions. Dashed lines represent either phonon propagator or impurity correlation function.

---

[u]The same types of higher order diagrams could also describe the electron–electron scattering if the photon propagator is used instead.

For illustrative purposes, let us calculate the conductivity by taking into account only the leading-order diagram in Fig. 2.5(a). The effects of disorder can be taken into account phenomenologically by introducing a nonzero quasiparticle width $\Gamma(\omega)$ in the Green function. Such a treatment is qualitatively equivalent to taking the self-energy corrections in Figs. 2.5(c) and 2.5(d) into account. Here we will ignore the vertex correction, see Fig. 2.5(b), which, however, could be important in the limit of small Γ and/or small frequencies. The retarded and advanced Green functions at a fixed spin are given by

$$G^{\mathrm{R/A}}(\omega, \mathbf{k}) = i\frac{[\hbar\omega \pm i\Gamma(\omega)]\gamma^0 - \hbar v_F(\mathbf{k} \cdot \boldsymbol{\gamma})}{[\hbar\omega \pm i\Gamma(\omega)]^2 - (\hbar v_F k)^2}. \tag{2.37}$$

The current–current correlator reduces to

$$\Pi_{ij}(\Omega + i0; \mathbf{0}) = e^2 v_F^2 \int\int d\omega d\omega' \frac{f^{\mathrm{eq}}(\omega') - f^{\mathrm{eq}}(\omega)}{\omega - \omega' - \Omega - i0}$$

$$\times \int \frac{d^2 k}{(2\pi)^2} \operatorname{tr}[\gamma_i A(\omega, \mathbf{k})\gamma_j A(\omega', \mathbf{k})], \tag{2.38}$$

where $f^{\mathrm{eq}}(\omega) = 1/[e^{(\hbar\omega - \mu)/T} + 1]$ is the equilibrium Fermi–Dirac distribution function and the quasiparticle spectral density is given by

$$A(\omega, \mathbf{k}) = \frac{1}{2\pi}[G^{\mathrm{R}}(\omega, \mathbf{k}) - G^{\mathrm{A}}(\omega, \mathbf{k})]$$

$$= \frac{\Gamma}{2\pi k}\left[\frac{k\gamma^0 - (\mathbf{k} \cdot \boldsymbol{\gamma})}{\hbar^2(\omega - v_F k)^2 + \Gamma^2(\omega)} + \frac{k\gamma^0 + (\mathbf{k} \cdot \boldsymbol{\gamma})}{\hbar^2(\omega + v_F k)^2 + \Gamma^2(\omega)}\right]. \tag{2.39}$$

By substituting this into Eq. (2.38), one finds that the only nontrivial components of the conductivity tensor are the diagonal ones, i.e., $\sigma_{ij}(\Omega) = \delta_{ij}\sigma(\Omega)$.

By assuming a small quasiparticle width $\Gamma(\omega)$ that has a weak dependence on frequency, i.e., $\Gamma(\omega) \approx \Gamma(\omega - \Omega)$, the explicit expression for $\sigma(\Omega)$ reads [237, 238]

$$\sigma(\Omega) = \frac{e^2}{8\pi} \int d\omega \frac{f^{\mathrm{eq}}(\omega - \Omega) - f^{\mathrm{eq}}(\omega)}{\Omega} \frac{(|\omega| + |\omega - \Omega|)[\omega^2 + (\omega - \Omega)^2]}{\omega(\omega - \Omega)}$$

$$\times \left[\frac{2\Gamma(\omega)}{(\hbar\Omega)^2 + 4\Gamma^2(\omega)} - \frac{2\Gamma(\omega)}{\hbar^2(\Omega + 2\omega)^2 + 4\Gamma^2(\omega)}\right]. \tag{2.40}$$

This result takes into account only the contribution of Dirac fermions with a single spin projection. After including both spins, therefore, the result in Eq. (2.40) should be doubled.

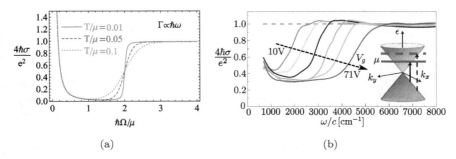

(a)   (b)

Fig. 2.6   (a) The real part of the optical conductivity $\sigma(\Omega)$ given by Eq. (2.40). The quasiparticle width is $\Gamma(\omega) = 0.01\hbar\omega/\mu$. Temperature is $T/\mu = 0.01$ (red solid lines), $T/\mu = 0.05$ (blue dashed lines), and $T/\mu = 0.1$ (green dotted lines). (b) The real part of the optical conductivity measured in graphene on a SiO_2/Si substrate at several gate voltages (dashed arrow) and $T = 45$ K. Inset shows the mechanism of the Pauli exclusion principle in graphene, where the transitions with $\hbar|\Omega| < 2|\mu|$ are suppressed. (Adapted from [239].)

The optical conductivity given in Eq. (2.40) is plotted in Fig. 2.6(a), where the quasiparticle width is assumed to depend linearly on frequency, i.e., $\Gamma(\omega) \propto \hbar\omega$. Such a dependence can be indeed expected in the Born approximation for weak scattering (see, e.g., [190, 238]). One can clearly see that the optical conductivity has a characteristic step-like feature at $\hbar|\Omega| = 2|\mu|$. This behavior can be seen most clearly from the expression for the conductivity (2.40) in the limit $\Gamma \to 0$ and $T \to 0$, which takes the following explicit form:

$$\sigma(\Omega) \simeq \frac{e^2}{\hbar}|\mu|\delta(\hbar\Omega) + \frac{e^2}{4\hbar}\theta\left(\frac{\hbar|\Omega|}{2} - |\mu|\right). \qquad (2.41)$$

Physically, the step function θ originates from the Pauli exclusion principle, which forbids transitions between fully occupied states. The same feature also occurs in 3D Dirac and Weyl semimetals, whose optical conductivity will be discussed in Sec. 7.3.1. The theoretical results in graphene are in a good agreement with the experimental data [173, 239] shown in Fig. 2.6(b), where the step-like feature also appears. Its position can be tuned by changing the gate voltage V_g, which regulates the electric chemical potential. It is worth noting that the conductivity at high frequencies $\sigma(\hbar\Omega \gg \mu) = e^2/(4\pi\hbar)$ is well reproduced at all values of the Fermi energy, which is in excellent agreement with the theoretical prediction (cf. Figs. 2.6(a) and 2.6(b)).

Let us note that graphene becomes an optically active medium in the presence of a background magnetic field. The high mobility of its charge carriers and the linear dispersion allow for a rather large Faraday rotation for the infrared light even at moderately strong magnetic fields [174]. Depending on the frequency, the typical value of the Faraday rotation angle can reach $0.1\,\mathrm{rad} \approx 6°$ [174], which is truly gigantic for a monolayer material. As will be discussed in Sec. 7.3.3, the Faraday and Kerr rotations in Dirac and Weyl semimetals are estimated to be large too. In addition, they are strongly affected by the topological properties of these materials and, in particular, the presence of the surface Fermi arc states.

Having established the basics of the linear response approach, where interaction effects are not crucial, it is interesting to also discuss the regime with a strong electron–electron interaction. When the rate of scattering of electrons on phonons, impurities, and defects is small compared to that of the electron–electron scattering, a collective electron flow forms. The transport properties of graphene in this regime resembles those of a viscous fluid described in terms of conventional hydrodynamics. The corresponding charge transport will be briefly considered in the following subsection.

2.5.2   *Hydrodynamic regime*

As was already mentioned in Sec. 2.5.1, the electron transport in solids under certain conditions is described by hydrodynamic equations. Originally, this regime was proposed by Radii Gurzhi in the 1960s [240, 241]. Gurzhi argued that electron quasiparticles in solids could have a sufficiently small electron–electron collision mean free path l_{ee} to form an *electron fluid*.

The hydrodynamic regime of electron transport is particularly suitable for graphene, where the effects of disorder are significantly reduced [190, 242, 243]. Even the scattering on acoustic phonons, which is dominant at high temperatures, can be weak in graphene. Indeed, the corresponding mean free path decreases with temperature as $\sim 1/T$, which is slower than its electron–electron counterpart $\sim 1/T^2$ (at least in the Fermi liquid regime). Therefore, it is possible to realize a hydrodynamic window, i.e., the range of parameters that allows for the formation of the electron fluid, in a finite range of temperatures [244].

The formation of the electron fluid leads to several interesting effects, such as the decrease of resistance with temperature (the Gurzhi effect), the characteristic quadratic dependence of resistivity on the width of the conducting channel (Poiseuille flow), a number of viscosity-related effects, and, possibly, even turbulence. Note that similar effects of *electron hydrodynamics*[v] could also be realized in 3D Dirac and Weyl semimetals, see Sec. 7.4 for more details. Here, we consider only a few characteristic manifestations of the electron hydrodynamic regime in graphene.

The hydrodynamic regime in graphene was studied theoretically in [245–260] and observed experimentally in [261–267]. It is important to emphasize that, unlike conventional metals, graphene allows one to realize the hydrodynamic regimes with either one or two fluid components. The corresponding phase diagram in the plane of the electric chemical potential μ and temperature T is given in Fig. 2.7. The one-component regime of the electron fluid is realized when the electric chemical potential exceeds temperature $|\mu| \gg T$. In this case, either electrons ($\mu > 0$) or holes ($\mu < 0$) dominate the transport. This Fermi liquid regime is similar to the one considered by Gurzhi in conventional solid-state systems. The two-component regime with the interacting electron and hole liquids takes place when $|\mu| < T$ and is known in the literature as the *Dirac fluid*. This regime relies on the particle–hole symmetry and is unique for systems with relativistic-like dispersion relation. Below we will briefly review both Fermi and Dirac fluid regimes.

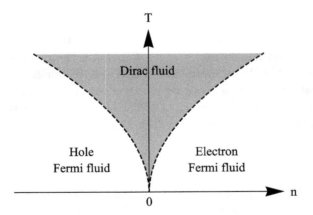

Fig. 2.7 The phase diagram of graphene as a function of the carrier density n and temperature T showing the Dirac (shaded region) and Fermi fluid regimes with the quantum critical point occurring at $n = T = 0$.

---

[v]Since electrons are electrically charged, taking into account Maxwell's equations is important in the self-consistent description. It is more appropriate, therefore, to call the corresponding framework electro-magneto-hydrodynamics.

From experimental viewpoint, electron hydrodynamics in graphene could be revealed by the negative nonlocal resistance and the formation of current vortices [249, 251–253, 260] or higher than ballistic conduction in samples with constrictions [254, 263]. Note that the profile of electric currents in the hydrodynamic regime can be visualized by measuring the stray magnetic fields [266] or the Hall field across the graphene channel [267]. One might also argue that the effects of the electron viscosity could be detected by observing the Dyakonov–Shur instability [268] or a certain characteristic frequency-dependent response in the Corbino disk [269].

2.5.2.1 *Fermi fluid*

While the detailed discussion of the hydrodynamic equations, i.e., the Navier–Stokes equation, continuity relations, etc., is postponed until Sec. 7.4, where Weyl and Dirac semimetals will be discussed, the key hydrodynamic features of the electron transport in graphene can be captured via the following steady-state linearized equations:

$$\eta_{\text{kin}}\Delta\mathbf{u} - \frac{\mathbf{u}}{\tau} = \frac{env_F^2}{w}\mathbf{E}, \qquad (2.42)$$

$$\boldsymbol{\nabla}\cdot\mathbf{u} = 0, \qquad (2.43)$$

where $\mathbf{u}$ is the fluid velocity, τ is the relaxation time due to collisions with phonons and impurities, $w = \epsilon + P$ is the enthalpy, ϵ is the energy density, and $P \simeq \epsilon/2$ is the pressure of a 2D relativistic-like electron gas. The kinematic viscosity η_{kin} can be re-expressed via the electron–electron scattering time τ_{ee} as $\eta_{\text{kin}} = v_F^2\tau_{\text{ee}}/4$ [270]. The electron–electron collision time τ_{ee} in the Fermi fluid scales as [271–273]

$$\tau_{\text{ee}} \sim \frac{\hbar\mu}{\alpha^2 T^2}, \qquad (2.44)$$

where $\alpha = e^2/(\hbar v_F \varepsilon_e)$ is the effective coupling constant and ε_e the dielectric permittivity of a substrate. Note that, in the Fermi fluid regime when the fluid velocity is small $u \ll v_F$, the heat transport can be usually neglected (see, e.g., [274]).

In addition to the equations of motion, the boundary conditions for the hydrodynamic flow should be specified. Their explicit form reads

$$[\partial_y u_x + \partial_x u_y]|_{\text{edge}} = \frac{1}{l_{\text{slip}}} u_{\parallel}|_{\text{edge}}, \qquad (2.45)$$

where $u_\parallel$ is the tangential component of the fluid velocity with respect to the edge of the sample and l_{slip} is the slip length. While the no-slip boundary conditions (i.e., when the fluid perfectly sticks to the surfaces) are obtained in the limit $l_{\text{slip}} \to 0$, the free-surface ones (i.e., when no stress is applied to the surface of the fluid) correspond to the limit $l_{\text{slip}} \to \infty$.

For illustrative purposes, let us demonstrate a few representative features of the electron hydrodynamics in graphene. As one can see by comparing Figs. 2.8(a) and 2.8(b), the viscous electron fluid allows for the formation of vortices and the sign change of the electric potential near the contacts.[w] Another interesting effect is the higher than ballistic conductivity of a graphene constriction with the opening at $|x| < w/2$, see Fig. 2.9. Qualitatively, this effect is explained by the fact that the electric current in the viscous regime concentrates primarily in the middle of the constriction, $J \propto \sqrt{(w/2)^2 - x^2}$, avoiding the edges where the momentum dissipation occurs. In the ballistic regime, on the other hand, the current profile is homogeneous that causes a more significant momentum dissipation.

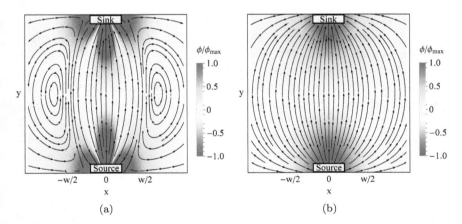

(a) (b)

Fig. 2.8   The electric current streamlines (black lines) and the electric potential map for the (a) viscous and (b) Ohmic flows.

---

[w]It is worth noting that the sign change of the electric potential is not necessarily accompanied by the formation of vortices or whirlpools but strongly depends on the geometry of the setup [251].

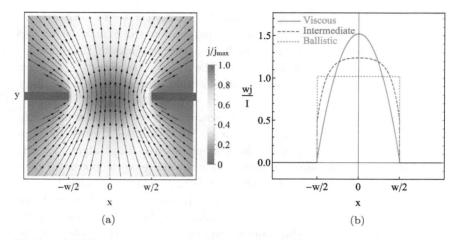

(a)   (b)

Fig. 2.9   (a) The electric current streamlines (black lines) and the current density in a graphene constriction. (b) Schematic electric current profile in the constriction for the viscous, intermediate, and ballistic transport regimes.

2.5.2.2   *Dirac fluid*

Let us turn to the *Dirac fluid* regime of hydrodynamic transport in graphene, which takes place at small charge density. The electrons and holes in the graphene plasma near the charge neutrality point $\mu = 0$ strongly interact through the weakly screened Coulomb interaction. This regime should be treated by using the framework of two-fluid hydrodynamics [245, 247, 248, 255–258]. Indeed, for systems with the relativistic-like dispersion relation, the separate conservation for the electron (e^-) and hole (h^+) numbers follows from the kinematic constraints of the following number-changing processes $e^- \to e^- + e^- + h^+$ and $h^+ \to h^+ + h^+ + e^-$. From the energy and momentum conservation, they are allowed only if all quasiparticles are collinear. The independent (approximate) conservation of the electron and hole numbers allows one to introduce the so-called imbalance mode for which the charge density is defined as the sum of the electron and hole components $\rho_{\text{imb}} = -e(n_e + n_h)$ rather than the difference, as for the electric charge $\rho = -e(n_e - n_h)$. As we will see in subsequent chapters, the presence of the additional mode is somewhat reminiscent to the chiral charge in Weyl and Dirac semimetals.

The equations of motion for the Dirac fluid are similar to those in the Fermi fluid regime (2.42) and (2.43) but should be written for both electron and hole fluids. Also, additional collision terms accounting for the interfluid scattering should be added to the Navier–Stokes and continuity equations. In this regime, the heat transport cannot be neglected. The corresponding steady-state linearized equations in the absence of magnetic fields read

$$\eta_{\text{kin}} \Delta \mathbf{u}_i - \frac{\mathbf{u}_i}{\tau_i} - \frac{\mathbf{u}_i - \mathbf{u}_j}{\tau_{\text{eh}}} = \frac{e_i n_i v_F^2}{w_i} \mathbf{E}, \tag{2.46}$$

$$\nabla \cdot \mathbf{u}_i = -\frac{n_e^{(1)} + n_h^{(1)}}{\tau_{\text{eh}}}, \tag{2.47}$$

$$-\kappa \nabla \cdot \left(\nabla T - \frac{T}{w_i} \nabla P_i \right) + w_i (\nabla \cdot \mathbf{u}_i) = -e n_i (\mathbf{E} \cdot \mathbf{u}_i) + Q_i, \tag{2.48}$$

where $i, j = (e, h)$ is the electron (e) and hole (h) index, $e_e = -e_h = e$, τ_i is the impurity relaxation time, τ_{eh} is the momentum relaxation time due to electron–hole scattering, and $n_i^{(1)}$ is the deviation of the corresponding carrier density from its equilibrium value. In addition, κ is the thermal conductivity and Q_i denote the heat sink rates for the electrons $(i = e)$ and the holes $(i = h)$.

The conservation law for the number of electrons and holes is, however, not exact. The imbalance mode decays due to higher-order processes with the characteristic time $\tau_{\text{imb}} \sim \hbar/(\alpha^4 T)$, which is long compared to the electron–electron collision time

$$\tau_{\text{ee}} \sim \frac{\hbar}{\alpha^2 T} \tag{2.49}$$

in a weak-coupling regime. This expression follows from the dimensional analysis because temperature is the only energy scale at the charge neutrality point.

While the electric charge transport is quite nontrivial in the Dirac fluid regime, one of its main manifestation is the unusual heat response. Indeed, the heat and charge transport channels are decoupled at the charge neutrality point. This leads to a few interesting effects. Among the most notable of them is a strong violation of the Wiedemann–Franz law (see also Sec. 7.2)

$$\kappa = \frac{\pi^2 T \sigma}{3 e^2} \tag{2.50}$$

reported in [261]. Here σ and κ are the electrical and thermal conductivities, respectively. [For simplicity, we ignored the tensor structure of thermal and electrical conductivities.] The physics behind this strong violation is simple and can be easily explained. While electrons and holes move in the opposite directions for the electric current, they move in the same direction for the thermal one. These two types of collective behavior are affected differently by disorder that ultimately leads to the breakdown of the Wiedemann–Franz law. Indeed, collisions between electrons and holes introduce a frictional dissipation that results in a finite conductivity even in the absence of disorder [275]. In contrast, the carriers of heat current are backscattered only on impurities and, consequently, the thermal conductivity is limited solely by disorder. In the end, the Lorenz ratio, defined as

$$\mathcal{L} = \frac{\kappa}{\sigma T}, \qquad (2.51)$$

exceeds its Fermi liquid value $\mathcal{L}_0 = \pi^2/(3e^2)$ as was theoretically predicted in [245, 276]. The experimental data obtained in [261] are shown in Fig. 2.10. It is remarkable that the peak value of the Lorenz ratio is rather large, $\mathcal{L}_{\max} \approx 22$. In agreement with the hydrodynamic regime, the breakdown of the Wiedemann–Franz law occurs in a finite window of temperatures and carrier densities. Note that the law remains approximately valid at low temperatures because there are fluctuations of the local chemical potential due to disorder-induced charge puddles that destroys the Dirac fluid. At high temperatures, the heat transfer to phonons becomes important and the Dirac fluid also deteriorates.

In summary, graphene provides one of the best examples of the hydrodynamic regime in solids. Moreover, not only the conventional manifestations

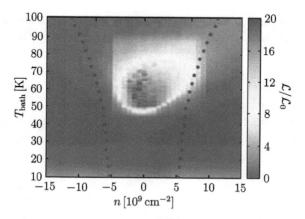

Fig. 2.10 The Lorenz ratio $\mathcal{L}/\mathcal{L}_0$ as a function of the carrier density n and the bath temperature T_{bath}. Green dotted lines show the corresponding minimal carrier density $n_{\min}(T)$ due to disorder, which is extracted from the conductivity measurements. (Adapted from [261].)

of hydrodynamic flow are realized, but also novel effects are predicted as well. The ability to tune graphene between different transport regimes makes it a perfect playground for testing various phenomena beyond the standard Drude picture. Interestingly, it has been suggested that the Dirac fluid can behave as a nearly perfect fluid [277] with the ratio of the shear viscosity η to the entropy density s approaching the conjectured absolute minimum, i.e., $\eta/s = \hbar/(4\pi)$, for strongly interacting systems [278]. Moreover, the range of parameters, which makes electronic preturbulence possible, was identified in [279, 280].

In addition to the exciting fundamental implications, graphene might also find practical applications in viscous electronics. It is possible, for example, that the formation of electron fluid might enhance conductive properties of nanodevices. The breakdown of the Wiedemann–Franz law could also be used for creating thermoelectric devices with high thermal conductivity. Other possible applications will definitely follow with the advancement of the field.

2.6 Landau levels and quantum oscillations

As will be discussed in this section, the presence of a background magnetic field affects both orbital motion and spin dynamics of electron quasiparticles in graphene, providing a direct way to probe their Dirac spinor structure.

To start with, let us consider the structure of the electron wave functions and the energy spectrum in a magnetic field. As usual, the interaction with electromagnetic field is introduced by replacing the momentum operator $-i\hbar\nabla$ in Hamiltonian (2.17) with the canonical momentum $\boldsymbol{\pi} = -i\hbar\nabla + e\mathbf{A}/c$, where $\mathbf{A}$ is the background vector potential and $-e$ $(e > 0)$ is the charge of the electron. One should also include the Zeeman energy associated with the intrinsic magnetic moment of quasiparticles $\mu_B B \Psi^\dagger \sigma_z \Psi$,[x] where $\mu_B = e\hbar/(2mc)$ is the standard Bohr magneton, B is the magnetic field strength (by assumption, the field is normal to the plane of graphene, $\mathbf{B} \parallel \hat{\mathbf{z}}$) and σ_z is the Pauli matrix acting in the spin space. Therefore, the effective Hamiltonian in a background constant magnetic field reads

$$H_{B,0} \simeq \int d^2r\,\bar{\Psi}(\mathbf{r})[v_F(\boldsymbol{\gamma}\cdot\boldsymbol{\pi}) - \mu\gamma^0 + \mu_B B\sigma_z \otimes \gamma^0]\Psi(\mathbf{r}), \qquad (2.52)$$

---

[x]For the sake of simplicity, we suppress the subscript s of the wave function Ψ in this Section.

where μ is the electric chemical potential. It is convenient to use the Landau gauge for the vector potential $\mathbf{A} = (0, Bx)$. The eigenvalues and the eigenstates of the above Hamiltonian can be found by solving the following Dirac equation:

$$\gamma^0[v_F(\boldsymbol{\gamma} \cdot \boldsymbol{\pi}) - \mu\gamma^0 + \mu_B B\sigma_z \otimes \gamma^0]\Psi(\mathbf{r}) = \epsilon_n \Psi(\mathbf{r}). \qquad (2.53)$$

This eigenvalue problem can be reduced to the harmonic oscillator problem. It will be considered in more detail for the 3D case in Sec. 6.4.1. Here, by omitting the details of derivation, let us present only the final results. The eigenvalues or, equivalently, the Landau level energies, are given by

$$\epsilon_{n,s} = -\mu_s \pm \frac{\sqrt{2}\hbar v_F}{l_{\mathrm{B}}}\sqrt{n}, \qquad (2.54)$$

where $l_{\mathrm{B}} = \sqrt{c\hbar/|eB|}$ is the magnetic length, $n = 0, 1, 2, \ldots$ labels the discrete Landau levels, and $\mu_s = \mu - s\mu_B B$ is the effective chemical potential for the quasiparticles with a fixed spin projection, $s = \pm$.

The corresponding expressions for the Landau level eigenstates with a given spin projection $(s = \pm)$ can be written as follows:

$$\Psi_{n,k,\zeta}(\mathbf{r}) = [(\epsilon_{n,s} + \mu_s)\gamma^0 - v_F(\boldsymbol{\gamma} \cdot \boldsymbol{\pi})]\mathcal{P}_\zeta \Phi_{n_\zeta,k}(\mathbf{r}), \qquad (2.55)$$

where $\mathcal{P}_\zeta = (1 + i\zeta\gamma_x\gamma_y)/2$ are the projectors onto the states with fixed pseudospins $(\zeta = \pm)$. The indices $n_+ = n - 1$ and $n_- = n$ label the orbital states for $\zeta = \pm$, respectively. One can verify that only one projection of pseudospin $(\zeta = -)$ is allowed in the lowest Landau level $(n = 0)$. The explicit form of the orbital part of eigenfunctions is given by

$$\Phi_{n,k}(\mathbf{r}) = \frac{1}{\sqrt{2\pi l_{\mathrm{B}}}} \frac{1}{\sqrt{2^n n! \sqrt{\pi}}} H_n(\xi) e^{-\xi^2/2} e^{ik_y y}, \qquad (2.56)$$

where $H_n(x)$ are the Hermite polynomials [281] and $\xi = (x + k_y l_{\mathrm{B}}^2)/l_{\mathrm{B}}$. By construction, functions $\Phi_{n,k}$ are the eigenstates of the operator $\boldsymbol{\pi}^2$ with eigenvalues $(2n + 1)\hbar^2/l_{\mathrm{B}}^2$. They also form an orthonormal and complete set of states (see also Eqs. (A.21) and (A.22) in Appendix A.2).

Note that eigenfunctions (2.56) depend on the y component of the wave vector k_y. Moreover, the value of k_y determines the location of the center of quasiparticle orbit in the x-direction, i.e., $x_c = -k_y l_{\mathrm{B}}^2$. Since the Landau level energies are independent of k_y, there is a large degeneracy of the corresponding states. It is not difficult to determine the number of degenerate states per unit area in coordinate space. To demonstrate this, let us assume that the sample has the dimensions L_x and L_y in the x- and y-directions, respectively. For the boundary conditions periodic in the y-direction, k_y becomes quantized $k_y = -2\pi j/L_y$, where j is an integer. Since k_y determines the location of the center of quasiparticle orbits, $x_c = -k_y l_B^2$, which should lie within the sample $0 \leq x_c \leq L_x$, one obtains the constraint $0 \leq j \leq L_x L_y/(2\pi l_B^2)$. This implies that the number of degenerate states in each Landau level is $N_{s,\varsigma} L_x L_y/(2\pi l_B^2)$, where $N_{s,\varsigma} = 4$ takes into account the degeneracy in the spin and valley degrees of freedom.[y]

A special role in the dynamics of Dirac fermions in a magnetic field is played by the lowest Landau level, which, for massless quasiparticles in graphene, describes states with zero energy.[z] It is important to note that these states are topologically protected. According to the Atiyah–Singer index theorem [283], their number is determined by the magnetic field flux. Indeed, for each valley and spin, there must exist exactly $[\Phi/\Phi_0]$ quantum states with the zero energy [284], where $\Phi = BL_x L_y$ is the total magnetic flux, $\Phi_0 = 2\pi \hbar c/e$ is the magnetic flux quantum, and [...] denotes the integer part. As will be shown in Chapter 6, the lowest Landau level plays a very important role in the transport properties of 3D Dirac and Weyl semimetals too, where it is responsible for the celebrated chiral anomaly.

It is instructive to compare the obtained results for the Landau levels in graphene with those for a 2D nonrelativistic electron gas (see, e.g., [285]). The Landau level energies in the latter case are given by

$$\epsilon_n^{(\mathrm{NR})} = \hbar \omega_c \left(n + \frac{1}{2} \right), \tag{2.57}$$

where $\omega_c = |eB|/(m^*c)$ is the cyclotron frequency and m^* is the effective mass. In contrast to the nonrelativistic spectrum (2.57) with an even spacing between the Landau levels, the spacing in the Dirac case given by Eq. (2.54) decreases with energy. Furthermore, the characteristic energy scale $\hbar \omega_c$ in a nonrelativistic gas is relatively small even when the magnetic field is strong and the effective electron mass is small. For example, by using the value of the effective electron mass in gallium arsenide (GaAs) $m^* \approx 0.068 m_e$ [286], one finds $\hbar \omega_c \approx 1.70\, B[\mathrm{T}]\,\mathrm{meV}$, where $B[\mathrm{T}]$ is the value of the magnetic field in teslas. For comparison, the characteristic

---

[y]Note that the Zeeman energy lifts slightly the degeneracy of the spin-up and spin-down states. Also, while there is only one pseudospin state in the lowest Landau level, the total degeneracy of the level is still four because of the "accidental" degeneracy of the electron and hole states with zero energy.

[z]In fact, the structure of the lowest Landau level in the Dirac spectrum could be interpreted as yet another consequence of the nontrivial Berry phase [282].

Landau energy scale in graphene is an order of magnitude larger, i.e.,
$\epsilon_L = \hbar v_F / l_B \approx 25.6\sqrt{B[\mathrm{T}]}$ meV.

Experimentally, the distinctive structure of Landau levels in graphene can be tested by numerous probes. As will be also discussed in the case of 3D Dirac and Weyl semimetals in Sec. 3.9, one of the most direct probes is the quantum oscillations in a magnetic field [287–289]. Conceptually, the oscillatory dependence of the density of states (DOS) can be achieved by sweeping the relative position of the Fermi level with respect to the Landau levels. This can be easily done, for example, by changing the external magnetic field or the gate voltage. Of course, the oscillations of DOS, in turn, cause oscillations of transport coefficients and thermodynamic characteristics [290]. The hallmark examples include the Shubnikov–de Haas oscillations of the electrical conductivity [167, 168] and the de Haas–van Alfven oscillations of the magnetization. The latter were studied theoretically in [291], but have not been observed directly. On the other hand, the measurements of the quantum capacitance confirm the oscillations of the DOS [290].

In order to understand the underlying mechanism behind quantum oscillations, let us compute the DOS in graphene. In the absence of a magnetic field, it is given by

$$\nu(\epsilon) = 4 \int \frac{d^2 k}{(2\pi)^2} \delta(\epsilon - \epsilon_\mathbf{k}) = 2\frac{|\epsilon|}{\pi \hbar^2 v_F^2}, \qquad (2.58)$$

where $\delta(x)$ is the Dirac δ-function, the factor 4 comes from the spin and valley degeneracy, and the energy dispersion relation $\epsilon_\mathbf{k} = \hbar v_F |\mathbf{k}|$ is used. This DOS is in a drastic contrast to the DOS of 2D nonrelativistic fermions with a quadratic dispersion relation $\epsilon_\mathbf{k} = \hbar^2 \mathbf{k}^2/(2m^*)$, which is independent of energy, i.e., $\nu(\epsilon) = m^*/(\pi \hbar^2)$.

In the presence of a magnetic field, the following DOS can be obtained (see also Sec. 3.9.1):

$$\nu(\epsilon) = 2\frac{|eB|}{\pi c \hbar} \sum_{n=0}^{\infty} \delta(\epsilon - \epsilon_n). \qquad (2.59)$$

For real samples, the δ-functions in the DOS are broadened because of the effects of disorder, temperature, etc. Phenomenologically, this can be captured by replacing the δ-functions in Eq. (2.59) with the Gaussian distributions, i.e.,

$$\delta(\epsilon - \epsilon_n) \to \frac{1}{\sqrt{2\pi}\Gamma} \exp\left[-\frac{(\epsilon - \epsilon_n)^2}{2\Gamma^2}\right], \qquad (2.60)$$

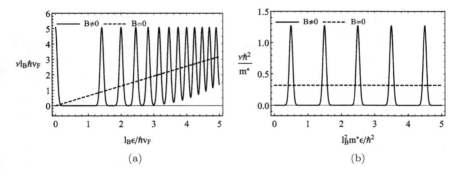

Fig. 2.11 Schematic DOS for the low-energy Dirac quasiparticles in (a) graphene and (b) nonrelativistic quasiparticles. While the results for graphene are shown for $\Gamma = 0.05\,\hbar v_F/l_B$, those for a 2D nonrelativistic gas are presented for $\Gamma = 0.05\,\hbar^2/(l_B^2 m^*)$.

where, for simplicity, the same constant value of the quasiparticle width Γ is used for all Landau levels.

Schematically, the DOS for graphene and a 2D gas of nonrelativistic electrons are plotted in Figs. 2.11(a) and 2.11(b), respectively. As was already discussed, the peaks in graphene's DOS are nonequidistant and oscillate around the mean value that linearly increases with ϵ.

The formation of Landau levels and the corresponding peaks in the DOS are manifested in many thermodynamic and transport properties. One of them, namely the unusual Hall response, is of particular interest. It will be discussed in detail in the following section.

2.7 Unconventional Hall effect

Historically, the relativistic-like Dirac nature of quasiparticles in graphene was first theoretically predicted in [212, 292]. It was confirmed experimentally through the observation of an unconventional or, as commonly called, "anomalous" quantization of the Hall conductivity [167, 168].[aa] The key features of the unconventional Hall effect will be presented in this section.

Let us start by reminding the key features about the integer quantum Hall effect in a 2D nonrelativistic electron gas in a strong magnetic field, which was discovered by Klaus von Klitzing, Gerhard Dora, and

---

[aa]Note that the term "anomalous Hall effect" is not perfect in the case of graphene, see the corresponding discussion in Sec. 6.3. In particular, it is not the same as the anomalous Hall effect in Weyl semimetals, which implies a nonzero Hall conductivity in the absence of a magnetic field.

Michael Pepper in 1980 [293]. In the experiment, they observed a series of Hall plateaus characterized by (i) the vanishing longitudinal conductivity $\sigma_{xx} = 0$ and (ii) the quantized values of Hall conductivity $\sigma_{xy} = Ne^2/(2\pi\hbar)$, where $N = 0, 1, 2, \ldots$. This quantization is remarkably exact with a relative precision of about 10^{-8} to 10^{-10} even in real materials with impurities and disorder [294]. Nowadays, the quantum Hall effect is used for reproducing the standard of electrical resistance by many metrology laboratories around the world. This effect also provides an extremely precise independent determination of the vacuum fine-structure constant α_0, which is one of the fundamental constants in physics. From a theoretical viewpoint, the quantization of the Hall conductivity is not accidental but has deep roots in topology. In particular, it is related to the Chern number [295].

It was experimentally observed [167, 168] that the Hall conductivity σ_{xy} in graphene has well-resolved plateaus $\sigma_{xy} = \pm 4(N + 1/2)e^2/(2\pi\hbar)$ where $N = 0, 1, 2, \ldots$, and the factor 4 counts the spin and valley degeneracy. The corresponding data are shown in Fig. 2.12(a). These are qualitatively different from the measurements in a conventional 2D nonrelativistic electron gas,[bb] whose conductivity $\sigma_{xy}^{(NR)} = 2Ne^2/(2\pi\hbar)$ is shown in Fig. 2.12(b). Note that, due to the anomalous shift of $1/2$, there is no quantum Hall

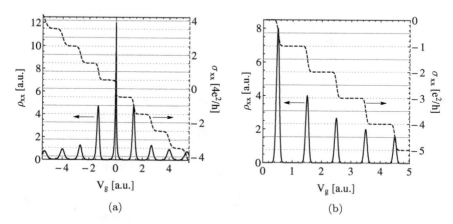

Fig. 2.12   Schematic dependencies of the quantum Hall conductivity in (a) graphene and in (b) a 2D nonrelativistic electron gas in a GaAs/AlGaAs heterostructure.

[bb]There is only the spin degeneracy for a nonrelativistic electron gas under consideration.

plateau at zero filling, i.e., $4(N + 1/2) \neq 0$, in graphene. This is the reason why such an unusual Hall response is also known as the *half-integer quantum Hall effect*.

The unconventional Hall plateaus in graphene are directly related to the Dirac nature of its quasiparticles as was theoretically predicted in [212, 292, 296]. This was confirmed experimentally in [167, 168]. In particular, the origin of the additional term $1/2$ in σ_{xy} can be traced back to the lowest Landau level, which has zero energy and is half-filled for the vanishing electric chemical potential (for the details of derivation, see, e.g., [212, 237]).

Alternatively, one can understand the anomalous quantization of Hall conductivity in terms of the Berry phase[cc] for the filled states [295, 297]:

$$\sigma_{xy} = \frac{e^2}{(2\pi)^2\hbar} \sum_n \oint_{\mathcal{C}} d\mathbf{k} \cdot \mathbf{A}_{n,\mathbf{k}} = -\frac{e^2}{(2\pi)^2\hbar} \sum_n \gamma_n(\mathcal{C}), \qquad (2.61)$$

where the sum runs over all occupied states and $\mathbf{A}_{n,\mathbf{k}} = -iu_{n,\mathbf{k}}^\dagger \nabla_{\mathbf{k}} u_{n,\mathbf{k}}$. If the state is trivial (completely filled), then its Berry phase is also trivial, i.e., $\mp 2\pi$. However, as was shown in Sec. 2.4, the phase is $\mp\pi$ in the case of relativistic-like quasiparticles in graphene. This explains the additional term $1/2$ in the Hall conductivity.

The subsequent experimental studies uncovered that, in sufficiently strong magnetic fields, additional quantum Hall plateaus at $2\pi\hbar\sigma_{xy}/e^2 = 0, \pm 1, \pm 3, \pm 4$ can be resolved [298–310]. This means that the fourfold degeneracy of the Landau levels is lifted and the Fermi energy falls in the gaps between the individual levels. Note that the Zeeman-type splitting partially lifts the degeneracy of the Landau levels. Because of the leftover $U_+(2) \times U_-(2)$ symmetry, however, it can only explain the plateaus at $2\pi\hbar\sigma_{xy}/e^2 = 0, \pm 4$. In particular, the plateau at zero filling appears because the zeroth Landau level is split by a dynamically enhanced Zeeman energy.

The breakdown of the leftover $U_+(2) \times U_-(2)$ symmetry and the complete lifting of the Landau level degeneracies is caused by many-body electron–electron interactions. In fact, the same interactions are also responsible for the dynamical enhancement of the Zeeman splitting. In order to illustrate the essence of spontaneous breaking of the internal symmetries and qualitatively describe the corresponding order parameters,

---

[cc]Note that [295, 297] appeared before the seminal paper by Berry [68].

a simple model will be considered below. We will assume that the electron–electron interaction is given by the screened Coulomb potential $U_C(\mathbf{r})$. The corresponding term in the Hamiltonian reads

$$H_C = \frac{1}{2} \int d^2 r d^2 r' \Psi_s^\dagger(\mathbf{r}) \Psi_s(\mathbf{r}) U_C(\mathbf{r} - \mathbf{r}') \Psi_s^\dagger(\mathbf{r}') \Psi_s(\mathbf{r}'). \tag{2.62}$$

In addition to the Coulomb interaction, there can exist many short-range four-fermion and six-fermion interaction terms [224], which break explicitly the U(4) spin–valley symmetry. Although they are expected to be small, such terms could play a very important role in triggering a preferred alignment of the otherwise highly degenerate ground state. Unfortunately, the rigorous theoretical analysis of the underlying dynamics is plagued with at least two serious difficulties: (i) a large number of possible symmetry-breaking terms and (ii) a strong renormalization of the interaction terms in the low-energy region. It is very challenging, therefore, to predict reliably the outcome of their competition and determine unambiguously the ground state. For illustrative purposes, therefore, all symmetry-breaking terms except for the Zeeman energy will be omitted below. While such a framework is oversimplified, it allows one to reveal and classify the most promising candidates for the electron ground states of the observed integer quantum Hall plateaus.

To start with, let us obtain the free Green function $S(\omega; \mathbf{r}, \mathbf{r}')$ for the Dirac quasiparticles in graphene (see also Appendix A.) We will use a mixed frequency-coordinate representation, which is defined by the following matrix element:

$$S(\omega; \mathbf{r}, \mathbf{r}') = i\langle \mathbf{r} | [(\hbar\omega + \mu_0 - \mu_B B \sigma_z)\gamma^0 - v_F(\boldsymbol{\pi} \cdot \boldsymbol{\gamma})]^{-1} | \mathbf{r}' \rangle. \tag{2.63}$$

By making use of this definition and utilizing the complete set of the orbital eigenstates in Eq. (2.56), one can easily derive an explicit form of the Green function in the Landau level representation. [The details of a similar derivation in the 3D case can be found in Appendix A.2]. The final expression for the Green function takes the following form:

$$S(\omega; \mathbf{r}, \mathbf{r}') = e^{i\Phi(\mathbf{r}, \mathbf{r}')} \bar{S}(\omega; \mathbf{r} - \mathbf{r}'), \tag{2.64}$$

where

$$\Phi(\mathbf{r}, \mathbf{r}') = -\frac{e}{\hbar c} \int_{\mathbf{r}'}^{\mathbf{r}} d\mathbf{r}'' \cdot \mathbf{A}(\mathbf{r}'') = -\frac{(x + x')(y - y')}{2 l_B^2} \tag{2.65}$$

is the celebrated Schwinger phase [311]. The translation-invariant part of the Green function is given by

$$\bar{S}(\omega; \mathbf{r}) = i \frac{e^{-r^2/(4l_{\mathrm{B}}^2)}}{2\pi l_{\mathrm{B}}^2} \sum_{s=\pm} \sum_{n=0}^{\infty} \frac{1}{(\hbar\omega + \mu_s)^2 - \epsilon_n^2}$$
$$\left\{ \gamma^0(\hbar\omega + \mu_s) \left[\mathcal{P}_- L_n \left(\frac{\mathbf{r}^2}{2l_{\mathrm{B}}^2} \right) + \mathcal{P}_+ L_{n-1} \left(\frac{\mathbf{r}^2}{2l_{\mathrm{B}}^2} \right) \right] \right.$$
$$\left. + \frac{iv_F}{l_{\mathrm{B}}^2} (\boldsymbol{\gamma} \cdot \mathbf{r}) L_{n-1}^1 \left(\frac{\mathbf{r}^2}{2l_{\mathrm{B}}^2} \right) \right\}, \tag{2.66}$$

where $\mu_s = \mu_0 - s\mu_B B$ is the effective chemical potential, $\mathcal{P}_\zeta = (1 + i\zeta\gamma_x\gamma_y)/2$ are the pseudospin projectors, and $L_n^\alpha(x)$ are the Laguerre polynomials [281].[dd] Interestingly, the inverse Green function has a very similar structure and exactly the same Schwinger phase, i.e., $S^{-1}(\omega; \mathbf{r}, \mathbf{r}') = e^{i\Phi(\mathbf{r},\mathbf{r}')}\bar{S}^{-1}(\omega; \mathbf{r}-\mathbf{r}')$, where the explicit expression for $\bar{S}^{-1}(\omega; \mathbf{r}-\mathbf{r}')$ can be obtained by multiplying the summand in Eq. (2.66) with $-[(\hbar\omega+\mu_s)^2 - \epsilon_n^2]$.

The translation invariant part of the full Green function $G(\omega; \mathbf{r})$ is determined by the following gap equation in the Hartree–Fock approximation:

$$\bar{G}^{-1}(\omega; \mathbf{r}) = \bar{S}^{-1}(\omega; \mathbf{r}) + e^2\gamma^0\bar{S}(\omega; \mathbf{r})\gamma^0 D(-\mathbf{r}), \tag{2.67}$$

where without the loss of generality we set $\mathbf{r}' = \mathbf{0}$. In the simplest approximation, $D(\mathbf{r})$ is the photon propagator describing the instantaneous Coulomb interaction.[ee] It can be conveniently rewritten as

$$D(\mathbf{r}) = \int \frac{d^2k}{(2\pi)^2} D(k) e^{i\mathbf{k}\cdot\mathbf{r}} = i \int_0^\infty \frac{dk}{2\pi} \frac{k J_0(kr)}{k + \Pi(0, k)}, \tag{2.68}$$

where $J_n(x)$ is the Bessel function of the first kind. Note that the screening effects are taken into account by including the polarization function $\Pi(0, k)$.

In order to solve the gap equation (2.67), one must use a suitable ansatz for the full Green function that captures the key aspects of the quasiparticle dynamics. The structure of this ansatz is usually determined by symmetry arguments and general physics insights. For example, by applying the ideas of magnetic catalysis [26, 227], one might expect a rise of excitonic ordering in the ground state [26, 312]. Similarly, one might argue that the spin

---

[dd]Note that $L_n^0(x) = L_n(x)$ and, by assumption, $L_{-1}^\alpha(x) \equiv 0$.
[ee]When the effects of retardation are taken into account, the last term in Eq. (2.67) will take the form of a convolution, rather than a product of two functions.

and pseudospin dynamics could induce some quantum Hall ferromagnetic ordering [313–318].

The ordering induced by the quasiparticle interactions can be described by the appropriate order parameters given by expectation values of composite field operators. In particular, the order parameters associated with the magnetic catalysis in graphene are given by $\langle \bar{\Psi}_s \gamma_z \gamma_5 \mathcal{P}_s \Psi_s \rangle$ and $\langle \bar{\Psi}_s \mathcal{P}_s \Psi_s \rangle$, where $\mathcal{P}_s = (1 + s\sigma_z)/2$ is the spin-projection operator. From a physics viewpoint, they describe the usual and valley-polarized charge density waves, respectively. Note that $\langle \bar{\Psi}_s \gamma_z \gamma_5 \mathcal{P}_s \Psi_s \rangle$ transforms as a singlet and $\langle \bar{\Psi}_s \mathcal{P}_s \Psi_s \rangle$ as a triplet under the $U_s(2)$ subgroup in each spin sector ($s = \pm$). The nonzero expectation values of these condensates are responsible for the generation of the Haldane and Dirac mass terms, $\Delta_s \bar{\Psi}_s \gamma_z \gamma_5 \mathcal{P}_s \Psi_s$ and $\tilde{\Delta}_{z,s} \bar{\Psi}_s \mathcal{P}_s \Psi_s$, respectively, in the low-energy effective Hamiltonian [211, 226, 319–326]. As will be shown in Sec. 3.1.1, the analog of the Haldane mass has a very different meaning in 3D systems, where it defines the momentum separation of the Weyl nodes in Weyl semimetals.

The order parameters associated with the quantum Hall ferromagnetism are the spin and pseudospin densities, i.e., $\langle \Psi_s^\dagger \sigma_z \Psi_s \rangle$ and $\langle \Psi_s^\dagger \gamma_z \gamma_5 \mathcal{P}_s \Psi_s \rangle$. In the effective Hamiltonian, they correspond to the two types of chemical potential terms given by $\mu_z \Psi_s^\dagger \sigma_z \Psi_s$ (singlet) and $\tilde{\mu}_s \Psi_s^\dagger \gamma_z \gamma_5 \mathcal{P}_s \Psi_s$ (triplet), respectively. Note that μ_z is exactly of the same type as the Zeeman energy and, therefore, could be interpreted as a dynamical correction to its bare value $\mu_B B$.

When the order parameters associated with both magnetic catalysis and quantum Hall ferromagnetism are allowed, the ansatz for the full quasiparticle Green function at a fixed spin ($s = \pm$) reads[ff]

$$G(\omega; \mathbf{r}, \mathbf{r}') = i\langle \mathbf{r} | [(\hbar\omega + \mu_s + \tilde{\mu}_s \gamma_z \gamma_5)\gamma^0 - v_F(\boldsymbol{\pi} \cdot \boldsymbol{\gamma})$$
$$- \tilde{\Delta}_{z,s} + \Delta_s \gamma_z \gamma_5]^{-1} | \mathbf{r}' \rangle. \tag{2.69}$$

The corresponding Landau level energies are given by the poles of Green's function, i.e.,

$$\epsilon_{0,\zeta} = -\mu_0 - \zeta \tilde{\mu}_0 + \Delta_0 + \zeta \tilde{\Delta}_{z,0}, \tag{2.70}$$

$$\epsilon_{n,\zeta} = -\mu_n - \zeta \tilde{\mu}_n \pm \sqrt{\frac{2n v_F^2 \hbar |eB|}{c} + (\Delta_n + \zeta \tilde{\Delta}_{z,n})^2}, \quad n \geq 1, \tag{2.71}$$

[ff]The full Green function can also be rewritten in the Landau level representation. The corresponding result can be found in [327].

where $\zeta = \pm$. In general, all masses and chemical potentials depend on the Landau level index n. For simplicity, the spin index s was omitted. It is clear that the presence of both singlet (i.e., Δ_n and μ_n) and triplet (i.e., $\tilde{\Delta}_{z,n}$ and $\tilde{\mu}_n$) parameters completely lifts the spin–valley degeneracy of the Landau levels.

The next step is to substitute the ansatz for the full Green function (2.69) into the gap equation (2.67). Projecting onto the individual Landau levels, a system of algebraic equations could be obtained and its solutions determined. The ground-state solution corresponds to the lowest free energy [211, 226, 319–328].

Experimentally, a lot of effort was devoted to revealing the underlying nature of the strongly insulating $\sigma_{xy} = 0$ quantum Hall state associated with the half-filled lowest Landau level [298–310], which is consistent with both charge density wave and antiferromagnet state. The progress was made by applying a tilted magnetic field to high-quality graphene devices [308, 309]. A careful analysis of the conductance and the bulk density of states [308–310] suggests that this insulating Hall state is not spin-polarized when the magnetic field is perpendicular to the plane of graphene. When an in-plane component of the magnetic field increases, however, the state gradually transforms into a fully polarized ferromagnetic state. In the intermediate regime, a canted antiferromagnetic state is presumably realized [225]. Such an interpretation is supported by the observation of a nonzero bulk gap that does not close and the edge states that become conducting with the increasing total magnetic field. Considering, however, how the arguments rely heavily on the properties of the edge states, the final conclusions should be still accepted with caution.

Here it is appropriate to underline that the unconventional quantum Hall effect is a very special regime of electronic transport. According to the celebrated bulk-boundary correspondence [329], the anomalous quantization of Hall conductance in the unconventional quantum Hall effect could be explained through the existence of nondissipative edge states at the interface between materials with different topological characteristics. As will be shown in Chapter 5, the Fermi arc surface states in Weyl semimetals have a similar topological origin rooted in the bulk-boundary correspondence.

Before finalizing this chapter, we discuss how the relativistic-like spectrum of electron quasiparticles in graphene allows one to probe some of the effects related to the formation of Landau levels even without a magnetic field. This can be achieved by mechanical deformations that are represented in terms of effective gauge fields.

2.8 Strain-induced pseudoelectromagnetic fields

As was shown in Sec. 2.2, the hopping of electrons on a honeycomb lattice is responsible for the emergence of Dirac quasiparticles in graphene. Therefore, it is natural to ask what happens if an external stress is applied and the lattice is deformed. While the hopping parameters change, it turns out that the Dirac structure of the low-energy effective Hamiltonian for electrons is preserved in a strained graphene. Moreover, smooth elastic deformations lead to the appearance of effective gauge fields. The latter interact with electron quasiparticles similarly to the usual electromagnetic fields but with different signs in K and K' valleys. Since the valley index plays the role of chirality in graphene, the fields of this type are known as *axial gauge fields*. In high energy physics, the examples of similar (albeit non-Abelian) fields are $W^{\pm}$ and Z_0 bosons that play an important role in the electroweak sector of the Standard Model.

The physical reasons for the appearance of axial gauge fields in a strained graphene are rooted in the linearity of the energy spectrum and the symmetry protection of the band touching points. In addition to elastic deformations, there could exist various defects connected with atomically sharp inhomogeneities such as disclinations, dislocations, etc., which can change the topology of the lattice and also induce gauge fields [330]. For some defects, the Dirac equation could still be applied for the description of quasiparticles in graphene (see, e.g., [331] and references therein). In what follows, however, only smooth elastic deformations and the corresponding strain-induced axial gauge fields will be considered.

Let us demonstrate how strains lead to the appearance of axial gauge fields. Microscopically, deformations modify the lattice spacings in graphene that, in turn, change the hopping parameters. In particular, the hopping between the nearest neighbors along the three directions $\boldsymbol{\delta}_1$, $\boldsymbol{\delta}_2$, and $\boldsymbol{\delta}_3$, see Eq. (2.4) and Fig. 2.1, change. Therefore, instead of a single parameter t in Eq. (2.5), one has to introduce three different hopping parameters t_i, where $i = 1, 2, 3$. Then, in the vicinity of K ($\zeta = +$) or K' ($\zeta = -$) point, the following low-energy Hamiltonian is obtained [332–334]:

$$H_\zeta = \zeta v_F \boldsymbol{\tau} \cdot \left(-i\hbar \boldsymbol{\nabla} + \zeta \frac{e}{c} \mathbf{A}^5 \right), \tag{2.72}$$

where the strain-induced gauge fields are

$$A_x^5 = \frac{c}{2ev_F}(t_2 + t_3 - 2t_1), \tag{2.73}$$

$$A_y^5 = \frac{\sqrt{3}c}{2ev_F}(t_2 - t_3). \tag{2.74}$$

Note that the modification of the quasiparticle velocities was ignored in Eq. (2.72). Such an approximation is reasonable because the terms $\propto (t_i - t_j)\nabla$ with $i \neq j$ are negligible when strains are weak and the momentum deviations from the Dirac points are small.[gg] It is important to note that mechanical deformations do not break the time-reversal symmetry, which interchanges valleys K and K'. Therefore, unlike the electromagnetic vector potential $\mathbf{A}$, the strain-induced field $\mathbf{A}^5$ couples to Dirac fermions in different valleys (equivalently, to quasiparticles of opposite chiralities) with opposite signs. It is important to emphasize that, unlike the gauge-dependent vector potential in electrodynamics, the strain-induced gauge field $\mathbf{A}^5$ could have direct observable implications. Beyond the leading order in small momenta and spatial derivatives, strains could also produce other effects in graphene. They include a tilt of Dirac cones, a position-dependent Fermi velocity, and a nonzero gap [335, 336]. While such higher-order effects are usually small, under right conditions they could have interesting physical implications.

In the case of weak deformations, the hopping constants can be rewritten in terms of the displacement vector $\mathbf{u}$ as follows:

$$t_i = t - \frac{\beta t}{a^2} \boldsymbol{\delta}_i (\boldsymbol{\delta}_i \cdot \nabla) \mathbf{u}, \tag{2.75}$$

where $\beta = -\partial \ln t / \partial \ln a$ is the electron Grüneisen parameter, which equals approximately 2 in graphene (see, e.g., [331, 332]). Then the axial gauge fields in Eqs. (2.73) and (2.74) can be expressed in terms of the linearized deformation tensor $u_{ij} = (\partial_i u_j + \partial_i u_j)/2$ [332, 337], i.e.,

$$A_x^5 \simeq -\frac{3c\beta t}{4ev_F}(u_{xx} - u_{yy}), \tag{2.76}$$

$$A_y^5 \simeq \frac{3c\beta t}{2ev_F} u_{xy}, \tag{2.77}$$

where some overall numerical prefactors, that depend on the model of chemical bonding, were omitted. In addition to $\mathbf{A}^5$, some strains could also generate the scalar potential $A_0 \propto u_{xx} + u_{yy}$, which is known as the deformation potential.

The strain-induced axial electric and magnetic fields are expressed through the axial gauge fields in the usual way $B_5 = \partial_x A_y^5 - \partial_y A_x^5$

[gg] In the case of large strains, the effective description of deformations in terms of gauge fields breaks down and a nonzero gap can open.

and $\mathbf{E}_5 = -\partial_t \mathbf{A}^5/c$. These *pseudoelectromagnetic* fields couple with opposite signs to electron quasiparticles in valleys K and K'. It was shown in [338, 339] that a nearly uniform pseudomagnetic field could be realized by applying normal forces to a graphene sample of certain shape, see Fig. 2.13(a). The field magnitude can be estimated as $B_5 \approx \beta \hbar c u_0/(eaL) \gtrsim 0.3\,\mathrm{T}$, where $u_0 \gtrsim 0.01$ is a typical value of stress and $L = 10\,\mu\mathrm{m}$ is the size of the sample. Since graphene can be stretched significantly,[hh] very high (albeit often nonuniform) pseudomagnetic fields could be obtained. For example, the pseudomagnetic field created by a spontaneous deformation of graphene nanobubbles on a platinum surface was estimated [341] to be approximately 300 teslas, which is significantly higher than the usual static magnetic field that can be achieved in a laboratory.

The pseudoelectromagnetic fields produce interesting observable effects and can be used to manipulate the electronic structure via strain engineering [342].[ii] In the case of a constant pseudomagnetic field, for example, the quasiparticle energy spectrum is given by the pseudo-Landau levels, see Fig. 2.13(b). A dispersionless zero-energy mode is clearly seen in the spectrum. By applying also a usual magnetic filed to such a system,

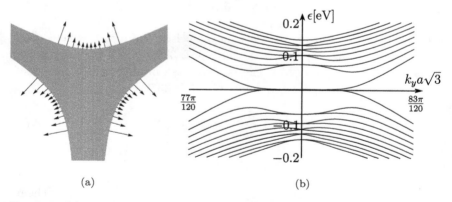

(a)   (b)

Fig. 2.13   (a) The shape of a graphene sample and the applied stress configuration that allows for a constant pseudomagnetic field. (b) The energy spectrum for graphene with sinusoidal ripples, where the change of the hopping parameters is quantified by $\delta t \sin(2\pi x/L_{\mathrm{r}})$. Averaged hopping is $t = 3\,\mathrm{eV}$, the width of the ripple is $L_{\mathrm{r}} = 1200a = 168\,\mathrm{nm}$, and the modulation of the hopping is $\delta t/t = 0.04$. (Adapted from [347].)

[hh] According to the *ab initio* calculations in [340], graphene can sustain reversible elastic deformations in excess of 20%.
[ii] This newly emerging field is known as straintronics.

one can produce a valley-polarized electron distribution in graphene. The strain-induced gauge fields can also strongly modify transport properties. For example, they could give rise to the valley quantum Hall effect, where the contributions from different valleys have opposite signs [338]. In suspended graphene with two contacts along the y-axis, the gauge field $\mathbf{A}^5 \propto -c\beta t u_{xx}/(ev_F)\hat{\mathbf{y}}$ emerging at the contacts strongly affects the transport properties of the device [343]. While weak strains reduce the transmission, sufficiently strong strains with $|eA_y^5/c| > \mu$ can block the transport completely.

It is worth mentioning that random strain-induced pseudomagnetic fields can mimic disorder effects. This is particularly important for graphene, where stochastic-like gauge fields are always generated by ripples. For a single ripple, the magnitude of the corresponding pseudomagnetic field can be estimated [344] as $B_5 \approx \hbar c L^2/(eaR^3) \approx 1.2\,\mathrm{T}$, where $L \approx 5\,\text{Å}$ is the height of a ripple and $R \approx 10\,\mathrm{nm}$ is its radius. The combined effects of disorder and ripples on the quasiparticle scattering in graphene were studied in [345, 346]. It was found that certain types of ripples create a long-range scattering potential, which is similar to Coulomb disorder. This ripple-induced potential explains a weak dependence of the mobility on the carrier concentration, as observed in experiment [75, 166, 168, 185]. Therefore, in addition to the extrinsic mechanism related to the charged disorder, the scattering on omnipresent ripples provides an alternative, intrinsic, mechanism that limits the charge mobility in graphene. For a more detailed review of the strain-induced gauge field effects, the reader is referred to [331].

Before finalizing this section, it is worth noting that deformation-induced gauge fields can also be realized in strained Weyl semimetals leading to very interesting physical effects as well. As will be discussed in Sec. 3.7, such fields couple to Weyl fermions of opposite chiralities with different signs, effectively realizing the axial analog of the usual vector field $\mathbf{A}$. As in graphene, the corresponding axial gauge field $\mathbf{A}^5$ leads to pseudo-electromagnetic (axial) fields $\mathbf{E}_5$ and $\mathbf{B}_5$. In 3D Weyl semimetals, however, there is no analog of ripple-induced random gauge fields.

As we will see, the electronic properties of graphene discussed in this chapter provide a useful benchmark for understanding 3D Dirac and Weyl materials in the rest of the book. As in graphene, the relativistic-like nature of the low-energy quasiparticles will play a very important role. Moreover, an additional spatial dimension enriches the properties of the corresponding systems and allows for new effects unavailable in graphene.

Chapter 3

Topology, Low-Energy Models, and Thermodynamic Properties

A great deal of my work is just playing with equations and seeing what they give.

Paul A. M. Dirac

Topology deals with the characteristics of systems that remain invariant under continuous deformations of model parameters. Therefore, a nontrivial topology often implies the robustness of physical properties with respect to perturbations, impurities, and disorder. The relevance of topology for condensed matter systems was understood a long time ago. In particular, it played a major role in the studies of numerous topological defects whose existence is often associated with spontaneous breaking of global symmetries. The magnetic domain walls, the quantization of circulation in superfluid ^{4}He, and the Abrikosov vortices in type-II superconductors are just a few renowned examples [348]. The existence of such defects is usually protected by one of the nontrivial homotopy groups $\pi_i(\mathcal{G}/\mathcal{H})$ (with $i = 0$ for domain walls, $i = 1$ for vortices, and $i = 2$ for point defects), where $\mathcal{G}$ is the symmetry group of the Hamiltonian and $\mathcal{H}$ is the symmetry subgroup of the system in the ground state. After the discovery of the integer quantum Hall effect in a two-dimensional (2D) electron gas in a strong magnetic field, it became clear that topology can be relevant also for electronic states in solids. In fact, the remarkable robustness of the experimentally observed quantization of the Hall conductivity is explained by the underlying topology [293, 295].

Over the last 10–15 years, it has become clear that topology can play an important role also in the characterization of *electronic bands* of solids even in the absence of a magnetic field. This is the case, for example, when a strong spin–orbit coupling causes the inversion of the normal order of the bands. Depending on the specific details of the band inversion, as well as the presence of additional discrete symmetries (e.g., time-reversal, parity-inversion, and crystal symmetries), various types of topological materials could be realized. The two subclasses that will be discussed in detail here are Dirac and Weyl semimetals. Some Dirac materials, for example, could be viewed as an intermediate state at the topological phase transition between topological and normal insulators, provided both time-reversal and parity-inversion symmetries are intact. Other band structures with Dirac points could be enforced by symmetries of crystalline lattices with non-symmorphic space groups. When either time-reversal or parity-inversion symmetry, or both are broken, a Weyl semimetal could be realized.

As we will discuss in detail, the Berry curvature plays a vital role in characterizing topological properties of Weyl semimetals. In this connection, let us note that the Weyl nodes in momentum space are monopoles of the Berry curvature with nonzero topological charges or the Chern numbers, which are also related to the chiralities of the nodes. This profound result can be traced back to the original analysis of Michael Berry in his seminal paper [68]. Additionally, in accordance with the Nielsen–Ninomiya theorem [51–53], the Weyl nodes in solids occur only in pairs of opposite chiralities. Thus, the total Chern number obtained after the integration over the whole Brillouin zone always vanishes. Because of their nonzero topological charges, the individual Weyl nodes are stable with respect to weak perturbations and can disappear only through a pairwise annihilation. This is in contrast to the usual topologically trivial Dirac points, where each point is composed of two Weyl nodes of opposite chiralities or, equivalently, opposite topological charges. As a result, the stability of Dirac points usually requires the presence of certain space groups.

One of the consequences of the nontrivial topological nature of Weyl and certain Dirac semimetals is the existence of unusual surface states. Such states, called Fermi arcs, have remarkable properties and will be discussed in detail in Chapter 5. In this chapter, however, the main focus will be on bulk topological properties and low-energy effective models of the electron states in Dirac and Weyl semimetals. Clearly, the relativistic-like energy spectrum of Dirac and Weyl semimetals should be imprinted in thermodynamic properties and quantum oscillations of the electrical conductivity and the magnetization in an external magnetic field, i.e., in the Shubnikov–de Haas (SdH) [349–352] and de Haas–van Alphen (dHvA) [353] effects, respectively. Similar methods are also useful for probing the surface–bulk orbits mediated by the Fermi arcs in Dirac and Weyl semimetals (see Sec. 5.6).

In the bulk probes, quantum oscillations provide detailed information about the fermiology of materials, i.e., the shape of the electronic Fermi surfaces.

Last but not least, we discuss the interpretation of strains in terms of effective pseudoelectromagnetic fields. In addition, a few interaction effects including the dynamical screening of the Coulomb potential as well as a renormalization of the Fermi velocity and the coupling constant in Dirac and Weyl semimetals will be discussed.

3.1 Band-touching points

The Bloch theorem [354] plays a fundamental role in the description of electron states in crystals. According to the theorem, the states are grouped into a countable set of energy bands. The latter are described by continuous periodic functions of the quasimomentum, defined in a compact Brillouin zone. In general, the energy bands may partially overlap or be separated by energy gaps. Their structure, supplemented by the relative position of the Fermi energy (or the electric chemical potential μ) with respect to the bands, determines whether the material is an insulator, a semiconductor, a semimetal, or a metal. Schematically, the corresponding types of band structures are shown in Fig. 3.1. The filled and (partially) empty energy bands in the vicinity of the Fermi level are known as the valence and conduction bands, respectively.

Insulators are materials where the electric chemical potential μ lies inside a sufficiently large gap Δ_g between the valence and conduction bands. The typical values of the gap in insulators are of the order of 5 eV or more. Because of a negligible amount of thermally excited states in the conductance band, the electrical conductivity of insulators is vanishingly small even at temperatures much higher than the room temperature $T_{room} \simeq 300$ K ≈ 26 meV. Semiconductors are qualitatively similar to insulators, but have smaller energy gaps, typically of the order of 1 eV or less, allowing for a substantial population of thermally induced electron states. Often semiconductors are also doped with donor- or acceptor-induced electron states within the gap, which supply additional charge carriers.[a] The relative ease of controlling the number density of charge carriers and,

---

[a]Note that a distinction between insulators and semiconductors based on the comparison of the size of the band gap with temperature is not precise. In addition, the doping of semiconductors usually plays a more important role than the temperature effects.

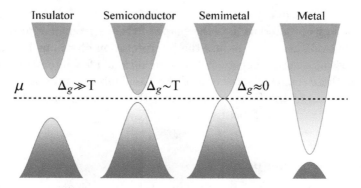

Fig. 3.1   Schematic band structure of insulators, semiconductors, semimetals, and metals. Because of a large band gap in insulators there are almost no thermally excited electrons in the conduction band. On the other hand, a relatively small gap in semiconductors allows for a substantial population of thermally excited electrons. In semimetals, the conduction and valence bands touch or have a small overlap and, therefore, the band gap vanishes. Finally, in metals, the Fermi level lies completely within the conduction band. Unlike semimetals, metals are usually characterized by a large density of states at the Fermi surface.

therefore, the conductive properties of semiconductors is what makes them invaluable in modern electronics.[b]

The defining feature of semimetals is the vanishing value of the gap that comes from the valence and conductance bands overlapping only slightly or touching at isolated points. Note that the latter is the case for Dirac and Weyl semimetals, which are the main focus of this book. Since the electric chemical potential lies in the overlap zone, there is a relatively small density of states at the Fermi surface. As a result, the electrical conductivity of semimetals is usually very small at low temperatures.[c] With increasing temperature, however, the conductivity of semimetals rapidly grows due to an increase of the number density of thermally excited electrons in the conductance band. This is in contrast to usual metals, which are characterized by a large density of states at the Fermi surface and, therefore, have a large electrical conductivity even at low temperatures. The large density

---

[b]It is curious to note how clueless Wolfgang Pauli was in 1931 when he wrote in a letter to Rudolf Peierls, "One shouldn't work on semiconductors, that is a filthy mess; who knows whether any semiconductors exist".

[c]Typically, the density of charge carriers in semimetals is more than four orders of magnitude smaller than that in normal metals ($10^{22} - 10^{23}$ cm^{-3}).

of states in metals comes from the fact that the electron chemical potential lies inside the conductance band.

For the first time, the energy bands with isolated touching points were studied by Conyers Herring.[d] In his paper [33], Herring considered the cases where the electron states of two bands have the same energy either accidentally or due to some symmetry arguments. In particular, he showed that in the vicinity of a discrete number of touching points, the energy dispersion could be linear in quasimomentum.

Let us consider the physics of isolated band-touching points in detail [24]. Since there are two degrees of freedom associated with a pair of touching bands, the most general Hermitian Hamiltonian for the electron states has the following form in momentum space:

$$H(\mathbf{k}) = \epsilon_0(\mathbf{k}) + \mathbf{a}(\mathbf{k}) \cdot \boldsymbol{\sigma}, \tag{3.1}$$

where $\boldsymbol{\sigma}$ are the pseudospin Pauli matrices[e] acting in the space of the two touching energy zones. The above Hamiltonian is fully defined by functions $\epsilon_0(\mathbf{k})$ and $\mathbf{a}(\mathbf{k})$ that depend on quasimomentum $\mathbf{k}$. If the zones touch at $\mathbf{k}_0$, then $\mathbf{a}(\mathbf{k}_0) = \mathbf{0}$. In the vicinity of $\mathbf{k}_0$, it is convenient to expand the Hamiltonian into a power series in $\delta\mathbf{k} = \mathbf{k} - \mathbf{k}_0$. To linear order in deviations, the result reads

$$H(\mathbf{k}) = \epsilon_0(\mathbf{k}_0) + \delta\mathbf{k} \cdot \partial_{\mathbf{k}}\epsilon_0(\mathbf{k}_0) + \sum_{i,j=x,y,z} a_{ij}\sigma_i\delta k_j + O(\delta k^2), \tag{3.2}$$

where $a_{ij} = \partial_{k_j} a_i(\mathbf{k})\big|_{\mathbf{k}=\mathbf{k}_0}$. The first term in the series plays a minor role as it simply redefines the electric chemical potential. The second term, on the other hand, carries an important information. As we will briefly discuss in Sec. 3.1.2, it introduces an overall tilt of the corresponding Weyl node and determines whether a type-I or type-II Weyl semimetal is realized [127].[f] While such a term is forbidden in high energy physics by the Lorentz symmetry, it is possible and, in fact, generically expected in condensed matter systems. The most crucial term in Eq. (3.2) is the third one. It defines an anisotropic Weyl Hamiltonian with a Weyl node of chirality $\lambda = \text{sign}[\det(a_{ij})] = \pm 1$. A few particular cases of Weyl Hamiltonians will be discussed below.

---

[d]Conyers Herring played a significant role in the development of the early solid state physics and is well known as the author of the orthogonal plane wave method [355, 356].
[e]In general, the pseudospin is an effective degree of freedom akin to a spin, but may have a different origin, e.g., associated with sublattices and/or valleys.
[f]It is worth noting that the tilt parameter was already present in the low-energy model in [102].

3.1.1 *Weyl and Dirac fermions*

In order to clarify the connection between the low-energy model in Eq. (3.2) and the canonical form of the Weyl Hamiltonian [31], it is instructive to concentrate on the third term in Eq. (3.2). Indeed, it is the only one containing a nontrivial dependence on the pseudospin degrees of freedom. By a suitable orthogonal transformation, this term can be diagonalized, i.e., $a_{ij} \rightarrow \lambda \hbar v_i \delta_{ij}$, where v_i are quasiparticle velocities along the principal axes. In the simplest case of an isotropic system, all three values of v_i are equal to a common Fermi velocity, $v_i = v_F$. Then the effective Hamiltonian takes the canonical Weyl form

$$H_W = \lambda \hbar v_F \left(\boldsymbol{\sigma} \cdot \mathbf{k} \right), \tag{3.3}$$

although, unlike its original relativistic version, it has v_F instead of the speed of light. As expected, the energy dispersion of the Weyl fermions in the effective model (3.3) is linear in momentum,

$$\epsilon = \pm \hbar v_F k. \tag{3.4}$$

Geometrically, this dispersion is a cone in the four-dimensional energy–momentum space.

Historically, the relativistic form of Hamiltonian (3.3) with v_F equal to the speed of light c was proposed by Weyl in order to describe massless spin-1/2 fermions. While none of the known elementary particles are truly Weyl fermions, it is worth mentioning that the Weyl equation still plays an essential role in the Standard Model of particles [42–44, 357], which relies on the chiral representations of the electroweak gauge group. Since the mass terms break the chiral symmetry, an additional mechanism is required to generate particle masses. They come from the spontaneous electroweak symmetry breaking via the Higgs mechanism.

Since electrons in solids are subject to a periodic lattice potential, there are additional constraints on the emerging chirality of quasiparticles, which are absent in high energy physics. Most crucially for the Weyl semimetals, it was proven by Holger Nielsen and Masao Ninomiya [51–53] that the chiral fermions in lattice systems should appear in pairs of opposite chirality. In particular, their theorem states that, for any local, Hermitian, and translation-invariant fermion action on a regular lattice in even dimensions, there must be an equal number of right- and left-handed fermion states. A simple intuitive argument for the Nielsen–Ninomiya theorem utilizing the concept of the Berry curvature will be provided at the end of Sec. 3.3.4.

In view of the Nielsen–Ninomiya theorem, a realistic model of a Weyl semimetal should always include an even number of Weyl nodes and the total chirality of all nodes must vanish. These constraints are satisfied in the following minimal low-energy model describing a Weyl semimetal with a single pair of Weyl nodes:

$$H_0 = \begin{pmatrix} \hbar v_F \boldsymbol{\sigma} \cdot (\mathbf{k} - \mathbf{b}) + b_0 & 0 \\ 0 & -\hbar v_F \boldsymbol{\sigma} \cdot (\mathbf{k} + \mathbf{b}) - b_0 \end{pmatrix}, \qquad (3.5)$$

which is paradigmatic and will be often used throughout this book.[g] Hamiltonian (3.5) describes two Weyl nodes of opposite chirality that are separated by $2\mathbf{b}$ in momentum and $2b_0$ in energy. Indeed, the energy dispersion of the corresponding quasiparticles is given by $\epsilon_\lambda = \lambda b_0 \pm \hbar v_F |\mathbf{k} - \lambda \mathbf{b}|$, where $\lambda = \pm$ is the node's chirality. The vector $\mathbf{b}$ is also known as the *chiral shift* [231]. Such a name is easy to justify since, as Eq. (3.5) implies, vector $\mathbf{b}$ shifts the locations of Weyl nodes away from the origin in momentum space. Moreover, the shifts have opposite signs for the fermions of left- and right-handed chiralities. The schematic illustration of the Weyl nodes in the two special cases, namely (i) $\mathbf{b} \neq \mathbf{0}$ and $b_0 = 0$ and (ii) $\mathbf{b} = \mathbf{0}$ and $b_0 \neq 0$, is given in Figs. 3.2(a) and 3.2(b), respectively. As will be discussed in more detail in Sec. 3.2, parameters $\mathbf{b}$ and b_0 are directly related to breaking of the time-reversal ($\mathcal{T}$) and parity-inversion ($\mathcal{P}$) symmetries, respectively.

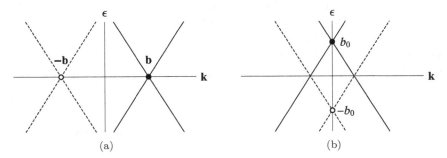

(a)   (b)

Fig. 3.2   Schematic representation of Weyl nodes in a Weyl semimetal with (a) $\mathbf{b} \neq \mathbf{0}$ and $b_0 = 0$ as well as (b) $\mathbf{b} = \mathbf{0}$ and $b_0 \neq 0$. Solid and dashed lines correspond to right- and left-handed quasiparticles. The left- and right-handed Weyl nodes are denoted by empty and filled dots, respectively.

---

[g]A more realistic periodic two-band model will be given in Sec. 3.5.

In order to demonstrate a close relation between the above minimal model of a Weyl semimetal and high energy physics, it is convenient to introduce the following four-dimensional Dirac matrices in the chiral representation:

$$\gamma^0 = \begin{pmatrix} 0 & -\mathbb{1}_2 \\ -\mathbb{1}_2 & 0 \end{pmatrix}, \quad \boldsymbol{\gamma} = \begin{pmatrix} 0 & \boldsymbol{\sigma} \\ -\boldsymbol{\sigma} & 0 \end{pmatrix}, \tag{3.6}$$

where $\mathbb{1}_2$ is the 2D unit matrix. By making use of these matrices, we can rewrite Hamiltonian (3.5) in a relativistic-like form,

$$H_0 = \hbar v_F \left(\boldsymbol{\alpha} \cdot \mathbf{k} \right) - (\mathbf{b} \cdot \boldsymbol{\alpha})\gamma_5 + b_0\gamma_5, \tag{3.7}$$

where $\boldsymbol{\alpha} = \gamma^0 \boldsymbol{\gamma}$ and

$$\gamma_5 = i\gamma^0\gamma_x\gamma_y\gamma_z = \begin{pmatrix} \mathbb{1}_2 & 0 \\ 0 & -\mathbb{1}_2 \end{pmatrix} \tag{3.8}$$

is the chirality matrix. By comparing it with the free Hamiltonian (3.5), it is clear that the two Weyl nodes, described by the diagonal blocks in Eq. (3.5), indeed correspond to opposite chiralities. By using the equation of motion, one could also verify that the chirality of quasiparticles with positive (negative) energies is equivalent (opposite) to their helicity, which is defined as the sign of the fermion's (pseudo-)spin projection on the direction of its momentum.

By using the same formalism, we can also discuss Dirac semimetals. As we mentioned before, a massless Dirac fermion is composed of a pair of right- and left-handed fermions with the same dispersion relations. Therefore, the low-energy Hamiltonian in the vicinity of a Dirac point can be obtained from Eq. (3.7) by setting $|\mathbf{b}| = b_0 = 0$, i.e.,

$$H_D = \hbar v_F \left(\boldsymbol{\alpha} \cdot \mathbf{k} \right). \tag{3.9}$$

The energy spectrum of this Hamiltonian, $\epsilon_D = \pm\hbar v_F|\mathbf{k}|$, is obtained from the Weyl spectrum in Fig. 3.3 by merging the Weyl nodes of opposite chiralities into a single Dirac point. Unlike the individual Weyl nodes, a generic Dirac point is not topologically protected. Therefore, unless there is another protection mechanism, interactions and perturbations could induce a Dirac mass term $-m\gamma^0$ in the Hamiltonian, which is equivalent to opening an energy gap between the valence and conductance bands. In fact, this is exactly what happens in nontopological Dirac semimetals, where a hybridization between the overlapping Weyl nodes is not forbidden. Sometimes the corresponding materials with a (small) nonzero gap in the spectrum are also called Dirac semimetals. In what follows, however, we will assume a more narrow definition of Dirac semimetals, where $m = 0$ is enforced by some mechanism.

3.1.2 Type-II Weyl fermions

Let us now discuss what happens when the first derivatives of $\epsilon_0(\mathbf{k})$ in Eq. (3.2) do not vanish at the band touching point. Note that, in particle physics, the presence of terms without the Pauli matrices and linear in momentum is strictly forbidden by the Lorentz symmetry. In condensed matter physics, however, there are no fundamental reasons that could prevent the appearance of such terms. By taking them explicitly into account, the effective Hamiltonian in the vicinity of a Weyl node could be written in the following form:

$$H_{\mathrm{II}} = (\mathbf{v} \cdot \mathbf{k}) + \sum_{i,j=x,y,z} a_{ij}\sigma_i k_j, \qquad (3.10)$$

where $\mathbf{v} = \partial_{\mathbf{k}}\epsilon_0(\mathbf{k})\big|_{\mathbf{k}\to\mathbf{k}_0}$. If the absolute value of $\mathbf{v}$ is sufficiently large, this Hamiltonian describes a *type-II Weyl semimetal* [127]. Its energy eigenvalues are given by

$$\epsilon(\mathbf{k}) = (\mathbf{v} \cdot \mathbf{k}) \pm \sqrt{\sum_{i,j=x,y,z} k_i [aa^{\mathrm{T}}]_{ij} k_j} \equiv T(\mathbf{k}) \pm U(\mathbf{k}). \qquad (3.11)$$

As is easy to check, the first term $T(\mathbf{k}) \equiv (\mathbf{v} \cdot \mathbf{k})$, which is linear in momentum, is responsible for tilting the Weyl cone. Most interestingly, it opens a qualitatively new possibility when the absolute value of $T(\mathbf{k})$ becomes larger than $U(\mathbf{k})$ for certain directions of $\mathbf{k}$. In this case, the Weyl cone becomes overtilted, as shown in Fig. 3.3(b). This has a profound effect on the occupation of quasiparticle states in the energy bands. As is clear from Fig. 3.3, the point-like Fermi surface at zero energy (in a type-I Weyl semimetal) turns into a pair of electron and hole pockets (in a type-II Weyl semimetal). Such a transformation due to tilting of the Weyl cone is an example of the celebrated Lifshitz transition [358], generically associated with a change of the Fermi surface topology. As we will briefly discuss later, the tilt profoundly affects the properties of type-II Weyl semimetals. As an example, the chiral anomaly can be realized in type-II Weyl semimetals only if the magnetic field points in the direction where $T(\mathbf{k})$ dominates.

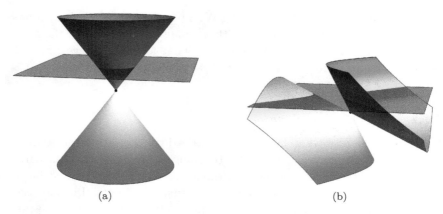

(a)   (b)

Fig. 3.3   Schematic energy spectra of (a) type-I and (b) type-II Weyl semimetals. The
Fermi level in both panels is denoted by the horizontal plane.

Also, there might exist additional surface states when the tilt is sufficiently
strong [359].

By equating energy in Eq. (3.11) to the electric chemical potential,
$\epsilon(\mathbf{k}) = \mu$, one derives the following equation defining the Fermi surface:

$$\sum_{i,j=x,y,z} k_i \left([aa^\mathrm{T}]_{ij} - v_i v_j\right) k_j + 2(\mathbf{v} \cdot \mathbf{k})\mu - \mu^2 = 0. \qquad (3.12)$$

This is an equation of a quadric surface in momentum space. The classification of all such surfaces is very well known in mathematics [360]. From
physics viewpoint, the closed Fermi surfaces (including the point-like one
in the limiting case $\mu \to 0$) are indicative of a type-I Weyl semimetal.
The open Fermi surfaces, on the other hand, correspond to type-II Weyl
semimetals.

For comparison, the schematic energy spectra in type-I and type-II Weyl
semimetals are shown in Figs. 3.3(a) and 3.3(b), respectively. It is clear
that the Fermi surface of a type-I Weyl semimetal is closed in general and
becomes point-like in the special case when $\mu = 0$. On the other hand,
the Fermi surface of a type-II Weyl semimetal is open. It delimits a pair
of extended electron and hole pockets at any value of μ, including $\mu = 0$.
Concerning the latter case, it should be noted that truly open Fermi surfaces cannot be realized in a compact Brillouin zone. In reality, the Fermi
surfaces in type-II Weyl semimetals are closed too, but they extend sufficiently far from the Weyl node, where the series expansion of the low-energy

Hamiltonian (3.10) breaks down. Therefore, in order to describe the corresponding electron- and hole-type pockets reliably, one needs a more realistic low-energy model whose range of validity covers at least the whole region of the pockets.

3.1.3 *Multi-Weyl fermions*

Another interesting type of Weyl fermions appears in the so-called multi-Weyl semimetals. They are materials with topological charges of Weyl nodes n greater than one. The examples include double-Weyl ($n = 2$) and triple-Weyl ($n = 3$) semimetals with quadratic and cubic energy dispersion relations, respectively. According to the theoretical analysis in [129], only the Weyl nodes with topological charges less than or equal to 3 are permitted by the possible crystallographic point symmetries.

The general forms of the low-energy Hamiltonians describing the double- and triple-Weyl fermions are given by

$$H_{\text{double}} = \left(A_1 k_+^2 + A_2 k_-^2 \right) \sigma_+ + \left(A_1^* k_-^2 + A_2^* k_+^2 \right) \sigma_- + \lambda \hbar v_F k_z \sigma_z, \quad (3.13)$$

and

$$H_{\text{triple}} = \left(B_1 k_+^3 + B_2 k_-^3 \right) \sigma_+ + \left(B_1^* k_-^3 + B_2^* k_+^3 \right) \sigma_- + \lambda \hbar v_F k_z \sigma_z, \quad (3.14)$$

respectively. Here $A_{1,2}$ and $B_{1,2}$ are some dimensionful coefficients and the shorthand notation $k_\pm = k_x \pm i k_y$ was used. (For specific examples of two-band periodic models of multi-Weyl semimetals, see also Sec. 3.5.) In the simplest case of a cubic lattice, the above low-energy Hamiltonians can be rewritten in the following explicit form:

$$H_{\text{double}} = \hbar v_F a \left(k_x^2 - k_y^2 \right) \sigma_x + 2 \hbar v_F a k_x k_y \sigma_y + \lambda \hbar v_F k_z \sigma_z, \quad (3.15)$$

and

$$H_{\text{triple}} = \hbar v_F a^2 \left(k_x^3 - 3 k_x k_y^2 \right) \sigma_x - \hbar v_F a^2 \left(k_y^3 - 3 k_x^2 k_y \right) \sigma_y + \lambda \hbar v_F k_z \sigma_z, \quad (3.16)$$

where a is the lattice constant. The corresponding energy dispersion relations read

$$\epsilon_{\text{double}} = \pm \hbar v_F \sqrt{k_z^2 + a^2 k_\perp^4}, \quad (3.17)$$

$$\epsilon_{\text{triple}} = \pm \hbar v_F \sqrt{k_z^2 + a^4 k_\perp^6}. \quad (3.18)$$

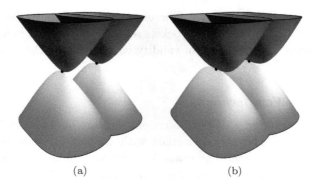

(a)   (b)

Fig. 3.4   Schematic energy spectra of (a) double-Weyl and (b) triple-Weyl semimetals.

As we see, these dispersions are anisotropic, with a linear dependence on k_z and a quadratic (double-Weyl) or cubic (triple-Weyl) dependence on $k_\perp = \sqrt{k_x^2 + k_y^2}$. This anisotropy is reflected in various physical properties of multi-Weyl semimetals. The dispersion relations of double- and triple-Weyl semimetals are shown schematically in Figs. 3.4(a) and 3.4(b), respectively.

By using the first-principles calculations, it was suggested that the double-Weyl nodes are realized in $HgCr_2Se_4$ [129, 361] and $SrSi_2$ [362].

3.2   Discrete symmetries

In condensed matter physics, symmetries play a crucial role in the classification of crystals and various phases of matter. In particular, ordered phases are characterized by order parameters that arise from spontaneous symmetry breaking. As will be clear from our discussion below, symmetries also play a very important role in topological semimetals. Among other things, they provide a protection for Dirac points and Weyl nodes, as well as help to classify Weyl semimetals.

It is instructive to start the discussion from the fundamental discrete symmetries, i.e., the parity-inversion $\mathcal{P}$, the time-reversal $\mathcal{T}$, and the charge-conjugation $\mathcal{C}$. As is well known [42–45], the Dirac Hamiltonian (3.9) is invariant with respect to all three of these symmetries, i.e.,

$$\hat{P}H(\mathbf{k})\hat{P}^{-1} = H(-\mathbf{k}), \quad \hat{P} = -\gamma^0 = \begin{pmatrix} 0 & \mathbb{1}_2 \\ \mathbb{1}_2 & 0 \end{pmatrix}, \tag{3.19}$$

$$\hat{T}H(\mathbf{k})\hat{T}^{-1} = H(-\mathbf{k}), \quad \hat{T} = \gamma_x\gamma_z\hat{K} = \begin{pmatrix} i\sigma_y & 0 \\ 0 & i\sigma_y \end{pmatrix}\hat{K}, \qquad (3.20)$$

$$\hat{C}H(\mathbf{k})\hat{C}^{-1} = -H(\mathbf{k}), \quad \hat{C} = i\gamma_y\gamma^0\hat{K} = \begin{pmatrix} -i\sigma_y & 0 \\ 0 & i\sigma_y \end{pmatrix}\hat{K}, \quad (3.21)$$

where $\hat{K}$ is the complex conjugation operator. It is important to note that $(\hat{P}\hat{T})^2 = -1$ for spin-1/2 fermions, which is the underlying reason for the celebrated *Kramer's theorem*. The latter states that eigenstates of a $\hat{P}\hat{T}$-invariant Hamiltonian are at least doubly degenerate [363]. When $\mathcal{T}$ symmetry is present, but $\mathcal{P}$ symmetry is absent (e.g., in an external electric field), the degeneracy is lifted except at the $\mathcal{T}$ *invariant momenta* (TRIM) in the Brillouin zone for which $\mathbf{k}$ and $-\mathbf{k}$ are the same up to a reciprocal lattice vector. The TRIM points include the center $\mathbf{k} = \mathbf{0}$ as well as several other special points (e.g., corners) in the Brillouin zone.

It is instructive to emphasize that the low-energy band structure with a Dirac point as in Eq. (3.9) cannot be realized in three-dimensional (3D) materials without fine-tuning. This is due to the fact that a crossing of two spin-degenerate bands is described by a rather general 4×4 effective Hamiltonian. Indeed, by imposing only the necessary constraints of the $\hat{P}\hat{T}$-invariance and tracelessness,[h] one can verify that the resulting effective Hamiltonian is a linear combination of the following five Hermitian matrices: γ^0, $\gamma_x\gamma_y\gamma_z$, and $\alpha_i = \gamma^0\gamma_i$ with $i = x, y, z$. In order to realize a Dirac point in momentum space, all five coefficient functions, multiplying these matrices, must vanish. However, in general, there are only three natural tuning parameters, namely the three components of momentum. Therefore, in contrast to the case of Weyl nodes, the appearance of Dirac points can be enforced only by additional constraints on the coefficient functions stemming, e.g., from crystal symmetries in Dirac semimetals.

In order to show how a Dirac point can arise, let us begin with the band inversion mechanism, which plays a principal role in the realization of topological insulators [98–101]. A natural order of bands at high symmetry points in the Brillouin zone is related to the order of the corresponding atomic orbitals. For example, usually, the energies of s-orbitals lie below those of p-orbitals. However, sometimes the order of orbitals may change

---

[h]The condition of tracelessness eliminates the term proportional to the unit matrix from the effective Hamiltonian. Since the only role of such a term is to shift the energy of the Dirac point, it can be always removed by choosing a suitable reference point.

leading to the band inversion. Clearly, the band inversion between two spin degenerate bands produces four-fold degenerate points in the energy spectrum. However, as noted by Shuichi Murakami [364, 365], the spin–orbit coupling will generically open a gap at the four-fold degenerate band crossing points. This is a common way to think about the realization of a topological insulator. However, a gap opening can be avoided if the spin-degenerate bands belong to different 2D irreducible representations of the little group at the crossing point. This requirement is crucial and was used in guiding the first theoretical predictions of the symmetry-protected Dirac semimetal phases in Na_3Bi [110] and Cd_3As_2 [111]. In both cases, the stability of the Dirac points is guaranteed by crystal symmetries [366, 367] and their formation is schematically depicted in Fig. 3.5. A more detailed discussion of the symmetry classification of Dirac semimetals will be given in Sec. 3.6.2.

In order to realize a Weyl semimetal phase, at least one of the symmetries $\mathcal{P}$ or $\mathcal{T}$ should be broken. By using the simplest Weyl Hamiltonian in Eq. (3.5) as an example, one can verify that a nonzero separation of the Weyl nodes in momentum (2**b**) breaks $\mathcal{T}$ symmetry. Similarly, a nonzero separation of the Weyl nodes in energy ($2b_0$) breaks $\mathcal{P}$ symmetry.[i] It is worth noting that, for the Weyl nodes away from the TRIM points, the presence of $\mathcal{T}$ symmetry implies that the minimal number of Weyl nodes is four. Indeed, this result stems from the fact that $\mathcal{T}$ symmetry transforms a Weyl node at $\mathbf{k}_0$ into a node at $-\mathbf{k}_0$ without changing its chirality. Since the chirality is directly related to the topological charge, which should

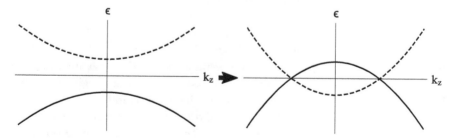

Fig. 3.5 Schematic representation of band inversion, where the trivial insulator band structure is depicted in the left panel. The Dirac phase occurs in the case of band inversion when the opening of a gap is forbidden by a discrete symmetry.

---

[i]The breakdown of $\mathcal{P}$ symmetry is also caused by the tilt of Weyl cones, discussed in Sec. 3.1.2.

vanish for the whole Brillouin zone, the minimal number of Weyl nodes in $\mathcal{T}$ symmetric but $\mathcal{P}$ broken Weyl semimetals must be indeed four. The Weyl nodes in the case with broken $\mathcal{T}$ and $\mathcal{P}$ symmetries are schematically shown in Figs. 3.2(a) and 3.2(b), respectively. Let us mention that the crystal symmetries could be also important for Weyl semimetals. In particular, they stabilize the multi-Weyl semimetals [129], which were briefly discussed in Sec. 3.1.3.

In addition to discrete symmetries, Weyl Hamiltonian possesses also nontrivial topological properties. To quantify them and understand their importance, we will introduce the concept of the Berry phase in the next Section.

3.3 Topology of electron states

As argued in Sec. 3.2, symmetries play a significant role in condensed matter systems. As became clear in the 1980s, topology is very important for the electron states in solids too. Indeed, starting with the discovery of the quantum Hall effect [293], it became evident that a topological order could be also used to distinguish different states of matter even in cases when no symmetry is spontaneously broken [368]. Often, a topological state is characterized by a long-range quantum entanglement. Since different topological states cannot be connected continuously without a phase transition, topology is imperative for their classification. Moreover, the use of topology language allows for a much deeper insight into the physical properties of novel materials and their stability with respect to small changes of parameters.

Historically, the nontrivial topological properties of electron states in solids were revealed in the analysis of semiclassical motion of charge carriers. Generally, the intuition about the motion of electrons in solids is based on considering them as classical particles with the energy dispersion defined by the corresponding band that move in response to electric and magnetic fields. This approach is used in the Drude model which provides a paradigmatic semiclassical description of the electron transport in solids that is both simple and surprisingly reliable. However, one phenomenon was completely enigmatic from the viewpoint of the semiclassical motion. Indeed, shortly after the discovery of the Hall effect [369], Edwin Hall found [370] that the Hall resistivity in magnetic materials is much larger than in nonmagnetic ones. This phenomenon is widely known as the anomalous Hall effect (AHE) [371].

The AHE did not get a clear theoretical explanation almost for a century. The reason is that, in order to understand the underlying physics, one requires the concepts from topology such as the Berry phase and the Berry curvature [68, 232], which were developed only in the second half of the 20th century. The major breakthrough in the theory of AHE was done by Robert Karplus and Joaquin Luttinger [372], who showed that the electrons may acquire an additional contribution to their group velocity perpendicular to an applied electric field. Such a contribution was dubbed the *anomalous velocity*. As will be shown below, since the anomalous velocity has its roots primarily in the band structure with a nonzero Berry curvature, it is usually referred to as the intrinsic contribution to the AHE. However, this may not be sufficient for explaining the AHE still [371]. As will be discussed in Sec. 6.3, the source of the AHE is particularly interesting in Weyl semimetals with broken $\mathcal{T}$ symmetry. There it is determined by the separation of the Weyl nodes in momentum space. In addition to the intrinsic contribution from the Berry curvature, the AHE may also have extrinsic contributions due to disorder effects connected with a spin-dependent scattering of charge carriers. The latter are referred to as the side jumps and the skew scattering.

3.3.1 *Berry phase*

In order to see how the anomalous velocity appears and what effects it causes in Weyl semimetals, it is instructive to begin our discussion with the notions of the *Berry phase* and the *Berry curvature*.[j] In 1984 Michael Berry studied [68] how a quantum-mechanical system evolves adiabatically along a closed loop in the parameter space. He found that, in addition to the conventional dynamical phase, the system acquires an extra contribution known as the Berry phase. The latter is of particular interest because it is determined by the geometry of the parameter space.

Let us show how the Berry phase appears. To start with, we assume that one-particle Hamiltonian H depends on parameter $\mathbf{R}(t)$, which evolves slowly in time making a loop, i.e., $\mathbf{R}(T) = \mathbf{R}(0)$. It is useful to introduce an instantaneous basis of eigenstates of $H(\mathbf{R})$ at a given value $\mathbf{R}(t)$, i.e.,

$$H\left(\mathbf{R}(t)\right)\left|n\left(\mathbf{R}(t)\right)\right\rangle = \epsilon_n\left(\mathbf{R}(t)\right)\left|n\left(\mathbf{R}(t)\right)\right\rangle. \tag{3.22}$$

---

[j]It is worth noting that the Berry phase and the Berry curvature are also often referred to as the geometric phase and the geometric curvature, respectively.

Let us assume that the system was prepared in state $|n\left(\mathbf{R}(0)\right)\rangle$ and then evolved adiabatically into state $|n\left(\mathbf{R}(t)\right)\rangle$ at time t, i.e.,

$$|\psi(t)\rangle = \exp\left[-\frac{i}{\hbar}\int_0^t \epsilon_n\left(\mathbf{R}(t')\right)dt'\right]\exp\left[i\gamma_n(t)\right]|n\left(\mathbf{R}(t)\right)\rangle. \qquad (3.23)$$

While the first exponential factor is the usual dynamical phase, the second factor with the phase $\gamma_n(t)$ has a geometrical origin. This extra phase $\gamma_n(t)$ is neither a function of $\mathbf{R}$ nor necessarily single-valued for a closed cycle, i.e., $\gamma(T) \neq \gamma(0)$.

The Schrödinger equation implies that $\gamma_n(t)$ satisfies the equation

$$\dot{\gamma}_n(t) = i\left\langle n\left(\mathbf{R}(t)\right)\right|\boldsymbol{\nabla}_\mathbf{R}\left|n\left(\mathbf{R}(t)\right)\right\rangle \cdot \dot{\mathbf{R}}(t). \qquad (3.24)$$

The total Berry phase accumulated along the loop $\mathcal{C}$ in a parameter space is

$$\gamma_n(\mathcal{C}) = -\oint_\mathcal{C} \mathbf{A}_n(\mathbf{R}) \cdot d\mathbf{R}, \qquad (3.25)$$

where the vector-valued function

$$\boldsymbol{\mathcal{A}}_n(\mathbf{R}) = -i\left\langle n(\mathbf{R})\right|\boldsymbol{\nabla}_\mathbf{R}\left|n(\mathbf{R})\right\rangle \qquad (3.26)$$

is called the *Berry connection*.[k] Sometimes, it is also referred to as the Berry vector potential. Since the definition of the geometric phase is reparameterization invariant, the dependence of $\mathbf{R}$ on t is omitted in Eq. (3.26) without loss of generality.

The normalization of $|n\rangle$ implies that $\langle n(\mathbf{R})|\boldsymbol{\nabla}_\mathbf{R}|n(\mathbf{R})\rangle$ is imaginary and, therefore, the Berry connection in Eq. (3.26) is real. It is easy to see, however, that $\boldsymbol{\mathcal{A}}_n(\mathbf{R})$ is gauge dependent and, therefore, is not a real observable. Indeed, by taking into account that the quantum wave function is defined locally up to an unobservable phase factor, one can perform the following phase transformation:

$$|n(\mathbf{R})\rangle \to e^{if(\mathbf{R})}|n(\mathbf{R})\rangle, \qquad (3.27)$$

which leads, in turn, to a different Berry connection,

$$\boldsymbol{\mathcal{A}}_n(\mathbf{R}) \to \boldsymbol{\mathcal{A}}_n(\mathbf{R}) + \boldsymbol{\nabla}_\mathbf{R}f(\mathbf{R}). \qquad (3.28)$$

---

[k]The use of the Berry connection in the description of quasiparticle motion was known for a long time. For example, it appeared in the coordinate operator $\hat{\mathbf{r}}$ already in [137]. Its topological consequences, however, were not recognized.

Naively, one might think that the gauge dependence could be always used to eliminate the Berry connection. However, the crucial observation made by Berry is that the cyclic evolution of the system along a closed path with $\mathbf{R}(T) = \mathbf{R}(0)$ requires that the basis $|n(\mathbf{R})\rangle$ must be single-valued, i.e.,

$$f\left(\mathbf{R}(T)\right) = f\left(\mathbf{R}(0)\right) + 2\pi m, \tag{3.29}$$

where m is an integer. In this case, $\gamma_n(\mathcal{C})$ is gauge invariant for a closed path. One also finds that the Berry phase depends only on the geometric aspects of the closed path $\mathcal{C}$ and is insensitive to the specific parameterization of $\mathbf{R}(t)$ on time.

3.3.2 *Berry curvature*

Although it was proven that the Berry phase is gauge invariant, it is useful to rewrite Eq. (3.25) in an explicitly gauge invariant form. By applying Stoke's theorem on the right-hand side of Eq. (3.25), the Berry phase could be expressed as a surface integral

$$\gamma_n(\mathcal{C}) = -\hbar \int_{\mathcal{C}} d\mathbf{S} \cdot \mathbf{\Omega}_n(\mathbf{R}), \tag{3.30}$$

where the Planck constant $\hbar$ is included for convenience, and

$$\mathbf{\Omega}_n(\mathbf{R}) = \frac{1}{\hbar} \left[\mathbf{\nabla_R} \times \boldsymbol{\mathcal{A}}_n(\mathbf{R})\right] \tag{3.31}$$

is the Berry curvature. By using the analogy with electrodynamics, this vector quantity can be viewed as an effective magnetic field in the parameter space. It can be also rewritten in terms of the Berry curvature tensor as $(\Omega_n)_i = \frac{1}{2}\epsilon_{ijl}(\mathcal{F}_n)_{jl}$, where ϵ_{ijk} is the Levi-Cività tensor and

$$(\mathcal{F}_n)_{jl} = \frac{1}{\hbar} \left[\partial_{R_j}(\mathcal{A}_n)_l(\mathbf{R}) - \partial_{R_l}(\mathcal{A}_n)_j(\mathbf{R})\right], \tag{3.32}$$

which, like the electromagnetic strength tensor, is explicitly gauge invariant.

It might be useful to derive another useful expression for the Berry curvature. Let us start by noting the following identities:

$$\gamma_n(\mathcal{C}) = -\mathrm{Im} \int_{\mathcal{C}} d\mathbf{S} \cdot \left[\mathbf{\nabla_R} \times \langle n(\mathbf{R})| \, \mathbf{\nabla_R}n(\mathbf{R})\rangle\right]$$

$$= -\mathrm{Im} \int_{\mathcal{C}} d\mathbf{S} \cdot \sum_{m \neq n} \left[\langle \mathbf{\nabla_R}n(\mathbf{R})| \, m(\mathbf{R})\rangle \times \langle m| \, \mathbf{\nabla_R}n(\mathbf{R})\rangle\right]. \tag{3.33}$$

By differentiating Eq. (3.22) with respect to $\mathbf{R}$, we also find that

$$\langle m| \, \boldsymbol{\nabla}_{\mathbf{R}} \, |n\rangle = \frac{\langle m| \, \boldsymbol{\nabla}_{\mathbf{R}} H \, |n\rangle}{\epsilon_n - \epsilon_m}, \quad m \neq n. \tag{3.34}$$

After substituting this into the expression for $\gamma_n(\mathcal{C})$ in Eq. (3.33) and comparing it with the definition in Eq. (3.30), the following new representation for the Berry curvature is derived:

$$\boldsymbol{\Omega}_n(\mathbf{R}) = \frac{1}{\hbar} \mathrm{Im} \sum_{m \neq n} \frac{[\langle n(\mathbf{R})| \, \boldsymbol{\nabla}_{\mathbf{R}} H(\mathbf{R}) \, |m(\mathbf{R})\rangle \times \langle m(\mathbf{R})| \, \boldsymbol{\nabla}_{\mathbf{R}} H(\mathbf{R}) \, |n(\mathbf{R})\rangle]}{[\epsilon_n(\mathbf{R}) - \epsilon_m(\mathbf{R})]^2}.$$

$$\tag{3.35}$$

This representation is convenient because it does not depend on the choice of eigenfunction gauge. In addition, the denominator in Eq. (3.35) shows that the main contribution comes from the energy bands that are close in energy. These properties make Eq. (3.35) particularly useful for numerical calculations.

It is worth noting that, so far, we considered only the case of nondegenerate energy levels. As discussed in [373–377], degenerate energy bands give rise to the Berry curvature with a non-Abelian structure. The latter may qualitatively affect the wave packet dynamics [378] and, in turn, some physical properties of Dirac semimetals. In particular, it is found that for parallel magnetic and electric fields, the trajectories of wave packets from different valleys (or, equivalently, Dirac points) can become spatially split in the plane perpendicular to the fields. The non-Abelian corrections allow for a spiral-like motion of wave packets on top of an almost linear separation as well as lead to a small oscillating chirality polarization of the packets. The physical origin of spiraling trajectories is connected with the precession of the magnetic moment. Although the effects related the non-Abelian structure could be experimentally accessible via certain local probes, unfortunately, this might be difficult to do in practice.

3.3.3 *Semiclassical motion*

At this point it is appropriate to discuss briefly why the Berry curvature is important for the electronic properties of solids. As a representative example, let us consider a semiclassical motion of charge carriers in crystals. The corresponding analysis is based on the semiclassical wave-packet approach developed by Ming-Che Chang and Qian Niu [379, 380] (see also review [377]). A general form of the electron Hamiltonian in a crystal, subject to an electromagnetic field, is given by [356, 381]

$$H = \frac{1}{2m} \left[-i\hbar \boldsymbol{\nabla} + \frac{e}{c} \mathbf{A}(\mathbf{r}) \right]^2 + V(\mathbf{r}) - e\phi(\mathbf{r}), \qquad (3.36)$$

where $-e$ is the electron charge, $\mathbf{A}(\mathbf{r})$ is the vector potential, $V(\mathbf{r})$ is the periodic potential of the crystalline lattice, and $\phi(\mathbf{r})$ is the electric potential. The wave packet for an electron in the nth energy zone is given by

$$|W\rangle = e^{-ie\mathbf{A}(\mathbf{r}_c)\cdot\mathbf{r}/(\hbar c)} \int \frac{d^3k}{(2\pi)^3} a\left(\mathbf{k}, t\right) |\psi_{n\mathbf{k}}\rangle. \qquad (3.37)$$

It is composed of the Bloch functions $|\psi_{n\mathbf{k}}\rangle = e^{i\mathbf{k}\cdot\mathbf{r}} u_{n\mathbf{k}}(\mathbf{r})$, where $u_{n\mathbf{k}}(\mathbf{r})$ has the lattice periodicity. By assumption, $a(\mathbf{k}, t)$ is a narrow distribution (e.g., a Gaussian one) that centers around the wave vector $\mathbf{k}_c(t)$ in momentum space and is normalized, $\int d^3k/(2\pi)^3 |a(\mathbf{k}, t)|^2 = 1$. The center of mass of the wave packet, or more precisely the expectation value of the coordinate for the wave packet function, is defined by

$$\langle W| \mathbf{r} |W\rangle = \mathbf{r}_c. \qquad (3.38)$$

Conceptually, the semiclassical wave-packet approach makes sense when the electromagnetic perturbations are smooth, varying on length scales much larger than the spatial spread of the wave packet. In this case, one can expand the electromagnetic fields in powers of $\mathbf{r} - \mathbf{r}_c$

$$\phi(\mathbf{r}) = \phi(\mathbf{r}_c) + ((\mathbf{r} - \mathbf{r}_c) \cdot \boldsymbol{\nabla}_{\mathbf{r}_c}\phi) + \cdots, \qquad (3.39)$$

$$\mathbf{A}(\mathbf{r}) = \mathbf{A}(\mathbf{r}_c) + ((\mathbf{r} - \mathbf{r}_c) \cdot \boldsymbol{\nabla}_{\mathbf{r}_c}\mathbf{A}) + \cdots. \qquad (3.40)$$

The wave packet dynamics can be obtained from the time-dependent variational principle at each order of the expansion in powers of $\mathbf{r} - \mathbf{r}_c$. To use such an approach, one should obtain the explicit form of the effective Lagrangian

$$L = \langle W| i\hbar\partial_t - H |W\rangle \qquad (3.41)$$

and then derive the equations of motion for the wave packet coordinate $\mathbf{r}_c$ and momentum $\mathbf{q}_c$ by using the variational principle.

To the first order in electromagnetic fields, one finds [381]

$$L = \hbar \left(\mathbf{k} \cdot \dot{\mathbf{r}}\right) - \epsilon_n(\mathbf{k}) - \hbar \left(\boldsymbol{\mathcal{A}}_n(\mathbf{k}) \cdot \dot{\mathbf{k}}\right) - \frac{e}{c} \left(\mathbf{A} \cdot \dot{\mathbf{r}}\right) + e\phi. \qquad (3.42)$$

Here, for brevity of notation, the subscript c is omitted in the center-of-mass quantities $\mathbf{r}_c$ and $\mathbf{q}_c$. In general, the quasiparticle energy $\epsilon_n(\mathbf{k})$ might include corrections due to interaction with a magnetic field. By making use of Eq. (3.42), the following equations of motion are obtained:

$$\dot{\mathbf{r}} = \frac{1}{\hbar}\boldsymbol{\nabla}_{\mathbf{k}}\epsilon_n(\mathbf{k}) + 0\hbar[\dot{\mathbf{k}} \times \boldsymbol{\Omega}_n(\mathbf{k})], \tag{3.43}$$

$$\hbar\dot{\mathbf{k}} = -e\mathbf{E} - \frac{e}{c}[\dot{\mathbf{r}} \times \mathbf{B}]. \tag{3.44}$$

Note that the Berry curvature $\boldsymbol{\Omega}_n$ plays the same role in momentum space as the magnetic field in coordinate space. The presence of the corresponding term with $\boldsymbol{\Omega}_n$ in Eq. (3.43) also has important physics implications. This can be seen already by considering the quasiparticle velocity in the simplest case without a magnetic field, $\mathbf{B} = \mathbf{0}$, i.e.,

$$\dot{\mathbf{r}} = \frac{1}{\hbar}\boldsymbol{\nabla}_{\mathbf{k}}\epsilon_n(\mathbf{k}) - e\,[\mathbf{E} \times \boldsymbol{\Omega}_n(\mathbf{k})]. \tag{3.45}$$

While the first term describes the standard group velocity defined by the energy dispersion, the second one is an unusual contribution perpendicular to the electric field $\mathbf{E}$. The appearance of this extra term demonstrates that the Berry curvature might be indeed responsible for the intrinsic contribution to the anomalous Hall effect. A more detailed derivation of the semiclassical equations of motion as well as the formulation of the kinetic theory for Dirac and Weyl semimetals will be given in Sec. 6.6. The anomalous Hall effect in Weyl semimetals with broken $\mathcal{T}$ symmetry will be discussed in Sec. 6.3.

3.3.4 *Weyl nodes as monopoles of Berry curvature*

Let us now calculate the Berry connection and the Berry curvature in the vicinity of a single Weyl node. As is easy to check, the eigenstates of momentum space Hamiltonian (3.3) with positive energy $\epsilon = \hbar v_F|\mathbf{k}|$ are given by

$$|\psi_+(\mathbf{k})\rangle_{\mathrm{N}} = \begin{pmatrix} \cos{(\theta/2)} \\ e^{i\varphi}\sin{(\theta/2)} \end{pmatrix}, \quad |\psi_-(\mathbf{k})\rangle_{\mathrm{N}} = \begin{pmatrix} -e^{-i\varphi}\sin{(\theta/2)} \\ \cos{(\theta/2)} \end{pmatrix}. \tag{3.46}$$

Here the subscript labels the chirality of Weyl fermions ($\lambda = \pm$) and the standard spherical coordinate system parameterization was used, i.e., $\mathbf{k} = k\,(\cos\varphi\sin\theta, \sin\varphi\sin\theta, \cos\theta)$, where $k = |\mathbf{k}|$. Note that the spinors in Eq. (3.46) do not depend on the absolute value of the wave vector k.

While the right-handed (left-handed) spinor is well defined in the whole northern hemisphere, it is singular at $\theta = \pi$, where the phase of its lower (upper) component is undefined. By making use of the definition of the Berry connection (3.26), it is easy to find that $\mathcal{A}_\varphi$ is the only nonvanishing component of the Berry connection,

$$\mathcal{A}_\theta^{\mathrm{N}} = -\frac{i}{k} \langle \psi_\lambda |_{\mathrm{N}} \, \partial_\theta \, |\psi_\lambda\rangle_{\mathrm{N}} = 0, \tag{3.47}$$

$$\mathcal{A}_\varphi^{\mathrm{N}} = -\frac{i}{k \sin\theta} \langle \psi_\lambda |_{\mathrm{N}} \, \partial_\varphi \, |\psi_\lambda\rangle_{\mathrm{N}} = \frac{\lambda}{2k} \tan\left(\frac{\theta}{2}\right). \tag{3.48}$$

By calculating its curl, one can derive the corresponding Berry curvature,

$$\boldsymbol{\Omega} = \lambda \frac{\mathbf{k}}{2\hbar k^3}. \tag{3.49}$$

Note that this expression has a monopole-like form. One must remember, however, that its validity breaks down along the direction $\theta = \pi$. Indeed, as one can verify, a more careful calculation of the curl gives

$$[\boldsymbol{\nabla}_{\mathbf{k}} \times \boldsymbol{\mathcal{A}}^{\mathrm{N}}] = \hbar\boldsymbol{\Omega} + 2\pi\lambda\hat{\mathbf{z}}\delta(k_x)\delta(k_y)\theta(-k_z). \tag{3.50}$$

The singularity along the negative k_z axis is called the Dirac string [382] and reflects a specific choice of the gauge for the wave functions in Eq. (3.46). While other gauges are possible, none of them results in a well-defined wave function without singularities in the whole momentum space.[1]

As usual, in order to define a nonsingular spinor in the southern hemisphere, one should use another gauge [383], e.g., by multiplying the wave functions in Eq. (3.46) by $\exp(-\lambda i\varphi)$. The new wave functions read

$$|\psi_+(\mathbf{k})\rangle_{\mathrm{S}} = \begin{pmatrix} e^{-i\varphi}\cos(\theta/2) \\ \sin(\theta/2) \end{pmatrix}, \quad |\psi_-(\mathbf{k})\rangle_{\mathrm{S}} = \begin{pmatrix} -\sin(\theta/2) \\ e^{i\varphi}\cos(\theta/2) \end{pmatrix}. \tag{3.51}$$

While these spinors are well defined in the whole southern hemisphere, they are singular at the north pole, $\theta = 0$. The corresponding Berry connection is given by

$$\mathcal{A}_\theta^{\mathrm{S}} = -\frac{i}{k} \langle \psi_\lambda |_{\mathrm{S}} \, \partial_\theta \, |\psi_\lambda\rangle_{\mathrm{S}} = 0, \tag{3.52}$$

$$\mathcal{A}_\varphi^{\mathrm{S}} = -\frac{i}{k \sin\theta} \langle \psi_\lambda |_{\mathrm{S}} \, \partial_\varphi \, |\psi_\lambda\rangle_{\mathrm{S}} = -\frac{\lambda}{2k} \cot\left(\frac{\theta}{2}\right). \tag{3.53}$$

---

[1]Note that the wave functions in Eq. (3.46) are defined on a 2D Bloch sphere spanned by the angular coordinates φ and θ, rather than the whole 3D momentum space.

As expected for a gauge-invariant quantity, the Berry curvature for the connection defined in Eqs. (3.52) and (3.53) is given by the same expression (3.49) as for the connection in Eqs. (3.47) and (3.48), although its validity breaks down along the direction $\theta = 0$. This can be verified by performing a more careful calculation of curl, which will produce a result similar to that in Eq. (3.50), but $\theta(-k_z)$ will be replaced by $\theta(k_z)$.

As one can see from Eq. (3.49), the Berry curvature for a generic Weyl node of chirality λ has a monopole structure with a singularity at $\mathbf{k} = \mathbf{0}$, where the two bands become degenerate. This point acts as a source or sink of the Berry curvature flux. Indeed, by rederiving the same result from Eq. (3.35), one could also see that the monopole singularity is connected with the crossing of two energy bands.

As is easy to verify, the Berry connections, defined in the northern and southern hemispheres, are related by the gauge transformation

$$\mathcal{A}^{\mathrm{N}} = \mathcal{A}^{\mathrm{S}} + \boldsymbol{\nabla}_{\mathbf{k}}\varphi = \mathcal{A}^{\mathrm{S}} + \lambda\frac{\hat{\boldsymbol{\varphi}}}{k\sin\theta}, \tag{3.54}$$

which is nonsingular in the region near the equator, $\theta = \pi/2$. In fact, this pair of connections is identical with the vector potential describing the Wu–Yang monopole [384] with charge $-\lambda/2$ and the total flux or, equivalently, the Berry phase $\gamma = -4\pi\lambda/2 = -2\pi\lambda$. In order to appreciate the significance of the nonsingular gauge transformation given by Eq. (3.54), it is instructive to calculate the Berry phase by using the definition in Eq. (3.30), i.e.,

$$\begin{aligned}
\gamma &= -\int_{\mathrm{N}} d\mathbf{S} \cdot [\boldsymbol{\nabla}_{\mathbf{k}} \times \mathcal{A}^{\mathrm{N}}] - \int_{\mathrm{S}} d\mathbf{S} \cdot [\boldsymbol{\nabla}_{\mathbf{k}} \times \mathcal{A}^{\mathrm{S}}] \\
&= -\oint_{\mathrm{equator}} d\mathbf{l} \cdot \mathcal{A}^{\mathrm{N}} + \oint_{\mathrm{equator}} d\mathbf{l} \cdot \mathcal{A}^{\mathrm{S}} \\
&= -\lambda \oint_{\mathrm{equator}} d\varphi = -2\pi\lambda.
\end{aligned} \tag{3.55}$$

As we see, the Berry phase is completely determined by a line integral of the gauge function along the equator. This is not accidental and has roots in the theory of fiber bundles that Wu and Yang used to describe a magnetic monopole without the introduction of the unphysical Dirac string.

By making use of the concept of the Berry curvature, one can also give a simple qualitative argument supporting the validity of the Nielsen–Ninomiya theorem in solids. The argument is built on the fact that each

Weyl node is a monopole of the Berry curvature in momentum space. The topological charge of such a monopole is determined by the chirality of the Weyl node λ. In other words, the flux of the Berry curvature through the sphere surrounding each Weyl node is proportional to λ and, therefore, the flux through the whole Brillouin zone is given by the sum of contributions from all Weyl nodes. Because of the periodicity of lattice models, however, the surface encompassing the whole Brillouin zone is topologically equivalent to a point. This means that the total flux from all Weyl nodes and, consequently, the total topological charge should vanish. For massless spin-1/2 quasiparticles, this implies that the Weyl nodes always come in pairs of opposite chirality.

As one can see, the topological invariant of Weyl semimetals is *not* determined by the integral over the whole Brillouin zone. The correct way to define such invariants is to integrate over closed surfaces encompassing the singular Weyl nodes. The corresponding flux of the Berry curvature defines the topological charge of Weyl nodes.

The distribution of the Berry curvature lines in the $k_x - k_z$ plane for a simple two-band model, defined by Eq. (3.93) later, is schematically shown in Fig. 3.6. As expected, the right- and left-handed Weyl nodes act as source and sink of the Berry curvature lines, respectively. In the vicinity of each node, one can easily recognize a monopole-like configuration analogous to that defined in Eq. (3.49).

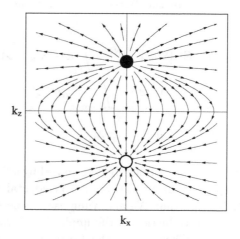

Fig. 3.6   The Berry curvature distribution in a realistic two-band model of Weyl semimetals (see Secs. 3.5 and 5.3) with the chiral shift parallel to $\hat{\mathbf{z}}$. Filled and empty points denote the right- and left-handed Weyl nodes.

Having considered the symmetry and topology protection in a noninteracting model, one can also ask about the effects of interactions in Dirac and Weyl semimetals and, in particular, their implications on the stability of Weyl nodes and Dirac points. The corresponding effects will be briefly considered in the following section.

3.4   Interaction effects

The effects of the Coulomb interaction on electron quasiparticles with a linear spectrum were first systematically considered by Abrikosov and Beneslavskii [34]. More recently, this issue was reexamined in more detail in [385–388]. In particular, in [388] the interaction corrections to the polarization function, the electron self-energy, and the vertex function were calculated up to second order in the coupling constant. In addition, the renormalization group analysis was performed for the running coupling, the Fermi velocity, and the quasiparticle residue. In this section, for simplicity, we will discuss only the key results obtained at the lowest order in coupling, which are sufficient to capture the gist of the underlying physics. In particular, the focus will be made on the polarization function and the self-energy. The corresponding results together with vertex corrections are fundamental for the renormalization group analysis in quantum field theory [42–45]. In Dirac and Weyl semimetals, however, the vertex corrections become relevant only at the second order in coupling [388]. To address the role of interaction effects, we consider the case of zero electric chemical potential and vanishing temperature.

Let us start with a discussion of dynamical screening of the Coulomb potential. In the momentum space representation,[m] the interaction of quasiparticles leads to the following modification of the Coulomb law:

$$V(\Omega, \mathbf{q}) = \frac{4\pi e^2}{\varepsilon_e q^2 + 4\pi\hbar\Pi_{00}(\Omega, \mathbf{q})}, \tag{3.56}$$

where ε_e is the background dielectric constant. To leading order in the coupling constant, the *polarization function* $\Pi_{00}(\Omega, \mathbf{q})$ is given by the following

---

[m]The screened Coulomb law in coordinate space can be easily obtained by taking the inverse Fourier transform. For example, in the case of a polarization function in the static limit $\Pi_{00}(0, \mathbf{0})$, the electron–electron interaction in Eq. (3.56) is described by the Yukawa potential $\propto e^2 \exp{(-r/r_{sc})}/r$, where $r_{sc} = 1/\sqrt{4\pi\hbar\Pi(0, \mathbf{0})}$ is the screening length.

diagram:

$$\Pi_{00}(\Omega, \mathbf{q}) = \text{(diagram)} \tag{3.57}$$

The corresponding analytical expression reads

$$\Pi_{00}(\Omega, \mathbf{q}) = -ie^2 \int \frac{d\omega d^3 k}{(2\pi)^4} \text{tr}\left[\gamma^0 S(\omega, \mathbf{k})\gamma^0 S(\omega - \Omega, \mathbf{k} - \mathbf{q})\right]. \tag{3.58}$$

The causal Green function (Feynman propagator) for the noninteracting quasiparticles, described by the relativistic-like Dirac Hamiltonian (3.9), reads (see Appendix A.1 for details)

$$S(\omega, \mathbf{k}) = i\frac{\hbar\omega\gamma^0 - \hbar v_F\left(\mathbf{k} \cdot \boldsymbol{\gamma}\right)}{\left[\hbar\omega + i0\,\text{sgn}\left(\omega\right)\right]^2 - (\hbar v_F k)^2}. \tag{3.59}$$

In order to utilize the standard Feynman diagram technique of QED, it is convenient to introduce a relativistic-like four-vector notation $\tilde{k}_\mu = (\hbar\omega, -\hbar v_F \mathbf{k})$. (Note that $\tilde{k}_\mu$ has the units of energy.) Then the expression for the polarization function can be rewritten as

$$\Pi_{00}(\tilde{q}) = 4i\frac{e^2}{\hbar^4 v_F^3} \int \frac{d^4 k}{(2\pi)^4} \frac{\tilde{k}_0(\tilde{k}_0 - \tilde{q}_0) + \tilde{\mathbf{k}} \cdot (\tilde{\mathbf{k}} - \tilde{\mathbf{q}})}{(\tilde{k}_0^2 - |\tilde{\mathbf{k}}|^2)[(\tilde{k}_0 - \tilde{q}_0)^2 - |\tilde{\mathbf{k}} - \tilde{\mathbf{q}}|^2]}. \tag{3.60}$$

After performing the Wick rotation, i.e., $\tilde{k}_0 \to i\tilde{k}_4$ and $\tilde{q}_0 \to i\tilde{q}_4$, and then integrating over $\tilde{k}_4$, one obtains

$$\Pi_{00}(\tilde{q}) = \frac{2e^2}{\hbar^4 v_F^3} \int \frac{d^3\tilde{k}}{(2\pi)^3} \frac{|\tilde{\mathbf{k}}| + |\tilde{\mathbf{k}} - \tilde{\mathbf{q}}|}{\tilde{q}_4^2 + (|\tilde{\mathbf{k}}| + |\tilde{\mathbf{k}} - \tilde{\mathbf{q}}|)^2} \left[1 - \frac{\tilde{\mathbf{k}} \cdot (\tilde{\mathbf{k}} - \tilde{\mathbf{q}})}{|\tilde{\mathbf{k}}||\tilde{\mathbf{k}} - \tilde{\mathbf{q}}|}\right]. \tag{3.61}$$

Formally, the integral in the last expression is logarithmically divergent. By noting, however, that the effective Dirac theory is valid only at sufficiently low energies, one can naturally regularize the divergent integral by introducing a suitable momentum cutoff Λ. Physically, the latter corresponds to the energy scale above which deviations from the linear dispersion relation become nonnegligible. In particular, it is clear that the upper limit for cutoff is always set by the size of the Brillouin zone, i.e., $\Lambda \lesssim \hbar v_F \pi/a$, where a is the lattice spacing. However, a more reasonable value of cutoff is usually much smaller than $\hbar v_F \pi/a$ for realistic band structures of Dirac and Weyl semimetals.[n] From the viewpoint of observables, a specific choice

---

[n] For example, in the model of a Dirac semimetal, discussed in Secs. 3.5 and 3.6, a natural choice of cutoff is the height of the dome in the energy spectrum between the two Dirac points, i.e., ϵ_0 (see also Fig. 3.8).

of the cutoff is almost irrelevant. Indeed, after fixing the value of cutoff Λ, one should perform a renormalization procedure, which amounts to introducing the "running" parameters of the coupling constant, the Fermi velocity, and others that depend on a renormalization scale. When observables are expressed in terms of renormalized low-energy parameters, however, a strong dependence on the energy cutoff Λ will be absent.

In order to perform the integration in Eq. (3.61) with an energy cutoff Λ, it is convenient to use the following prolate spheroidal coordinates:

$$\tilde{k}_x = \frac{1}{2}|\tilde{\mathbf{q}}|\sinh u \sin\theta \cos\varphi, \tag{3.62}$$

$$\tilde{k}_y = \frac{1}{2}|\tilde{\mathbf{q}}|\sinh u \sin\theta \sin\varphi, \tag{3.63}$$

$$\tilde{k}_z = \frac{1}{2}|\tilde{\mathbf{q}}|(\cosh u \cos\theta - 1), \tag{3.64}$$

where $0 \le u < \infty$, $0 \le \theta < \pi$, and $0 \le \varphi < 2\pi$. After integrating over φ and u, the result reads

$$\Pi_{00}(\tilde{q}) = e^2 \frac{|\tilde{\mathbf{q}}|^2}{24\pi^2\hbar^4 v_F^3} \ln\left\{ \frac{\left[\tilde{q}_4^2 + (2\Lambda + |\tilde{\mathbf{q}}|)^2\right]\left[\tilde{q}_4^2 + (2\Lambda - |\tilde{\mathbf{q}}|)^2\right]}{\left(\tilde{q}_4^2 + |\tilde{\mathbf{q}}|^2\right)^2} \right\}$$
$$- e^2 \frac{|\tilde{\mathbf{q}}|^3}{8\pi^2\hbar^4 v_F^3} \int_{-1}^{1} dy\, y \left(1 - \frac{y^2}{3}\right) \frac{2\Lambda + y|\tilde{\mathbf{q}}|}{\tilde{q}_4^2 + (2\Lambda + y|\tilde{\mathbf{q}}|)^2}. \tag{3.65}$$

Here the new integration variable $y = \cos\theta$ was introduced. While the first term in Eq. (3.65) contains a logarithmic divergence at large Λ, the second term is finite. By assuming that $\hbar|\Omega| \ll \Lambda$ and $\hbar v_F q \ll \Lambda$, the following approximate result can be obtained [34, 388]:

$$\Pi_{00}(\Omega, \mathbf{q}) \approx \frac{\alpha\varepsilon_e q^2}{6\pi^2\hbar} \left[\ln\left(\frac{\Lambda}{\hbar v_F q}\right) - \frac{1}{2}\ln\left(\frac{v_F^2 q^2 - \Omega^2}{4v_F^2 q^2}\right)\right]. \tag{3.66}$$

It should be emphasized that the effective coupling constant $\alpha = e^2/(\varepsilon_e \hbar v_F)$ and the Fermi velocity v_F on the right-hand side of the last expression are the "bare" parameters defined at the cutoff scale Λ. After substituting $\Pi_{00}(\Omega, \mathbf{q})$ into the expression for the potential in Eq. (3.56), one obtains the renormalized potential at a reference momentum scale q_*

$$V(\Omega, \mathbf{q}_*) = \frac{4\pi\hbar\alpha(q_*)v_F(q_*)}{(q_*)^2} \simeq \frac{4\pi\hbar\alpha v_F}{(q_*)^2\left[1 + \frac{2\alpha}{3\pi}\ln\left(\frac{\Lambda}{\hbar v_F q_*}\right)\right]}, \tag{3.67}$$

where $\alpha(q_*)$ and $v_F(q_*)$ are the renormalized parameters defined at scale q_*. As usual in the renormalization theory, their values should be fixed experimentally. In order to determine the running of both coupling constant $\alpha(q)$ and Fermi velocity $v_F(q)$, the knowledge of the polarization function alone is not sufficient. As will be shown below, one also needs to calculate radiative corrections to the self-energy.

To leading order in coupling, the radiative correction to the *self-energy* is given by the following one-loop diagram:

$$\Sigma\left(\Omega,\mathbf{q}\right) = \text{} \tag{3.68}$$

The corresponding expression reads

$$\Sigma\left(\Omega,\mathbf{q}\right) = -i \int \frac{d\omega d^3 k}{(2\pi)^4} \gamma^0 S(\omega,\mathbf{k})\gamma^0 D(\omega - \Omega, \mathbf{k} - \mathbf{q}), \tag{3.69}$$

where the photon propagator

$$D\left(\Omega,\mathbf{q}\right) = \frac{4\pi e^2}{\varepsilon_e} \frac{i}{q^2} \tag{3.70}$$

corresponds to the unscreened Coulomb potential. It is important to emphasize that there is no frequency dependence in Eq. (3.70) because the retardation effects are negligible for quasiparticles in Weyl and Dirac semimetals. This is in contrast to truly relativistic systems, such as QED plasma, where a frequency-dependent photon propagator must be used. With that being said, in a more refined treatment beyond the leading order approximation, one needs to use the screened Coulomb potential in Eq. (3.56), which introduces a weak frequency dependence.

As in the calculation of the polarization function, it is convenient to perform the Wick rotation before integrating over $\tilde{k}_4$. The result reads

$$\Sigma\left(\tilde{q}\right) = -\frac{2\pi e^2}{\varepsilon_e \hbar^3 v_F^2} \int \frac{d^3 \tilde{k}}{(2\pi)^3} \gamma^0 \frac{\left(\tilde{\mathbf{k}} - \tilde{\mathbf{q}}\right) \cdot \boldsymbol{\gamma}}{|\tilde{\mathbf{k}}||\tilde{\mathbf{k}} - \tilde{\mathbf{q}}|} \gamma^0. \tag{3.71}$$

By switching to spherical coordinates and using the symmetry with respect to rotations of $\tilde{\mathbf{k}}$ about $\tilde{\mathbf{q}}$, Eq. (3.71) can be rewritten in the following form:

$$\Sigma\left(\tilde{q}\right) = -\frac{e^2}{2\pi\varepsilon_e \hbar^3 v_F^2 |\tilde{\mathbf{q}}|^2} \int_0^\Lambda d\tilde{k} \int_0^\pi d\theta \frac{\sin\theta \left(|\tilde{\mathbf{q}}|^2 + |\tilde{\mathbf{k}}||\tilde{\mathbf{q}}|\cos\theta\right)}{\sqrt{|\tilde{\mathbf{k}}|^2 + |\tilde{\mathbf{q}}|^2 + 2|\tilde{\mathbf{k}}||\tilde{\mathbf{q}}|\cos\theta}} (\tilde{\mathbf{q}} \cdot \boldsymbol{\gamma}). \tag{3.72}$$

After integrating over momentum, the approximate low-energy result for the self-energy reads [388]

$$\Sigma\left(\Omega, \mathbf{q}\right) = i\frac{2\alpha}{3\pi}\left[\frac{4}{3} + \ln\left(\frac{\Lambda}{\hbar v_F q}\right)\right]\hbar v_F\left(\mathbf{q}\cdot\boldsymbol{\gamma}\right). \tag{3.73}$$

Note that the self-energy does not lead to a renormalization of the wave function because there is no term proportional to γ^0. Physically, this is related to the fact that the interaction potential is nonretarded. Since the self-energy modifies the inverse Green function, i.e.,

$$G^{-1}\left(\Omega, \mathbf{q}\right) = S^{-1}\left(\Omega, \mathbf{q}\right) + i\Sigma\left(\Omega, \mathbf{q}\right), \tag{3.74}$$

it is straightforward to verify that the result in Eq. (3.73) gives a radiative correction to the Fermi velocity, $v_F \to v_F(q) = v_F + \delta v_F$, where δv_F is determined from the equation $\hbar\delta v_F\left(\mathbf{q}\cdot\boldsymbol{\gamma}\right) = -i\Sigma\left(\Omega, \mathbf{q}\right)$. Then, by making use of the result in Eq. (3.73), the renormalized Fermi velocity at the reference momentum scale q_* reads

$$v_F(q_*) \simeq v_F\left[1 + \frac{2\alpha}{3\pi}\ln\left(\frac{\Lambda}{\hbar v_F q_*}\right)\right] \simeq v_F\left(\frac{\Lambda}{\hbar v_F q_*}\right)^{\frac{2\alpha}{3\pi}}. \tag{3.75}$$

Note that the second form of the expression for $v_F(q_*)$ can be obtained by solving the leading-order renormalization group equation [388]. As is easy to check, it agrees with the series representation obtained at the leading order in coupling. Because of the positive sign in front of the logarithmic term, the Fermi velocity grows with decreasing momentum q_*. By combining relation (3.75) with that in Eq. (3.67), the following expression for the renormalized coupling is obtained:

$$\alpha(q_*) \simeq \alpha\left[1 - \frac{4\alpha}{3\pi}\ln\left(\frac{\Lambda}{\hbar v_F q_*}\right)\right] \simeq \frac{\alpha}{1 + \frac{4\alpha}{3\pi}\ln\left(\frac{\Lambda}{\hbar v_F q_*}\right)}. \tag{3.76}$$

While the first representation in the form of a series follows from the perturbative analysis at the leading order in coupling, the other is an "improved" expression obtained by solving the renormalization group equation [388]. The two representations are equivalent, of course, at the leading order. Unlike the Fermi velocity, the coupling constant becomes smaller with decreasing the momentum scale.[o]

---

[o] A similar running behavior is also realized in QED, where it describes the screening of a bare charge by vacuum fluctuations [42–45].

    The dependence of the coupling constant and the Fermi velocity on the momentum scale is an important property, which can be straightforwardly tested in experiments. For example, as suggested in [388], this could be done by measuring the temperature dependence of the optical conductivity and the plasmon frequency (see also Sec. 7.3 and Chapter 8, respectively). The corresponding general idea is simple. One starts with a certain experimental setup, where, e.g., temperature and/or doping level are fixed. This defines the momentum scale q_* and the parameters $\alpha(q_*)$ and $v_F(q_*)$. By varying experimental conditions, a different momentum scale q can be probed and the corresponding coupling constant $\alpha(q)$ and the Fermi velocity $v_F(q)$ can be extracted. After repeating a series of such experiments, the running behavior could be verified.

    The dependence of the coupling constant $\alpha(q)$ and the Fermi velocity $v_F(q)$ on momentum q is shown in Figs. 3.7(a) and 3.7(b), respectively. As one can see from Fig. 3.7(a), the screening effects in the Dirac semimetals Cd_3As_2 and Na_3Bi, as well as the Weyl semimetal TaAs (see Appendix D for the characteristic values for numerical parameters) are small. A similarly weak dependence of the Fermi velocity on the momentum scale is evident from Fig. 3.7(b). The smallness of the interaction effects can be mainly attributed to the relatively large dielectric permittivities of these materials. It is worth noting, however, that the interaction effects can be enhanced substantially in materials with multiple band touching points. This stems from the fact that contributions from all Dirac points or pairs of Weyl nodes add up in the polarization function. Such an enhancement mechanism could

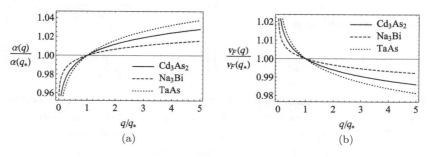

Fig. 3.7   (a) Coupling constant $\alpha(q)$ and (b) Fermi velocity $v_F(q)$ as functions of momentum scale q. The results for the Dirac semimetals Cd_3As_2 and Na_3Bi are shown with solid and dashed lines, while the results for the Weyl semimetal TaAs are represented by dotted lines. We used the following parameters: (i) $v_F(q_*) = 1.5 \times 10^8$ cm/s and $\varepsilon_e = 36$ for Cd_3As_2, (ii) $v_F(q_*) = 8 \times 10^7$ cm/s and $\varepsilon_e = 120$ for Na_3Bi, (iii) $v_F(q_*) = 4.3 \times 10^7$ cm/s and $\varepsilon_e = 93$ for TaAs.

be important, for example, in the Weyl semimetal TaAs, which has 12 pairs of Weyl nodes (see Sec. 4.2.2.2).

Here it might be instructive to briefly compare the results in Dirac semimetals with those obtained in QED [42–45]. One of the biggest differences is the renormalization of the Fermi velocity, which is in drastic contrast to the constant speed of light in QED. Technically, this is related to the fact the Lorentz symmetry is explicitly broken in solids. Also, because of substantial retardation effects in a relativistic treatment, the whole structure of divergencies and their cancelation is different in QED. For example, unlike in Dirac semimetals, a logarithmic divergence in the relativistic vertex function appears already at the first order in the coupling constant. Additional quantitative differences are related to the fact that the coupling constant $\alpha = c\alpha_0/(\varepsilon_e v_F)$, where $\alpha_0 = e^2/(\hbar c)$ is the vacuum fine-structure constant, is much larger in solids. While large values of the coupling constant provide an opportunity to study nonperturbative regimes unavailable in a relativistic QED plasma, they also limit the applicability of the well-developed perturbative techniques.

Before finalizing this section, it might be also useful to compare the interaction effects in Dirac and Weyl semimetals with those in conventional metals [136, 138, 235]. As was already discussed at the beginning of this chapter, the latter are characterized by large Fermi surfaces and parabolic dispersion relations. As a result, there are specific Fermi liquid renormalization effects in normal metals that are determined by the electron density. The effects of this type are relevant for doped Weyl and Dirac semimetals but of little importance at low doping.

3.5   Effective models of Weyl semimetals

Until now, the discussions were mostly limited to idealized relativistic-like models. In order to make a better connection to experimental observations, however, it is instructive to consider more realistic models of Weyl and Dirac semimetals. In this section, we review two types of such models. We start with a simple two-band model realized on a cubic lattice and then consider a Weyl semimetal phase induced in a heterostructure.

3.5.1   *Two-band model*

A simple two-band model is defined by the following Hamiltonian [389, 390]:

$$\mathcal{H}(\mathbf{k}) = -\mu + (\mathbf{d} \cdot \boldsymbol{\sigma}). \tag{3.77}$$

Here $\boldsymbol{\sigma} = (\sigma_x, \sigma_y, \sigma_z)$ is the vector of Pauli matrices, μ is the electric chemical potential, and $\mathbf{d}$ is a periodic vector function in momentum space. The explicit form of $\mathbf{d}$ is defined by the following components:

$$d_1 = \Lambda \sin(ak_x), \tag{3.78}$$

$$d_2 = \Lambda \sin(ak_y), \tag{3.79}$$

$$d_3 = t_0 + t_1 \cos(ak_z) + t_2 [\cos(ak_x) + \cos(ak_y)], \tag{3.80}$$

where Λ, t_0, t_1, and t_2 are material-dependent parameters and a is the lattice spacing of a cubic lattice. It is interesting to note that the same Hamiltonian as in Eq. (3.77) can be also used to describe the multi-Weyl semimetals (see Sec. 3.1.3), provided functions d_1 and d_2 are modified appropriately.[P] The quasiparticle energy spectrum of Hamiltonian (3.77) reads

$$\epsilon_{\mathbf{k}} = -\mu \pm |\mathbf{d}|. \tag{3.81}$$

A slice of the corresponding dispersion (at $k_x = k_y = 0$) is shown in Fig. 3.8. As one can see from the figure, the model Hamiltonian in Eq. (3.77) describes a Weyl semimetal with a pair of Weyl nodes separated in momentum space by $2|b_z|$, provided $|t_0 + 2t_2| \leq |t_1|$. Analytically, the value of the chiral shift b_z is given by

$$b_z = -\frac{1}{a} \arccos\left(\frac{-t_0 - 2t_2}{t_1}\right). \tag{3.82}$$

To make a connection to a linearized relativistic-like model, let us derive the low-energy dispersion in the vicinity of Weyl nodes. The corresponding result reads

$$\epsilon_{\mathbf{k},\lambda} = -\mu \pm a \sqrt{\Lambda^2 k_\perp^2 + t_1^2 \left[1 - \frac{(t_0 + 2t_2)^2}{t_1^2}\right] (k_z - \lambda b_z)^2}, \tag{3.83}$$

where $\lambda = \pm$ denotes the chirality. The linearized spectra for chiral quasiparticles are shown by the dashed lines in Fig. 3.8. As is clear from the figure, two separate Fermi surfaces, that form around the Weyl nodes of opposite chiralities at small chemical potential μ, should merge when

---

[P]In the case of double-Weyl semimetals, for example, $d_1 = \Lambda\left[\sin^2(ak_x) - \sin^2(ak_y)\right]$ and $d_2 = 2\Lambda \sin(ak_x)\sin(ak_y)$. Similarly, in the case of triple-Weyl semimetals, $d_1 = \Lambda\left[\sin^3(ak_x) - 3\sin(ak_x)\sin^2(ak_y)\right]$ and $d_2 = -\Lambda\left[\sin^3(ak_y) - 3\sin(ak_y)\sin^2(ak_x)\right]$.

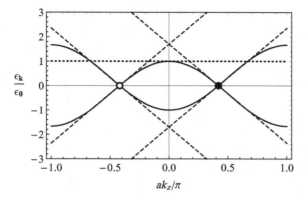

Fig. 3.8   The energy spectrum (3.81) of the two-band periodic model (3.77). The filled and empty points denote the right- and left-handed Weyl nodes. The approximate linear dispersion relation in the vicinity of Weyl nodes is shown by dashed lines. The horizontal dotted line corresponds to the energy at which the two separate Fermi surfaces merge, i.e., $\epsilon_0 = \lim_{\mathbf{k}\to 0} \epsilon_{\mathbf{k}}$. The representative set of model parameters corresponding to a Weyl semimetal phase $t_1 = 4\,t_0$, $t_2 = -t_0$, and $\Lambda = t_0$ was used.

$\mu = \epsilon_0$, where $\epsilon_0 = |t_0 + t_1 + 2t_2|$ is the height of the dome between the Weyl nodes. This energy scale is shown by the dotted horizontal line in Fig. 3.8. As expected, deviations from the approximate linear dispersion relation (3.83) become noticeable outside a close vicinity of Weyl nodes. Such deviations also suggest that the concept of quasiparticle chirality gradually deteriorates.

3.5.2   *Realization of Weyl semimetals in heterostructures*

In this subsection, we discuss a proposal to realize Weyl semimetals in heterostructures. Historically, the first model of this type was proposed by Anton Burkov and Leon Balents in [391] and played an important role in the theoretical investigations of Weyl semimetals. Therefore, it is instructive to discuss this model in more detail. The model describes a heterostructure composed of alternating layers of ordinary and magnetically doped topological insulators. The corresponding second-quantized Hamiltonian reads

$$H = \sum_{i,j}\left[\hbar v_F\left(k_y\tau_z\otimes\sigma_x - k_x\tau_z\otimes\sigma_y\right)\delta_{i,j} + m\mathbb{1}_2\otimes\sigma_z\delta_{i,j} + \Delta_S\tau_x\otimes\mathbb{1}_2\delta_{i,j}\right.$$
$$\left. + \frac{\Delta_D}{2}\tau_+\otimes\mathbb{1}_2\delta_{j,i+1} + \frac{\Delta_D}{2}\tau_-\otimes\mathbb{1}_2\delta_{j,i-1}\right]c^\dagger_{\mathbf{k}_\perp,i}c_{\mathbf{k}_\perp,j},\qquad(3.84)$$

where $\mathbf{k}_\perp = (k_x, k_y)$ is the 2D momentum, $\boldsymbol{\sigma}$ denotes the Pauli matrices acting on the spin degree of freedom, $\boldsymbol{\tau}$ corresponds to the Pauli matrices in the pseudospin space, indices i and j label layers of topological insulators. The first term in Eq. (3.84) describes surface states of individual topological insulator layers. The second term corresponds to an exchange spin splitting of the surface states, which can be induced by doping each layer with magnetic impurities. The remaining terms describe tunneling between the top and bottom surfaces within the same layer of a topological insulator (Δ_S) or between the top and bottom surfaces of the neighboring layers (Δ_D). The corresponding heterostructure is schematically presented in Fig. 3.9(a). Hamiltonian (3.84) can be rewritten in momentum space as follows:

$$H(\mathbf{k}) = \hbar v_F \left(k_y \tau_z \otimes \sigma_x - k_x \tau_z \otimes \sigma_y \right) + m \mathbb{1}_2 \otimes \sigma_z$$

$$+ \left[\Delta_S + \Delta_D \cos{(k_z d)} \right] \tau_x \otimes \mathbb{1}_2 - \Delta_D \sin{(k_z d)} \tau_y \otimes \mathbb{1}_2, \quad (3.85)$$

where d is the superlattice period of the heterostructure in the z-direction. By using the transformation $\mathbb{1}_2 \otimes \sigma_\pm \to \tau_z \otimes \sigma_\pm$ and $\tau_\pm \otimes \mathbb{1}_2 \to \tau_\pm \otimes \sigma_z$ and partially diagonalizing the result in the pseudospin space, one finds that $H(\mathbf{k}) = \mathrm{diag}\left[H_+(\mathbf{k}), H_-(\mathbf{k}) \right]$, where

$$H_s(\mathbf{k}) = \hbar v_F \left(k_y \sigma_x - k_x \sigma_y \right) + M_\zeta(k_z) \sigma_z. \quad (3.86)$$

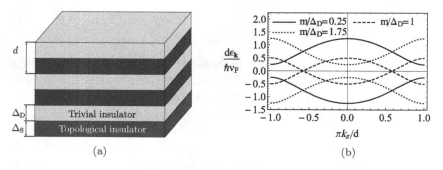

(a)   (b)

Fig. 3.9   (a) The layered heterostructure described by the effective Hamiltonian (3.84). Here d is the period of the resulting superlattice heterostructure in the z-direction, as well as Δ_S and Δ_D denote the tunneling between the top and bottom surfaces within the same layer of topological insulator and between the top and bottom surfaces of neighboring layers, respectively. (b) The low-energy branches ($\zeta = -$) of spectrum (3.87) for $\Delta_S = \Delta_D/2 = \hbar v_F/d$. Here $m = 0.25\,\Delta_D$ (solid line), $m = \Delta_D$ (dashed line), and $m = 1.25\,\Delta_D$ (dotted line).

Here $\zeta = \pm$ and $M_\zeta(k_z) = m + \zeta\sqrt{\Delta_S^2 + \Delta_D^2 + 2\Delta_S\Delta_D \cos{(k_z d)}}$. The corresponding energy spectrum

$$\epsilon_{\mathbf{k},\zeta} = \pm\sqrt{\hbar^2 v_F^2 k_\perp^2 + [M_\zeta(k_z)]^2} \tag{3.87}$$

describes two Weyl nodes located at $\mathbf{k}_0^\pm = (0, 0, \pi/d \pm b_z)$, where

$$b_z = \frac{1}{d}\arccos\left(1 - \frac{m^2 - (\Delta_S - \Delta_D)^2}{2\Delta_S\Delta_D}\right). \tag{3.88}$$

As is easy to see, a Weyl semimetal phase exists for the intermediate values of m satisfying $(\Delta_S - \Delta_D)^2 < m^2 < (\Delta_S + \Delta_D)^2$. To illustrate this graphically, we show the lower energy branch ($\zeta = -$) of the energy spectrum (3.87) in Fig. 3.9(b) for a few values of m. The other model parameters were set as follows: $\Delta_S = \Delta_D/2 = \hbar v_F/d$. In agreement with the criterion above, the model describes a minimal Weyl semimetal phase with a single pair of Weyl nodes only at intermediate values of m. At small and large values of m, on the other hand, insulating phases are realized. It is should be added, however, that one of the insulating phases is trivial but the other is a quantum anomalous Hall insulator. The latter is topological and occurs at large values of magnetization m, i.e., $m^2 > (\Delta_S + \Delta_D)^2$. A trivial insulator phase is realized when $m^2 < (\Delta_S - \Delta_D)^2$.

While the heterostructure model discussed in this subsection increased interest to Weyl semimetals in the condensed matter community, it was not realized experimentally so far. One of the possible reasons might be connected with the difficulty of controlling precisely the preparation of layered samples.

3.6 Effective low-energy model and topological properties of Dirac semimetals

In this section, an effective low-energy model that describes a realistic Dirac semimetal will be considered. In general, such models are obtained by using symmetry arguments to restrict the form of the $\mathbf{k} \cdot \mathbf{p}$ Hamiltonian and then adjusting the parameters to fit the low-energy quasiparticle spectrum [392, 393] (this approach is known as the method of invariants). A few such models will be discussed below and their relation to atomic orbitals will be presented in Sec. 4.2. Another approach to obtain effective models is to employ the multi-band Kane model with the subsequent downfolding to the low-energy basis.

According to their topological properties, Dirac semimetals can be split into two distinct classes [367, 394]. Dirac semimetals similar to Na_3Bi and Cd_3As_2 belong to the first class, in which Dirac points occur in pairs separated in momentum space along a rotation axis. It is worth emphasizing that topological properties of such Dirac semimetals are nontrivial because they possess a $\mathbb{Z}_2$ topological invariant [367, 394–396] which protects Dirac points in the bulk. On the other hand, Dirac semimetals of the second class, which are characterized by a single isolated Dirac point at a $\mathcal{T}$ invariant momentum, have trivial topological properties.

3.6.1 Low-energy model

By following [110], let us consider the following low-energy effective Hamiltonian for electron excitations in the Dirac semimetal A_3Bi ($A =$ Na,K,Rb):

$$H(\mathbf{k}) = C(\mathbf{k}) + H_{4\times4}, \tag{3.89}$$

where

$$H_{4\times4} = \begin{pmatrix} M(\mathbf{k}) & Ak_+ & 0 & B^*(\mathbf{k}) \\ Ak_- & -M(\mathbf{k}) & B^*(\mathbf{k}) & 0 \\ 0 & B(\mathbf{k}) & M(\mathbf{k}) & -Ak_- \\ B(\mathbf{k}) & 0 & -Ak_+ & -M(\mathbf{k}) \end{pmatrix}. \tag{3.90}$$

Interestingly, the same general form is also valid in the case of the low-energy Hamiltonian for one of the structures of Cd_3As_2 [111]. The diagonal elements of the Hamiltonian are determined by two quadratic functions of the momentum, $C(\mathbf{k}) = C_0 + C_1 k_z^2 + C_2(k_x^2 + k_y^2)$ and $M(\mathbf{k}) = M_0 - M_1 k_z^2 - M_2(k_x^2 + k_y^2)$. The off-diagonal elements are defined by functions Ak_+ and $B(\mathbf{k}) = \alpha k_z k_+^2$, where $k_\pm = k_x \pm i k_y$. This model was derived by using the $\mathbf{k} \cdot \mathbf{p}$ approximation and taking into account $\mathcal{T}$, $\mathcal{P}$, and crystal space group ($P6_3/mmc$) symmetries of A_3Bi ($A =$ Na, K, Rb).

The energy spectrum of Hamiltonian (3.89) is given by

$$\epsilon_{\mathbf{k}} = C(\mathbf{k}) \pm \sqrt{M^2(\mathbf{k}) + A^2 k_+ k_- + |B(\mathbf{k})|^2}. \tag{3.91}$$

The energy bands intersect at two points, $\mathbf{k}_0^\pm = (0, 0, \pm\sqrt{m})$, where $m \equiv M_0/M_1$. Because of the double degeneracy, these intersections are Dirac points rather than Weyl nodes. Indeed, by linearizing the Hamiltonian $H_{4\times4}$ in vicinity of one of these points (e.g., at $\mathbf{k}_0^-$ with $\mathbf{k} = \mathbf{k}_0^- + \delta\mathbf{k}$) and performing a unitary transformation defined by matrix $U = \text{diag}\left[\sigma_x, \mathbb{1}_2\right]$, we obtain the following Dirac Hamiltonian in the chiral representation:

$$H_{4\times4}^{\text{lin}} = \begin{pmatrix} A\left(\tilde{\mathbf{k}} \cdot \boldsymbol{\sigma}\right) & B^*(\mathbf{k}) \\ B(\mathbf{k}) & -A\left(\tilde{\mathbf{k}} \cdot \boldsymbol{\sigma}\right) \end{pmatrix}, \tag{3.92}$$

where, by definition, $\tilde{\mathbf{k}} = (k_x, k_y, 2\delta k_z \sqrt{M_0 M_1}/A)$. By setting $\sqrt{M_0 M_1} \approx A/2$, which is equivalent to neglecting the anisotropy, we find that Eq. (3.92) is indeed very similar to the standard Dirac Hamiltonian in the chiral representation, albeit with an unusual mass term $B(\mathbf{k})$. Unlike the conventional Dirac mass, the latter vanishes at the Dirac points.

When the mass function $B(\mathbf{k})$ is neglected, the Hamiltonian $H_{4\times4}$ takes a block diagonal form $H_{4\times4}(\alpha = 0) = H_{2\times2}^+ \oplus H_{2\times2}^-$, where

$$H_{2\times2}^\pm = \begin{pmatrix} M(\mathbf{k}) & A(\pm k_x + ik_y) \\ A(\pm k_x - ik_y) & -M(\mathbf{k}) \end{pmatrix}. \tag{3.93}$$

Each of the two Hamiltonians, $H_{2\times2}^+$ and $H_{2\times2}^-$, describes a pair of Weyl nodes of opposite chiralities located at $\mathbf{k}_0^\pm$. Because of the opposite signs of the k_x terms in the upper block $H_{2\times2}^+$ and the lower block $H_{2\times2}^-$, the chiralities of the states in the two blocks are opposite near each Weyl node. Therefore, the complete Hamiltonian $H_{4\times4}(\alpha = 0)$ describes two superimposed copies of Weyl semimetals with two pairs of overlapping nodes and interchanged chiralities (see also the schematic illustration in Fig. 3.10). As will be shown below, however, the overlapping Weyl nodes of opposite chiralities cannot annihilate by forming topologically trivial Dirac points because of additional discrete symmetries of Hamiltonian (3.89).

3.6.2 *Topological protection*

Having defined an effective low-energy model of Dirac semimetals, let us discuss the symmetry and topological reasons for the stability of Dirac points. We will start our presentation from a general classification of 3D Dirac semimetals. This will be followed then by an overview of symmetries of the effective Hamiltonian (3.89) in Sec. 3.6.2.2.

3.6.2.1 *General classification*

As was discussed in Sec. 3.2, an interplay of various symmetries is crucial for the stability of Dirac points. The key role is played by $\mathcal{P}$ and $\mathcal{T}$ symmetries. If a system has both symmetries, then its energy bands are doubly degenerate at every value of the momentum. The formation of band cross-ings is, generically, allowed only under specific conditions [364, 397, 398]. For example, a Dirac point can emerge at a TRIM point, when the system undergoes a transition from a normal to a topological insulator. However, such a point is not stable and requires fine tuning of material parame-ters. The situation drastically changes if there exists an additional uniaxial rotational symmetry that could stabilize Dirac points. Qualitatively, this is possible because the valence and conduction bands can have different rotational eigenvalues that provide the necessary mechanism for stabilizing the Dirac points.

By following [367], let us consider a generalized Dirac Hamiltonian in momentum space

$$H(\mathbf{k}) = \sum_{n,m=0}^{3} a_{nm}(\mathbf{k})\,(\tau_n \otimes \sigma_m), \qquad (3.94)$$

where $\boldsymbol{\tau}$ and $\boldsymbol{\sigma}$ are the Pauli matrices acting in the orbital and (pseudo)spin spaces, respectively, $\tau_0 = \sigma_0 = \mathbb{1}_2$, and $a_{nm}(\mathbf{k})$ are some real functions of wave vector $\mathbf{k}$. In addition to symmetries (3.19) and (3.21), there could also exist a rotational symmetry. Under spatial rotations, the Hamiltonian transforms as follows:

$$\hat{C}_n H(\mathbf{k}) \hat{C}_n^{-1} = H(R_n \mathbf{k}), \qquad (3.95)$$

where R_n is an orthogonal matrix that describes the rotation by an angle $2\pi/n$ about a certain axis. For the sake of concreteness, let us assume that the rotation is about the z-axis. The Hamiltonian $H(\mathbf{k}_{\parallel})$ must commute with $\hat{C}_n$ for any momenta $\mathbf{k}_{\parallel}$ along the rotational axis because $R_n \mathbf{k}_{\parallel} = \mathbf{k}_{\parallel}$. This allows us to significantly simplify Eq. (3.94) at $k_x = k_y = 0$, i.e.,

$$H(k_z) = c_0(k_z) + c_1(k_z)\,(\tau_0 \otimes \sigma_z) + c_2(k_z)\,(\tau_z \otimes \sigma_0) + c_3(k_z)\,(\tau_z \otimes \sigma_z). \qquad (3.96)$$

The presence of $\mathcal{P}$ and $\mathcal{T}$ symmetries implies the double degeneracy of the states. For the Hamiltonian in Eq. (3.96), this is the case when only one of the functions $c_i(k_z)$, with $i = 1, 2, 3$, is nonzero. Since the (pseudo)spins of degenerate bands should be opposite for the same orbital, one must set $c_1(k_z) = 0$. In addition, depending on the structure of the $\mathcal{P}$ operator, the remaining coefficient, i.e., either $c_2(k_z)$ or $c_3(k_z)$, should be an even or odd function of k_z. When the coefficient is even, e.g., $c_3(k_z) = M_0 - M_1 k_z^2$, the corresponding dispersion relation describes a Dirac semimetal with two Dirac points at $k_z^\pm = \pm\sqrt{M_0/M_1}$ along the rotational axis. As was shown in Sec. 3.6.1, this is indeed the case for Dirac semimetals A_3Bi (A = Na, K, Rb) and Cd$_3$As$_2$, whose Dirac points are protected by $\hat{C}_3$ [110] and $\hat{C}_4$ [111] symmetries, respectively. On the other hand, when $c_2(k_z)$ or $c_3(k_z)$ in Eq. (3.96) is an odd function of k_z, there is a single Dirac point at $k_z = 0$ or another TRIM point. It was predicted theoretically that β-cristobalite BiO$_2$ [366] and distorted spinels [399] belong to this class of Dirac semimetals.

After clarifying the role of symmetries in Dirac semimetals, let us now briefly discuss the topological features of Dirac points [367]. In this connection, it is worth emphasizing that topology is usually irrelevant for Dirac semimetals. There is a class of materials, however, where Dirac points have nontrivial topological structure. For example, the Dirac semimetals A_3Bi (A = Na,K,Rb) have the $\hat{C}_3$ symmetry, which ensures the existence of a 2D topological $\mathbb{Z}_2$ invariant defined in the $k_z = 0$ plane. Note that this invariant is similar to that in 3D topological insulators [88, 400] (see also [98–101]). In the presence of both $\mathcal{P}$ and $\mathcal{T}$ symmetries, the $\mathbb{Z}_2$ invariant can be related to the parity of the valence and conduction bands [89]. Similarly, in $\mathcal{P}$ symmetric Dirac semimetals with the $\hat{C}_4$ and $\hat{C}_6$ symmetries, the mirror Chern number can be used in addition to the $\mathbb{Z}_2$ invariant. As will be discussed in Secs. 5.2 and 5.4, the existence of the unusual surface states in Dirac semimetals is related to either the $\mathbb{Z}_2$ invariant or the mirror Chern number.

3.6.2.2 *Dirac semimetals as $\mathbb{Z}_2$ Weyl semimetals*

The nontrivial topological properties of A_3Bi (A = Na, K, Rb) semimetals can be inferred from the symmetries of the effective Hamiltonian (3.89). To start with, let us first discuss its symmetries in the simplest case when the mass function $B(\mathbf{k}) = \alpha k_z k_+^2$ vanishes. Then, as is easy to check, the corresponding block-diagonal Hamiltonian $H_{4\times4}(\alpha = 0) = H_{2\times2}^+ \oplus H_{2\times2}^-$ with $H_{2\times2}^\pm$ defined in Eq. (3.93) possesses a continuous $U_+(1) \times U_-(1)$ symmetry. This symmetry is generated by independent phase transformations of the up and down spinors of the block Hamiltonians $H_{2\times2}^+$ and $H_{2\times2}^-$, respectively. Since Dirac nodes of $H_{4\times4}(\alpha = 0)$ are composed of two superimposed pairs of Weyl nodes, the corresponding Dirac material can be viewed as a $\mathbb{Z}_2$ Weyl semimetal [395].

As is well known in relativistic quantum mechanics, a constant $B(\mathbf{k})$ gives rise to a nonzero Dirac mass term and breaks the continuous "chiral" symmetry $U_+(1) \times U_-(1)$ down to a diagonal subgroup $U_{em}(1)$, generated by the same phase transformations for both up and down spinors. In Hamiltonian (3.89), however, the mass function has a very special dependence on momentum, i.e., $B(\mathbf{k}) = \alpha k_z k_+^2$, which vanishes at the Dirac points. As will be argued below, this has important consequences for the effective Hamiltonian, which reveals additional discrete symmetries.

As expected in the case of a Dirac semimetal, Hamiltonian (3.89) is invariant under $\mathcal{P}$ and $\mathcal{T}$ symmetries given by Eqs. (3.19) and (3.20), respectively, where $\hat{P} = \mathbb{1}_2 \otimes \sigma_z$ and $\hat{T} = -i\tau_y \otimes \mathbb{1}_2$. It is interesting, however, that there is an additional symmetry of the effective low-energy model (3.89), the so-called *up–down parity* [395], which is defined by

$$\hat{U}_\chi H(\mathbf{k}) \hat{U}_\chi^{-1} = H(\mathbf{k}), \tag{3.97}$$

where $\hat{U}_\chi = \hat{U}\hat{\Pi}_{k_z}$, $\hat{U} = \tau_z \otimes \mathbb{1}_2$, and $\hat{\Pi}_{k_z}$ is the operator that changes the sign of the z component of momentum, $k_z \to -k_z$. For the existence of this symmetry, it is crucial that $B(\mathbf{k})$ is an odd function of k_z. By noting that the operator $\hat{U}_\chi$ has two eigenvalues ± 1, the eigenstates of Hamiltonian (3.89) can be separated into two sectors with fixed eigenvalues of $\hat{U}_\chi$. However, since the $\mathcal{T}$ symmetry operator does not commute with the up–down parity, $\mathcal{T}$ symmetry is broken in each sector. In other words, each subspace of eigenstates with a fixed eigenvalue of $\hat{U}_\chi$ defines a copy of a Weyl semimetal with broken $\mathcal{T}$ symmetry and a pair of Weyl nodes at $\mathbf{k}_0^\pm$. Overall, of course, $\mathcal{T}$ symmetry is preserved when both sectors are taken into account. Thus, the presence of the up–down parity symmetry means that the Dirac semimetals A_3Bi (A = Na, K, Rb) are, in fact, $\mathbb{Z}_2$ Weyl semimetals. This explains, among other things, why these materials have nontrivial topological properties, revealed for example by the existence of the surface states resembling Fermi arcs (see also Sec. 5.4). The structure of the energy spectrum and the surface states of $\mathbb{Z}_2$ Weyl semimetals is schematically illustrated in Fig. 3.10.

In addition to topology and symmetry characteristics of Weyl and Dirac semimetals, there are many other effects that occur when the corresponding models are weakly perturbed. As in the case of graphene, various lattice distortions lead to the modification of the effective low-energy Hamiltonian in Dirac and Weyl semimetals. Despite being model-dependent, new strain-induced terms could be also described in terms of effective gauge potentials and lead to pseudoelectromagnetic fields. The latter will be discussed in the following section, where a simple lattice model will be introduced.

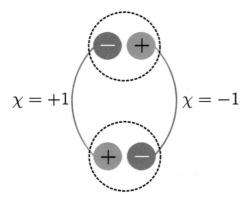

Fig. 3.10   Schematic illustration of the structure of the energy spectrum and the surface states in $\mathbb{Z}_2$ Weyl semimetals. Each copy of a $\mathcal{T}$ symmetry broken Weyl semimetal contains well-separated Weyl nodes of positive (red filled circles) and negative (blue filled circles) chiralities as well as a Fermi arc connecting them. These copies occupy the same position in momentum space and lead to trivial Dirac nodes (marked by dashed circles) and two Fermi arc-like states.

3.7   Pseudoelectromagnetic fields

In this section, we discuss the role of deformations in Dirac and Weyl semimetals. These materials not only mimic the properties of truly relativistic matter but also allow for the realization of novel quantum phenomena that cannot be easily observed in high energy physics.[q] One of such exotic phenomena is the possibility of synthetic axial gauge fields that can be effectively induced by certain elastic deformations of solids. As we discussed in Sec. 2.8, this phenomenon was investigated in graphene theoretically and experimentally. The underlying mechanism is simple to understand after noting that elastic deformations change the lattice constants and, in turn, alter the electron hopping parameters on the lattice. Using the same line of arguments, the appearance of strain-induced gauge fields was also predicted for Weyl semimetals [401–403]. It is worth noting that this finding can be also inferred from the fact that Weyl nodes are topologically protected (see Sec. 3.1). Indeed, from a general symmetry approach, strain-induced small perturbations of the Weyl Hamiltonian

---

[q]Although the $W^{\pm}$ and Z gauge bosons in the Standard Model are described by axial vector fields, it is practically impossible to realize them experimentally as classical fields.

can be represented in the following form:

$$\delta H_\lambda = g_0 \mathbb{1}_2 \sum_i u_{ii} + \lambda g_1 \mathbb{1}_2 \sum_{i,j} u_{ij} \frac{b_i b_j}{b^2} + \lambda g_2 \sum_{i,j} \sigma_i u_{ij} b_j$$
$$+ \lambda g_3 \mathbb{1}_2 \sum_{i,j} k_i u_{ij} b_j + g_4 \sum_{i,j} k_i u_{ij} \sigma_j + \lambda g_5 \sum_{i,j,l,n} \epsilon_{ijl} \sigma_i k_j u_{ln} b_n,$$

$$(3.98)$$

where g_i with $i = 0, \ldots, 5$ are coupling constants, $\boldsymbol{b}$ is the chiral shift, b_0 is the separation of the Weyl nodes in energy, $u_{ij} = (\partial_i u_j + \partial_j u_i)/2$ is the symmetrized strain tensor, and $\mathbf{u}$ is the displacement vector. Let us discuss the physical meaning of the terms in Eq. (3.98). The first term is a deformation potential, which is induced by a change of the volume. The pseudoscalar gauge field $A_0^5 \propto g_1 b_0 \sum_{i,j} u_{ij} b_i b_j / b^2$ is given by the second term. In essence, this is a correction to the Weyl node separation in energy b_0. The third term describes the *axial gauge field* $A_i^5 \propto g_2 \sum_j u_{ij} b_j$, which can be viewed as a strain-modulated correction to the chiral shift $\mathbf{b}$. There are also three terms that depend on momentum. In particular, the term with g_3 corresponds to a tilt of Weyl cones. The fifth term makes the Fermi velocity anisotropic. Finally, the last term in Eq. (3.98) can be identified with the pseudo-Zeeman effect.

It is clear that, by construction, small perturbations of the Weyl Hamiltonian can be represented by shifts of the band touching points. By generalizing this argument to smooth time- and coordinate-dependent deformations, the overall effect can be described by effective vector potentials. The latter are axial as they couple to quasiparticles from Weyl nodes of opposite chirality with opposite signs. This possibility gave birth to the new field of *straintronics* in 3D materials, where the electron transport is controlled by deformations. Numerous early studies indicate that this direction of research is holding a great potential for applications.

3.7.1 *Strains as artificial gauge fields*

In order to show how strain-induced gauge fields arise, let us employ the following simple cubic lattice model, which describes electron states in the s and p orbitals [390, 403–405]:

$$H_{\text{latt}} = \sum_{j,l} \hat{c}_j^\dagger \left(it\gamma^0 \gamma_j - r\gamma^0 \right) \hat{c}_{j+l} + (m + 3r) \sum_j \hat{c}_j^\dagger \gamma^0 \hat{c}_j$$
$$+ \sum_j d_z \hat{c}_j^\dagger \gamma^0 \gamma_z \gamma_5 \hat{c}_j + \sum_j d_0 \hat{c}_j^\dagger \gamma_5 \hat{c}_j + h.c., \qquad (3.99)$$

where the index j denotes the positions of atoms, l labels the six nearest neighbors, a is the lattice constant, d_0 and d_z are the parameters corresponding to $\mathcal{P}$ and $\mathcal{T}$ symmetry breaking terms, respectively. In this model, a Weyl phase is realized when $d_z > m$. (To be specific, we assume that $d_z + m > 0$.) While parameter t describes hopping between different s and p states, parameter r corresponds to hopping between the same states. The difference between the on-site energies of the s and p states is parameterized by m. By performing the Fourier transform and a unitary transformation, one obtains

$$H_{4\times4} = \mathbb{1}_2 \otimes [t\sigma_x \sin(ak_x) + t\sigma_y \sin(ak_y) + \sigma_z d_z] + \tau_z \otimes \sigma_z M(\mathbf{k})$$

$$+ \tau_x \otimes \sigma_z t \sin(ak_z) + \tau_x \otimes \mathbb{1}_2 d_0, \tag{3.100}$$

where $M(\mathbf{k}) = m + 3r - r\sum_i \cos(ak_i)$. The energy spectrum of this Hamiltonian has four branches. However, only two of them intersect and produce Weyl nodes. By projecting out the high energy modes and expanding in small k, one finds

$$H_{2\times2} = -\frac{2d_0 tak_z}{m + d_z} + ta(\sigma_x k_x + \sigma_y k_y)$$

$$+ \frac{d_z^2 - d_0^2 - m^2 - t^2 a^2 k_z^2}{m + d_z} \sigma_z. \tag{3.101}$$

As is easy to see, this is similar to a single (up or down) block Hamiltonian in the Dirac semimetal model in Eq. (3.93). Therefore, it describes a Weyl semimetal with two Weyl nodes located at $\mathbf{k}_0^{\pm} = \pm\mathbf{b} = \mp\sqrt{d_z^2 - d_0^2 - m^2}/(at)\hat{\mathbf{z}}$ when $|d_z| > \sqrt{d_0^2 + m^2}$ and separated in energy by $2b_0 = -4d_0 atb_z/(m + d_z)$. In the vicinity of each Weyl node, where $\mathbf{k} = \delta\mathbf{k} + \lambda\mathbf{b}$, one has

$$H_\lambda = \lambda b_0 + v(\sigma_x \delta k_x + \sigma_y \delta k_y) + \lambda v_z \delta k_z \sigma_z, \tag{3.102}$$

where $\lambda = \pm$ denotes the Weyl node's chirality, $v = at$, and $v_z = 2v\sqrt{d_z^2 - d_0^2 - m^2}/(d_z + m)$.

Now let us show how strains affect the low-energy properties of Weyl semimetals. For this purpose, we return to the tight-binding model (3.99), where small deformations are included in the first term by changing the hopping parameters as

$$t\gamma_j \to t\gamma_j(1 - \beta u_{jj}) + t\beta \sum_{l \neq j} u_{jl}\gamma_l, \tag{3.103}$$

$$r \to r_j \approx r(1 - \beta u_{jj}). \tag{3.104}$$

Here, the first equation describes the modification of hopping between the different orbitals and the second equation accounts for the change of a bond length [404]. The effect of deformation is parameterized by the Grüneisen parameter $\beta = -(a/t)\partial t/\partial a$. Note that strain-induced potentials could be also obtained using the symmetry arguments, as was done in [406].

It is clear that the electronic properties of Dirac and Weyl semimetals strongly depend on the properties of the underlying lattices. Repeating the same steps that lead to Hamiltonian (3.102), one finds that strains produce the following additional terms:

$$H_\lambda^{(\text{str})} = \lambda\beta b_0 u_{zz} \mathbb{1}_2 - \lambda\beta v b_z \left(u_{zx}\sigma_x + u_{zy}\sigma_y\right)$$

$$+ \beta \left[\frac{2v^2 b_z^2}{m + d_z} u_{zz} - r \sum_{j=x,y,z} u_{jj} \right] \sigma_z. \tag{3.105}$$

In the vicinity of Weyl nodes, the complete Hamiltonian $H_\lambda + H_\lambda^{(\text{str})}$ reads

$$H_\lambda + H_\lambda^{(\text{str})} = \lambda \left(b_0 - eA_0^5\right) \mathbb{1}_2 + v\sigma_x \left(\delta k_x + \lambda\frac{e}{c\hbar}A_x^5\right)$$

$$+ v\sigma_y \left(\delta k_y + \lambda\frac{e}{c\hbar}A_y^5\right) + \lambda v_z \left(\delta k_z + \lambda\frac{e}{c\hbar}A_z^5\right)\sigma_z, \tag{3.106}$$

where

$$A_0^5 = -\frac{1}{e}b_0\beta u_{zz}, \tag{3.107}$$

$$A_x^5 = -\frac{c\hbar}{e}\beta b_z u_{zx}, \tag{3.108}$$

$$A_y^5 = -\frac{c\hbar}{e}\beta b_z u_{zy}, \tag{3.109}$$

$$A_z^5 = -\frac{c\hbar}{e}\beta b_z u_{zz} - \frac{c\hbar}{e}\beta\frac{r}{v_z} \sum_{j=x,y,z} u_{jj}. \tag{3.110}$$

As we see from Eq. (3.106), strains enter the low-energy Hamiltonian as effective axial or, equivalently, pseudoelectromagnetic gauge fields. It is crucial that, unlike their usual electromagnetic counterparts, the strain-induced fields couple to quasiparticles from the Weyl nodes of opposite chiralities with different signs. This is especially fascinating from the viewpoint of high energy physics, where the experimental realization of classical background axial fields is impossible in practice.

When strains are spatially nonuniform and/or slowly varying in time, deformations of Weyl semimetals can be effectively described in terms of pseudoelectromagnetic fields

$$\mathbf{B}_5 = \mathbf{\nabla} \times \mathbf{A}^5, \tag{3.111}$$

$$\mathbf{E}_5 = -\mathbf{\nabla} A_0^5 - \frac{1}{c}\partial_t \mathbf{A}^5. \tag{3.112}$$

From the mathematical point of view, a constant pseudomagnetic field $\mathbf{B}_5$ breaks both $\mathcal{T}$ and $\mathcal{P}$ symmetries. On the other hand, a constant pseudoelectric field $\mathbf{E}_5$ preserves them. As we discuss below, in order to generate $\mathbf{B}_5$, it is sufficient to apply static strains to Weyl and Dirac semimetals. Concerning the pseudoelectric fields, one of the ways to generate it is to create a time-dependent chiral shift [405], which could indeed be realized in experiment by applying circularly polarized light [407]. Time-dependent deformations leading to pseudoelectric fields could be also induced by acoustic waves [408–411].

Pseudoelectromagnetic fields lead to a variety of interesting effects [412]. Among them are the strain-induced chiral magnetic effect and negative pseudomagnetic resistivity in Weyl semimetals [405, 408, 413–415], the chiral torsional effect [416–418], quantum oscillations in pseudomagnetic fields [419], unusual collective modes [420–423], axial analogs of the chiral separation and anomalous Hall effects [415], the lensing of Weyl quasiparticles [424–426], emergent gravity [402, 427], etc. It is curious that strains are not the only way to create effective pseudoelectromagnetic gauge fields. As an example, nontrivial magnetization textures can be also used to produce effective axial gauge fields [425, 428, 429] in Weyl and Dirac semimetals.

3.7.2 *Explicit examples of strains*

It is instructive to consider a few explicit examples of deformations that produce pseudoelectromagnetic fields. As suggested by Eqs. (3.107)–(3.110), both chiral shift $\mathbf{b}$ and strain tensor u_{ij} play important roles in the underlying mechanism behind effective pseudoelectromagnetic fields.

The first example is a long wire of a Weyl semimetal with torsion [408, 430]. For concreteness, let us assume that the axis of torsion is along the z-axis and $\mathbf{b} \parallel \hat{\mathbf{z}}$. Then, the displacement vector field is given by [431]

$$\mathbf{u} = \frac{\theta}{L}\,[\mathbf{r} \times \mathbf{z}], \tag{3.113}$$

where θ is the twist angle, L is the length of the wire, and $\mathbf{r}$ is the coordinate with respect to the center of the wire. The nontrivial components of the symmetrized strain tensor are

$$u_{xz} = \frac{\theta}{2L}y, \quad u_{yz} = -\frac{\theta}{2L}x. \qquad (3.114)$$

By using Eqs. (3.107)–(3.110), one concludes that the strain-induced gauge fields are given by $A_i^5 \propto \sum_{j=x,y,z} u_{ij}b_j$ and $A_0^5 = 0$. In vector notation, $\mathbf{A}^5 \propto \theta b_z/(2L)\,(y, -x, 0)$ and the corresponding pseudomagnetic field reads

$$\mathbf{B}_5 \propto -\frac{\theta b_z}{L}\hat{\mathbf{z}}. \qquad (3.115)$$

The generation of a pseudomagnetic field by torsion is schematically illustrated in Fig. 3.11(a).

Another potentially viable way to generate $\mathbf{B}_5$ is to bend a slab of a Weyl semimetal [419, 430]. In the case of bending about the y-axis, provided the undeformed lattice is located in the $x - y$-plane, the nonzero components of deformation vector are [431]

$$u_x = \frac{2u_0}{d}xz, \quad u_z = -\frac{u_0}{d}\left(x^2 + D_\mathrm{L}z^2\right), \qquad (3.116)$$

where u_0 is the maximum stress, d is the thickness of the slab, and D_L is a certain function of the Lamé coefficients. In order to have a nontrivial pseudomagnetic field, the chiral shift should point in the x-direction. Then $\mathbf{A}^5 \propto 2u_0 b_x z\hat{\mathbf{x}}/d$, $A_0^5 = 0$, and the pseudomagnetic field reads

$$\mathbf{B}_5 \propto \frac{2u_0 b_x}{d}\hat{\mathbf{y}}. \qquad (3.117)$$

Note that its direction is along the bending axis.

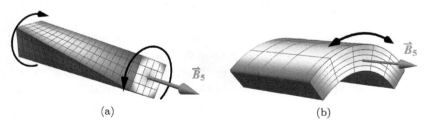

Fig. 3.11   Schematic representation of pseudomagnetic fields induced by (a) torsion and (b) bending.

Since deformations are usually relatively small in crystals, it is important to provide estimates of pseudomagnetic fields in real materials. For the Dirac semimetal Cd_3As_2, for example, the strength of the strain-induced pseudomagnetic field could range from about $B_5 \approx 0.3$ T in twisted nanowires [408] to $B_5 \approx 15$ T in bent thin films [419]. Such pseudomagnetic fields are clearly quite strong.

It is worth noting that the description of strains as effective pseudo-electromagnetic fields is not universal. For example, higher-order corrections [430], deviations from the linearized dispersion relation [432, 433], as well as a nonlinearity of the energy spectra in multi-Weyl semimetals [434] lead to the appearance of terms that cannot be interpreted as pseudoelec-tromagnetic fields. The new terms could be momentum-dependent leading to a Dirac cone tilt [430] and an anisotropic Fermi velocity [427]. Furthermore, uniaxial strains could change the band structure and even open a gap in Dirac and Weyl semimetals, as was argued theoretically, for example, in the case the Dirac semimetal Na_3Bi [110, 435] and tested experimentally in the Dirac semimetal $ZrTe_5$ [436].

Having discussed low-energy effective models of Dirac and Weyl materials as well as their topological properties, let us now turn to thermodynamic properties. As we argue, the latter also reveal characteristic features associated with relativistic-like quasiparticles. In addition, we discuss how the Fermi surface of Dirac and Weyl materials can be probed via quantum oscillations.

3.8 Electronic thermodynamic properties

Thermodynamic properties are among the simplest characteristics of solid state materials. In this section, we discuss thermodynamic manifestations of low-energy quasiparticles of Dirac and Weyl semimetals with a relativistic-like dispersion. As in the rest of the book, we will concentrate only on the electron contribution to thermodynamic characteristics. By taking into account that multiple Weyl nodes contribute additively to most thermodynamic properties, it will be sufficient to analyze the effects of a single Weyl node.

Because of a relativistic-like linear dispersion relation of electron quasiparticles, the thermodynamic properties of Dirac and Weyl semimetals are similar to those of an ultrarelativistic fermion gas (see, e.g., [80]). The main difference from the ultrarelativistic case is connected with the quasiparticle velocity v_F, which is much smaller than the speed light. Let us start with the specific heat.[r] It is defined as follows:

$$C = T \left(\frac{\partial s}{\partial T} \right)_n, \tag{3.118}$$

where s is the entropy density and the partial derivative is calculated at a fixed quasiparticle density n. The entropy density itself is defined by

$$s = - \left(\frac{\partial \Omega}{\partial T} \right)_\mu, \tag{3.119}$$

where Ω is the grand canonical potential per unit volume. In the absence of external fields, it is given by

$$\Omega = T \sum_{e,h} \int \frac{d^3 p}{(2\pi\hbar)^3} \ln \left[1 - f^{eq}(\epsilon_\mathbf{p})\right], \tag{3.120}$$

which includes both electron and hole contributions. It should be taken into account that the electric chemical potential for holes has the opposite sign, $\mu \to -\mu$. By using the equilibrium Fermi–Dirac distribution for quasiparticles

$$f^{eq}(\epsilon_\mathbf{p}) = \frac{1}{e^{(\epsilon_\mathbf{p} - \mu)/T} + 1}, \tag{3.121}$$

where $\epsilon_\mathbf{p} = v_F p$ and $p = \hbar k$, we derive an explicit expression for the density of the grand canonical potential

$$\Omega = \frac{T^4}{\pi^2 \hbar^3 v_F^3} \sum_{e,h} \mathrm{Li}_4 \left(-e^{\mu/T} \right) = -\frac{1}{24\pi^2 \hbar^3 v_F^3} \left(\mu^4 + 2\pi^2 \mu^2 T^2 + \frac{7\pi^4 T^4}{15} \right),$$

where $\mathrm{Li}_m(x)$ is the polylogarithm function. Note that in the derivation we used several table integrals given in Appendix B. By calculating the partial derivative with respect to temperature, we then obtain the expression for the entropy density

$$s = \frac{T}{6\hbar^3 v_F^3} \left(\mu^2 + \frac{7\pi^2 T^2}{15} \right). \tag{3.122}$$

As expected, it agrees with the entropy of an ultrarelativistic electron gas, provided v_F is replaced with the speed of light.

---

[r]Note that pressure does not depend on temperature for strictly periodic crystals. In such a case, the specific heat at fixed pressure C_p and volume C_v are the same. Therefore, to simplify the notation, we omit the corresponding subscripts.

Since the specific heat is calculated at a fixed density n, it is convenient to utilize the following alternative representation:

$$C = T \left(\frac{\partial s}{\partial T} \right)_{\mu} - T \left(\frac{\partial s}{\partial \mu} \right)_{T} \left(\frac{\partial n}{\partial T} \right)_{\mu} \left(\frac{\partial n}{\partial \mu} \right)_{T}^{-1}, \qquad (3.123)$$

where the last term is due to the dependence of density on temperature at a fixed chemical potential. The quasiparticle number density per Weyl node is given by (see also Sec. 6.6)

$$n = \frac{\rho}{-e} = \int \frac{d^3 p}{(2\pi\hbar)^3} f^{\mathrm{eq}}(\epsilon_{\mathbf{p}}) - (\mu \to -\mu) = \frac{\mu \left(\mu^2 + \pi^2 T^2 \right)}{6\pi^2 \hbar^3 v_F^3}, \qquad (3.124)$$

where ρ is the electric charge density. Then, by making use of Eq. (3.123), we obtain the final expression for the specific heat per Weyl node

$$C = \frac{T}{6\hbar^3 v_F^3} \left(\mu^2 + \frac{7\pi^2 T^2}{5} \right) - \frac{2\pi^2 \mu^2 T^3}{3\hbar^3 v_F^3 \left(3\mu^2 + \pi^2 T^2 \right)}. \qquad (3.125)$$

It can be shown that the same expression remains valid also in the limit $\mu \to 0$ when the Fermi-liquid effects are taken into account [437].

As is easy to check, the expression for the specific heat in Eq. (3.125) has a very different temperature dependence at low and high temperatures. At small temperature ($T \ll |\mu|$), one finds

$$C \simeq \frac{\mu^2 T}{6\hbar^3 v_F^3}, \qquad (3.126)$$

which resembles a linear scaling in usual metals in the free electron gas approximation, $C \propto k_F^3 T / |\mu|$ (note that the Fermi momentum is $k_F = |\mu|/v_F$). In the opposite limit of a small chemical potential ($|\mu| \ll T$), which can be also realized in Dirac and Weyl semimetals, one derives

$$C \simeq \frac{7\pi^2 T^3}{30\hbar^3 v_F^3}. \qquad (3.127)$$

In this regime, the temperature scaling of the specific heat is very different from that of an electron contribution in conventional metals. Instead, the cubic dependence on temperature in Eq. (3.127) resembles the low-temperature limit of the phonon specific heat $C_{\mathrm{ph}} \simeq 2\pi^2 T^3 / (15\hbar^3 v_s^3)$, where v_s is the speed of sound. This similarity to the phonon contribution is a hallmark signature of relativistic-like quasiparticles. It implies that the unambiguous separation of the electron and phonon contributions may be difficult in Weyl and Dirac semimetals, especially when the chemical potential is vanishingly small, $\mu \to 0$.

By making use of the quasiparticle distribution function in Eq. (3.121), one can also calculate the energy density per Weyl node,

$$\epsilon = \sum_{e,h} \int \frac{d^3 p}{(2\pi\hbar)^3} \epsilon_{\mathbf{p}} f^{\text{eq}}(\epsilon_{\mathbf{p}}) = \frac{1}{8\pi^2\hbar^3 v_F^3} \left(\mu^4 + 2\pi^2 \mu^2 T^2 + \frac{7\pi^4 T^4}{15} \right).$$

(3.128)

By comparing this relation with Eq. (3.122), we see that the standard relation for an ultrarelativistic fermion gas, $\Omega = -\epsilon/3$, is satisfied. Furthermore, by using the definition for the electron pressure $P = -\Omega$, the equation of state for free Dirac quasiparticles is reproduced, $P = \epsilon/3$.

In the following section, we discuss the magnetization, which is another important thermodynamic characteristics affected by the relativistic-like nature of quasiparticles in Weyl and Dirac semimetals.

3.9 Magnetic oscillations

Quantum oscillations in an applied magnetic field is one of the oldest methods to investigate the fermiology of materials [287–289]. The oscillatory pattern of the density of states is manifested in many physical properties of materials, including the electrical conductivity (the Shubnikov–de Haas effect) and the magnetization (the de Haas–van Alphen effect). According to the Onsager relation, the period of the oscillations $T_{1/B}$ encodes the information about the extremal cross-section area of the Fermi surface normal to the magnetic field S_F, i.e.,

$$T_{1/B} = \frac{2\pi e}{\hbar c} \frac{1}{S_F}.$$

(3.129)

The phase of oscillations may additionally allow one to distinguish relativistic-like quasiparticles in Weyl and Dirac semimetals from those with a parabolic energy spectrum in conventional solids.

3.9.1 *Quantum oscillations of density of bulk states*

Quantum oscillations in Weyl and certain Dirac semimetals could be highly convoluted due to the presence of the surface Fermi arc states. Indeed, as will be discussed in Sec. 5.6.1, the Fermi arcs allow for unconventional surface–bulk orbits in an external magnetic field leading to an additional period of oscillations. For the sake of clarity, in this section, we will concentrate exclusively on the bulk states, assuming that they are described by the following low-energy Weyl Hamiltonian:

$$H = -\mu + \lambda \hbar v_F \boldsymbol{\sigma} \cdot \left(\mathbf{k} - \lambda \mathbf{b} + \frac{e}{\hbar c} \mathbf{A} \right), \tag{3.130}$$

where $\mathbf{A}$ is the vector potential and $-e$ is the electron charge ($e > 0$). While the chiral shift $\mathbf{b}$ is important for determining the separation between the Weyl nodes, it does not affect the calculation of the density of states (DOS). Therefore, it is justified to set $\mathbf{b} = \mathbf{0}$ in the rest of this section. Note that the generalization to the case of multiple Weyl nodes is trivial since the corresponding contributions to the DOS are additive.

Without loss of generality, an external magnetic field $\mathbf{B}$ is assumed to point in the $+z$-direction and is described by the vector potential $\mathbf{A} = (0, Bx, 0)$. The corresponding energy spectrum is given by Landau levels and has a form similar to that in gapped graphene (see Sec. 2.7), but with the gap Δ replaced by $\hbar v_F k_z$,

$$\epsilon_{n=0} = -\mu - s_B \lambda \hbar v_F k_z, \tag{3.131}$$

$$\epsilon_{n>0} = -\mu \pm \sqrt{\hbar^2 v_F^2 k_z^2 + 2n\epsilon_L^2}, \tag{3.132}$$

where $s_B = \text{sgn}(eB)$, $\epsilon_L = \hbar v_F / l_B$ is the Landau energy scale, $l_B = \sqrt{\hbar c / |eB|}$ is the magnetic length, and $n = 0, 1, 2, \ldots$ is the Landau level index. A more rigorous derivation of the energy spectrum in an external magnetic field will be postponed until Secs. 6.2.3.2 and 6.4.1.

The density of states is defined in terms of the retarded Green's function as follows:

$$\nu(\mu) = -\frac{1}{\pi} \text{Im} \int \frac{d^3 k}{(2\pi)^3} \text{tr} \left[(-i) G^R(0, \mathbf{k}) \right]. \tag{3.133}$$

By using Green's function in the clean limit (see Appendix A.2 and Eq. (6.32)), the DOS for a single Weyl node reads [438]

$$\nu(\mu) = \frac{1}{4\pi^2 l_B^2} \int dk_z \sum_{n=0}^{\infty} \left(n + \frac{1}{2} \right) \{ \delta(\mu - \epsilon_n) + \delta(\mu + \epsilon_n)$$

$$- \delta(\mu - \epsilon_{n+1}) - \delta(\mu - \epsilon_{n+1}) \}. \tag{3.134}$$

After integrating over k_z, one arrives at the following result:

$$\nu(\mu) = \frac{1}{4\pi^2 \hbar v_F l_B^2} \left(1 + 2 \sum_{n=1}^{N_\mu} \frac{|\mu|}{\sqrt{\mu^2 - 2n\epsilon_L^2}} \right), \qquad (3.135)$$

where $N_\mu = \left[\mu^2 l_B^2 / (2\hbar^2 v_F^2) \right]$ is the highest occupied Landau level and $[\ldots]$ denotes the integer part. In the limit $B \to 0$, the DOS in Eq. (3.135) reduces to the well-known result for a single Weyl node

$$\lim_{B \to 0} \nu(\mu) = \int \frac{d^3 k}{(2\pi)^3} \left[\delta\left(\mu - \hbar v_F k\right) + \delta\left(\mu + \hbar v_F k\right) \right] = \frac{\mu^2}{2\pi^2 \hbar^3 v_F^3}. \quad (3.136)$$

The DOS for both cases, $B \neq 0$ and $B = 0$, is shown in Fig. 3.12(a). When a magnetic field is nonzero, the DOS takes a nonzero constant value $|eB|/(4\pi^2 \hbar^2 c v_F)$ for $\mu < \epsilon_L$ and exhibits a series of sawtooth-like peaks originating from the Landau levels at larger μ. Note that the peaks in Fig. 3.12(a) are not equidistant because of the relativistic-like dispersion relation that produces the Landau level energies scaling as $\sqrt{n}$. This is in stark contrast to an equidistant spacing for the conventional fermions with a nonrelativistic quadratic dispersion relation, whose Landau level energies scale linearly with n [287].

In order to calculate the sum in Eq. (3.135), it is convenient to use the following Poisson formula [287, 288]:

$$\frac{1}{2} f(0) + \sum_{n=1}^{\infty} f(n) = \int_0^\infty dx f(x) + 2 \operatorname{Re} \sum_{l=1}^{\infty} \int_0^\infty dx f(x) e^{2\pi i l x}. \quad (3.137)$$

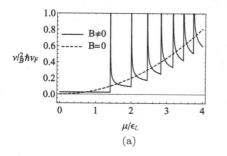
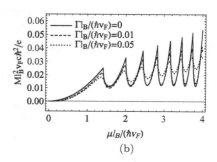

Fig. 3.12 (a) The density of states of a clean Weyl semimetal in (solid line) and without (dashed line) a magnetic field, see Eqs. (3.135) and (3.136), respectively. (b) The magnetization M given by Eq. (3.147) as a function of an electric chemical potential μ. The disorder strength is $\Gamma l_B/(\hbar v_F) = 0$ (solid line), $\Gamma l_B/(\hbar v_F) = 0.01$ (dashed line), and $\Gamma l_B/(\hbar v_F) = 0.05$ (dotted line).

Then, one finds that the oscillating part of the DOS is given by

$$\nu_{\text{osc}}(\mu) = \frac{\mu^2}{2\pi^2\hbar^3 v_F^3} \sum_{l=1}^{\infty} \int_0^1 \frac{dx}{\sqrt{x}} \cos\left[\pi l \frac{\mu^2 l_B^2}{\hbar^2 v_F^2}(1-x)\right]$$

$$= \frac{\mu}{\pi^2\hbar^2 v_F^2 l_B} \sum_{l=1}^{\infty} \frac{1}{\sqrt{2l}} \left[\cos\left(\frac{\pi l \mu^2 l_B^2}{\hbar^2 v_F^2}\right) \mathcal{C}\left(\frac{\sqrt{2l}\mu l_B}{\hbar v_F}\right)\right.$$

$$\left. + \sin\left(\frac{\pi l \mu^2 l_B^2}{\hbar^2 v_F^2}\right) \mathcal{S}\left(\frac{\sqrt{2l}\mu l_B}{\hbar v_F}\right)\right]$$

$$\approx \frac{\mu}{2\pi^2 l_B \hbar^2 v_F^2} \sum_{l=1}^{\infty} \frac{1}{l} \sin\left(\frac{\pi l \mu^2 l_B^2}{\hbar^2 v_F^2} + \frac{\pi}{4}\right). \tag{3.138}$$

In the derivation, we made use of the Fresnel integrals $\mathcal{C}(x)$ and $\mathcal{S}(x)$

$$\mathcal{C}(x) = \int_0^x dt \cos\left(\frac{\pi}{2}t^2\right) \simeq \frac{1}{2} + \frac{1}{\pi x} \sin\left(\frac{\pi}{2}x^2\right) + O\left(\frac{1}{x^3}\right), \tag{3.139}$$

$$\mathcal{S}(x) = \int_0^x dt \sin\left(\frac{\pi}{2}t^2\right) \simeq \frac{1}{2} - \frac{1}{\pi x} \cos\left(\frac{\pi}{2}x^2\right) + O\left(\frac{1}{x^3}\right). \tag{3.140}$$

By assuming the weak magnetic field limit, we also approximated the Fresnel cosine and sine integrals by their asymptotes at large values of argument $x \propto \mu l_B \gg 1$.

It is instructive to spell out explicitly the difference between the case of relativistic-like fermions in Weyl or Dirac semimetals and usual non-relativistic quasiparticles with a quadratic (parabolic) dispersion relation $\epsilon^{(\text{par})} = \hbar^2 k^2/(2m)$. In the latter case, the energy dispersion in a magnetic field is given by

$$\epsilon_n^{(\text{par})} = \frac{\hbar^2}{ml_B^2}\left(n + \frac{1}{2}\right) + \frac{\hbar^2 k_z^2}{2m}. \tag{3.141}$$

This leads to the following DOS:

$$\nu_{\text{osc}}^{(\text{par})}(\mu) = \frac{m}{2\pi^2\hbar^2 l_B} \sum_{l=1}^{\infty} \frac{(-1)^l}{\sqrt{l}} \sin\left(\frac{2\pi lm\mu l_B^2}{\hbar^2} + \frac{\pi}{4}\right), \tag{3.142}$$

and, in the limit of the vanishing magnetic field,

$$\lim_{B\to 0} \nu^{(\text{par})}(\mu) = \frac{m^{3/2}\sqrt{\mu}}{\sqrt{2}\pi^2\hbar^3}. \tag{3.143}$$

By comparing the zero magnetic field DOS expressions in Eqs. (3.136) and (3.143), we see a major difference in their dependence on the electric chemical potential μ. Indeed, while the DOS is quadratic in μ for the relativistic-like case, it has a square-root dependence for conventional electron quasiparticles with a quadratic dispersion. The results at a nonzero magnetic field in Eqs. (3.138) and (3.142) are also very different. In particular, they differ by a phase shift of π, which stems from the absence of factor $(-1)^l$ in Eq. (3.138). At its root, this is intimately connected with the effect of the Berry curvature.

In order to get a better understanding of the phase shift in quantum oscillations, let us employ a semiclassical approach. In such a framework, the oscillations are described by the Lifshitz–Onsager quantization condition

$$\frac{S_\mathrm{F}}{l_\mathrm{B}^2} = 2\pi \left(n + \frac{1}{2} + \beta + \delta \right), \tag{3.144}$$

where β is the phase connected with the Berry curvature [282] and δ is an additional phase due to other effects such as, for example, the warping of the Fermi surface. While the latter is absent in the case of 2D cylindrical Fermi surfaces, $\delta = \pm 1/8$ in the 3D case [439].[s] As one can show explicitly, the Berry curvature effects lead to the phase shift $\beta = \pm 1/2$ for materials with a relativistic-like spectrum [282, 377, 440]. On the other hand, the trivial Berry curvature in the case of parabolic bands produces $\beta = 0$.

As suggested by the analysis in [440], which relies on a two-band model similar to that discussed in Sec. 3.5.1 (see also Sec. 5.3), the phase shift could be also sensitive to μ. It was argued, in particular, that its value gradually grows with μ as the Fermi level changes from the charge neutrality point to the point of the *Lifshitz transition* [358], where the two separate chiral sheets of the Fermi surface start to merge. Afterwards, the value of β quickly decreases and approaches $\beta = 0$, which is the standard value for parabolic bands. Thus, from the viewpoint of quantum oscillations, the most crucial distinction between relativistic-like and nonrelativistic energy dispersion relations is connected with the phase shift of oscillations. This is true also for 2D fermions (see the corresponding discussion in Chapter 2).[t]

---

[s]The $+$ and $-$ signs of phase δ correspond to holes and electrons, respectively.
[t]It is worth noting, however, that the phase of oscillations might depend on different factors and should be used with caution when distinguishing Dirac and Weyl semimetals from conventional materials. In particular, a phase offset might be enforced by symmetries rather than a relativistic-like nature of quasiparticles [441].

3.9.2 de Haas–van Alphen effect

Let us now discuss a canonic example of quantum oscillations, namely, the oscillations of magnetization, which is known as the de Haas–van Alphen effect. The magnetization $\mathbf{M}$ can be calculated by taking the first derivative of the grand thermodynamic potential density Ω with respect to an applied magnetic field, $\mathbf{M} = -\partial_{\mathbf{B}}\Omega$, where it is convenient to define the thermodynamic potential in terms of the density of states $\nu(\epsilon)$ as follows [291, 442]:

$$\Omega = -T \int_{-\infty}^{\infty} d\epsilon\, \nu(\epsilon) \ln\left[2\cosh\left(\frac{\epsilon - \mu}{2T}\right)\right]. \qquad (3.145)$$

By following the approach of [291, 438], we separate the "vacuum" or topological contribution, which stems from the filled states and does not depend on μ, from the "matter" contribution in the thermodynamic potential $\Omega = \Omega^{(\text{vac})} + \Omega^{(\text{mat})}(\mu)$. In the limit of vanishing temperature, the latter reads

$$\Omega^{(\text{mat})}(\mu) = \frac{1}{8\pi^2 l_{\text{B}}^2 \hbar v_F}\left\{\mu^2 + \frac{2}{l_{\text{B}}^2}\sum_{n=1}^{N_\mu}\left[\mu l_{\text{B}}\sqrt{\mu^2 l_{\text{B}}^2 - 2n\hbar^2 v_F^2}\right.\right.$$

$$\left.\left. - 2n\hbar^2 v_F^2 \ln\left(\frac{\mu l_{\text{B}} + \sqrt{\mu^2 l_{\text{B}}^2 - 2n\hbar^2 v_F^2}}{\sqrt{2n}\hbar v_F}\right)\right]\right\}. \qquad (3.146)$$

Since the vacuum part $\Omega^{(\text{vac})}$ does not contribute to the magnetization [438], the latter comes from the matter part alone. In the absence of intrinsic anisotropies, the magnetization points in the same direction as the magnetic field and its magnitude is given by

$$M(\mu) = \frac{e\mu^2}{8\pi^2 \hbar^2 c v_F} + \frac{e}{4\pi^2 \hbar^2 c v_F l_{\text{B}}^2}\text{Re}\sum_{n=1}^{\infty}\left[\mu l_{\text{B}}\sqrt{\mu^2 l_{\text{B}}^2 - 2n\hbar^2 v_F^2}\right.$$

$$\left. - 4n\hbar^2 v_F^2 \ln\left(\frac{\mu l_{\text{B}} + \sqrt{\mu^2 l_{\text{B}}^2 - 2n\hbar^2 v_F^2}}{\sqrt{2n}\hbar v_F}\right)\right]. \qquad (3.147)$$

The dependence of the magnetization on the chemical potential is shown in Fig. 3.12(b), where quantum oscillations are evident from a series of nonequidistant peaks. (Note that, in order to plot the results, nonzero values of the quasiparticle width were used.) The oscillations go away at a sufficiently high magnetic field or low μ. Note that, for $\mu l_{\text{B}} < \sqrt{2}$, only the lowest Landau level is partially filled. In such a case, $M(\mu) \approx e\mu^2/(8\pi^2\hbar^2 c v_F)$.

By using the same method as in the analysis of the DOS, one can rewrite the magnetization by using the Poisson summation formula (3.137). The final result reads

$$\Omega_{\text{osc}}^{(\text{mat})}(\mu) = \frac{\mu^4}{12\pi^2\hbar^3 v_F^3} \sum_{k=1}^{\infty} {}_2F_3\left(1,1;\frac{5}{4},\frac{7}{4},2; -\frac{\pi^2 k^2 \mu^4 l_B^4}{4\hbar^4 v_F^4}\right), \qquad (3.148)$$

where $_nF_m(x_1, x_2, \ldots, x_n; y_1, y_2, \ldots, y_m; z)$ is the generalized hypergeometric function [443]. This is analogous to Eq. (3.138) for the DOS. The corresponding magnetization reads

$$M = \frac{e\mu^4 l_B^2}{6\pi^2\hbar^4 c v_F^3} \sum_{k=1}^{\infty} \left[{}_2F_3\left(1,1;\frac{5}{4},\frac{7}{4},2; -\frac{\pi^2 k^2 \mu^4 l_B^4}{4\hbar^4 v_F^4}\right)\right.$$
$$\left. - {}_1F_2\left(1;\frac{5}{4},\frac{7}{4}; -\frac{\pi^2 k^2 \mu^4 l_B^4}{4\hbar^4 v_F^4}\right)\right]. \qquad (3.149)$$

Let us note that the analytical expressions for the quantum oscillations and the DOS were derived in the clean limit, which assumes that the quasiparticle width vanishes, $\Gamma \to 0$. As will be discussed in detail in Sec. 7.1, disorder can be taken into account via the self-energy corrections in Green's function $G^R(0, \mathbf{k})$. In view of the definition in Eq. (3.133), this will also modify the DOS. Qualitatively, a nonzero quasiparticle width will be manifested as broadening of the δ-functions in Eq. (3.134) and flattening of the peaks in Fig. 3.12(a). Phenomenologically, the effects of disorder can be also modeled semirigorously by (i) introducing the Dingle factor $\exp\left[-2\pi l \mu \Gamma l_B^2/(\hbar^2 v_F^2)\right]$ in the Poisson resummation expressions (3.137) for the DOS or (ii) replacing the δ-functions with Lorentz distributions in Eq. (3.134). Then, a corrected expression for the magnetization can be obtained by following the same steps as in the clean limit. The corresponding results for the magnetization are shown in Fig. 3.12(b) for several choices of constant Γ, which confirms that the peaks become smoother and less pronounced with increasing the quasiparticle width. It might be appropriate to mention, however, that the effects of realistic disorder potentials could be much more complicated.

The oscillating part of the magnetization for semimetals with a relativistic-like dispersion relation can be also obtained from the Lifshitz–Kosevich formula [287, 444]

$$M_{\text{osc}} = -\left(\frac{e}{2\pi\hbar c}\right)^{3/2} \frac{\hbar^2 S_F}{\pi^2 m^*} \sqrt{\frac{B}{\partial^2 S_F/\partial k_z^2}} \sum_{l=1}^{\infty} \frac{1}{l^{3/2}} R_T R_D R_S$$

$$\times \sin\left[2\pi l\left(\frac{c\hbar S_{\text{extr}}}{2\pi e B} - \beta - \frac{1}{2}\right) + 2\pi\delta\right], \tag{3.150}$$

where S_F is the extremal cross-section of the Fermi surface normal to the direction of the field, $m^* = \hbar^2/(2\pi)\,[\partial S/(\partial\epsilon)]_{\epsilon=\mu}$ is the cyclotron mass in the semiclassical approximation, and S_{extr} is the area covered by the cyclotron orbit. In the case of a linear dispersion relation, the cyclotron mass is $m^* = \mu/v_F^2$. The terms R_T, R_D, and R_S correspond to the temperature, Dingle, and spin damping factors that contribute to the net "blurring" of the oscillation phase. They are given by

$$R_T = \frac{2\pi^2 clTm^*}{e\hbar B \sinh\left[2\pi^2 clTm^*/(e\hbar B)\right]}, \tag{3.151}$$

$$R_D = \exp\left(-\frac{2\pi^2 clT_D m^*}{e\hbar B}\right), \tag{3.152}$$

$$R_S = \cos\left(\frac{l\pi g m^*}{2m_e}\right). \tag{3.153}$$

Here g is the g-factor of charge carriers ($g \approx 2$ for relativistic fermions) and T_D is the effective Dingle temperature, which is determined by disorder, i.e., $T_D = \Gamma/\pi$. Such a representation is particularly handy when one needs to estimate the effects of temperature and/or disorder.

Experimentally, quantum oscillations were studied in many Dirac [57, 445–448] and Weyl [65, 66, 449–456] semimetals. It is worth noting, however, that the analysis of quantum oscillations in real materials is a very complicated task. Indeed, in addition to the multiple pairs of Dirac points and Weyl nodes, additional energy bands with parabolic dispersion relations might be also present. Despite numerous complications, the analysis of quantum oscillations is an invaluable tool in determining the shape of the Fermi surface. The representative results for the Weyl semimetal TaP [454] are shown in Fig. 3.13(a). By comparing the band structure calculations

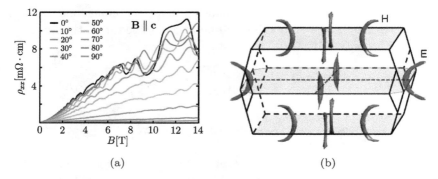

Fig. 3.13   (a) The quantum oscillations of resistivity (the Shubnikov–de Haas oscillations) in TaP as a function of a magnetic field at different angles in plane defined by the [001] ($\theta = 0$) and [100] ($\theta = 90°$) basis vectors. (b) The reconstructed Fermi surface where E and H denote electron (blue) and hole (red) pockets.[u] (Adapted from [454].)

with the full angular dependence of quantum oscillation frequencies, the Fermi surface shape could be determined. As one can see from Fig. 3.13(b), one finds that the Fermi surface of TaP has nontrivial crescent-like pockets. Note that separate pockets correspond to distinct sheets of the Fermi surface near different pairs of Weyl nodes.

3.9.3   *Specific heat in a magnetic field*

Let us now investigate the effects of a nonzero magnetic field on the specific heat in Weyl and Dirac semimetals (see also [438]). The specific heat is defined as

$$C = \left(\frac{\partial \epsilon}{\partial T} \right)_n.$$
(3.154)

The energy density per Weyl node reads[v]

$$\epsilon = \int_{-\infty}^{\infty} dx\, \nu(x) x f^{\text{eq}}(x).$$
(3.155)

---

[u]By using the classification in Sec. 4.3.3, the electron pockets correspond to W1-type Weyl nodes and the hole ones to W2-type Weyl nodes.
[v]Since an ultraviolet divergent part of the energy density does not depend on temperature, its subtraction is not crucial for the calculation of the specific heat.

By making use of the DOS $\nu(x)$ given in Eq. (3.135), one derives

$$C = \frac{1}{16\pi^2\hbar v_F l_B^2 T} \int_{-\infty}^{\infty} dx \frac{x}{\cosh^2\left(\frac{x-\mu}{2T}\right)} \left(\frac{x-\mu}{T} + \frac{d\mu}{dT}\right)$$

$$\times \left(1 + 2\sum_{n=1}^{N_x} \frac{|x|}{\sqrt{x^2 - 2n\epsilon_L^2}}\right), \tag{3.156}$$

where $N_x = [x^2 l_B^2/(2\hbar^2 v_F^2)]$ is the highest occupied Landau level, [...] denotes the integer part, and the following expression was used:

$$\frac{\partial f^{\text{eq}}(x)}{\partial T} = \frac{1}{4T\cosh^2\left(\frac{x-\mu}{2T}\right)} \left(\frac{x-\mu}{T} + \frac{d\mu}{dT}\right). \tag{3.157}$$

The second term in the parentheses, which accounts for a temperature dependence of the electric chemical potential, can be found by requiring a constant density of quasiparticles $dn/dT = 0$. By using the definition of n,

$$n = \int_{-\infty}^{\infty} dx\, \nu(x) f^{\text{eq}}(x), \tag{3.158}$$

we derive

$$\frac{d\mu}{dT} = -\int_{-\infty}^{\infty} dx \frac{1}{\cosh^2\left(\frac{x-\mu}{2T}\right)} \frac{x-\mu}{T} \left(1 + 2\sum_{n=1}^{N_x} \frac{|x|}{\sqrt{x^2 - 2n\epsilon_L^2}}\right)$$

$$\times \left[\int_{-\infty}^{\infty} dx \frac{1}{\cosh^2\left(\frac{x-\mu}{2T}\right)} \left(1 + 2\sum_{n=1}^{N_x} \frac{|x|}{\sqrt{x^2 - 2n\epsilon_L^2}}\right)\right]^{-1}. \tag{3.159}$$

In the limit of small temperature ($T/\epsilon_L \ll 1$) and zero chemical potential, the $d\mu/dT$ correction could be ignored. Then, the specific heat is given primarily by the LLL contribution

$$C_{n=0} \approx \frac{1}{16\pi^2\hbar v_F l_B^2 T^2} \int_{-\infty}^{\infty} dx \frac{x^2}{\cosh^2\left(\frac{x}{2T}\right)} = \frac{T|eB|}{12\hbar^2 c v_F}. \tag{3.160}$$

It is worth emphasizing that, unlike the zero-field result in Eq. (3.125), the LLL contribution to the specific heat is linear in temperature even at $\mu = 0$. It also increases linearly with a magnetic field. These properties caused by a high sensitivity of electron–hole plasma to magnetic fields allow one to separate the electron and phonon contributions to the specific heat.

The specific heat in a magnetic field as well as its asymptotes are shown in Fig. 3.14(a) at $\mu = 0$. As one can see, the linear LLL contribution

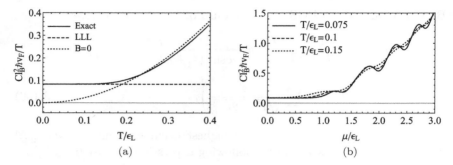

Fig. 3.14 (a) The specific heat in Dirac and Weyl semimetals as a function of temperature. The exact solution given in Eq. (3.156) is shown by the solid line. The LLL contribution as well as the $B = 0$ result (3.125) are shown by the dashed and dotted lines, respectively. (b) The quantum oscillations of the specific heat as a function of electric chemical potential at a few values of temperature. Here, we set $T/\epsilon_L = 0.075$ (solid line), $T/\epsilon_L = 0.1$ (dashed line), and $T/\epsilon_L = 0.15$ (dotted line).

dominates at small T. At larger temperatures, the effects of magnetic field are weak and the specific heat is well described by the $B = 0$ result in Eq. (3.125). The quantum oscillations of the specific heat with μ are shown in Fig. 3.14(b). As expected, temperature smears peaks that become barely distinguishable at large T.

While useful in weakly correlated Weyl semimetals, thermodynamic measurements are particularly important for investigating strongly correlated systems. Indeed, the Kondo temperature of typical heavy-fermion systems is usually smaller than the Debye temperature. Furthermore, the Fermi velocity is strongly renormalized and can be $100 - 1000$ times smaller than in conventional Weyl semimetals. Therefore, the electron contribution to the specific heat, which scales as T^3/v^3 where v standing for renormalized quasiparticle velocity, is strongly enhanced and could dominate over the phonon one.

An observation of a relativistic-like scaling of the specific heat with temperature $C \propto T^3/v^3$ was used to identify experimentally the signatures of the Weyl–Kondo semimetal phase [437] in strongly correlated heavy-fermion materials such as $Ce_3Bi_4Pd_3$ [457] and YbPtBi [458]. In addition to a relativistic-like dependence of the specific heat, the negative magnetoresistivity and the anomalous Hall effect (see also Chapter 6) were also observed in YbPtBi [458]. The experimental data for $Ce_3Bi_4Pd_3$ [457] and YbPtBi [458] are presented in Figs. 3.15(a) and 3.15(b), respectively. To separate the electron contribution in the specific heat, the phonon part was

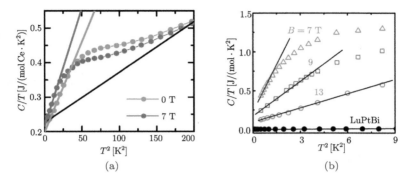

Fig. 3.15 (a) The dependence of the specific heat to temperature ratio C/T on temperature squared in $Ce_3Bi_4Pd_3$ with (blue) and without (red) a magnetic field. Solid lines correspond to analytical fits in Eq. (3.127). The black solid line corresponds to the phonon contribution determined in $La_3Bi_4Pt_3$. (Adapted from [457].) (b) The dependence of the specific heat to temperature ratio C/T on temperature squared in YbPtBi for several values of a magnetic field. Solid black lines correspond to analytical fits in Eq. (3.127). For comparison, the data for nonmagnetic material LuPtBi are given by black dots. (Adapted from [458].)

determined in the isostructural nonmagnetic material $La_3Bi_4Pt_3$ in the left panel. As one can see from both panels in Fig. 3.15, the specific heat in heavy-fermion materials is well-fitted with a relativistic-like $\sim T^3$ dependence of the specific heat. Moreover, its electron origin is supported by strong deviations from the results for isostructural materials.

Before concluding this section, it is worthwhile mentioning that a novel type of quantum oscillations could be realized in Weyl and Dirac semimetals in the absence of a background magnetic field. As suggested in [419], the necessary key ingredient that replaced the magnetic field is a strain-induced pseudomagnetic field discussed in Sec. 3.7. Since each Weyl node contributes additively, the oscillations induced by a strain-induced pseudomagnetic field are similar to those produced by a usual magnetic field. Moreover, when both magnetic and pseudomagnetic fields are applied, the Weyl nodes of opposite chirality will experience different effective fields leading to a splitting of quantum oscillation peaks.

Chapter 4

Ab Initio Calculations and Materials Realizations

I do not fear computers. I fear lack of them.

Isaac Asimov

The rapid increase of the computational power at the beginning of the 21st century as well as the development of effective first-principles calculation methods and software packages strongly improved the quality and significance of numerical calculations in science. Nowadays, in condensed matter physics, various properties of materials are routinely studied by using the *ab initio* simulations, which provide a powerful supplemental tool for the conventional theoretical and experimental research. Numerical studies and modeling of topological phases of matter are particularly fruitful because their nontrivial band topology leads to an intrinsic robustness with respect to disorder and numerical errors [17, 459].

The first-principles calculations have already demonstrated the unprecedented predictive power that helped to guide experimental efforts in discovering materials with novel interesting physical properties. In particular, it is the *ab initio* calculations that lead to significant breakthroughs in studies of two-dimensional (2D) and three-dimensional (3D) topological insulators as well as various types of Dirac and Weyl semimetals. In fact, many of their materials realizations were first predicted by numerical methods and only later discovered experimentally. It is worth emphasizing, however, that the *ab initio* calculations are not the panacea. They have some inherent limitations and drawbacks, and, most importantly, they should not be viewed as a substitute for real empirical data. Therefore, a cross-verification between first-principles calculations and experimental data remains important in the search for novel materials.

In regards to Dirac and Weyl semimetals, as it turns out, the prospective compounds should have the following key properties: (i) appropriate orbitals giving overlapping valence and conduction bands and (ii) suitable crystal symmetries providing the necessary protection of the band touching points. Often, these conditions are met when the chemical composition of 3D topological materials includes heavy elements, e.g., bismuth, antimony, or arsenic from the pnictogen family. Interestingly, some of the Dirac semimetals, such as Cd_3As_2, were synthesized a long time ago and were well known for their excellent transport properties. For a long time, however, the experimental detection of Dirac points and/or Weyl nodes was practically impossible without reliable direct methods that would be able to map precisely the band structure.

Among such methods, arguably, the most valuable is the *angle-resolved photoemission spectroscopy* (ARPES) [460–464]. The core principles of ARPES were established more than a hundred years ago by Albert Einstein [465] in his groundbreaking paper on the photoelectric effect. The ARPES is able to map the bulk and surface band dispersions with a relatively good precision and, in recent years, it became one of the ubiquitous experimental tools in the investigation of various topological materials. Another very useful technique, which is often used for studying the surface states in Weyl and Dirac semimetals, is the *scanning tunneling microscopy/spectroscopy* (STM/STS) [466–468]. Unlike ARPES, these methods rely more heavily on ancillary theoretical analysis. However, they are able to provide additional information on the scattering channels and the selection rules for surface quasiparticles.

4.1   *Ab initio* methods

The *ab initio* methods play an important role in the development of condensed matter physics. In practice, these methods utilize a wide range of approximations that were honed over decades by trial and error. The corresponding compilation of approximations and various numerical techniques is usually referred to as the electronic band structure calculations. In what follows, the key aspects of the *ab initio* methods will be briefly outlined.

Let us begin with a general discussion of electron models in solids. The existing wide range of models can be roughly split into three categories: (i) free-electron models [469], (ii) tight-binding models [470], (iii) microscopic models with various interaction effects. In the first type of models, the conducting electrons are treated as a gas of free particles with an effective mass determined by the mean value of the crystal potential energy.

This approach was first put forward by Paul Drude who studied the electrical conductivity of metals in 1900 [471, 472]. The use of the classical statistics for quasiparticles was a major limitation in the original model. In 1928, its formulation was greatly improved by Arnold Sommerfeld [473], who included quantum effects of quasiparticles subject to the Pauli exclusion principle by using the Fermi–Dirac statistics. Surprisingly, such a simple model was quite successful in explaining the generic features of many properties of metals, including the specific heat, the electrical and thermal conductivities, the Wiedemann–Franz law, the Hall effect, etc. Upon a closer examination, however, it was found that the free-electron models usually are not accurate. They also fail completely in describing the electronic properties of semimetals, semiconductors, and insulators.

In tight-binding models, as the name suggests, electrons are assumed to be tightly bound to ions. In its simplest formulation, the model takes into account only the nearest-neighbor and on-site interactions. The corresponding tight-binding Hamiltonian reads

$$H = \sum_{\langle i,j \rangle} t_{ij} \hat{c}_i^\dagger \hat{c}_j + \sum_i u_i \hat{c}_i^\dagger \hat{c}_i + \text{h.c.}, \tag{4.1}$$

where $\langle i,j \rangle$ denotes the summation over nearest neighbors, $\hat{c}_i$ and $\hat{c}_i^\dagger$ are the annihilation and creation operators, t_{ij} is the hopping matrix, and u_i is the on-site potential. It is worth noting that the quasiparticle energy spectrum in graphene is described really well by a tight-binding model of this type [195] (see Chapter 2).

The third class of models combines various approaches inspired by a microscopic description of solids with approximations of a varying degree that take into account electron–ion and electron–electron interactions. The corresponding examples include the Hartree–Fock approximation, the density functional theory, the quantum Monte–Carlo methods, etc. Often, the description can be simplified by using the Born–Oppenheimer approximation, which assumes that the dynamics of ions and electrons are separable. In the special case when ions are treated as static, the electron dynamics is governed by the following Schrödinger equation[a]:

$$-\frac{\hbar^2}{2m} \sum_{i=1}^N \Delta_i \Psi + \sum_{i=1}^N U_{\text{ion}}(\mathbf{r}_i)\Psi + \sum_{i<j} \frac{e^2}{|\mathbf{r}_i - \mathbf{r}_j|}\Psi = \epsilon\Psi, \tag{4.2}$$

---

[a]Relativistic effects usually become important for electrons in heavy atoms. Thus, the Dirac equation or the Schrödinger equation with relativistic corrections should be used for the description of such electrons.

where $\Psi = \Pi_{i=1}^{N}\psi_i$ is the wave function of N electrons and $U_{\text{ion}}(\mathbf{r}_i)$ is the electron potential energy in the field of ions. The last term on the left-hand side describes the Coulomb interaction between electrons.

Given that (i) the number of electrons in typical macroscopic samples is of the order of $N \sim 10^{23}$ and (ii) the computational time grows exponentially with N, it is impossible to analyze the dynamics of electrons in solids by solving the microscopic equation (4.2) directly. The prime difficulty in the analysis of the corresponding Schrödinger equation is the electron–electron interaction. As was first suggested by Douglas Hartree [474], a useful simplification can be obtained by utilizing the one-electron Schrödinger equation, in which the electron–electron interactions are taken into account through an effective "mean field" potential

$$U_{\text{ee}}(\mathbf{r}) = \int d^3 r' \sum_i \frac{e^2 \left|\psi_i\left(\mathbf{r}'\right)\right|^2}{\left|\mathbf{r} - \mathbf{r}'\right|}. \tag{4.3}$$

In other words, it is assumed that each electron moves in an effective potential produced by all other electrons. Unfortunately, this Hartree approximation does not account for the Pauli principle. This problem was later resolved in the *Hartree–Fock approximation* [475], which generalizes the Hartree approach by including an exchange interaction energy. The latter arises from the antisymmetric nature of the electron wave function, given by the Slater determinant [476]. The corresponding energy functional reads

$$\epsilon = \sum_{i,s_1} \int d^3 r_1 \left[\psi_i^\dagger(\mathbf{r}_1, s_1) \frac{-\hbar^2}{2m} \Delta \psi_i(\mathbf{r}_1, s_1) + U_{\text{ion}} \left|\psi_i(\mathbf{r}_1, s_1)\right|^2 \right]$$

$$+ \frac{1}{2} \sum_{i,j,s_1,s_2} \int d^3 r_1 \int d^3 r_2 \frac{e^2}{\left|\mathbf{r}_1 - \mathbf{r}_2\right|} [\left|\psi_i(\mathbf{r}_1, s_1)\right|^2 \left|\psi_j(\mathbf{r}_2, s_2)\right|^2$$

$$- \psi_i^\dagger(\mathbf{r}_1, s_1)\psi_j^\dagger(\mathbf{r}_2, s_2)\psi_i(\mathbf{r}_2, s_2)\psi_j(\mathbf{r}_1, s_1)], \tag{4.4}$$

where $\sum_{s_l}$ with $l = 1, 2$ denotes the summation over spin projections. The first term in the second square brackets is known as the Coulomb integral and the second is the exchange integral. The corresponding equations of motion for the one-electron wave function $\psi_i(\mathbf{r})$ can be obtained by using the variation principle. However, when trying to solve the problem with a large number of particles, one finds that the calculation of the Coulomb and exchange integrals is very expensive numerically. Therefore, additional approximations are highly desirable in practice.

4.1.1 *Density functional theory*

One of the most popular and versatile first-principles methods available in condensed matter physics is the *density functional theory* (DFT) [477–479]. As suggested by its name, the properties of a many-electron system are determined by the electron density[b]

$$n(\mathbf{r}) = N \int \ldots \int d^3r_1 \ldots d^3r_N \Psi^\dagger(\mathbf{r}_1, \ldots, \mathbf{r}_N)\delta\,(\mathbf{r} - \mathbf{r}_1)\,\Psi(\mathbf{r}_1, \ldots, \mathbf{r}_N). \tag{4.5}$$

The other quantity of principal importance is the free energy $F[n]$, which is a functional of the electron density. The DFT method evaluates the electronic structure by taking into account the external potential $U_{\text{ext}}[n]$ and the electron–electron potential $U_{\text{ee}}[n]$. While the former stems from the static Coulomb potential of the lattice of ions, which is determined by the materials structure and composition, the latter captures the electron–electron interactions. Were the potential and kinetic functionals known, the free energy $F[n] = T[n] + U_{\text{ext}}[n] + U_{\text{ee}}[n]$ would be completely fixed for any given external potential $U_{\text{ext}}(\mathbf{r})$. Unfortunately, this is usually not the case because $U_{\text{ee}}[n]$ and $F[n]$ are not known exactly. In particular, the exact functionals for the exchange and the correlation energies in a realistic system are usually too complicated or known only approximately.

The simplest DFT formulation is the *local density approximation* (LDA), which assumes that the free energy functional depends only on the local electron density but not on its derivatives. For example, the Coulomb part of the energy functional in this approximation is given by

$$\frac{1}{2} \int d^3r_1 d^3r_2 \frac{e^2 n(\mathbf{r}_1)n(\mathbf{r}_2)}{|\mathbf{r}_1 - \mathbf{r}_2|}. \tag{4.6}$$

In order to improve the accuracy of DFT calculations, it is advantageous to introduce the Kohn–Sham equation [480] for the one-electron wave function

$$-\frac{\hbar^2}{2m}\Delta\psi_i(\mathbf{r}) + \left(U_{\text{ext}}(\mathbf{r}) + \int d^3r' \frac{e^2 n(\mathbf{r}')}{|\mathbf{r} - \mathbf{r}'|} + \frac{\delta\mathcal{E}_{\text{xc}}[n]}{\delta n}\right)\psi_i(\mathbf{r}) = \epsilon_i\psi_i(\mathbf{r}), \tag{4.7}$$

---

[b]While this statement might seem surprising, it can be shown that two different external potentials never result in the same electron density. Therefore, since many-electron systems differ only by the external potentials and the number of particles, the properties of these systems can be indeed determined by the electron density.

where $\mathcal{E}_{xc}[n]$ denotes the exchange-correlation part of the potential energy functional.[c] The local electron density is then expressed self-consistently in terms of the electron wave functions $\psi_i(\mathbf{r})$ as follows: $n(\mathbf{r}) = \sum_i |\psi_i(\mathbf{r})|^2$. A more refined approximation can be obtained by including gradients of the density and is known as the *generalized gradient approximation* (GGA).

As was already stated above, the DFT is particularly suitable for studying topological phases of matter. Indeed, many modern *ab initio* methods describe quite well weakly correlated electronic materials and can easily incorporate the spin–orbit coupling [459]. Moreover, the evaluation of topological properties is relatively robust and is not very susceptible to numerical errors. Nevertheless, some technical issues and complications that require a more careful treatment may still arise. For example, the calculation of the Berry curvature generally requires additional adjustments such as dense momentum meshes, extra gauge-fixing conditions, the use of symmetry arguments, etc.

No matter which algorithm is used, numerical methods always have discretization errors due to a finite resolution in sampling of 3D momenta (or coordinates). Even for the finest sampling grids, numerical errors make it difficult to identify unambiguously gapless Dirac points or Weyl nodes. Generally, the use of LDA or GGA tends to underestimate the value of a band gap. While this is not crucial for identifying topological insulators, errors in finding the gap could lead to an incorrect determination of non-topological Dirac semimetals. Furthermore, by definition, the LDA and GGA cannot describe strongly correlated systems.

Taking into account the limitations of the DFT, it is often beneficial to use it in conjunction with other methods such as the pseudopotential approach [482], the methods of orthogonalized plane waves [355] and projector augmented waves [483], etc. The basic idea behind the pseudopotential method is to replace the strong ionic potential with a residual potential that accounts for the screening effects by the core electrons. Since there is a high degree of freedom in the definition of pseudopotentials, the main criterion for choosing them is the simplicity of implementation. As for the orthogonalized plane waves and the projector augmented waves methods, their underlying idea is to separate a rapidly oscillating part of the wave function in the vicinity of lattice ions from the smooth part in the space

---

[c]For example, one of the simplest approximations for the exchange-correlation part of the potential energy functional is obtained by using the Thomas–Fermi theory, where $\mathcal{E}_{xc}[n] \propto n^{4/3}$ (see, e.g., [478, 481]).

between the lattice sites. While the latter is relatively easy to determine numerically, the calculation near the ions requires a much larger computation effort.

4.1.2 *Topological aspects and surface states*

Even when the band structure is determined, the numerical study of the corresponding topological properties is not simple. For example, as we already emphasized before, the determination of the Berry curvature requires a fine momentum mesh. Moreover, since the Berry connection is not gauge invariant, its numerical calculation is impossible without fixing the phases of wave functions in the whole Brillouin zone, which is usually difficult in conventional numerical methods. In order to relatively quickly and efficiently determine the topological numbers, the Wilson loop method [459, 484–486] can be used. While this method is usually applied for investigating topological insulators, it can also be used to calculate topological numbers in Weyl semimetals, including the mirror Chern numbers and $\mathbb{Z}_2$ indices for the mirror and glide mirror planes, respectively. Unlike other approaches, the Wilson loop method does not rely on gauge fixing and can be used for systems with broken $\mathcal{P}$ symmetry.

The Wilson loop method greatly benefits from the use of the *maximally localized Wannier functions* (MLWFs) [487–489]. Compared to the Bloch functions $\psi(\mathbf{r})$, the Wannier functions $w(\mathbf{R}, \mathbf{r})$ [490] are localized in coordinate space

$$w(\mathbf{R}, \mathbf{r}) = \int \frac{d^3 k}{(2\pi)^3} e^{i\mathbf{k} \cdot \mathbf{R}} \psi(\mathbf{r}), \qquad (4.8)$$

but they are not the eigenstates of the Hamiltonian. The MLWFs are a specific type of the Wannier functions with the minimized second-moment spread of each Wannier charge density about its center. The MLWFs can provide an insight into the nature of bonding and are useful for electronic-structure calculations. Furthermore, they can be used in constructing the corresponding tight-binding Hamiltonians and in studying ballistic transport.

An efficient approach to calculate the Fermi surface averages and spectral properties of solids is the interpolation technique [489, 491, 492]. It starts from performing a low-resolution first-principles calculation of the electronic structure, e.g., by using the DFT. Then, the obtained Bloch functions are mapped onto the MLWFs and the tight-binding Hamiltonian is derived. Finally, the resulting MLWFs can be used to interpolate the eigenfunctions and other momentum-space quantities on a much finer grid compared to the initial first-principles calculations. This method is very helpful for determining topological properties of materials because there is a close connection between the charge centers of the MLWFs and the Berry phases of the Bloch functions as they are carried around the Brillouin zone.

An important feature of Weyl and certain Dirac semimetals is the existence of special topologically protected surface states. While their appearance will be discussed in Chapter 5, it is worth outlining briefly how they can be studied numerically. The surface states can be investigated by using two main methods [459]. Firstly, the surface states can be calculated by utilizing a supercell[d] of a semimetal slab surrounded by vacuum [493]. While such a method is straightforward and takes into account various properties of the surface, it is rather demanding computationally because it requires a supercell and a vacuum region of large sizes. In addition, it is necessary to eliminate possible coupling between the boundaries, which could deform the surface states. The second method for studying surface states relies on Green's functions built from the MLWFs [494]. Since the MLWFs are maximally localized, the Hamiltonian of a finite system can be approximated as a block tridiagonal matrix. Then an iteration method [495–497] can be utilized to calculate the projected Green functions on the surfaces. Unlike the first method, such an approach is not able to take into account fine details of the surface, but it is faster and simpler.

4.1.3 *Ab initio software packages*

Before discussing the *ab initio* results, it might be appropriate to list the most popular software packages usually employed in numerical investigations of Weyl and Dirac semimetals. The DFT calculations are often performed in the Vienna *ab initio* simulation package (VASP) [498–501], which also includes post-DFT corrections such as hybrid functionals, many-body perturbation theory, and dynamical electronic correlations in the random phase approximation. Another popular package is OpenMX [502], which makes possible to perform nanoscale material simulations based on the DFT, norm-conserving pseudopotentials, and pseudo-atomic localized

---

[d]The supercell is a cell that contains many unit cells of a crystal. Supercells are useful when the physical properties of a solid are not determined by a single unit cell, e.g., as in finite crystals.

basis functions. The electronic structure calculations of periodic solids in the DFT approach can also be performed using the software package WIEN2k [503, 504]. A variety of solid-state properties, including ground-state calculations, structural optimization, quantum transport, as well as response and spectroscopic properties, can be calculated via Quantum ESPRESSO *ab initio* package [505–507]. The DFT calculations can also be performed by using the full-potential local-orbital (FPLO) code [508, 509]. A surge of interest to novel topological materials has led to the development of specialized numerical software packages such as WannierTools [510]. Quantum transport properties, especially in finite systems, can also be studied numerically by using the KWANT toolbox [511].

4.2 First-principles results

A few examples of the *ab initio* results for Dirac and Weyl semimetals will be discussed in this section. It is instructive to start by recalling some general criteria for the realization of topological materials with relativistic-like dispersion relations. As suggested in [5, 512], a large atomic number and a small electronegativity[e] difference of constituents are general prerequisites of a gapless energy spectrum. Both factors increase the tendency for a pair of adjacent bands to get inverted. This is also supported by the empirical data since all experimentally verified Dirac and Weyl semimetals contain heavy atoms in their composition. Considering that the band crossings are not always stable against opening a gap (see Chapter 3), they should be protected by either a crystal symmetry (Dirac points) or topology (Weyl nodes). To be of practical interest, the crossing points should lie near or, ideally, at the Fermi level. In Weyl semimetals, there is an additional requirement of $\mathcal{P}$ or $\mathcal{T}$ symmetry breaking that leads to a splitting of Dirac points into pairs of Weyl nodes.

4.2.1 *Dirac semimetals*

Historically, the realization of a Dirac semimetal phase was first proposed at a quantum critical point where a topological insulator transforms into a normal insulator. Among the known materials of this type, there is a number of alloys, including $Bi_{1-x}Sb_x$ [89, 513], with the band gap vanishing at $x \approx 0.03 - 0.04$, $TlBiSe_{2-x}S_x$ [514–516], $TlBiTe_{2-x}S_x$ [517],

---

[e]Electronegativity measures the tendency of atoms to attract a bonding pair of electrons.

$Bi_{2-x}In_xSe_3$ [518, 519], $Pb_{1-x}Sn_xSe$ [520], and $Pb_{1-x}Sn_xTe$ [521]. However, the realization of a Dirac semimetal phase in such alloys requires fine tuning of the elemental composition, which is often hard to achieve with sufficient precision. Therefore, the most promising candidates for Dirac semimetals are stoichiometric materials where Dirac points are stabilized by symmetries. Such Dirac semimetals are called *symmetry protected*.

As was already mentioned in Sec. 3.2, it was first predicted theoretically by using symmetry considerations and the *ab initio* calculations that symmetry-protected Dirac points could be realized in the metastable β-cristobalite AO_2 [366] with $A = As, Sb, Bi$, the distorted spinels $BiASiO_4$ [522] with $A = Zn, Ca, Mg$, as well as the alkali pnictides A_3Bi [110] with $A = Na, K, Rb$, and Cd_3As_2 [111]. The physics and chemistry principles behind the underlying structure of Dirac semimetals are discussed in [512] and supplemented by representative results of the DFT calculations for three different types of materials. The first type corresponds to Dirac semimetals obtained from charge-balanced semiconductors. The second type covers Dirac semimetals whose stacked hexagonal structures have weak interlayer coupling. The corresponding Dirac nodes come from the orbital degeneracies. Finally, an essential feature of the third type is nonsymmorphic symmetries that include space groups with glide planes and screw axes. Since the double group C_{2v} has only one irreducible representation [523] in the presence of spin–orbit coupling, Dirac nodes are not symmetry protected in such materials and a higher symmetry, e.g., C_3 or C_4, is required. The materials of the first class include the ZrBeSi and LiGaGe families, where BaAgBi and YbAuSb are among the most promising candidates, respectively, the pyrite family ($PtBi_2$), and the Bi_2Te_2Se family ($SrSn_2As_2$). As a representative example of the second class, one can identify $BaGa_2$, but it has other states at the Fermi level. As for the third class, the most promising candidates are HfI_3 and $TlMo_3Te_3$.

Let us discuss the *ab initio* calculations in Dirac semimetals by using trisodium bismuthide, Na_3Bi, as a representative example [110]. This compound was synthesized and structurally characterized almost a century ago [524]. It belongs to the alkali pnictides A_3Bi family with $A = Na, K, Rb$. The crystal structure of Na_3Bi is shown in Fig. 4.1(a), where Na and Bi atoms form simple honeycomb lattice layers stacking along the c-axis ([001]) with additional Na atoms inserted at the interlayer positions on top of Bi atoms. The lattice has $P6_3/mmc$ symmetry in the Hermann–Mauguin notation. The Brillouin zone of Na_3Bi is hexagonal and schematically shown in Fig. 4.1(b).

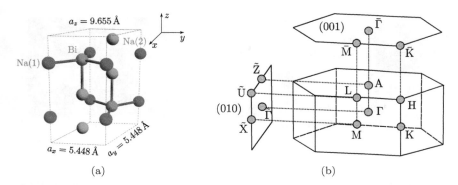

Fig. 4.1 (a) Lattice structure of the Dirac semimetal Na_3Bi, which has $P6_3/mmc$ symmetry. Here a_x, a_y, and a_z are the lattice constants. Blue and green spheres denote the Na atoms and the red spheres correspond to the Bi atoms. (b) The bulk Brillouin zone as well as its projections onto (001) and (010) planes.

The band structure of Na_3Bi was determined [110] by using the GGA for the exchange-correlation functional and then was cross-verified by using the hybrid functional method, which is defined in [525]. The valence and conduction bands are dominated by the Bi-6p and Na-3s states showing a relatively large band inversion of the order of 0.5 eV. The Fermi surface of Na_3Bi has two Dirac points at $\mathbf{k}_0^{\pm} = \pm 0.26\pi/a_z$ along the $\Gamma - A$ line. These points are protected by the threefold rotation symmetry. The breaking of this symmetry by compression along the y-axis, for example, leads to a topologically nontrivial insulating state.

Next, let us discuss the projected surface states. As in topological insulators [92], they can be determined from the surface Green's function of a semi-infinite system. Usually, one uses the *ab initio* results to obtain the MLWFs [487, 488] and then constructs the surface Green's function by employing an iterative approach defined in [495, 496]. The corresponding projected surface density of states for the [010] surface as well as the Fermi surface calculated in [110] are shown in Figs. 4.2(a) and 4.2(b), respectively. The surface states are unusual and, as will be discussed in Sec. 5.4, consist of a pair of Fermi arcs, whose existence can be explained by topological arguments. In essence, the Fermi arcs are open segments of the Fermi surface that connect the projections of the Weyl nodes and are related to nontrivial topological properties of the band structure. Their existence was first suggested by the *ab initio* calculations in the context of Weyl semimetals in [102]. Without going into details, let us note that the surface states have

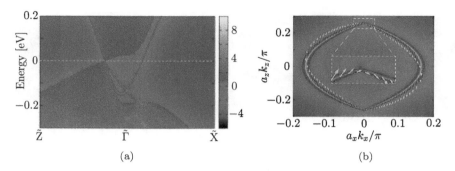

Fig. 4.2 (a) The projected surface density of states for the (010) surface of the Dirac semimetal Na_3Bi. The green dashed line corresponds to the Fermi level. (b) The Fermi surfaces and their spin textures (green arrows) for the (010) surface of Na_3Bi. (Adapted from [110].)

an interesting helical spin texture and discontinuous derivative of velocity at Dirac points.

Last but not least, let us comment on the derivation of the effective Hamiltonian given in Eqs. (3.89) and (3.90). Similarly to topological insulators [526], the effective Hamiltonian can be obtained by using the theory of invariants where the symmetries define the form of the Hamiltonian and the numerical values of its parameters are obtained by fitting the *ab initio* (or empirical) data. By noting that the dominant contribution to the low-energy spectrum comes from the Bi-$6p$ and Na-$3s$ states, it is convenient to introduce the following superpositions:

$$|S^\pm\rangle = \frac{|Na, s\rangle \pm |\widetilde{Na}, s\rangle}{\sqrt{2}}, \tag{4.9}$$

$$|P_i^\pm\rangle = \frac{|Bi, p_i\rangle \mp |\widetilde{Bi}, p_i\rangle}{\sqrt{2}}, \tag{4.10}$$

where the tilde denotes the $\mathcal{P}$ inversion partner state, the signs $\pm$ label parity, and $i = x, y, z$. When the spin–orbit coupling is included, one should use a new set of states defined in terms of the eigenstates of the total angular momentum: $|S_{1/2}^\pm, \pm 1/2\rangle$, $|P_{1/2}^\pm, \pm 1/2\rangle$, $|P_{3/2}^\pm, \pm 1/2\rangle$, and $|P_{3/2}^\pm, \pm 3/2\rangle$. Here subscript is the total angular momentum and its projection is denoted by the second entry in the wave function. Fortunately, only four of these states are needed for the description of Dirac points [110, 527], namely $|S_{1/2}^+, \pm 1/2\rangle$ and $|P_{3/2}^-, \pm 3/2\rangle$. In fact, this was already taken into account in the effective Hamiltonian given in Eqs. (3.89) and (3.90), which is defined in the basis of states $|S^+_{1/2}, 1/2\rangle$, $|P^-_{3/2}, 3/2\rangle$, $|S^+_{1/2}, -1/2\rangle$, $|P^-_{3/2}, -3/2\rangle$ and respects $\mathcal{T}$, $\mathcal{P}$, and C_3 rotational symmetry. By fitting the parameters of the effective Hamiltonian to the *ab initio* numerical data, the following values of model parameters were obtained in [110]:

$$C_0 = -0.06382 \text{ eV}, \quad C_1 = 8.7536 \text{ eV} \cdot \text{Å}^2, \quad C_2 = -8.4008 \text{ eV} \cdot \text{Å}^2,$$

$$M_0 = -0.08686 \text{ eV}, \quad M_1 = -10.6424 \text{ eV} \cdot \text{Å}^2, \quad M_2 = -10.3610 \text{ eV} \cdot \text{Å}^2,$$

$$A = \hbar v_F = 2.4598 \text{ eV} \cdot \text{Å}. \tag{4.11}$$

It is worth noting that another way to derive the effective Hamiltonian is to use the $\mathbf{k} \cdot \mathbf{p}$ expansion [392, 393]. The basic idea of the method is to treat the Hamiltonian at the Γ point ($\mathbf{k} = \mathbf{0}$) as an exact one and consider its dependence on crystal momentum $\mathbf{k}$ as a perturbation $\sim \mathbf{k} \cdot \mathbf{p}$, where $\mathbf{p} = -i\hbar \mathbf{\nabla}$. Such an approach with the subsequent downfolding of the resulting Kane model was used in [111] to obtain an effective Hamiltonian for the Dirac semimetal Cd_3As_2, whose quasiparticles were known to have a remarkably high mobility since the 1970s [528]. The effective Hamiltonian as well as the properties of Cd_3As_2 are qualitatively similar to those of Na_3Bi. Interestingly, Cd_3As_2 can be realized in two crystal structures known as structure I and structure II [111], where structure II has the body-centered tetragonal symmetry $I4_1cd$. The absence of the inversion symmetry in the structure II lifts the degeneracy for the states away from C_{4v} symmetry-protected Dirac points[f] but does not lead to the formation of Weyl nodes. The Weyl nodes separated in momentum space can be introduced either by lowering the crystal symmetry from C_{4v} to C_4 or by breaking $\mathcal{T}$ symmetry. Moreover, unlike Na_3Bi, Cd_3As_2 is stable in air under ambient conditions, which highly increases its practical value.[g]

In passing, we note that type-II Dirac semimetals, where Dirac cones are tilted, were theoretically predicted in the transition metal dichalcogenide $PtSe_2$ [529], the transition-metal icosagenides XA_3 [530] with $X = V, Nb, Ta$ and $A = Al, Ga, In$, the ternary wurtzite $CaAgBi$ materials family [531], $AMgBi$ [532] with $A = K, Rb, Cs$, and YPd_2Sn family [533].

---

[f]The point group C_{4v} is equivalent to C_4 with the addition of four mirror planes parallel to the axis of rotation.
[g]Note, however, that since Cd and As are highly toxic, a special care is required during the synthesis of Cd_3As_2.

4.2.2 Weyl semimetals

In this subsection, let us discuss the *ab initio* predictions for material realizations of Weyl semimetals. Since $\mathcal{T}$ and/or $\mathcal{P}$ symmetries should be broken in Weyl semimetals, it is convenient to separate them into the two main classes: magnetic (with broken $\mathcal{T}$ symmetry) and non-centrosymmetric (with broken $\mathcal{P}$ symmetry).

4.2.2.1 Weyl semimetals with broken $\mathcal{T}$ symmetry

The first theoretical proposal to realize Weyl fermions in solids was made in [102], where the magnetic pyrochlores $A_2Ir_2O_7$ with $A = $ Y, Eu, Nd, Sm, Pr were proposed as candidate materials. The *ab initio* calculations based on the LDA+U approximation[h] predicted an "all-in/all-out" spin structure. It was found that depending on the Coulomb energy parameter U, a few phases can be realized. What is important, a Weyl semimetal phase can appear for intermediate values of U. The nontrivial topological properties of this phase were confirmed by determining the surface states, which, as will be discussed in Chapter 5, have the characteristic Fermi arc form. However, despite exhaustive experimental studies, no direct evidence for Weyl nodes has been found so far in the pyrochlore iridates family of compounds.

Other materials, such as YbMnBi$_2$ were also proposed [534] to host a (type-II) Weyl semimetal phase. However, a later analysis [535] suggested a Dirac phase with a small gap. Another potentially viable proposal [536, 537] is related to the cobalt-based magnetic Heusler compounds ACo_2X with $A = $ V, Zr, Ti, Nb, Hf and $X = $ Si, Ge, Sn, as well as VCo$_2$Al and VCo$_2$Ga. Numerical results suggest several Weyl nodes, which are, however, separated in energy. By tuning the Fermi level, it is possible to realize a minimal Weyl semimetal with only two Weyl nodes. The antiferromagnetic half Heusler compounds GdPtBi and NdPtBi were also predicted [538–540] to host a Weyl semimetal phase with four to six pairs of Weyl nodes when $\mathcal{T}$ symmetry is explicitly broken by an external magnetic field. The latter pushes valence and conduction bands in the opposite directions allowing for the band inversion. These theoretical findings appear to agree with the experimental transport measurements. As predicted in [541, 542], the

---

[h]The LDA+U approximation includes an additional Hubbard-like term U, which models a strong on-site Coulomb interaction of localized electrons.

antiferromagnetic materials Mn_3Ge and Mn_3Sn are also Weyl semimetals with several Weyl points. In [541, 542], the results of the *ab initio* band structure analysis are supplemented with the calculation of the Berry curvature, whose monopole-like structure strongly supports the identification of band crossings with Weyl nodes. Other promising magnetic Weyl phases were predicted in $Co_3Sn_2S_2$ [543–546] and $Co_3Sn_2Se_2$ [544]. These materials have a quasi-2D structure of stacked kagome lattices. Band structure analysis revealed six Weyl nodes with a relatively large separation in momentum, which can reach up to a one-third of the Brillouin zone.

Conceptually, an external magnetic field provides a direct route to create Weyl semimetals by splitting each of the Dirac points into two Weyl nodes of opposite chirality via the Zeeman coupling [547] as well as the electron–electron [548] and exchange interactions [110, 549, 550]. Indeed, such a possibility is relatively obvious since a magnetic field breaks explicitly $\mathcal{T}$ symmetry. Concerning the role of interactions, it is worth noting that the corresponding effect was first discussed in a quantum electrodynamic plasma in [231, 551]. Indeed, by calculating the electron self-energy to linear orders in the coupling constant and an external magnetic field, it was shown that radiative corrections generate a chiral shift that splits a Dirac point into a pair of Weyl nodes separated in momentum space.

By using the *ab initio* calculations [552–554] and ARPES measurements [553, 554], it was shown that $EuCd_2As_2$ is a Weyl semimetal with a single pair of Weyl nodes in the Brillouin zone either in a magnetic field or when alloyed with barium at the europium sites. While the low-energy Hamiltonian of these material looks similar to that in the Dirac semimetals A_3Bi with $A = Na, K, Rb$ and Cd_3As_2, the ferromagnetic exchange splitting is large enough to annihilate one pair of Weyl nodes. In addition, the remaining pair is predicted to be close to the Fermi level. These properties make $EuCd_2As_2$ a promising candidate for testing various effects related to the chiral shift in the simplest form. Moreover, it was suggested [554] that due to ferromagnetic fluctuations Weyl fermions emerge already in the paramagnetic phase of $EuCd_2As_2$.

4.2.2.2 *Weyl semimetals with broken $\mathcal{P}$ symmetry*

Let us discuss another group of Weyl semimetals, where $\mathcal{T}$ symmetry is intact but $\mathcal{P}$ symmetry is broken. From a general topological viewpoint (see Sec. 3.2), the minimal number of Weyl nodes in $\mathcal{T}$ symmetric materials equals four (see the discussion in Sec. 3.2).[i] Of course, the presence of additional symmetries could increase this number.

The first proposals to realize $\mathcal{P}$ symmetry broken Weyl semimetals were made in [555, 556] shortly after the ideas of semimetals with broken $\mathcal{T}$ symmetry appeared. In particular, it was argued that one can use a normal-topological insulator multilayer structure, similar to that discussed at the end of Sec. 3.5.2. The two terms that respect all symmetries of the model but break $\mathcal{P}$ symmetry are the electrostatic potential difference between the top and bottom surfaces of a topological insulator layer in each unit cell $V_1 \tau_z \otimes \mathbb{1}_2$ and the momentum-independent spin–orbit interaction term $V_2 \tau_y \otimes \sigma_z$. A Weyl semimetal phase was also predicted to occur in noncentrosymmetric topological insulators LaBiTe$_3$ and LuBiTe$_3$ by doping them with Sb, i.e., LaBi$_{1-x}$Sb$_x$Te$_3$ and LuBi$_{1-x}$Sb$_x$Te$_3$ [557] or under pressure in BiTeI [557], as well as in Te and Se [558]. Unfortunately, these proposals were not experimentally confirmed.

A stoichiometric and pressure-free realization of a Weyl semimetal phase with broken $\mathcal{P}$ symmetry was proposed in the transition metal monopnictides TaAs, TaP, NbAs, and NbP [112–118]. The crystal structure and the Brillouin zone with the corresponding Weyl nodes for this family of materials are shown in Figs. 4.3(a) and 4.3(b), respectively.[j] Note that the space group of TaAs-like materials contains two mirror planes M_x and M_y (green planes in Fig. 4.3(b)) and two glide mirror planes M_{xy} and M_{-xy}, which are normal to the horizontal plane and pass through the corners of the Brillouin zone. As one can see, the nodal structure of this family is quite complicated and contains 24 Weyl nodes. Moreover, the nodes are split into two sets separated in energy, which are dubbed as W1 and W2 in the literature [5, 20, 113, 115, 117, 118].[k] In the case of the Weyl semimetal TaAs, for example, the energies of W1 and W2 nodes lie about 26 meV and 14 meV below the Fermi energy, respectively.

In order to provide a symmetry perspective on the existence of Weyl nodes, the Wilson loop analysis can be performed (for details, see [17, 112]).

---

[i] Actually, the minimal number of Weyl nodes is equal to two when the nodes are located at a $\mathcal{T}$ invariant momentum (e.g., $\mathbf{k} = \mathbf{0}$).

[j] Note that the results for the chirality of Weyl nodes differ in the literature. While the chirality of Weyl nodes determined in [112, 115, 116] is shown in Fig. 4.3(b), the chirality of the Weyl nodes near the horizontal plane is opposite in [113, 117].

[k] According to the standard convention, W1 denotes a pair of opposite chirality Weyl nodes with the smallest separation in momentum space.

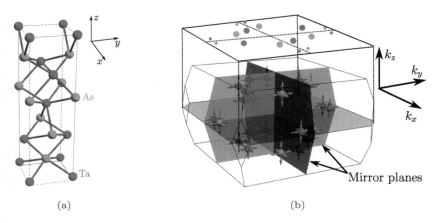

Fig. 4.3 (a) Lattice structure of the Weyl semimetal TaAs, which has nonsymmorphic space group $I4_1md$. (b) The bulk Brillouin zone as well as its projections onto (001) plane. The right- and left-handed Weyl nodes are denoted by red and blue colors, respectively. Arrows signify whether the Weyl node is a source or sink of the Berry curvature. Green planes denote the mirror planes.

The appearance of Weyl points is inferred by analyzing the mirror Chern numbers for mirror planes and $\mathbb{Z}_2$ indices for the glide mirror planes. It was shown [112] that while the mirror Chern numbers are nonzero, the $\mathbb{Z}_2$ indices are trivial in TaAs. This suggests that the Weyl nodes should be split with respect to the mirror planes and there should be the Fermi arc surface states.

The presence of 24 Weyl nodes can be deducted from the symmetry properties of the TaAs-family of materials. Indeed, let us consider the right-handed Weyl node at a given point in the Brillouin zone (k_x, k_y, k_z). In view of the mirror symmetries, there are three other Weyl nodes at $(-k_x, k_y, k_z)$, $(k_x, -k_y, k_z)$, and $(-k_x, -k_y, k_z)$, where the first two nodes have the opposite chirality. Since $\mathcal{T}$ symmetry is not broken, these four Weyl nodes should have four $\mathcal{T}$ partners of the same chirality situated at $\mathbf{k} \to -\mathbf{k}$. Then, the eight Weyl nodes have other eight partners due to the C_4 rotation symmetry, which preserves chirality. In total, there should be eight W1 (because $k_z = 0$) and sixteen W2 Weyl nodes. The precise positions of the Weyl nodes follow from the analysis of the Berry curvature distribution.

As will be discussed in detail in Chapter 5, the hallmark feature of any Weyl semimetal is the presence of nontrivial surface states, which are known as the Fermi arcs. Numerically, the surface states can be obtained via the Green's function method. It is worth noting that the shape of the

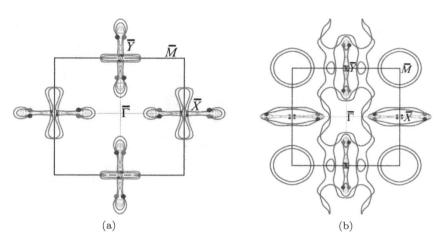

(a)   (b)

Fig. 4.4   Fermi surface projection for the (001) plane of the Weyl semimetal TaP with broken $\mathcal{P}$ symmetry for (a) P and (b) Ta terminations. Blue and green dots describe Weyl nodes of positive and negative chirality, respectively. Larger dots denote multiple Weyl nodes of the same chirality projected onto the same point in the surface Brillouin zone. (Adapted from [115].)

Fermi arcs depends on whether the crystal has an anion (As or P) or a cation (Ta or Nb) termination. In particular, the Fermi surface projections for the (001) surface of TaP with the P and Ta terminations are shown in Figs. 4.4(a) and 4.4(b), respectively. As one can see, while the clearest representation of the nontrivial states is obtained for the anion termination of the slab, there are many additional trivial projections for the cation termination.

Since the results for TaP are clear and resemble qualitatively those for other TaAs-like materials, let us discuss the shape of the Fermi arcs by using TaP with the P termination as a representative example. As one can see from Fig. 4.4, the Fermi surface projection for the (001) surface of TaP is composed of spoon-like and bowtie-like parts. Since the projections of the W2 Weyl nodes on the surface in the spoon-like part has effectively doubled topological charge, it is not surprising that two Fermi arcs form its tip. Similarly to the surface states in Na$_3$Bi presented in Fig. 4.2, the spin texture of the Fermi arcs is also helical. As for the Fermi arcs connecting the W1 nodes, they are assumed to be represented by very short parts of the Fermi surface projection between the Weyl nodes.

In general, the connectivity of the Fermi arcs depends on the material composition and the termination type (see, e.g., [559]). Therefore, it is important to formulate qualitative criteria that allow one to identify the surface states as Fermi arcs [117]. (i) A Fermi arc should be an open curve, which connects Weyl nodes of opposite chirality. (ii) The number of Fermi arcs is equal to the topological charge of Weyl nodes or their projections on the surface Brillouin zone. (iii) For Weyl node projections with topological charges greater than one, the Fermi arcs should form a kink when attaching to such projections. Note that the kink on the rotation axis might be present for some Dirac semimetals too, e.g., Na_3Bi [110]. (iv) The sum of the Fermi velocities signs corresponding to the surface states for a closed loop in the surface Brillouin zone should be nonzero. If this is the case, then the loop surrounds the Weyl node and the corresponding chiral state corresponds to the Fermi arc. While the last criterion requires the most effort, it allows for a more rigorous evidence for a Weyl semimetal phase, especially when the experimental resolution is low. It is worth noting that these criteria are not absolute and, due to various experimental limitations, some topologically trivial materials could also satisfy them. For example, photoemission from certain regions of the Fermi surface might be suppressed giving rise to apparently disjoined Fermi contours. The kink in the case of multiple Fermi arcs can be easily washed out due to an insufficient experimental resolution. Therefore, the experimental data alone might be insufficient to prove the existence of Weyl nodes and Fermi arcs. It is always desirable to have a cross-verification between the *ab initio* calculations and the experimental results.

Before describing the experimental results, let us briefly discuss the type-II Weyl semimetals. *Ab initio* calculations suggest that WTe_2 [127], $MoTe_2$ [560–562], AP_2 [563] with $A = Mo, W$, $Mo_xW_{1-x}Te_2$ [564], Ta_3S_2 [565], and noncentrosymmetric $TaIrTe_4$ [566] contain type-II Weyl nodes. An interesting proposal was made in [567], where a number of phases in RAlGe (R stands for a rare-earth element) were predicted. It was found that, under a suitable choice of rare-earth elements, type-I, type-II, $\mathcal{P}$, and $\mathcal{T}$ symmetry broken Weyl semimetals can be realized. For example, LaAlGe was predicted to be a $\mathcal{P}$ symmetry broken type-II Weyl semimetal, which has a total of forty Weyl nodes of three types. The energy spectra of WTe_2 and $MoTe_2$ as well as $TaIrTe_4$ are simpler and contain eight and four Weyl nodes, respectively. In addition, they tend to have long ($\sim 1/3$ of the Brillouin zone) Fermi arcs.

4.3　Experimental signatures

As usual, experiments play a decisive role in the discovery of novel materials. Fortunately, the hallmark features of Weyl and Dirac semimetals could be probed by several methods, including angle-resolved photoemission spectroscopy, scanning tunneling microscopy/spectroscopy, quantum magnetooscillations, and various transport measurements. The first method provides the most direct evidence of the relativistic-like dispersion relations and the Fermi arc surface states. Therefore, it will be the main focus of this section. In addition, the STM results will also be discussed briefly here. Transport measurements, on the other hand, will be presented in Chapters 6 and 7. It is worth noting that such measurements provide an important insight into the anomalous properties of Dirac and Weyl semimetals, including a negative magnetoresistivity due to the chiral anomaly, various optical signatures, etc.

4.3.1　*Overview of experimental techniques*

Given its experimental importance, let us first discuss the key physical principles of ARPES [460–464]. The schematics of an ARPES experiment is shown in Fig. 4.5(a), where the crystal is irradiated by a monochromatic beam of photons, which is usually supplied by a synchrotron. Due to the photoeffect, electrons escape from the sample and their angular distribution and kinetic energy are analyzed. By using the relation $p = \sqrt{2m_e \epsilon_{\mathrm{kin}}}$, the photoelectron momentum $\mathbf{p}$ can be determined. In the absence of interactions, the kinetic energy ϵ_{kin} of the photoelectron is related to the binding energy of the electron in a solid as follows:

$$\epsilon_{\mathrm{kin}} = \hbar\omega - \phi - |\epsilon_{\mathrm{B}}|, \qquad (4.12)$$

where ω is the frequency of radiation, ϕ is the electron work function (i.e., the energy required to remove the electron from the material to vacuum), and ϵ_{B} is the binding energy. The energetics of the photoemission process is schematically shown in Fig. 4.5(b). For an ideal crystal, the parallel component of momentum is conserved, i.e.,

$$\mathbf{p}_{\|} = \hbar\mathbf{k}_{\|} \quad \rightarrow \quad \hbar k_{\|} = \sqrt{2m\epsilon_{\mathrm{kin}}} \sin\theta, \qquad (4.13)$$

where $\mathbf{k}_{\|}$ is the component of the crystal momentum parallel to the surface in the extended zone scheme. As for the normal component of the crystal momentum, it is *a priori* not known since the translation symmetry is broken along the surface normal. This difficulty is avoided in 2D systems,

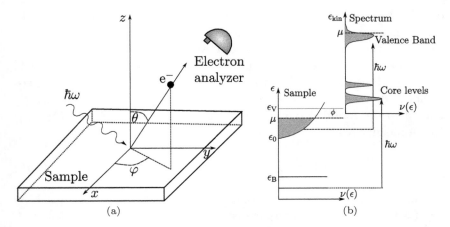

Fig. 4.5 (a) Schematics of the ARPES experiment, where photoelectrons emitted from a sample by incident photons with energy $\hbar\omega$ are collected by the electron analyzer. The latter determines not only the kinetic energy, but also the angle distribution of electrons. (b) Energetics of the photoemission process, where the left and right sides correspond to the electron energy distribution function in terms of the binding ϵ_B and kinetic ϵ_{kin} energies, respectively. Here ϵ_0 denotes the bottom of the valence band, ϵ_V denotes the energy of vacuum, and ϕ is the work function.

where the mapping of the band energy is trivially obtained just from $\mathbf{p}_\parallel$. Indeed, the ARPES played a crucial role in the experimental identification of surface states in topological insulators [97–99].

In 3D systems, the corresponding mapping can be done by using certain assumptions about the electron dispersion in the final state. For example, the electron energy in the free electron model equals [568]

$$\epsilon = \epsilon_{kin} + \phi = \frac{\hbar^2 (\mathbf{k}_\parallel^2 + \mathbf{k}_\perp^2)}{2m} - |\epsilon_0|. \qquad (4.14)$$

Here ϵ_0 denotes the position of the valence band bottom measured from the Fermi level (see Fig. 4.5(b)). Then, by using Eq. (4.13), the normal component of the electron momentum is given by

$$\hbar k_\perp = \sqrt{2m \left(\epsilon_{kin} \cos^2 \theta + V_0\right)}, \qquad (4.15)$$

where $V_0 = \phi + |\epsilon_0|$ is the inner potential, which corresponds to the energy of the bottom of the valence band with respect to vacuum. The inner potential can be obtained by comparing numerical calculations and experimental data or by using the periodicity of the Brillouin zone. In the latter case, one should measure the kinetic energy of photoelectrons emitted along the surface normal. It is worth noting also that the ARPES can be generalized to include the measurements of the electron spin.

A more formal description of photoemission can be found in [460–464]. The key observation is that the intensity measured in ARPES experiments is determined by the spectral function $A(\omega, \mathbf{k})$, which is directly related to the retarded (advanced) Green's function $G^{\mathrm{R}}(\omega, \mathbf{k})$ $(G^{\mathrm{A}}(\omega, \mathbf{k}))$ (see Appendix A.1),

$$A(\omega, \mathbf{k}) = \frac{1}{2\pi} \left[G^{\mathrm{R}}(\omega, \mathbf{k}) - G^{\mathrm{A}}(\omega, \mathbf{k}) \right]_{\mu=0}. \tag{4.16}$$

The momentum resolved density of states (DOS) is then given by $\nu(\omega, \mathbf{k}) = \mathrm{tr}\left[A(\omega, \mathbf{k})\right]$. Its spin counterpart equals $\nu_i(\omega, \mathbf{k}) = \mathrm{tr}\left[\sigma_i A(\omega, \mathbf{k})\right]$ with $i = x, y, z$.

Measurements of the DOS are fundamental for the STM/STS technique [466–468]. Technically, the STM is realized by placing a metal tip above the material surface and measuring the tunneling current as a function of applied voltage. This allows one to build a spatial map of the local density of states (LDOS). On the other hand, the STS probes the LDOS as a function of energy by changing applied voltage. The Fourier transform of the STS map is heuristically approximated by the *joint DOS* (JDOS) and the *spin-selective scattering probability* (SSP) (see, e.g., [569–571]),

$$\mathrm{JDOS}(\omega, \mathbf{q}) \propto \int \frac{d^3k}{(2\pi)^3} \nu_0\left(\omega, \mathbf{k} + \mathbf{q}\right) \nu_0\left(\omega, \mathbf{k}\right), \tag{4.17}$$

$$\mathrm{SSP}(\omega, \mathbf{q}) \propto \frac{1}{2} \sum_{i=x,y,z} \int \frac{d^3k}{(2\pi)^3} \nu_i\left(\omega, \mathbf{k} + \mathbf{q}\right) \nu_i\left(\omega, \mathbf{k}\right). \tag{4.18}$$

The STM/STS studies allow one to investigate *quasiparticle interference* (QPI) patterns that appear due to the interference of incident and reflected waves for surface defects or impurities. Combined with theoretical calculations, QPI measurements provide information on the surface states, including the band structure as well as possible scattering channels. This makes QPI patterns a useful tool in the study of various topological materials. For example, the QPI results in topological insulators demonstrate [97–99] that the backscattering of surface quasiparticles is forbidden for nonmagnetic disorder in agreement with the spin-momentum locking of the conductive electron states. As will be discussed below, the QPI is also useful in the study of Weyl and Dirac semimetals, where it can provide information on the spin structure of surface states. Unlike the ARPES, STM/STS measurements are indirect and should be always supplemented with theoretical calculations.

It should be mentioned that the description of the QPI in terms of the JDOS and/or SSP is not always correct. Indeed, the impurity landscape in the SSP is effectively replaced with a single scattering center. In addition, there might be additional issues related to the use of a simplified approach in multiband systems. A more precise description of the QPI is given by the following Fourier transformed LDOS difference [571, 572]:

$$\Delta\nu(\omega, \mathbf{q}) = \frac{i}{2\pi}\text{tr}\left[Q(\omega, \mathbf{q}) - Q^*(\omega, -\mathbf{q})\right], \tag{4.19}$$

$$Q(\omega, \mathbf{q}) = -\int \frac{d^3k}{(2\pi)^3}\text{tr}\left[G(\omega, \mathbf{k} + \mathbf{q})T(\omega, \mathbf{k} + \mathbf{q}, \mathbf{k})G(\omega, \mathbf{k})\right], \tag{4.20}$$

where $G(\omega, \mathbf{k}+\mathbf{q})$ is Green's function and $T(\omega, \mathbf{k}+\mathbf{q}, \mathbf{k})$ is the T-matrix connected with the disorder potential [235]. For example, for a single impurity, one has

$$T(\omega, \mathbf{k}_1, \mathbf{k}_2) = \delta_{\mathbf{k}_1, \mathbf{k}_2}\left[1 - U\int \frac{d^3k}{(2\pi)^3}(-i)G(\omega, \mathbf{k})\right]^{-1}U, \tag{4.21}$$

where U is the impurity potential (see also Sec. 7.1.5). By comparing the results obtained from Eq. (4.19) and the JDOS (4.17), the qualitative difference in the QPI patterns was found in [572].

4.3.2 *Dirac semimetals*

Let us begin with the review of experimental data for Dirac semimetals obtained by the ARPES methods. The bulk and surface dispersion relations of the Dirac semimetals Na_3Bi and Cd_3As_2 were reported in [104, 105] and [106–108], respectively. The characteristic ARPES data for the Fermi surface projections on the (001) plane and the bulk dispersions of the Dirac semimetal Na_3Bi are presented in the top and bottom panels of Fig. 4.6, respectively. The ARPES spectrum near the Dirac point in the Dirac semimetal Cd_3As_2 is shown in Fig. 4.7. It is clear that the energy spectrum of quasiparticles is linear in the vicinity of the band touching point and the value of the electric chemical potential measured from the Dirac point is quite large, $\mu \approx 200$ meV.

By comparing the *ab initio* results in Fig. 4.2 with the experimental data in Fig. 4.6, it is clear that there is a good qualitative agreement between

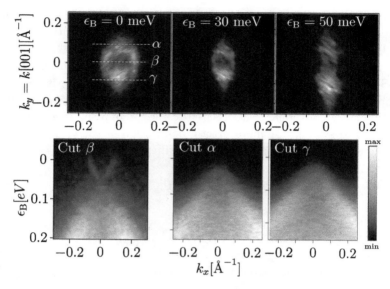

Fig. 4.6   (Top panel) The Fermi surface projections of the Dirac semimetal Na_3Bi on the (001) plane for three binding energies ϵ_B at the photon energy $\hbar\omega = 55$ eV. White dotted lines denote the constant k_y cuts. (Bottom panel) The bulk dispersion for the three cuts identified in the top left panel. (Adapted from [105].)

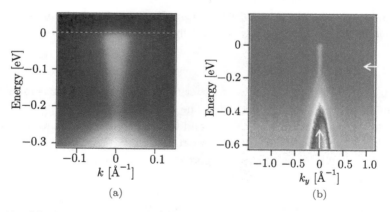

Fig. 4.7   (a) The ARPES data for the Dirac semimetal Cd_3As_2 measured at photon energy $\hbar\omega = 22$ eV and temperature $T = 15$ K along the $(-\pi, -\pi)-(0, 0)-(\pi, \pi)$ momentum space direction. (Adapted from [107].) (b) The ARPES of the Dirac semimetal Cd_3As_2 measured with photon energy $\hbar\omega = 90$ eV and temperature $T = 0.9$ K along the cut going through one of the $k_x = k_z = 0$ points. (Adapted from [108].)

the two. One can easily resolve two arc-like branches of surface states for small binding energies as well as a relativistic-like linear dispersion relation for the bulk electrons in the vicinity of Dirac points. In order to further clarify whether the observed arc-like states correspond to the surface or the bulk, photons of different energies can be used. While the arc-like states remained well-resolved for $\hbar\omega = 10 - 58$ eV, the cone-like states showed a strong dependence on the photon energy, which agrees with their 3D bulk nature [105]. It is worth noting that Na_3Bi and Cd_3As_2 were also studied by using the STM/STS techniques in [570, 573], which confirmed the key qualitative conclusions of the ARPES measurements.

As to the type-II Dirac semimetals, shortly after their theoretical prediction, the ARPES results verified the presence of type-II Dirac nodes in $PtTe_2$ [574], $PtSe_2$ [575], and $PdTe_2$ [576]. It is worth noting, however, that the corresponding Dirac points are buried deeply below the Fermi level. Therefore, additional efforts are needed either to find new material realizations or to move Dirac points closer to the Fermi level. The latter can be possibly achieved by doping.

4.3.3 *Weyl semimetals*

In conjunction with *ab initio* predictions, the ARPES technique played a crucial role in identifying Weyl semimetals. An excellent sensitivity of the high-resolution ARPES to surface phenomena makes it an ideal tool for observing the characteristic Fermi arc surface states in Weyl semimetals. In fact, the ARPES was used to confirm experimentally the realization of a Weyl semimetal phase in the transition metal monopnictides [116, 117, 119–126].

The ARPES data from [119] for the (001) surface of TaAs is presented in Fig. 4.8. It is remarkable that the experimental results are in good qualitative agreement with the *ab initio* calculations, whose representative example is shown in Fig. 4.4. Unfortunately, the experimental resolution is not sufficient to reliably investigate the surface states in the bowtie-like part of the Fermi surface projection. On the other hand, the Fermi arcs in the spoon-like part are well separated. The high-resolution image in Fig. 4.8(a) shows the presence of two crescent-shaped Fermi arcs. Their topologically nontrivial nature is supported by the peculiar shape of these states and the fact that their termination points appear to coincide with the surface projections of the W2 Weyl nodes. The possibility of a closed contour formation is excluded by the signs of surface state velocities, which are evident

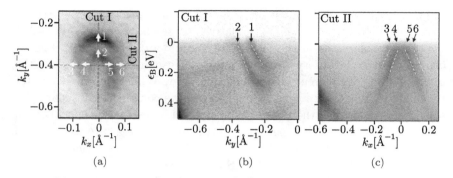

(a)   (b)   (c)

Fig. 4.8   (a) A region of the Fermi surface projected onto the (001) plane showing two
crescent-shaped Fermi arcs in the Weyl semimetal TaAs. The results are obtained for
the binding energy $\epsilon_B = 0$ with the incident photon energy $\hbar\omega = 90$ eV. Blue dashed
lines correspond to the positions of the energy dispersion cuts in panels (b) and (c),
respectively. White arrows with numbers mark the positions of the surface states in
these panels. (b) The dispersion relation of the surface states for the cut I (vertical)
in panel (a). There are two states with the same sign of the Fermi velocity. (c) The
dispersion relation of the surface states for the cut II (horizontal) in panel (a). There
are four distinctive features that correspond to two Fermi arcs entering and leaving the
cut. (Adapted from [119].)

from Figs. 4.8(b) and 4.8(c). In the case of usual surface states, one should
expect different signs or slopes of the states for a closed contour. However, this is clearly not the case and the two velocities have the same sign.
Recall that the presence of two Fermi arcs is required because the effective
topological charge of the Weyl nodes projection onto the surface is ± 2.

In addition to the surface analysis, it is important to establish the presence of Weyl nodes in the bulk energy spectrum. Experimentally, this can
be done via the soft X-ray ARPES. The corresponding results for TaAs [119]
are presented in Fig. 4.9. By choosing appropriate cuts of the Brillouin
zone, both W1 and W2 types of Weyl nodes could be, in principle, investigated. In agreement with the surface data in Fig. 4.8, the W2 Weyl nodes
have dispersion relations linear in k_y and are well separated in momentum space by $\Delta k_y \approx 0.08$ Å^{-1} (see Fig. 4.9(a)). Furthermore, the data
in Fig. 4.9(b) reconfirm a linear dispersion relation in the vicinity of Weyl
nodes. Unfortunately, the resolution of the currently available ARPES data
is insufficient to rigorously show the separation of the W1 Weyl nodes.
Nevertheless, the results in Fig. 4.9(c) still suggest a relativistic-like linear
dispersion relation, which agrees with the *ab initio* predictions. It is worth
noting also that the location of Weyl nodes could be reliably pinpointed

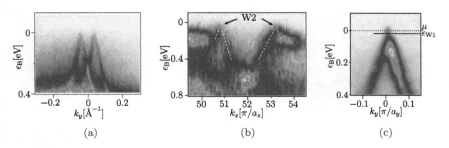

Fig. 4.9   The bulk energy dispersion for the W2 Weyl nodes as a function of (a) k_y and (b) k_z obtained with the incident photon energy $\hbar\omega = 650$ eV. White dashed lines in panel (b) are guides to the eye that show two linearly dispersing Weyl nodes. The bulk dispersion relation for the W1 Weyl node is shown in panel (c). Despite the experimental resolution is not sufficient to resolve the weakly separated W1 Weyl nodes, the relativistic-like linear energy spectrum is clearly evident. (Adapted from [119].)

by requiring that the results obtained in both ultraviolet and soft X-ray ARPES experiments are in good agreement.

The results for all TaAs-like materials are qualitatively similar to those presented in Figs. 4.8 and 4.9. As shown in [116], the key difference between TaAs, TaP, NbAs, and NbP is related primarily to the value of the spin–orbit coupling. The strength of the latter decreases from its largest value in TaAs down to its smallest value in NbP. As expected from the *ab initio* calculations, the separation between the Weyl nodes scales accordingly with the spin–orbit coupling. This fact makes TaAs one of the best materials among the transition metal monopnictides for the observation of the Weyl nodes separation.

Another aspect of the surface states is related to the termination of a crystal. Indeed, as follows from the numerical simulations presented in Fig. 4.4, the anion type of crystal termination (e.g., P or As atoms) could significantly affect the shape of the Fermi surface projection. The corresponding predictions were checked in NbP [559], where the shapes of probed states for the cation type (Nb) termination appeared to be notably different from those of the anion termination in Fig. 4.8. In addition, the possibility of a different connectivity pattern of the Fermi arcs was also suggested. The *in situ* manipulation of the Fermi arcs connectivity via the doping of the surface with potassium was reported in NbAs [577]. Similarly, the surface states of the magnetic Weyl semimetal $Co_3Sn_2S_2$ also strongly depend on the crystal termination [544, 578].

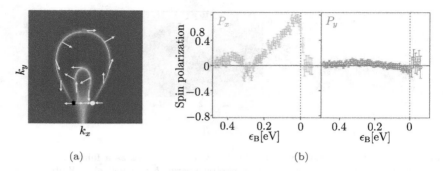

(a)   (b)

Fig. 4.10   (a) Schematic illustration of the spin structure of the crescent-like Fermi arcs connecting the W2 Weyl nodes. The results for the Fermi arcs are obtained numerically. (b) Representative data for the in-plane spin polarization P_x and P_y measured experimentally for the momenta denoted by the red square in panel (a). (Adapted from [126].)

Let us also discuss the spin texture of the Fermi arcs reported in [126, 579]. By using a well-resolved spoon-like part of the surface projection of the Fermi surface, it was shown that the spin on the crescent-like Fermi arcs rotates in the opposite direction along the arc, e.g., counterclockwise for the clockwise movement along the arc. The schematics of the spin polarization and a representative example of spin-resolved experimental data are presented in Figs. 4.10(a) and 4.10(b). As one can see from Fig. 4.10(b), the total spin polarization can reach up to about 80% and lies completely in the surface plane.

Last but not least, we consider the STM/STS measurements in Weyl semimetals. As was discussed at the end of Sec. 4.3.1, this technique allows one to obtain the information on the scattering channels as well as their dependence on the (pseudo)spin degrees of freedom. This is particularly useful for investigating surface states. Theoretically, the use of the QPI patterns, which arise due to the interference of scattered quasiparticles, for identification of the Fermi arcs in Weyl semimetals was proposed in [571, 572, 580]. It was suggested than the inter-arc scattering in Weyl semimetals is much more intense than the intra-arc one. This can be attributed to the quantum interference effects. Almost simultaneously with the theoretical predictions, the experiments were performed in transition metal monopnictides [570, 581–583]. A characteristic example of the STS data for TaAs [581] is shown in Fig. 4.11. From this data three major QPI patterns can be identified: (i) ellipse patterns at G_0 and $G_{\pm Y}$, (ii) half a

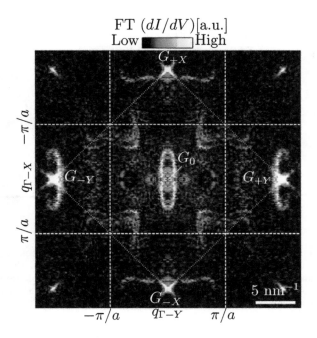

Fig. 4.11 The Fourier transform of the dI/dV map of TaAs with a few As vacancies. The QPI patterns appear on both G_0 and Bragg peaks at $G_{\pm X}$ and $G_{\pm Y}$. (Adapted from [581].)

bowtie patterns at $G_{\pm X}$, and (iii) quarter-square patterns at the intersections of the white dashed lines.

In order to explain the corresponding data, the DFT and SSP calculations of the surface states were performed, with the corresponding results shown in Figs. 4.12(a) and 4.12(b). By comparing Figs. 4.11 and 4.12(b), one can identify the ellipse-like structure with the blue feature in the SSP pattern, stemming from the trivial state transitions. Furthermore, the quarter-square patterns are in good agreement with the green SSP patterns, which also originate from the trivial states, albeit with a large momentum transfer. Interestingly, the intraband yellow SSP feature is invisible in the experimental results. On the other hand, a weak signal related to the contribution from the Fermi arc scattering vector Q_1 in Fig. 4.12(a) can be observed as small hump-like features along the x-axis.

Similar results were also reported for NaP [583]. However, the scattering that arises from the W2 Fermi arcs is suppressed, owing to their low spectral weight and a different orbital character of scattering. It is notable

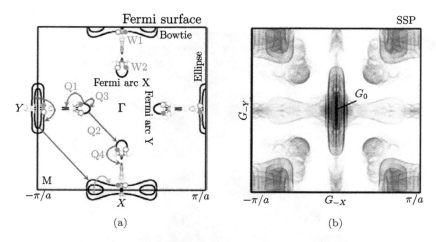

Fig. 4.12   (a) Surface states calculated via the DFT theory with the possible scattering vectors. Here Q_i, $i = 1, \ldots, 4$ indicate scattering vectors related to the Fermi arcs. Blue, green, and yellow vectors are related to the topologically trivial states. (b) The SSP patterns obtained via the DFT theory. Here the colors indicate the contributions related to the specific scattering vectors in panel (a). (Adapted from [581].)

that the QPI measurements give insights not only into the spin-dependent scattering of the Fermi arcs, but also provide important information about the delocalization of the surface states into the bulk, which occurs at the projections of the Weyl nodes onto the surface [581].[1] In addition, unlike the ARPES, the STS measurements are equally suitable for studying Weyl semimetals with broken $\mathcal{T}$ symmetry.

Last but not least, let us briefly mention type-II Weyl semimetals, where Weyl nodes are tilted. The ARPES results show the existence of the type-II Weyl nodes and Fermi arcs in $MoTe_2$ [562, 584–587], $Mo_xW_{1-x}Te_2$ [588, 589], $TaIrTe_4$ [590, 591], $LaAlGe$ [592], and WP_2 [593]. It is worth noting that, in general, it is hard to observe type-II Weyl nodes in $Mo_xW_{1-x}Te_2$ in view of their small separation and the fact that the Fermi level lies below the nodes.[m] The problem can be overcome, however, by using optical pumping [588]. A similar problem also arises in $TaIrTe_4$, although the lengths of its Fermi arcs are much larger [590].

---

[1]The delocalization of the Fermi arcs will be discussed more rigorously in Chapter 5.
[m]Recall that ARPES maps only the filled electron states.

Chapter 5

Surface Fermi Arcs

To them, I said, the truth would be literally nothing but the shadows of the images.

Plato

All crystals encountered in nature are, of course, finite. It is of interest, therefore, to discuss how an abrupt change of the periodic atomic lattice near the surface of a crystal affects an electron band structure. In this connection, one should note that there exist global bulk effects due to a finite size as well as local effects at the surface itself. First of all, in finite samples, the quasiparticle energy spectra become discrete. Since the distance between energy levels quickly decreases with the sample size, the corresponding effects are usually negligible for sufficiently large crystals, but may become important in thin films.

The other important effect is connected with the lattice potential near the crystal boundaries, which differs from that in the bulk. In particular, the modified boundary potential can give rise to additional states localized at the surfaces. Their energies usually lie in the bulk energy gap that allows for a clear separation of these states from their bulk counterparts. Since the surface states are very interesting from both theoretical viewpoint and practical applications, their formation and key properties in crystals were already extensively studied in the 20th century (see, e.g., [594]). In this chapter, we will discuss primarily the nontrivial properties of surface states in Weyl and Dirac semimetals as well as compare them with surface states in other materials.

As we will explain in detail below, the most prominent distinction between the surface states in Weyl and certain Dirac semimetals and those in other materials is related to their nontrivial topological properties.

These topological states connect the projections of the bulk Weyl nodes of opposite chiralities onto the spatial surfaces. Geometrically, they are open segments of the Fermi surface in the surface Brillouin zone and, therefore, are called the *surface Fermi arcs*.[a] This unique feature distinguishes the surface Fermi arc states in Weyl semimetals from the conventional surface states in other solids. Indeed, the latter usually form closed curves in the surface Brillouin zone. The unique shape of the Fermi arcs makes them one of the hallmark features in the experimental verification of Dirac and Weyl semimetals especially by using the ARPES techniques. Allegorically speaking, the observation of the surface Fermi arcs in the ARPES data could be seen as the master's signature on a painting showing the portrait of Weyl and certain Dirac semimetals (see Sec. 4.3).

The notable features of the surface Fermi arc states are their 1D chiral character and a linear dispersion relation. By using the analogy with the quantum Hall effect (QHE), one might argue that such states should be immune to the effects of disorder and, therefore, have a profound effect on the transport properties. However, because of a subtle interplay between the surface and bulk states in Dirac and Weyl semimetals, the corresponding properties of the Fermi arc states appear to be quite different from those of the edge states in QHE. For example, they are affected by disorder and produce dissipative transport. The effect of an external magnetic field on the Fermi arcs is also interesting, resulting in a novel type of closed *surface-bulk orbits*, in which the Fermi arcs on the opposite crystal surfaces are connected via the bulk states.

5.1   Tamm–Shockley states

Historically, the presence of electron states bound to the surface was established in the seminal work of Igor Tamm in 1932 [595]. He considered the case of a semi-infinite crystal with a deformed potential in its surface layer.

To get a clear perspective and for pedagogical reasons, we start by briefly reviewing the underlying physics of conventional surface states. For our purposes, it is sufficient to consider the simplest Kronig–Penney model of a 1D crystal. The corresponding Schrödinger equation reads

$$\frac{\hbar^2}{2m}\frac{d^2}{dy^2}\psi(y) + [\epsilon - V(y)]\,\psi(y) = 0, \tag{5.1}$$

---

[a]The surface Fermi arc states in Weyl semimetals should not be confused with the Fermi arcs in high-T_c superconductors, which are bulk states.

where m is the electron mass, $\psi(y)$ is the wave function, ϵ is the energy, and $V(y)$ is the potential energy. Assuming an infinite crystal, we use the Kronig–Penney model with the following periodic potential:

$$V(y) = aV_1 \sum_{n=-\infty}^{\infty} \delta(y - na), \tag{5.2}$$

where a is the lattice spacing and V_1 is a constant that determines the strength of the potential.

According to the Bloch theorem [470], a solution to the Schrödinger equation (5.1) with the periodic potential (5.2) should take the following form:

$$\psi(y) = A_+ u_k(y)e^{iky} + A_- u_{-k}(y)e^{-iky}, \tag{5.3}$$

which will be also suitable for solving the problem with boundaries. Here $u_k(y)$ and $u_{-k}(y)$ are periodic functions, $A_\pm$ are some constants, and k is the wave vector. For an infinite crystal, the normalizability of the wave function requires that k should be real. It is straightforward to find (see, e.g., [594, 596, 597]) that the energy spectrum for potential (5.2) is determined by

$$\cos(ka) = \frac{ma^2 V_1}{\hbar^2} \frac{\sin(\xi a)}{\xi a} + \cos(\xi a), \tag{5.4}$$

where $\xi = \sqrt{2m\epsilon}/\hbar$. Real solutions for the wave vector exist only when the absolute value of the expression on the right-hand side of Eq. (5.4) is less than 1. As is easy to check, this condition restricts the energies to lie within a specific set of allowed energy bands. The formation of such energy bands, separated by finite gaps, is the consequence of the periodicity of the electron potential. Note that there are no real solutions for k in the forbidden energy bands (i.e., within the energy band gaps). This is consistent with the fact that no normalizable solutions exist for those energies.

The situation drastically changes in semi-infinite crystals. In this case, the electron potential reads

$$U(y) = \theta(-y)V_0 + \theta(y)aV_1 \sum_{n=-\infty}^{\infty} \delta(y - na), \tag{5.5}$$

where $\theta(y)$ is the Heaviside step function. Here it is assumed that vacuum with a constant potential energy V_0 is located at $y < 0$ and the crystal occupies the half-space at $y > 0$. The same ansatz as in Eq. (5.3) can be used for the solution at $y > 0$. In the vacuum region, on the other hand, the solution takes the form

$$\psi(y < 0) = Ce^{\xi_0 y}, \tag{5.6}$$

where $\xi_0 = \sqrt{2m(V_0 - \epsilon)}/\hbar$. When studying the quasiparticle spectrum in the crystal, it is sufficient to consider only the solutions with energies less than V_0, which cannot propagate in vacuum. By solving the Schrödinger equation at $y > 0$ as before, and then matching the wave functions and their first derivatives at the boundary $y = 0$, one finds that the spectrum contains the same allowed energy bands as in the infinite crystal. In addition, however, there are solutions that correspond to complex wave vectors $k = i\kappa + n\pi/a$ with $\kappa > 0$ and $n = 0, 1, 2, \dots$. These solutions are normalizable since they are given by Eq. (5.3) with $A_- = 0$. Their wave functions are peaked near the boundary $y = 0$ and have a damped oscillatory behavior in the bulk as a function of y. The corresponding relative probability density $|\psi(y)|^2/|\psi(0)|^2$ is schematically depicted in Fig. 5.1(a).

By matching the wave functions at the boundary and using Eq. (5.4), one obtains the following relation:

$$\xi a \cot(\xi a) = \frac{V_0}{V_1} - \xi_0 a, \tag{5.7}$$

which determines the surface state energies. In the model at hand, the condition for the existence of the surface solutions is given by the inequality

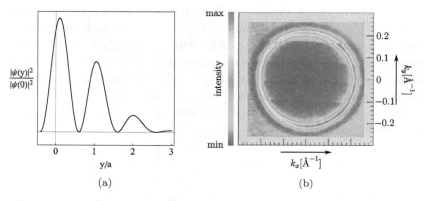

Fig. 5.1 (a) Schematic relative probability density for a Tamm surface state in the first forbidden energy gap. (b) ARPES spectrum of the surface states for copper. Here the surface normal is along the z-axis. (Adapted from [598].)

$V_1 > \hbar^2 \xi_0/(am)$, which follows from the assumption $\kappa > 0$. If the value of V_0 is sufficiently large, a unique real solution $\xi = \xi_n$ exists in each interval between $n\pi/a$ and $(n+1)\pi/a$ where n is a positive integer.[b] As for the lowest energy surface solution with $\xi_0 < \pi/a$, it exists only if the following additional condition is satisfied: $V_0/V_1 - \sqrt{2mV_0}a/\hbar < 1$.

Formally, the dispersion relation in Eq. (5.4) remains valid even for the surface states, albeit with complex wave vectors. This implies, therefore, that the energies of such states lie in the forbidden bands. While here we discussed only a simplified 1D model, the same qualitative conclusion can be also reached in the case of 3D crystals and more realistic potentials (see, e.g., [594, 596]).

Another important step in the development of the surface states theory was made by William Shockley in 1939 [599]. Unlike Tamm, who explicitly distorted the potential of the surface lattice cell, Shockley assumed that the crystal potential remains periodic through the entire crystal, including the outermost cells. Historically, these states were called the Shockley states and were thought to be different from the Tamm surface states. By omitting the long and vibrant history of their development,[c] we should mention that there is no real physical distinction between these states, especially for non-ideal surfaces. Therefore, in the modern literature, the surface states are often called the *Tamm–Shockley states*.

It is instructive to overview briefly the effects of the surface states on the physical properties of materials as well as their observational signatures. As is clear, the presence of the surface states could affect substantially transport properties (e.g., electrical and thermal conductivities). Also, the surface states could significantly influence the reflection of light. In semiconductors, the modification of the energy bands near the surface could be utilized for designing various heterojunctions, where surface states provide barrier layers. In addition, the surface states may change chemical properties of crystals, which could be of great value, e.g., for the surface catalysis of chemical reactions.

Experimentally, in addition to various indirect transport studies (the so-called field-induced experiments), one of the most straightforward ways

---

[b]As is clear from Eq. (5.6), the necessary condition for the state to be localized on the surface is $\epsilon < V_0$. Therefore, there exists a maximum value of integer n above which the solutions ξ_n become part of the continuum spectrum of delocalized states in vacuum.
[c]For an in-depth discussion, the interested reader is referred to the book by Davison and Stęślicka [594].

to study the surface states in crystals is the angle resolved photoemission spectroscopy and the scanning tunneling spectroscopy discussed in Sec. 4.3.1. In particular, the ARPES technique measures the distribution of electrons in momentum space and, therefore, provides a detailed information on the dispersion relation of the surface states. As an example, the ARPES spectrum for copper is shown in Fig. 5.1(b). Note that the surface states for copper form almost an exact circle. As will be argued in Sec. 5.2 and then shown explicitly in Sec. 5.3, this is in drastic contrast with the surface states in Weyl and certain Dirac semimetals, where the surface states have the shape of open segments or arcs and are characterized by a nontrivial topology.

5.2   Fermi arcs and bulk-boundary correspondence

5.2.1   *Topological protection of Fermi arcs*

As has been already discussed in Chapter 3, one of the most profound distinction of Weyl semimetals from conventional metals or semiconductors is the nontrivial topology of their electron states. The same applies also to the surface states. Therefore, instead of using the conventional paradigm outlined in Sec. 5.1, where the surface states come from the change of the crystal potential, we will approach the problem of surface states in the topological materials from a different direction. Let us recall that the Weyl nodes can be viewed as the monopoles of the Berry curvature whose topological charge is directly related to their chirality. As one might anticipate, the abrupt change of the topological characteristics at a surface of a Weyl semimetal has profound implications. Indeed, simple arguments of continuity suggest that the Berry curvature flux cannot simply disappear at the semimetal's boundary. There should exist some states that "shunt" the Weyl nodes. These states are the *surface Fermi arcs*, which were first discussed in [102]. It is important to emphasize that the existence as well as the properties of the Fermi arcs are determined primarily by the chiral nature of bulk states. Thus, the corresponding relation between bulk and surface states is known as the *bulk-boundary correspondence*. As will be explicitly shown below, unlike the Tamm–Shockley states, the Fermi arcs are open segments of the Fermi surface connecting the projections of the bulk Weyl nodes onto the surface Brillouin zone [102, 103]. On the other hand, their shape and connectivity, i.e., the way how different Weyl nodes are connected by the Fermi arcs, depend on the surface properties.

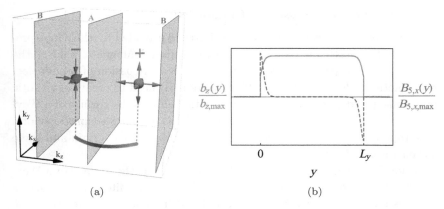

Fig. 5.2   (a) Schematic illustration of the Weyl nodes of opposite chiralities in momentum space. Red and blue spheres correspond to the left- and right-handed nodes. Arrows
denote the Berry curvature field. The surface of the semimetal is in the $x-z$ plane. The
topological charge, which is quantified by the Berry curvature flow, is unity for the green
plane A and equals zero for the gray planes B. (b) Schematic profile of the chiral shift
b_z (red solid line) and the pseudomagnetic field $B_{5,x}$ (blue dashed line) in a slab of a
Weyl semimetal. Here L_y is the width of the slab.

The change of the Fermi arc pattern on the surfaces with different terminations was demonstrated both theoretically [115, 600] and experimentally [544, 577, 578].

In order to illustrate the idea of the bulk-boundary correspondence, let
us consider a simple example of a Weyl semimetal with broken $\mathcal{T}$ symmetry and a single pair of Weyl nodes. The corresponding schematic illustration is shown in Fig. 5.2(a), where the red and blue spheres denote the
right- and left-handed Weyl nodes and the arrows show the Berry curvature
field. As is clear, the total flux of the Berry curvature through a surface
in momentum space placed between the Weyl nodes, e.g., the green plane
A in Fig. 5.2(a), is nonzero and is characterized by the unit topological
(Chern) charge. Such a plane can be viewed as a 2D Chern insulator with
chiral edge state [97, 98]. Since the total Berry curvature flux vanishes, the
topological charge for planes placed outside the Weyl nodes is zero (see,
e.g., the gray planes B in Fig. 5.2(a)). Therefore, each such plane corresponds to a usual (nontopological) 2D insulator, which has no edge states.
By stacking all 2D planes together, one finds that the chiral edge states
form a composite surface state only between the Weyl nodes. This is the
Fermi arc that connects the Weyl nodes of opposite chirality.

5.2.2   *Fermi arcs as pseudo-Landau levels*

Another useful perspective on the Fermi arcs is provided by their interpretation as the zeroth Landau level states in a pseudomagnetic field at the surface [413, 601] (see also Sec. 3.7 for a detailed discussion of pseudoelectromagnetic fields). Strictly speaking, there is no need to create any exotic conditions in order to realize pseudomagnetic fields on the surface of a Weyl semimetal. These fields appear automatically as an inherent property of a finite topological material. Below we address briefly their origin and some of their implications.

A nonvanishing pseudomagnetic field localized at the surface of a Weyl semimetal is related to an abrupt change of the chiral shift $\mathbf{b}$ at the surface of a crystal. Indeed, for a finite crystal of width L_y in the y-direction, the local chiral shift can be modeled as $\mathbf{b}(y) \approx [\theta(y) - \theta(y - L_y)]\mathbf{b}$. Here, for the sake of concreteness, it is assumed that the chiral shift in the bulk is parallel to the surfaces, $\mathbf{b} = b_z\hat{\mathbf{z}}$. Then, by making use of the definition $\mathbf{B}_5 = [\boldsymbol{\nabla} \times \mathbf{A}^5] = -\hbar c [\boldsymbol{\nabla} \times \mathbf{b}(y)]/e$ (see Sec. 3.7), one finds that the x component of the pseudomagnetic field is nonzero and localized mostly on the semimetal surface, i.e.,

$$B_{5,x} \simeq \frac{\hbar c b_z}{e} [\delta(y) - \delta(y - L_y)]. \qquad (5.8)$$

The profiles of the pseudomagnetic field and the chiral shift are schematically shown in Fig. 5.2(b). The pseudomagnetic field also leads to the formation of pseudo-Landau levels.[d] Most interestingly, the energy of the zeroth Landau level states is given by $\epsilon_0(k_x) = \text{sgn}(B_{5,x}) \hbar v_F k_x$, which reproduces exactly the dispersion relation of the Fermi arcs given in Eq. (5.17). It was also verified numerically that the wave functions of these pseudo-Landau level states are localized on the surfaces and quickly decrease as a function of the distance into the bulk of semimetal. This reconfirms an alternative interpretation of the Fermi arcs as the zeroth pseudo-Landau level states.

By making use of the above interpretation in terms of the lowest pseudo-Landau level, it is instructive to discuss the degeneracy of the Fermi arcs. Let us start by noting that, for a given surface (e.g., at $y = 0$) and a fixed value of k_x, the Dirac equation with a nonuniform pseudomagnetic field

---

[d]The pseudo-Landau levels are similar to the usual Landau levels discussed in Sec. 6.4.1.

$\mathbf{B}_5 = B_{5,x}(y)\hat{\mathbf{x}}$ can be viewed as a 2D problem (in the $y - z$ plane) with a parity-odd mass $m = \hbar v_F k_x$. Therefore, for each chiral sector $\lambda = \pm$, the energy of the zeroth Landau level state with $\epsilon_0(k_x) = \hbar v_F k_x$ can be determined by using the Aharonov and Casher approach [602]. The result of the analysis agrees with the Atiyah–Singer index theorem and, therefore, is robust with respect to a specific choice of a pseudomagnetic field profile. The degeneracy of the corresponding zeroth level can be also calculated, i.e.,

$$N_{\text{arc}} = 2\left[\frac{\Phi}{\Phi_0}\right] = 2\left[\frac{b_z L_z}{2\hbar\pi}\right], \tag{5.9}$$

where $\Phi = \int d\mathbf{S}_{yz} \cdot \mathbf{B}_5 = c b_z L_z/e$ is the pseudomagnetic field flux through the $y - z$ plane, $\Phi_0 = 2\pi\hbar c/e$ is the flux quantum, the overall factor of 2 accounts for two chiralities, and $[\ldots]$ denotes an integer part. It should be emphasized that, in this derivation, we assume that the top and bottom surfaces are decoupled. As suggested in Fig. 5.2(b), this is justified since the pseudomagnetic field is strongly localized near the boundaries. This allows us to consider each boundary separately where the 2D surface $\mathbf{S}_{yz}$ encompass only one of them.[e] It is useful to mention also that the result for the degeneracy is consistent with the direct calculation of the arc length in momentum space, i.e., $N_{\text{arc}} = L_z \int_{-b_z}^{b_z} dk_z/(2\pi\hbar)$.

The Fermi arcs play an important role in the anomalous transport in background electromagnetic and pseudo-electromagnetic fields [601]. They are crucial for explaining an apparent nonconservation of the bulk electric charge in the covariant picture and for providing a bridge between the covariant and consistent anomaly frameworks. A detailed discussion of the corresponding anomalous transport will be given in Sec. 6.6.2. Among the other interesting effects is the unusual cubic scaling of conductivity with the crystal width predicted in [603] in the ultraquantum regime.

Before concluding this section, we note that the interpretation of the Fermi arcs as the zeroth pseudo-Landau levels reinforces their topological origin. It is the power of topology that, in turn, explains the existence and robustness of such states in real materials. The shape and many other specific properties of the surface Fermi arc states depend on specific models of Weyl and Dirac semimetals. A few of them will be discussed below.

---

[e]The net flux of the pseudomagnetic field through the cross-section of the whole sample is zero.

5.3 Minimal model of Fermi arcs

In order to clarify the concept of the Fermi arcs, let us derive them analytically by employing a simple continuum model of a Weyl semimetal with broken $\mathcal{T}$ symmetry and two Weyl nodes (see, e.g., [604]). In essence, such a model contains half of the degrees of freedom of the Dirac semimetal model presented in Sec. 3.6.1. Its momentum space, the corresponding Hamiltonian reads

$$H(\mathbf{k}) = \gamma \left(k_z^2 - m\right) \sigma_z + \hbar v_F \left(k_x \sigma_x + k_y \sigma_y\right), \qquad (5.10)$$

where m and γ are positive constants, v_F is the Fermi velocity, σ is the vector of the Pauli matrices acting in the (pseudo)spin space, and $\mathbf{k}$ is the momentum. As is easy to check, the bulk energy spectrum of the model is given by

$$\epsilon_{\mathbf{k}} = \pm\sqrt{\gamma^2(k_z^2 - m)^2 + \hbar^2 v_F^2(k_x^2 + k_y^2)}. \qquad (5.11)$$

The two Weyl nodes of opposite chiralities are located at $\mathbf{k}_0^{\pm} = (0, 0, \pm\sqrt{m})$. The corresponding bulk wave functions are given by

$$\psi = N e^{i\mathbf{k}\cdot\mathbf{r}} \begin{pmatrix} 1 \\ \frac{\hbar v_F k_+}{\gamma(k_z^2 - m) + \epsilon_{\mathbf{k}}} \end{pmatrix}, \qquad (5.12)$$

where N is the normalization factor and $k_{\pm} = k_x \pm i k_y$. The bulk energy spectrum at $k_y = 0$ is shown in Fig. 5.3(a), where the presence of two Weyl nodes is evident.

Now let us consider the surface states in a semi-infinite semimetal. For concreteness, we assume that the semimetal is in the upper half-plane $y > 0$ and vacuum is in the lower one $y < 0$. The energy spectrum in the vacuum region is gapped and the value of the gap is sufficiently large to prevent electrons from escaping from the semimetal. For simplicity, vacuum is modeled by the same type Hamiltonian as in Eq. (5.10), but with m replaced by $-\tilde{m}$, where $\tilde{m} \to \infty$. As is obvious from Eq. (5.11), parameter $\tilde{m}$ defines the energy gap in vacuum. Therefore, an infinitely large value of $\tilde{m}$ confines the bulk quasiparticle to the semimetal.

By noting that the boundary breaks the translation invariance in the y direction, it is necessary to use the operator form $k_y = -i\partial_y$ in Hamiltonian (5.10). Then, by solving the eigenvalue problem $H\psi = \epsilon_s\psi$ in the two distinct regions, the following wave function on the semimetal side is

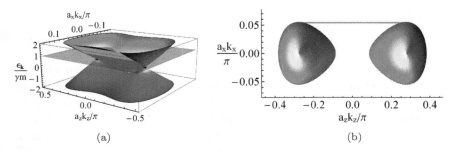

Fig. 5.3 (a) The bulk (5.11) and Fermi arc (5.17) (green plane) states for the effective model of Weyl semimetal (5.10). For bulk states, $k_y = 0$. The brown plane represents a fixed Fermi energy, $\epsilon_\mathbf{k} = 0.9\,\gamma m$. (b) The bulk sheets of the Fermi surface connected by the Fermi arc at $\epsilon_\mathbf{k} = 0.9\,\gamma m$. The numerical values of model parameters are given in Eq. (5.21).

obtained:

$$\psi_{y>0} = N_1 e^{ik_x x + ik_z z - \kappa y} \begin{pmatrix} 1 \\ \frac{\hbar v_F (k_x - \kappa)}{\gamma(k_z^2 - m) + \epsilon_s} \end{pmatrix}, \tag{5.13}$$

where

$$\kappa^2 = k_x^2 + \frac{\gamma^2 \left(k_z^2 - m\right)^2 - \epsilon_s^2}{\hbar^2 v_F^2}. \tag{5.14}$$

Similarly, on the vacuum side, one has

$$\psi_{y<0} = N_2 e^{ik_x x + ik_z z + \tilde{\kappa} y} \begin{pmatrix} 1 \\ \frac{\hbar v_F (k_x + \tilde{\kappa})}{\gamma(k_z^2 + \tilde{m}) + \epsilon_s} \end{pmatrix} \approx N_2 e^{ik_x x + ik_z z + \gamma \tilde{m} y} \begin{pmatrix} 1 \\ 1 \end{pmatrix}, \tag{5.15}$$

where

$$\tilde{\kappa}^2 = k_x^2 + \frac{\gamma^2 \left(k_z^2 + \tilde{m}\right)^2 - \epsilon_s^2}{\hbar^2 v_F^2} \approx \frac{\gamma^2 \tilde{m}^2}{\hbar^2 v_F^2}. \tag{5.16}$$

By matching the wave functions (5.13) and (5.15) at the boundary $y = 0$, it is straightforward to find that $N_1 = N_2$ and the energy spectrum for the surface states is given by

$$\epsilon_s = \hbar v_F k_x. \tag{5.17}$$

Note that the quasiparticle velocity equals $\mathbf{v}_s = \partial \epsilon_s / (\hbar \partial \mathbf{k}) = v_F \hat{\mathbf{x}}$. It is clear that the surface quasiparticle with the energy dispersion relation in Eq. (5.17) could be described by the following effective surface

Hamiltonian[f]:

$$H_{\text{arcs}} = \hbar v_F k_x. \tag{5.18}$$

It should be noted that the above effective Hamiltonian is incomplete and does not contain information about the Fermi arc length. Moreover, as we will discuss in Sec. 5.5, the effective Hamiltonian for the surface Fermi arcs should be used with great caution since, unlike topological insulators, the surface states in Weyl and Dirac semimetals coexist with bulk states. From the wave function matching and Eq. (5.14), one derives the following relation:

$$\frac{\hbar v_F k_x - \sqrt{\hbar^2 v_F^2 k_x^2 + \gamma^2 \left(k_z^2 - m\right)^2 - \epsilon_s^2}}{\gamma \left(k_z^2 - m\right) + \epsilon_s} = 1, \tag{5.19}$$

which defines the shape of the Fermi arcs in the surface Brillouin zone at fixed energy ϵ_s.

It is easy to see that Hamiltonian (5.18) describes 1D chiral electron states. Indeed, the quasiparticle velocity $\mathbf{v}_s$ is positive for all states and the Hamiltonian does not depend on k_z at all. It is important to note that the energy spectrum (5.17) and the surface Hamiltonian (5.18) are valid only for $k_z^2 < m$. Otherwise, the surface state (5.13) is not localized on the surface.

The overall constant N_1 can be straightforwardly obtained from the normalization condition $\int_{-\infty}^{0} dy\, \psi_{y<0}^{\dagger} \psi_{y<0} + \int_{0}^{\infty} dy\, \psi_{y>0}^{\dagger} \psi_{y>0} = 1$. It is given by $N_1 = \sqrt{\kappa}$, where $\kappa = \gamma(m - k_z^2)/(\hbar v_F) > 0$. Thus, the surface state wave function (5.13) reads

$$\psi_{y>0} = \sqrt{\kappa}\, e^{ik_x x + ik_z z - \kappa y} \begin{pmatrix} 1 \\ 1 \end{pmatrix}. \tag{5.20}$$

It is important to note that the dispersion relation for the Fermi arcs in Eq. (5.17) has the opposite overall sign if the semimetal occupies the lower half-space and vacuum is in the upper half-space. Therefore, in a sufficiently thick slab-shaped crystal, such that the states on different surfaces can be considered as independent, the surface energies and quasiparticle velocities have opposite signs on top and bottom surfaces. As will be shown in Sec. 5.6, this property plays a crucial role in the formation of the surface–bulk orbits in a magnetic field.

---

[f]For a general approach to derive the effective surface Hamiltonian, see, e.g., [101].

The bulk energy spectrum (5.11) of Hamiltonian (5.10) as well as the Fermi arc states with the dispersion relation (5.17) are schematically shown in Fig. 5.3. To present the results, the following numerical parameters reminiscent to those in Na_3Bi [110] (see Eq. (4.11)) were used

$$m = -M_0/M_1 = 0.0082 \, \text{Å}^{-2}, \quad \gamma = -M_1 = 10.6424 \, \text{eV} \cdot \text{Å}^2,$$

$$\hbar v_F = A = 2.4598 \, \text{eV} \cdot \text{Å}, \quad a_x = 5.448 \, \text{Å}, \quad a_z = 9.655 \, \text{Å}. \quad (5.21)$$

Here a_x and a_z are the lattice constants along the x- and z-directions, respectively. Note that, at constant energy, the Fermi arc surface states are attached exactly in the same way as predicted in [103]. In particular, these states are straight lines in the surface Brillouin zone and they are tangential to the projections of the bulk Fermi surfaces. It is important that the Fermi arcs connect separate sheets of the bulk Fermi surfaces associated with opposite chirality quasiparticles. Therefore, as argued in [103], the arcs are crucial for sustaining a common equilibrium chemical potential at the opposite chirality Weyl nodes that formally appear to be disconnected.

It is worth noting that, the Fermi arcs in real materials are usually not straight in surface Brillouin zone. More often than not, they have a nonzero curvature (see, e.g., the theoretical analysis in a low-energy effective model in [605]) and could even be spiraling in the vicinity of Weyl nodes [606]. The latter behavior is argued to be the result of a band bending near the crystal surface. As is clear, additional deviations from the simplified model considerations above occur when the crystals are sufficiently small (or thin) and the energy spectrum discretization becomes nonnegligible. The Fermi arcs in such a regime were studied theoretically in [607–612].

5.4 Fermi arcs in Dirac semimetals

As was discussed in Sec. 3.6.2, the Dirac materials A_3Bi ($A = $ Na,K,Rb) are, in fact, $\mathbb{Z}_2$ Weyl semimetals. Therefore, for each of the $\mathbb{Z}_2$ sectors, the presence of the Fermi arcs could be demonstrated by using the topological arguments given in Sec. 5.2. It should be noted, however, that, unlike the stability of the bulk Dirac nodes, the topological protection of the surface Fermi arcs in Dirac semimetals is not absolute. Indeed, according to [613], one can add some additional terms to the effective Hamiltonian that respect the initial symmetries and preserve the Dirac nodes in the bulk but merge pairs of Fermi arcs into closed ring-like structures. Such terms, however, are of higher order in the wave vector and are absent in the model derived on the basis of the *ab initio* calculations presented in [110, 111]. Therefore, in what follows, they will be ignored.

The complete low-energy model of A_3Bi (A = Na,K,Rb) is established in Sec. 3.6.1. For illustrative purposes, however, it is reasonable to consider only the simplified case where the momentum-dependent gap term in Eq. (3.90) vanishes $B(\mathbf{k}) = 0$. (The results for $B(\mathbf{k}) \neq 0$ can be obtained in the same way, albeit are much bulkier [605].) Then the low-energy Hamiltonian (3.89) in Sec. 3.6.1 is block diagonal and each of its blocks can be considered independently. By performing a unitary transformation with matrix $U_y = (\mathbb{1}_2 + i\sigma_y)/\sqrt{2}$, the following model Hamiltonian is obtained:

$$H_{2\times2}^+ = C(\mathbf{k})\mathbb{1}_2 + \left[\gamma\left(k_z^2 - m\right) - M_2(k_x^2 + k_y^2)\right]\sigma_x - \hbar v_F k_x \sigma_z - \hbar v_F k_y \sigma_y,$$
(5.22)

where, in order to make connections with the results in Sec. 5.3, we used the notation $\gamma = -M_1$ and $\hbar v_F = A$. The results for the lower block described by $H_{2\times2}^-$ can be obtained by replacing $k_x \to -k_x$. Similarly to Sec. 5.3, we assume that the surface of the semimetal is at $y = 0$. The semimetal itself is in the upper half-space $(y > 0)$ and vacuum is in the lower half-space $(y < 0)$. Since the boundary breaks the translation symmetry in the y-direction, one should use the operator form for the corresponding momentum $k_y \to -i\partial_y$. We also set $m \to -\tilde{m}$ with $\tilde{m} \to \infty$ in the vacuum region. The general form of the solution inside the semimetal reads

$$\psi_{y>0}(y) = \sum_{i=1}^{2}\begin{pmatrix} a_i \\ b_i \end{pmatrix} e^{-\kappa_i y},$$
(5.23)

where κ_i are the roots of the characteristic equation

$$\left[C_2(k_x^2 - \kappa^2) + C_1 k_z^2 + C_0 - \epsilon_s\right]^2 - \left[M_2(\kappa^2 - k_x^2) + \gamma\left(k_z^2 - m\right)\right]^2$$
$$+ \hbar^2 v_F^2(\kappa^2 - k_x^2) = 0$$
(5.24)

with a positive real part. When the real part is negative, the wave functions are not normalizable. As for the purely imaginary roots, they correspond to bulk states. The requirement of the normalizability, identified by the non-vanishing real part of $\kappa_{1,2}$ (see also a discussion after Eq. (5.18)), restricts the range of energies and momenta and limits the length of the Fermi arcs.

This can be seen explicitly from the following explicit expressions for κ_i:

$$\kappa_1 = \sqrt{k_x^2 - \frac{X + \sqrt{X^2 + Y}}{2(M_2^2 - C_2^2)}}, \quad \kappa_2 = \sqrt{k_x^2 - \frac{X - \sqrt{X^2 + Y}}{2(M_2^2 - C_2^2)}}, \qquad (5.25)$$

where

$$X = 2C_2 \left(C_1 k_z^2 + C_0 - \epsilon_s\right) + 2\gamma M_2 \left(k_z^2 - m\right) - \hbar^2 v_F^2, \qquad (5.26)$$

$$Y = 4(M_2^2 - C_2^2)[\left(C_1 k_z^2 + C_0 - \epsilon_s\right)^2 - \gamma^2 \left(k_z^2 - m\right)^2]. \qquad (5.27)$$

The spinor components a_i and b_i of the corresponding nontrivial solution satisfy the constraint

$$\frac{b_i}{a_i} = Q_i = \frac{C_2(k_x^2 - \kappa_i^2) + C_1 k_z^2 + C_0 - \epsilon_s - \hbar v_F k_x}{M_2(k_x^2 - \kappa_i^2) - \gamma(k_z^2 - m) + \hbar v_F \kappa_i}. \qquad (5.28)$$

In order to find the solution in vacuum ($y < 0$), one should replace $m \to -\tilde{m}$ and take the limit $\tilde{m} \to \infty$ in Eq. (5.23). This leads to

$$\psi_{y<0}(y) \simeq \frac{\tilde{a}_1}{\sqrt{\gamma\tilde{m}}} \begin{pmatrix} 1 \\ -1 \end{pmatrix} e^{\tilde{\kappa}_1 y} + \frac{\tilde{a}_2}{\sqrt{\gamma\tilde{m}}} \begin{pmatrix} 1 \\ 1 \end{pmatrix} e^{\tilde{\kappa}_2 y}, \qquad (5.29)$$

where, for convenience, we took the overall constants to be inversely proportional to $\sqrt{\gamma\tilde{m}}$.

By matching the wave functions at $y = 0$ and taking the limit $\tilde{m} \to \infty$, one finds that $a_1 = -a_2 \neq 0$ and $Q_1 = Q_2$. In addition, since Hamiltonian (5.22) is quadratic in k_y, one should include the matching equations for the wave functions derivatives. The corresponding equations determine coefficients $\tilde{a}_{1,2}$ on the vacuum side, whose form, however, is not crucial for the Fermi arcs. Thus, the condition $Q_1 = Q_2$ amended with the requirement of positive real parts of κ_i defines the Fermi arc surface states in the simplified model (5.22).

It is instructive to briefly discuss the case where the roots κ_i are complex numbers. According to the classification in [610, 614], the states with zero imaginary part of κ_i are called type B and exhibit an exponential decay in the bulk. On the other hand, type A states, whose characteristic roots have a nonvanishing imaginary part, have an oscillatory behavior with an exponentially decaying envelope. It is argued that type A states have a much smaller penetration depth compared to their type B counterparts. Note, however, that the presence of type A states depends on the model details. For example, such states are completely absent in a simple model discussed in Sec. 5.3 as well as in the low-energy model of Na_3Bi (see Sec. 3.6).

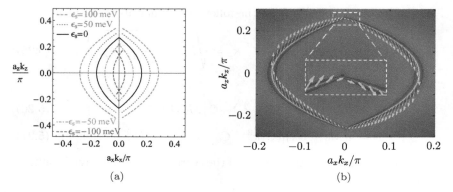

Fig. 5.4   (a) The dispersion relation of the Fermi arcs obtained from the condition $Q_1 = Q_2$ for several values of energy at $B(\mathbf{k}) = 0$. (b) The Fermi arcs obtained via the *ab initio* calculations for Na$_3$Bi. Green arrows show the spin texture of the Fermi arcs. (Adapted from [110]).

In order to visualize the Fermi arcs, we use the numerical values of parameters for Na$_3$Bi determined via the *ab initio* calculations in [110], which are given in Eq. (4.11) as well as lattice constants defined in Eq. (5.21). The results obtained by solving the Fermi arc equation $Q_1 = Q_2$ for $B(\mathbf{k}) = 0$ and those calculated by more sophisticated numerical methods in [110] are shown in Figs. 5.4(a) and 5.4(b), respectively. The Fermi arcs at small energies ($\epsilon_s \approx 0$) are in good agreement with the numerical results. Note that the surface states obtained here correspond to type B, i.e., the wave functions of these states decay exponentially without oscillations.

Before concluding this subsection, let us mention that the model of a semi-infinite crystal becomes unreliable for sufficiently thin samples of Dirac semimetals. In such a case, one needs to carefully reanalyze the problem of the surface states in a slab of finite thickness. While such calculations can be straightforwardly performed in the same way as in the semi-infinite case outlined above, the final expressions are much more cumbersome. (For the corresponding analytical calculations in the case of Dirac, Weyl, and nodal line semimetals, see [610].) Among the most distinctive features of the surface states in a finite slab is the opening of the energy gap between the Fermi arcs on the top and bottom surfaces, which exponentially decreases with the slab's width.

Having considered the models for the Fermi arc surface states, let us discuss their manifestation in various properties of Weyl and certain Dirac semimetals. It is natural to expect that the surface states could have a number of qualitative observational effects. For example, they modify the surface transport properties and, as we will see in Sec. 5.6, lead to nontrivial phenomena in a magnetic field.

5.5 Transport properties of Fermi arcs

In this section, we consider the effects of Fermi arcs on transport properties. While such effects are already expected for the conventional Tamm–Shockley surface states (see Sec. 5.1), the topological stability of the Fermi arcs at a moderate amount of disorder and their unusual dispersion relations may suggest that a qualitatively different type of the surface transport in Weyl and certain Dirac semimetals is possible. Moreover, by taking into account that the Fermi arcs are described by an effective Hamiltonian of 1D chiral fermions,[g] the transport properties could be very unusual. Indeed, as is known, the 1D chiral quasiparticles cannot backscatter. A classical example is the QHE [286, 293, 615, 616], where the longitudinal electrical conductivity σ_{xx} vanishes due to the absence of backscattering for 1D chiral fermions. In addition, similarly to the Fermi arcs, the quantum Hall edge states are also topologically protected [286, 616]. Other examples of nontrivial surface states in condensed matter systems include the Dirac cones on the surface of 3D topological insulators [97–101] and Majorana modes in topological superconductors [100, 617–622]. A similarity with these systems suggests that the electron surface transport due to the surface Fermi arcs might be also nondissipative. There is an important difference too. Indeed, the Fermi arcs coexist with the gapless bulk spectrum. As will be shown below, this fact plays a crucial role in the transport properties of the Fermi arcs in the presence of disorder. Note that general issues connected with the formulation of effective surface theories in gapless systems were also discussed in [572, 623].

5.5.1 *Disorder effects in effective 1D model*

In order to discuss the effects of disorder on the surface states in Weyl semimetals, let us utilize the model defined in Sec. 5.3. Recall that the corresponding model describes a semi-infinite Weyl semimetal with broken

---

[g]For curved Fermi arcs like those displayed in Fig. 5.4, there is always a possibility to introduce a coordinate along the arc and a surface coordinate perpendicular to the arc. Then the effective Hamiltonian describes the energy dispersion along the latter coordinate.

$\mathcal{T}$ symmetry and two Weyl nodes located at $\mathbf{k}_0^{\pm} = (0, 0, \pm\sqrt{m})$. The Weyl nodes are connected by the Fermi arc with the linear energy dispersion $\epsilon_s = \hbar v_F k_x$. For simplicity, we assume that the interaction of electrons with impurities is described by a superposition of local potentials

$$U(\mathbf{r}) = \sum_j u(\mathbf{r} - \mathbf{r}_j) = \sum_j u_0 \delta(\mathbf{r} - \mathbf{r}_j), \tag{5.30}$$

where u_0 is the strength of disorder potential and $\sum_j$ runs over all positions of impurities $\mathbf{r}_j$. The corresponding term in the Hamiltonian reads

$$H_{\text{dis}} = \int d^3 r \psi^{\dagger}(\mathbf{r}) U(\mathbf{r}) \psi(\mathbf{r}). \tag{5.31}$$

As is obvious, the disorder breaks the translation invariance. However, the latter is effectively restored after averaging over the positions of impurities (such an approach is known as the crossed diagram technique [136, 138, 139, 236])

$$\langle A(\mathbf{r}) \rangle_{\text{dis}} = \frac{1}{V} \sum_j \int d^3 r_j A(\mathbf{r}_j), \tag{5.32}$$

where $A(\mathbf{r})$ is an arbitrary function and V is the volume of the system. In the case of the local disorder potential (5.30), one can obtain the following average values:

$$\langle u(\mathbf{r} - \mathbf{r}_j) \rangle_{\text{dis}} = u_0 n_{\text{imp}}, \tag{5.33}$$

$$\langle u(\mathbf{r} - \mathbf{r}_j) u(\mathbf{r}' - \mathbf{r}_j) \rangle_{\text{dis}} = u_0^2 n_{\text{imp}} \delta(\mathbf{r} - \mathbf{r}'), \tag{5.34}$$

where $n_{\text{imp}} = N_{\text{imp}}/V$ is the concentration and N_{imp} is the total number of impurities.

For our transport calculations, it will be convenient to use the Kubo linear response theory. The conductivity tensor is defined in terms of the retarded current–current correlation function, i.e.,

$$\sigma_{nm}(\omega; \mathbf{r}, \mathbf{r}') = -\frac{i\hbar}{\omega} \Pi_{nm}(\omega; \mathbf{r}, \mathbf{r}'), \tag{5.35}$$

where the current–current correlator itself is given by the following ground state expectation value:

$$\Pi_{nm}(t - t'; \mathbf{r}, \mathbf{r}') = i\theta(t - t') \langle [j_n(t, \mathbf{r}), j_m(t', \mathbf{r}')] \rangle. \tag{5.36}$$

Here $j_n(t, \mathbf{r})$ is the nth spatial component of the electric current operator and $[\ldots, \ldots]$ denotes a commutator.

Before studying directly the transport properties of the Fermi arcs, it is instructive to point out that, naively, might expect a nondissipative transport for such surface states. As has already been mentioned at the beginning of Sec. 5.5, the Fermi arc electron states can be formally viewed as 1D chiral fermions described by the Hamiltonian $H_{\text{arcs}} = \hbar v_F k_x$. In absence of other gapless degrees of freedom, the transport due to such chiral states should be immune to disorder. Indeed, by assuming a quenched 1D disorder in Eqs. (5.30) and (5.31), one can easily obtain the following exact eigenfunction:

$$\psi_{1\text{D}}(x) = e^{i\left[\epsilon x - \sum_j u_0 \int_{x_0}^{x} dz \delta(z - x_j)\right]/(\hbar v_F)}, \tag{5.37}$$

where ϵ is the quasiparticle energy. The corresponding Green's function for the 1D chiral states can be obtained exactly, i.e.,

$$G_{1\text{D}}(i\omega_n; x, x') = i \int \frac{d\epsilon}{2\pi\hbar} \frac{e^{i\left\{\epsilon(x - x') - \sum_j u_0 \left[\theta(x - x_j) - \theta(x' - x_j)\right]\right\}/(\hbar v_F)}}{i\hbar\omega_n + \mu - \epsilon}. \tag{5.38}$$

Here $\omega_n = (2n + 1)\pi T/\hbar$ is the fermionic Matsubara frequency and n is an integer. This leads to the following one-loop correlation function (5.36):

$$\Pi_{1D}(i\Omega_n; \mathbf{r}, \mathbf{r}') = e^2 v_F^2 \frac{T}{\hbar} \sum_{l=-\infty}^{\infty} \langle G_{1\text{D}}(i\omega_l; \mathbf{r}, \mathbf{r}') G_{1\text{D}}(i\omega_l - i\Omega_n; \mathbf{r}', \mathbf{r}) \rangle_{\text{dis}}, \tag{5.39}$$

where $\Omega_n = 2n\pi T/\hbar$ denotes the bosonic Matsubara frequency. As one can see from the nonperturbative result (5.38), the disorder potential appears only as the phase $\sum_j u_0 \left[\theta(x - x_j) - \theta(x' - x_j)\right]$ in the exponent of Green's function. It is important that this phase changes the overall sign under the interchange $x \leftrightarrow x'$. Therefore, the resulting current–current correlation function (5.39) does not depend on disorder at all. In other words, the transport of 1D chiral fermions in nondissipative.

Because of the presence of the gapless bulk states, however, the transport properties of the Fermi arcs cannot be described accurately by using the 1D surface chiral fermions alone. In essence, the dynamics of the corresponding surface states in Weyl semimetals does not decouple from the gapless bulk states. This implies that the 1D kinematic constraint on the scattering of chiral fermions is lost and the surface Fermi arc transport becomes dissipative [624]. The statement will be supported below by analyzing the conductivity in the simplified model given in Sec. 5.3. In doing so, we will try to emphasize the underlying physics that supports the non-decoupling of the surface and bulk states, as well as causes dissipation of the Fermi arc transport. While mostly semirigorous, these qualitative results are also confirmed by the direct numerical calculations performed in [625, 626].

5.5.2   *Surface–bulk transitions*

The physical argument that provides evidence for the dissipative nature of the Fermi arc states is based on a perturbative treatment in the Born approximation. It is convenient to start the analysis by writing down the *Lippmann–Schwinger equation* for the scattered wave function $\psi(\mathbf{r})$ in the presence of disorder

$$\psi(\mathbf{r}) = \psi^{(0)}(\mathbf{r}) - i \int d^3\mathbf{r}' S^R(\mathbf{r}, \mathbf{r}') U(\mathbf{r}') \psi(\mathbf{r}'). \tag{5.40}$$

Here $\psi^{(0)}(\mathbf{r})$ describes an incident wave and

$$S^R(\omega, \mathbf{r}, \mathbf{r}') = i \sum_\epsilon \frac{\psi(\mathbf{r})\psi^\dagger(\mathbf{r}')}{\hbar\omega - \epsilon + i0} \tag{5.41}$$

is the retarded propagator in the clean limit, where the sum runs over the complete set of eigenstates. By including both surface and bulk states, the retarded propagator can be represented as a sum of two terms

$$S^R(\omega, \mathbf{r}, \mathbf{r}') = \int \frac{d^2 k_\parallel \, e^{i\mathbf{k}_\parallel \cdot (\mathbf{r}_\parallel - \mathbf{r}_\parallel')}}{(2\pi)^2} S_s^R(\omega, \mathbf{k}_\parallel; y, y')$$

$$+ \int \frac{d^3 k \, e^{i\mathbf{k} \cdot (\mathbf{r} - \mathbf{r}')}}{(2\pi)^3} S_b^R(\omega, \mathbf{k}). \tag{5.42}$$

Here subscript $\parallel$ denotes the components of vectors parallel to the surface. In the model defined in Sec. 5.3, the explicit expressions for the surface and bulk[h] Green's functions are given by

$$S_s^R(\omega, \mathbf{k}_\parallel; y, y') = i\frac{\kappa(k_z)}{\hbar} \frac{(\mathbb{1}_2 + \sigma_x) e^{-(y+y')\kappa(k_z)}}{\omega - v_F k_x + i0}, \tag{5.43}$$

$$S_b^R(\omega, \mathbf{k}) = i\frac{\hbar\omega\mathbb{1}_2 + \gamma(k_z^2 - m)\sigma_z + \hbar v_F(k_x\sigma_x + k_y\sigma_y)}{\hbar^2\omega^2 - \epsilon_\mathbf{k}^2 + i0\,\mathrm{sgn}(\omega)}, \tag{5.44}$$

---

[h]Note that, for the sake of simplicity, the modification of bulk states near the surface is neglected and the translation-invariant Green's function of bulk states is employed.

where $\kappa(k_z) = \gamma(m - k_z^2)/(\hbar v_F)$ and $\epsilon_\mathbf{k} = \pm \hbar v_F \sqrt{[\kappa(k_z)]^2 + k_x^2 + k_y^2}$. Recall that the Fermi arcs have a finite extent in momentum space. Therefore, the surface Green's function is defined only for $|k_z| < \sqrt{m}$.

A perturbative solution to the Lippmann–Schwinger equation can be obtained by replacing the wave function $\psi(\mathbf{r}')$ with $\psi^{(0)}(\mathbf{r}')$ on the right-hand side. By using Eqs. (5.40) and (5.42), we found that the wave function of a scattered surface state $\psi_s^{(0)}(\mathbf{r})$ is composed of both surface and bulk contributions, i.e.,

$$\psi^{(1)}(\mathbf{r}) = \psi_s^{(1)}(\mathbf{r}) + \psi_b^{(1)}(\mathbf{r}), \tag{5.45}$$

where

$$\psi_s^{(1)}(\mathbf{r}) \simeq -i \int d^3 r' S_s^R(\mathbf{r}, \mathbf{r}') U(\mathbf{r}') \psi_s^{(0)}(\mathbf{r}')$$

$$= -i u_0 \sum_j S_s^R(\mathbf{r}, \mathbf{r}_j) \psi_s^{(0)}(\mathbf{r}_j), \tag{5.46}$$

$$\psi_b^{(1)}(\mathbf{r}) \simeq -i \int d^3 r' S_b^R(\mathbf{r}, \mathbf{r}') U(\mathbf{r}') \psi_s^{(0)}(\mathbf{r}')$$

$$= -i u_0 \sum_j S_b^R(\mathbf{r}, \mathbf{r}_j) \psi_s^{(0)}(\mathbf{r}_j). \tag{5.47}$$

It is important to note that the bulk part of the scattered wave function $\psi_b^{(1)}(\mathbf{r})$ has the same form, but Green's function $S_s^R(\mathbf{r}, \mathbf{r}')$ is replaced with $S_b^R(\mathbf{r}, \mathbf{r}')$ on the right-hand side.

While the scattered wave functions can be obtained straightforwardly by using Green's functions in Eqs. (5.43) and (5.44), here we discuss only a few qualitative results. One of them is the power-law dependence of $\psi_s^{(1)}(\mathbf{r})$ on the z and y coordinates. In particular, by taking the limit of large z, one obtains the following wave function of the scattered state:

$$\psi_s^{(1)}(\mathbf{r}) \simeq 4i\sqrt{m}\gamma u_0 \sum_j \frac{\theta(x - x_j) \cos[\sqrt{m}(z - z_j)]}{\pi \hbar^2 v_F^2 (z - z_j)^2} \sqrt{\kappa(k_z)}$$

$$\times e^{ik_x x + ik_z z_j - \kappa(k_z) y_j} \begin{pmatrix} 1 \\ 1 \end{pmatrix}. \tag{5.48}$$

In essence, as one can see, such an outgoing wave is a superposition of two waves with the limiting values of the momentum $k_z = \pm\sqrt{m}$. Moreover, far from the scattering impurities, its amplitude decreases as $1/z^2$. The dependence on y has the same power-law character. It is worth noting

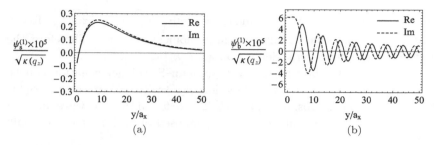

Fig. 5.5   (a) The upper component of the surface part of scattered wave function $\psi_s^{(1)}$ as a function of y/a. (b) The upper component of the bulk part of scattered wave function $\psi_b^{(1)}$ as a function of y. In both panels $\kappa(q_z) = \gamma(m - q_z^2)/(\hbar v_F)$, $\mathbf{q} = (\omega/v_F, 0, 0)$ is the wave vector of an impurity, $\mathbf{r} = (5a_x, y, 5a_z)$, and a single impurity is located at $\mathbf{r}_j = (0, 0, 0)$. The numerical values of the model parameters are given in Eq. (5.21). In addition, $u_0 = 0.1$ eV $\cdot$ Å^3 and $\hbar\omega = 0.5$ eV.

that the power-law rather than an exponential dependence on the spatial coordinates appears after the integration over k_z in Eq. (5.42).

The dependence of the upper components of the surface and bulk parts of the scattered wave function on y, i.e., the coordinate that quantifies the propagation into the bulk, is presented in Fig. 5.5 for the case of a single impurity. It is evident from Fig. 5.5(a) that the surface part of the scattered wave function, $\psi_s^{(1)}(\mathbf{r})$, falls off as a power-law function of y rather than an exponent. Most importantly, there are also outgoing waves propagating into the bulk, i.e., $\psi_b^{(1)}(\mathbf{r})$ shown in Fig. 5.5(b). This implies that the surface Fermi arc states are coupled with the bulk ones and that a part of the arc state leaks into the bulk during each scattering on an impurity. In the analysis of the surface transport, therefore, it is crucial to take into account the *surface–bulk transitions*. The corresponding bulk part of the scattered surface state is free from the 1D kinematic constraints and contributes to dissipation. In the end, one finds that, unlike truly 1D chiral fermions, the Fermi arc transport is affected by disorder.

Before concluding this subsection, let us also discuss the evidence from numerical simulations that illustrates the effect of disorder on the Fermi arcs and supports the dissipative nature of the surface states in Weyl semimetals. By using a simple lattice model of a Weyl semimetal subject to a random disorder potential, it was shown in [625] that a sufficiently strong disorder potential can dissolve completely the surface Fermi arcs. While there is a straight Fermi arc in the clean limit, disorder tends to blur the surface states with increasing the disorder strength. The numerical results in [626] also suggest that the Fermi arcs lose their topological robustness due to a hybridization with nonperturbative bulk rare states [627].[i] It is noted, however, that the surface velocity might still keep its chiral nature.

5.5.3 *Electrical conductivity*

As argued in Sec. 5.2.2, the electric transport due to the Fermi arcs should be dissipative. Therefore, in order to qualitatively study the surface conductivity, one can apply the standard perturbative methods.

5.5.3.1 *Kubo formalism for surface transport*

The electrical conductivity is given by the general expression (5.35), where the current–current correlator is defined in Eq. (5.36) and the components of the electric current operator $\hat{\mathbf{j}} = -e\partial_{\mathbf{k}}H$ are

$$\hat{j}_x = -ev_F\sigma_x, \tag{5.49}$$

$$\hat{j}_z = -\frac{2e\gamma k_z}{\hbar}\sigma_z. \tag{5.50}$$

By taking into account these definitions and the matrix structure of the surface state propagator (5.43), it is straightforward to show that three out of the total four components of the conductivity tensor vanish, i.e., $\sigma_{xz} = -\sigma_{zx} = \sigma_{zz} = 0$. The remaining nontrivial component of the direct current (DC) conductivity tensor reads

$$\sigma_{xx}(\mathbf{r}) = -\lim_{\Omega\to 0}\frac{i\hbar}{\Omega}\Pi_{xx}(\Omega;\mathbf{r}). \tag{5.51}$$

In coordinate space, the corresponding correlator is given in terms of the advanced and retarded surface Green's functions $G_s^{A/R}$

$$\Pi_{xx}(\Omega;\mathbf{r}) \simeq -e^2v_F^2\Omega\int\frac{d\omega}{2\pi}\frac{\partial f^{eq}(\hbar\omega)}{\partial\omega}\int d^3r'$$
$$\times \mathrm{tr}\left[\langle\sigma_x G_s^A(\omega;\mathbf{r},\mathbf{r}')\sigma_x G_s^R(\omega;\mathbf{r}',\mathbf{r})\rangle_{\mathrm{dis.}}\right], \tag{5.52}$$

where $f^{eq}(\epsilon) = 1/\left[1 + e^{(\epsilon-\mu)/T}\right]$ is the equilibrium Fermi–Dirac quasiparticle distribution function. Since here we are interested in the DC limit, it is sufficient to keep only the terms up to linear order in Ω.

---

[i]The bulk rare states will be discussed in Sec. 7.1.5.

Note that the full advanced and retarded propagators $G_s^{A/R}(\omega; \mathbf{r}, \mathbf{r}')$ should be calculated in the presence of disorder. They can be formally represented by the following infinite series of diagrams:

$$G \equiv \longrightarrow + \longrightarrow + \longrightarrow + \cdots, \qquad (5.53)$$

where the solid lines correspond to the bare Green's function S calculated in the absence of impurities and the dashed lines with the crosses at the ends represent the interaction with the impurity potential.

In the presence of disorder, the diagrammatic representation of the current–current correlation function is given by

$$\Pi_{xx} = \qquad . \qquad (5.54)$$

The corresponding analytic expression reads

$$\Pi_{xx}(\Omega; \mathbf{r}) = -e^2 v_F^2 \Omega \int \frac{d\omega}{2\pi} \frac{\partial f^{eq}(\hbar\omega)}{\partial \omega} \int d^3r' d^3r_1 d^3r_2$$
$$\times \operatorname{tr}[\sigma_x \left\langle G_s^A(\omega; \mathbf{r}, \mathbf{r}_1) \right\rangle_{\mathrm{dis}} \Lambda_x(\omega; \mathbf{r}_1, \mathbf{r}', \mathbf{r}_2)$$
$$\times \left\langle G_s^R(\omega; \mathbf{r}_2, \mathbf{r}) \right\rangle_{\mathrm{dis}}]. \qquad (5.55)$$

Here, as usual in the 3D analysis, we omitted the diagrams with crossing lines in the correlator. Such an approximation is reasonable because the surface states are not decoupled from the 3D bulk states and, in fact, have a 3D-like dynamics. In Eq. (5.55), the double solid lines represent the disorder-averaged quasiparticle propagators $\langle G \rangle_{\mathrm{dis}}$. In the quenched rainbow or first-order Born approximation, the latter is given by

$$\langle G \rangle_{\mathrm{dis}} \equiv \longrightarrow = \longrightarrow + \longrightarrow + \longrightarrow . \qquad (5.56)$$

Note that the second term on the right-hand side, describing the linear correction from the impurity potential, can be omitted. Indeed, as will be shown below, its effect is simply to shift the value of the electric chemical potential, $\mu^* = \mu - n_{\mathrm{imp}} u_0$.

In the same ladder approximation, the diagrammatic representation of the vertex function takes the form

$$\Lambda_x\left(\omega; \mathbf{r}_1, \mathbf{r}', \mathbf{r}_2\right) \equiv \quad = \circ + \quad , \qquad (5.57)$$

and the corresponding explicit expression reads

$$\Lambda_x\left(\omega; \mathbf{r}_1, \mathbf{r}', \mathbf{r}_2\right) = \sigma_x\delta(\mathbf{r}_1 - \mathbf{r}')\delta(\mathbf{r}' - \mathbf{r}_2) - D(\mathbf{r}_1, \mathbf{r}_2)\int d^3r'_1 d^3r'_2$$

$$\times \left\langle G_s^{\mathrm{A}}(\omega; \mathbf{r}_1, \mathbf{r}'_1)\right\rangle_{\mathrm{dis}} \Lambda_x\left(\omega; \mathbf{r}'_1, \mathbf{r}', \mathbf{r}'_2\right) \left\langle G_s^{\mathrm{R}}(\omega; \mathbf{r}'_2, \mathbf{r}_2)\right\rangle_{\mathrm{dis}}, \qquad (5.58)$$

where

$$D(\mathbf{r}_1, \mathbf{r}_2) = n_{\mathrm{imp}} u_0^2 \delta\left(\mathbf{r}_1 - \mathbf{r}_2\right) \qquad (5.59)$$

is the impurity correlation function (5.34).

Since the y coordinate plays a special role for the surface states, it is convenient to use a mixed coordinate and momentum space representation. The corresponding expression for the conductivity of the surface states can be written as follows (see also Eq. (5.51)):

$$\sigma_{xx}(\mathbf{q}_\parallel, y) = -\lim_{\Omega \to 0} \frac{i\hbar}{\Omega}\Pi_{xx}(\Omega; \mathbf{q}_\parallel, y). \qquad (5.60)$$

In response to spatially homogeneous fields, momentum $\mathbf{q}_\parallel$ should be set to zero. Then the corresponding current–current correlator reads

$$\Pi_{xx}(\Omega; 0, y) = -e^2 v_F^2 \Omega \int \frac{d\omega}{2\pi}\frac{\partial f^{\mathrm{eq}}(\hbar\omega)}{\partial \omega} \int \frac{d^2 k_\parallel}{(2\pi)^2} \int dy' dy_1 dy_2$$

$$\times \mathrm{tr}[\sigma_x \left\langle G_s^{\mathrm{A}}(\omega; \mathbf{k}_\parallel, y, y_1)\right\rangle_{\mathrm{dis}} \Lambda_x\left(\omega; \mathbf{k}_\parallel, y_1, y', y_2\right)$$

$$\times \left\langle G_s^{\mathrm{R}}(\omega; \mathbf{k}_\parallel, y_2, y)\right\rangle_{\mathrm{dis}}], \qquad (5.61)$$

where the vertex function satisfies the following equation:

$$\Lambda_x\left(\omega; \mathbf{k}_\parallel, y_1, y', y_2\right) = \sigma_x\delta(y_1 - y')\delta(y' - y_2) - \int \frac{d^2 l_\parallel}{(2\pi)^2} \int dy'_1 dy'_2$$

$$\times D(\mathbf{k}_\parallel - \mathbf{l}_\parallel, y_1, y_2) \left\langle G_s^{\mathrm{A}}(\mathbf{l}_\parallel, \omega; y_1, y'_1)\right\rangle_{\mathrm{dis}}$$

$$\times \Lambda_x\left(\omega; \mathbf{l}_\parallel, y'_1, y', y'_2\right) \left\langle G_s^{\mathrm{R}}(\omega; \mathbf{l}_\parallel, y'_2, y_2)\right\rangle_{\mathrm{dis}}. \qquad (5.62)$$

As one can see from Eqs. (5.61) and (5.62), both correlator and vertex function are determined by the disorder-averaged full quasiparticle propagator.

5.5.3.2 *Disorder-averaged propagator*

Let us start with the disorder-averaged full Green's function. Henceforth, only the retarded Green's function will be considered and, therefore, the superscript R will be omitted. In the end, the advanced Green's function can be obtained simply by inverting the sign of the imaginary term in the denominator. As usual in the studies of weakly disordered systems, it is sufficient to expand the full surface Green's function only up to the second order in the disorder strength, i.e.,

$$-iG(\omega, \mathbf{k}_{\parallel}, \mathbf{k}'_{\parallel}; y, y') \approx S^{(0)} + S^{(1)} + S^{(2)}, \tag{5.63}$$

where, for the sake of brevity, the arguments of $S^{(n)}$ are omitted,

$$S^{(0)} = (2\pi)^2 \delta(\mathbf{k}_{\parallel} - \mathbf{k}'_{\parallel})(-i)S_{\mathrm{s}}(\omega, \mathbf{k}_{\parallel}; y, y'), \tag{5.64}$$

and $S_{\mathrm{s}}(\omega, \mathbf{k}_{\parallel}; y, y')$ is the surface Green's function given by (5.43) in the clean limit.

The first-order correction takes the form

$$S^{(1)} = -\sum_j \int dy'' S_{\mathrm{s}}(\omega, \mathbf{k}_{\parallel}; y, y'') u_0 \delta(y'' - y_j)$$

$$\times e^{-i\mathbf{r}_j \cdot (\mathbf{k}_{\parallel} - \mathbf{k}'_{\parallel})} S_{\mathrm{s}}(\omega, \mathbf{k}'_{\parallel}; y'', y'). \tag{5.65}$$

After averaging over the impurities by using the prescription in Eq. (5.32) and then integrating over y'', one derives

$$\langle S^{(1)} \rangle_{\mathrm{dis}} = -n_{\mathrm{imp}} u_0 \frac{e^{(y+y')\kappa(k_z)}}{2\kappa(k_z)} (2\pi)^2 \delta(\mathbf{k}_{\parallel} - \mathbf{k}'_{\parallel})$$

$$\times S_{\mathrm{s}}(\omega, \mathbf{k}_{\parallel}; y, y') S_{\mathrm{s}}(\omega, \mathbf{k}_{\parallel}; y', y). \tag{5.66}$$

In terms of the self-energy, which enters the full Green's function as an additional term in the denominator $\hbar\omega + \mu \to \hbar\omega + \mu - \Sigma(\mathbf{k}_{\parallel})$, such a correction is nothing else but a constant shift

$$\Sigma^{(1)}(\mathbf{k}_{\parallel}) = n_{\mathrm{imp}} u_0. \tag{5.67}$$

This shift can be effectively absorbed into the electric chemical potential and is not important qualitatively.

Since the Fermi arcs are coupled to bulk states, there are two nontrivial contributions at the second order in powers of the disorder strength. They are given by the one-loop diagrams represented by the last term in Eq. (5.56) where the fermion line in the loop corresponds to either surface or bulk Green's function. In the first case, we have

$$S_s^{(2)} = i \sum_j \int dy_1 dy_2 \int \frac{d^2 q_\parallel}{(2\pi)^2} S_s(\omega, \mathbf{k}_\parallel; y, y_1)$$

$$\times u_0 \delta(y_1 - y_j) e^{-i\mathbf{r}_j \cdot (\mathbf{k}_\parallel - \mathbf{q}_\parallel)} S_s(\omega, \mathbf{q}_\parallel; y_1, y_2)$$

$$\times u_0 \delta(y_2 - y_j) e^{-i\mathbf{r}_j \cdot (\mathbf{q}_\parallel - \mathbf{k}_\parallel')} S_s(\omega, \mathbf{k}_\parallel'; y_2, y'). \tag{5.68}$$

After averaging over the impurities, one obtains

$$\langle S_s^{(2)} \rangle_{\text{dis}} = \frac{i n_{\text{imp}} u_0^2}{2\pi \hbar v_F} (2\pi)^2 \delta(\mathbf{k}_\parallel - \mathbf{k}_\parallel') e^{(y+y')\kappa(k_z)} S_s(\omega, \mathbf{k}_\parallel; y, y') S_s(\omega, \mathbf{k}_\parallel; y', y)$$

$$\times \left[\sqrt{m} - \frac{k_z^2 - m}{\sqrt{k_z^2 - 2m}} \arctan\left(\frac{\sqrt{m}}{\sqrt{k_z^2 - 2m}} \right) \right]. \tag{5.69}$$

This leads to the following imaginary term in the self-energy that depends only on k_z:

$$\Sigma_s^{(2)}(\mathbf{k}_\parallel) = -i \frac{\kappa(k_z) n_{\text{imp}} u_0^2}{\pi \hbar v_F} \left[\sqrt{m} - \frac{k_z^2 - m}{\sqrt{k_z^2 - 2m}} \arctan\left(\frac{\sqrt{m}}{\sqrt{k_z^2 - 2m}} \right) \right]. \tag{5.70}$$

This nonzero imaginary part of the self-energy implies that surface quasiparticles have a nonvanishing width $\Gamma_s(k_z) = i\Sigma_s^{(2)}(k_z)$ which depends on k_z.

Let us now turn to the bulk mediated one-loop diagram in Eq. (5.56). At the quadratic order in powers of the disorder strength, the corresponding contribution reads

$$\langle S_b^{(2)} \rangle_{\text{dis}} = i \left\langle \sum_j \int dy_1 dy_2 \int \frac{d^2 q_\parallel}{(2\pi)^2} S_s(\omega, \mathbf{k}_\parallel; y, y_1) \right.$$

$$\times u_0 \delta(y_1 - y_j) e^{-i\mathbf{r}_j \cdot (\mathbf{k}_\parallel - \mathbf{q}_\parallel)} S_b(\omega, \mathbf{q}_\parallel; y_1 - y_2)$$

$$\left. \times u_0 \delta(y_2 - y_j) e^{-i\mathbf{r}_j \cdot (\mathbf{q}_\parallel - \mathbf{k}_\parallel')} S_s(\omega, \mathbf{k}_\parallel'; y_2, y') \right\rangle_{\text{dis}}. \tag{5.71}$$

After integrating over $\mathbf{q}_\parallel$, one derives the following bulk-mediated contribution to the imaginary part of the self-energy:

$$\mathrm{Im}[\Sigma_\mathrm{b}^{(2)}(\omega)] = -n_\mathrm{imp} u_0^2 \frac{|\hbar\omega + \mu|}{4\pi\hbar^2 v_F^2 \sqrt{\gamma}} [\sqrt{\gamma m + |\hbar\omega + \mu|}$$

$$- \theta\left(\gamma m - |\hbar\omega + \mu|\right) \sqrt{\gamma m - |\hbar\omega + \mu|}], \qquad (5.72)$$

which depends only on ω. This leads to an additional energy-dependent contribution to the quasiparticle width, namely, $\Gamma_\mathrm{b}(\omega) = -\mathrm{Im}[\Sigma_\mathrm{b}^{(2)}(\omega)]$.

Finally, by collecting all self-energy corrections, the full disorder-averaged Green's function for the surface quasiparticles is obtained

$$\langle G(\omega, \mathbf{k}_\parallel; y, y')\rangle_\mathrm{dis} \simeq \frac{i\kappa(k_z)(\mathbb{1}_2 + \sigma_x)e^{-(y+y')\kappa(k_z)}}{\hbar\omega + \mu - \hbar v_F k_x + i\Gamma(\omega, k_z)}, \qquad (5.73)$$

where $\Gamma(\omega; k_z) = \Gamma_\mathrm{s}(k_z) + \Gamma_\mathrm{b}(\omega)$ is the total Fermi arc quasiparticle width, which accounts for scattering into both surface and bulk states.

Last but not least, let us briefly comment on the role of the vertex function in Eq. (5.62). By taking into account that the right-hand side of Eq. (5.62) does not depend on momentum,[j] one concludes that the vertex function depends only on ω and the spacial coordinates y_1 and y_2. For simplicity, the corresponding Fredholm integral equation of the second kind will be treated perturbatively in the following subsection. This approach should be reliable for sufficiently large values of the energy or the electric chemical potential.

5.5.3.3 *Analysis of Fermi arc conductivity*

Here we present the explicit results for the surface Fermi arc conductivity and discuss their implications. It is convenient to re-express Green's functions in terms of the spectral function $A(\omega, \mathbf{k})$

$$\langle G(i\omega_l, \mathbf{k}_\parallel; y, y')\rangle_\mathrm{dis} = i \int_{-\infty}^{\infty} \hbar d\omega \frac{A(\omega, \mathbf{k}_\parallel; y, y')}{i\hbar\omega_l + \mu - \hbar\omega}. \qquad (5.74)$$

---

[j]This follows from the fact that the disorder correlation function $D(\mathbf{k}_\parallel - \mathbf{l}_\parallel, y_1, y_2) = n_\mathrm{imp} u_0^2 \delta(y_1 - y_2)$ does not depend on momentum.

The spectral function for the surface states itself is defined by

$$A(\omega, \mathbf{k}_\parallel; y, y') = \frac{1}{2\pi} \left[\langle G^{\mathrm{R}}(\omega, \mathbf{k}_\parallel; y, y') \rangle_{\mathrm{dis}} - \langle G^{\mathrm{A}}(\omega, \mathbf{k}_\parallel; y, y') \rangle_{\mathrm{dis}} \right]_{\mu=0}$$

$$= \frac{1}{\pi} \frac{(\mathbb{1}_2 + \sigma_x) \kappa(k_z) e^{-(y+y')\kappa(k_z)} \Gamma(\omega, k_z)}{\hbar^2 (\omega - v_F k_x)^2 + \Gamma^2(\omega, k_z)}, \tag{5.75}$$

which is nonzero only for $|k_z| < m$.

By using the correlator in Eq. (5.61) and the spectral representation in Eq. (5.74), we can rewrite the surface DC conductivity (5.60) as follows:

$$\sigma_{xx}^{\mathrm{tot}}(y) = \lim_{\Omega \to 0} \frac{i}{\Omega} e^2 v_F^2 \hbar^2 T \sum_{l=-\infty}^{\infty} \int \frac{d^2 k_\parallel}{(2\pi)^2} \int\int\int dy' dy_1 dy_2 \int\int d\omega d\omega'$$

$$\times \frac{\mathrm{tr} \left[\sigma_x A(\omega, \mathbf{k}_\parallel; y, y_1) \Lambda_x \left(\omega; \mathbf{k}_\parallel, y_1, y', y_2 \right) A(\omega', \mathbf{k}_\parallel; y_2, y) \right]}{(i\hbar\omega_l + \mu - \hbar\omega)(i\hbar\omega_l - \hbar\Omega - i0 + \mu - \hbar\omega')}. \tag{5.76}$$

By expanding the vertex function in Eq. (5.62) to the second order in the disorder strength, one obtains the following leading and subleading contributions to the conductivity tensor:

$$\sigma_{xx}^{(0)}(y) = \frac{e^2}{4\pi^2} \int_{-\sqrt{m}}^{\sqrt{m}} dk_z \int d\omega \frac{v_F \hbar \kappa(k_z) e^{-2(y+y')\kappa(k_z)}}{4T\Gamma(\omega, k_z) \cosh^2 \left(\frac{\hbar\omega - \mu}{2T} \right)} \tag{5.77}$$

and

$$\sigma_{xx}^{(1)}(y) = \frac{e^2 \hbar n_{\mathrm{imp}} u_0^2}{16\pi^3} \int \frac{d\omega}{4T \cosh^2 \left(\frac{\hbar\omega - \mu}{2T} \right)} \int_{-\sqrt{m}}^{\sqrt{m}} \int_{-\sqrt{m}}^{\sqrt{m}} dk_z dl_z$$

$$\times \frac{e^{-2y\kappa(k_z)} \kappa(l_z) \kappa^2(k_z)}{\Gamma(\omega, k_z) \Gamma(\omega, l_z) [\kappa(l_z) + \kappa(k_z)]}, \tag{5.78}$$

respectively. The resulting surface conductivity is shown as a function of the spatial coordinate y (recall that y is perpendicular to the surface) and the electric chemical potential μ in Figs. 5.6(a) and 5.6(b), respectively.

As expected, the Fermi arc conductivity is dominated by the surface state contribution. This is clear from the y dependence of the local conductivity in Fig. 5.6(a), which is maximal at $y = 0$ and rapidly decreases with increasing the distance from the surface into the bulk of

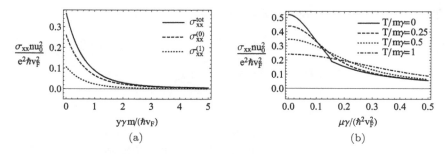

Fig. 5.6   (a) The surface conductivity as a function of the coordinate perpendicular to the surface y at $\mu = 50$ meV. (b) The surface conductivity at $y = 0$ as a function of the electric chemical potential μ for several values of temperature. In addition to the numerical parameters in Eq. (5.21), the impurity parameters $u_0 = 0.1$ eV $\cdot$ Å^3 and $n_{\text{imp}} = 10^{-3}$ Å^{-3} were used in both panels.

the semimetal. One can also see that the subleading quadratic correction given by Eq. (5.78) is nonnegligible at small y. A more detailed analysis reveals that perturbative corrections tend to increase with decreasing the electric chemical potential μ. Therefore, for sufficiently small μ, many terms in the perturbative expansion should be included.

As is evident from Fig. 5.6(b), unlike the bulk conductivity, which usually grows with the chemical potential μ, its surface counterpart decreases with μ. In retrospect, this should not be surprising since the quasiparticle width of the Fermi arcs grows with increasing the density of gapless bulk states. In other words, the effect is driven primarily by the increase of the phase space for the surface–bulk scattering. Another interesting feature of the surface conductivity is a characteristic kink at $\mu = \gamma m$, which is partially blurred by the effects of nonzero temperature. It is easy to verify that the kink is related to the Van Hove singularity [628], which is associated with the Lifshitz transition [358]. It occurs at a sufficiently large value of μ when the disjoined sheets of the Fermi surface around the opposite chirality Weyl nodes start to merge.[k] Let us note that, because of the perturbative treatment of the vertex, the results presented in Fig. 5.6(b) should be viewed only as qualitative.

Since the presence of bulk states is crucial for the Fermi arc dissipation, it is important to understand at least conceptually the regime of a small

---

[k]The existence of such a Lifshitz transition is evident from Fig. 5.3(a).

or even vanishing electric chemical potential. In such a case, there exist several nonperturbative effects that could produce a nonzero density of bulk states [627, 629, 630]. For example, depending on the nature of impurities, a nonvanishing density of bulk states can be induced by the nonperturbative rare region effects [627] (short-range impurities) or the electron and hole puddles [629, 630] (charged impurities). Additionally, as suggested by the numerical simulations in [626], the nonperturbative bulk rare states can hybridize with the Fermi arcs and, as a result, force them to extend into the bulk. Generically, therefore, the presence of disorder appears to always make the Fermi arc transport dissipative. It is worth pointing out that the electron transport in the bulk is also affected by the nonperturbative effects. Some of them will be discussed in Sec. 7.1.5.

Before finalizing this section, let us briefly discuss a different regime of the surface electron transport in Weyl semimetals, in which the sample width is much smaller than the relaxation length for the surface–bulk scattering. It was found in [631] that the Fermi arc surface states might contribute to the longitudinal conductivity at the same order of magnitude as the bulk. Moreover, it was argued that the large contribution of the surface states could occur even for $\mathcal{P}$ symmetric Weyl semimetals in which the scattering rate between the states related by $\mathcal{T}$ symmetry is small enough. Then, in terms of the surface transport, such a system can be described as a sum of two $\mathcal{T}$ symmetry broken subsystems.

So far we considered Fermi arc properties without any magnetic field. However, as was shown in previous chapters, its presence leads to several interesting effects and allows one to probe relativistic-like nature of bulk quasiparticles in Weyl and Dirac semimetals. Therefore, it is interesting to investigate what happens to the Fermi arc surface states when an external magnetic field is applied to the semimetal.

5.6 Fermi arcs in magnetic field

The effects of magnetic field could reveal observable signatures of the quantum anomaly, which, as will be discussed in Chapter 6, plays a profound role in the bulk transport in Dirac and Weyl semimetals. Among the most fascinating effects related to the interplay of the Fermi arcs and magnetic field are quantum oscillations of a novel type associated with closed surface–bulk orbits, the quantum Hall effect, and a nontrivial optical response mediated by the Fermi arcs.

5.6.1 *Surface–bulk quantum oscillations*

As was discussed in the previous sections, the surface Fermi arc states link separate chiral pockets of the Fermi surface in Weyl and certain Dirac semimetals. This gives rise to an unconventional *fermiology* of such materials. In addition to the spectroscopic methods such as the ARPES discussed in Chapter 4, one of the most direct and time-honored methods to investigate the Fermi surface structure relies on the quantum oscillations of the density of states (DOS) in a magnetic field (see, e.g., [287, 288]).

Generically, the quantum oscillations are connected with closed orbits in momentum space that quasiparticles in a magnetic field are forced to follow. The existence of such oscillations is manifested in many observable properties, including the electrical conductivity, the magnetization, etc. Naively, since the Fermi arcs form open contours in momentum space, one might suggest that they do not contribute to any type of quantum oscillations. However, as was argued in [632], new types of closed orbits can be realized in a magnetic field when both surface and bulk states are taken into account. Indeed, as we will explain below, the quasiparticles in a magnetic field can slide along the Fermi arcs on the surface as well as move through the bulk via the Landau-level (LL) states. Therefore, closed magnetic orbits, which include the surface states on both surfaces connected by the bulk modes, can form and give rise to a unique type of quantum oscillations. Of course, Dirac and Weyl semimetals have also the usual quantum oscillations connected with the bulk states themselves. These oscillations were discussed in Sec. 3.9.

Let us now describe the mechanism of the surface–bulk quantum oscillations. For simplicity, it will be assumed that a Weyl semimetal with broken $\mathcal{T}$ symmetry has the form of a slab of thickness L in the y-direction. Also, by assumption, the Weyl nodes are located at $\mathbf{k}_0^{\pm} = \pm\mathbf{b} = \pm(0, b_y, b_z)$, where b_y and b_z are the two components of the chiral shift that are perpendicular and parallel to the slab's surface, respectively. As follows from Eq. (5.17) (see also Sec. 5.3 for a semi-infinite case and [607] for a slab geometry) the quasiparticle velocity of the Fermi arc states is $\mathbf{v}_b = v_F\hat{\mathbf{x}}$ on the bottom surface ($y = 0$) and $\mathbf{v}_t = -\mathbf{v}_b$ on the top surface ($y = L$). Note that the latter property is also valid in the more realistic case of curved Fermi arcs discussed in Sec. 5.4 because the arcs at the top and bottom surfaces are related by the reflection $k_x \rightarrow -k_x$. It is convenient to write the external magnetic field in the form $\mathbf{B} = B\hat{\mathbf{n}}$, where the unit vector $\hat{\mathbf{n}} = (\sin\theta\sin\varphi, \cos\theta, \sin\theta\cos\varphi)$ is determined by the direction of

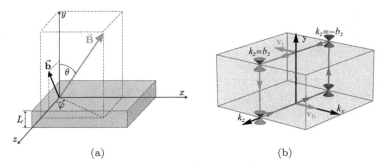

Fig. 5.7 (a) Model setup for a slab of a Weyl semimetal with broken $\mathcal{T}$ symmetry in an external magnetic field **B**. (b) Schematic illustration of closed orbits in mixed coordinate-momentum space composed of Fermi arcs and bulk states.

the field. The schematic setup for the slab in the magnetic field is shown in Fig. 5.7(a).

Let us illustrate the closed orbits in Fig. 5.7(b) for the simplest case of straight Fermi arcs. In a semiclassical description, the Lorentz force causes the z component of momentum to change, i.e.,

$$\hbar \partial_t k_z = -\frac{e}{c} \left[\mathbf{v}_{\mathrm{b,t}} \times \mathbf{B} \right]_z = \mp \frac{e}{c} v_F B \cos \theta, \qquad (5.79)$$

where the sign $-$ $(+)$ corresponds to the bottom (top) surface. From a physics viewpoint, Eq. (5.79) describes the sliding of the quasiparticles along the bottom Fermi arc from the right-handed Weyl node at $k_z = b_z$ to the left-handed one at $k_z = -b_z$.[1] At the left-handed Weyl node, the Fermi arc couples to bulk states. Here, the single-band semiclassical description breaks down and one needs to employ the quantum-mechanical treatment. Qualitatively, however, it is essential only that the surface electrons start to propagate through the bulk. The motion through the bulk is made possible by the gapless lowest Landau level (LLL). When the quasiparticle reaches the top surface, it is transferred to a surface state and slides along the top Fermi arc in the opposite direction, i.e., from the left-handed Weyl node at $k_z = -b_z$ to the right-handed one at $k_z = b_z$. Afterwards, the quasiparticle goes through the bulk via an LLL state and eventually closes the orbit.

---

[1]Since only the projections of Weyl nodes onto the surface determine the length and shape of the Fermi arcs, the y component of the chiral shift is not important for the surface–bulk oscillations. This might not be the case for very thin films [633].

By applying the semiclassical quantization condition for the closed orbits depicted in Fig. 5.7(b), one obtains the following energy levels $\epsilon_n t = 2\pi\hbar(n + \alpha)$, where n is an integer and α denotes a phase shift due to quantum effects. The time t needed to complete the orbit consists of two parts: the propagation time through the bulk t_{bulk} and the time of sliding along the Fermi arcs t_{arcs}. Qualitatively, they can be estimated as $t_{\mathrm{bulk}} = 2L/(v_F \cos\theta)$ and $t_{\mathrm{arcs}} = 4\hbar b_z/(v_F |eB| \cos\theta)$, respectively. Thus, the semiclassical energy spectrum reads

$$\epsilon_n = \frac{\pi\hbar v_F(n + \alpha)}{L/\cos\theta + 2\hbar b_z/(|eB| \cos\theta)}. \qquad (5.80)$$

By fixing the energy as $\epsilon_n = \mu$, the maxima of the DOS oscillations are achieved at the following discrete values of the inverse magnetic field strength:

$$\frac{1}{B_n} = \frac{e}{2\hbar c b_z} \left[\frac{\pi\hbar v_F(n + \alpha)\cos\theta}{\mu} - L \right]. \qquad (5.81)$$

Therefore, the period of oscillations is given by

$$T_{1/B} = \frac{e\pi v_F \cos\theta}{2c\mu b_z}. \qquad (5.82)$$

Since the right-hand side in Eq. (5.81) should be positive, there is a saturation value of the magnetic field $B_{\mathrm{sat}} = B_{n_{\min}}$, where $n_{\min} = [\mu L/(\pi\hbar v_F \cos\theta) - \alpha + 1]$ with $[\ldots]$ denoting the integer part of the expression. The peaks of the DOS persist only for $B < B_{\mathrm{sat}}$. In general, as argued in [634], Eq. (5.81) should include the contribution due to the surface enclosed by the Fermi arcs S_{s} as well as additional corrections in the bulk term. The corresponding expression obtained via the phase space quantization reads

$$\frac{1}{B_n} = \frac{e}{c\hbar S_{\mathrm{s}}} \left[2\pi(n + \alpha)\cos\theta - L\left(2\mathbf{b} \cdot \hat{\mathbf{n}} + \frac{\mu}{\hbar v_F} \right) \right]. \qquad (5.83)$$

Among the key features of a more refined treatment in [634] is the possibility of quantum oscillations even at $\mu = 0$ and the cancelation of the L-dependence for certain orientations of the magnetic field.

From the experimental viewpoint, it is important to keep in mind that, because of the gapless energy spectrum in Weyl semimetals, the oscillations due to the Fermi surface of bulk quasiparticles are present at any nonzero μ [635] (see also Sec. 3.9). The model results for the total DOS $\nu(\mu)$ that includes the contributions from both bulk states and surface–bulk closed

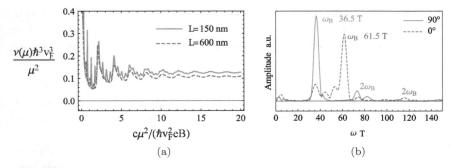

(a)   (b)

Fig. 5.8 (a) The oscillations of the total density of states that include the contributions from both bulk states and closed surface–bulk orbits defined in Eqs. (5.84) and (5.85), respectively. Red solid and blue dashed lines correspond to different slab thicknesses $L = 150$ nm and $L = 600$ nm. In order to plot the results, a small nonzero quasiparticle width was introduced. (b) The experimentally obtained Fourier transform of resistivity oscillations for Cd_3As_2 for two orientations of magnetic field, $\theta = 0°$ (blue dashed line) and $\theta = 90°$ (red line). The oscillations related to the surface–bulk orbits appear only in the former case. (The data provided by P. J. W Moll were used.)

orbits are presented in Fig. 5.8(a). The analytical expression for the bulk DOS for two Weyl nodes at energy ϵ is given by

$$\nu(\epsilon)_{\text{bulk}} = \frac{|eB|}{2\pi^2 \hbar^2 c v_F} \left(1 + 2 \sum_{n=1}^{N_\epsilon} \frac{|\epsilon|}{\sqrt{\epsilon^2 - 2n v_F^2 \hbar |eB|/c}} \right), \qquad (5.84)$$

where $N_\epsilon = [\epsilon^2 c/(2\hbar v_F^2 |eB|)]$ and $[\ldots]$ denotes the integer part of the expression (see Sec. 3.9 for a detailed derivation). The corresponding expression for the DOS of the hybrid surface–bulk closed orbits reads

$$\nu(\epsilon)_{\text{orbt}} = \frac{|eB|}{2\pi^2 L \hbar c} \sum_{n=n_{\text{min}}}^{\infty} [\delta (\epsilon - \epsilon_n) + \delta (\epsilon + \epsilon_n)], \qquad (5.85)$$

where the energies ϵ_n are given by Eq. (5.80).

As we see from Eq. (5.85), the DOS for the hybrid surface–bulk closed orbits $\nu(\epsilon)_{\text{orbt}}$ is inversely proportional to the slab thickness L. Therefore, unlike conventional bulk oscillations $\nu(\epsilon)_{\text{bulk}}$, whose amplitude is almost insensitive to the thickness of the slab, the surface–bulk oscillations are strongly suppressed at large L. Another distinctive feature of the surface–bulk oscillations is the unconventional dependence on the direction of the magnetic field, which is captured by $\cos \theta$ in Eq. (5.82). It is worth noting, however, that such a dependence is slightly modified when the electron–electron interactions are taken into account. Indeed, as was shown in [607], the interactions induce a dynamical correction to the chiral shift vector $\delta b\,\hat{n}$ along the magnetic field, where the magnitude δb is proportional to the interaction constant and to the strength of the magnetic field. Such a correction can be easily included in the analysis by replacing the bare arc length $2b_z$ with the renormalized value $2\sqrt{(b_z)^2 + 2b_z\delta b \cos\varphi \sin\theta + (\delta b \sin\theta)^2}$ in Eqs. (5.81) and (5.82). One finds, therefore, that the interaction effects cause a nontrivial dependence of the oscillation period on the azimuthal angle φ.

For completeness, let us also mention that the surface–bulk orbits can be also modified by another quantum effect proposed in [632]. In particular, it was argued that the tunneling from the Fermi arcs into bulk states can happen even before the surface quasiparticles reach the ends of the arcs. This effect can be captured qualitatively by reducing the length of the Fermi arcs by about $\propto \sqrt{\hbar c/(eB)}$.

In the presence of a strong in-plane component of the magnetic field, a 2D Fermi surface of a square shape (in the $k_x - k_z$ plane) could form in thin films of Dirac and Weyl semimetals [633, 637]. It originates from a combination of the chiral bulk channels near the boundary and the surface Fermi arcs. While one side of the contour is controlled by the Fermi arc length, the other is determined by the slab width and the in-plane component of the magnetic field $\propto L_y B_z$. The presence of these closed contours causes quantum oscillations determined by the following quantization rule [633]:

$$\frac{1}{B_y} = \frac{2\pi e(n + \alpha)}{2b_z L_y eB_z + c\hbar\delta S},\qquad(5.86)$$

where B_y is the component of the magnetic field normal to the surface, n is an integer, δS is a correction to the Fermi surface area associated with the curvature of the Fermi arcs, and $B_y \ll B_z$. As one can verify, the period of these quantum oscillations depends on the azimuthal angle even in the noninteracting case.

It was also predicted that, in thin films of Weyl semimetals with broken $\mathcal{P}$ symmetry, the Fermi arcs could hybridize with bulk states and form a figure-eight-shaped contour [609] instead of the square-shaped [633, 637] one in $\mathcal{P}$ symmetric Weyl semimetals. This unusual type of contour, where the Fermi arcs on the top and the bottom surfaces cross in the surface Brillouin zone, can be distinguished from a closed contour without self-intersections [633, 637] by investigating the edge transport in a sample of finite thickness and width, with a magnetic field applied normal to the surface. While the opposite edges are counterpropagating for a simple contour, the figure-8 contour could allow for the edge selection, where the current-carrying edge is determined by the orientation of the chiral shift rather than by the sign of the voltage. This effect is related to the formation of two wide and narrow types of channels. While wide channels are composed of parts of the figure-8 contours with intersections, narrow channels do not contain crossings.

For completeness, let us also discuss briefly the case of Dirac semimetals. In general, since Dirac semimetals are usually topologically trivial, one should not expect any surface–bulk orbits. However, as was discussed in Sec. 5.4, certain Dirac semimetals such as Na_3Bi or Cd_3As_2 are topologically nontrivial and do possess well-defined Fermi arcs. Strictly speaking, these semimetals can be classified as $\mathbb{Z}_2$ Weyl semimetals and, therefore, they possess a double set of the surface–bulk orbits. Nevertheless, there are additional complications when a sufficiently strong magnetic field is applied. Generically, such a field breaks the discrete rotation symmetry responsible for the protection of the Dirac points. This, in turn, leads to mixing of the pairs of orbits [632]. Usually, mixing becomes important when the magnetic field exceeds a certain critical value determined by material parameters. In the supercritical regime, closed Fermi arc orbits disconnected from the bulk develop on the top and bottom surfaces. They lead to quantum oscillations that are qualitatively the same as in nontopological materials. Because of this, such oscillations cannot be considered as a true hallmark feature of the Fermi arcs in Dirac semimetals.

Experimentally, the oscillations of resistivity due to the surface–bulk orbits were measured in thin slabs of Dirac semimetal Cd_3As_2 [636]. One of the most striking features in the experimental data is an additional characteristic frequency ω_S in the oscillations of resistivity that appears when the magnetic field is normal to the surface ($\theta = 0$). As one can see from Fig. 5.8(b), the surface component has a higher frequency and disappears when the magnetic field is parallel to the slab. Moreover, the characteristic $1/\cos\theta$ dependence of ω_S was also explicitly demonstrated. It is fair to mention, however, that a similar dependence on angle θ is also valid for closed orbits originating from the topologically trivial surface states, as well as from pairs of surface–bulk orbits that are strongly mixed in the bulk. These alternative scenarios are at odds, however, with the following two empirical facts: (i) the amplitude of oscillations decreases exponentially with the slab width, see Fig. 5.9(a), and (ii) the evidence of oscillations is relatively robust with respect to the surface disorder [636]. Moreover, the hybrid surface–bulk character of orbits is supported by the absence of

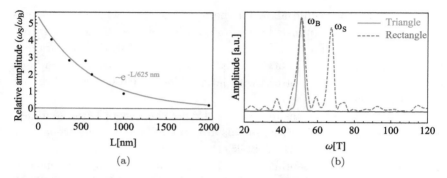

Fig. 5.9   (a) Relative amplitude of the surface oscillations compared to the bulk oscillations as a function of the sample thickness. (b) Fourier transform spectrum of the quantum oscillations in triangular (red line) and rectangular (blue dashed line) samples. (The data provided by P. J. W. Moll were used.)

the corresponding frequency for a triangle geometry, see Fig. 5.9(b). This is easy to understand in view of a destructive interference between different quantum paths. All these observations provide a strong support for the hybrid surface–bulk orbits, where both Fermi arcs and bulk states are involved.

5.6.2   *3D quantum Hall effect*

As is well known, 2D systems in a magnetic field allow for the QHE with the quantized Hall conductance [286, 293, 295, 615, 638]. As expected, such an effect is absent for 3D systems, where the additional dimension prevents the quantization. However, as was shown in [639], the presence of the Fermi arcs in Weyl and Dirac semimetals could make it possible to realize an unconventional 3D QHE in Weyl semimetals. In essence, such a Fermi arc QHE stems from the closed surface–bulk orbits discussed in Sec. 5.6.1. However, in order to describe this effect, a more complicated model of the surface should be employed. In particular, an anisotropy and a nonzero curvature of the Fermi arcs are crucial ingredients for the Fermi arc QHE. The corresponding minimal model reads

$$H(\mathbf{k}) = C_1 k_y^2 \mathbb{1}_2 + C_2(k_x^2 + k_z^2)\mathbb{1}_2 + \gamma\left(k^2 - m\right)\sigma_z + \hbar v_F\left(k_x\sigma_x + k_y\sigma_y\right).$$

$$(5.87)$$

This model has two Weyl nodes located at $\mathbf{k}_0^\pm = \pm\sqrt{m}\hat{\mathbf{z}}$ and the bulk energy spectrum $\epsilon_{\mathbf{k}} = C_1 k_y^2 + C_2(k_x^2 + k_z^2) \pm \sqrt{\gamma^2\left(k^2 - m\right)^2 + \hbar^2 v_F^2\left(k_x^2 + k_y^2\right)}$.

For concreteness, let us assume that $\gamma > |C_1|$. The Fermi arc surface states can be obtained by using the same method as in Sec. 5.3. The corresponding effective surface Hamiltonian reads [639]

$$H_{\text{arcs}}(k_x, k_z) = C_1 m + \zeta \hbar v_0 k_x + C_- \left(k_x^2 + k_z^2\right), \qquad (5.88)$$

where $\zeta = \pm$ for the bottom and top surfaces of the slab, respectively, $v_0 = v_F \sqrt{\gamma^2 - C_1^2}/\gamma$, and $C_- = C_2 - C_1$. The Fermi arc wave functions are normalizable only for

$$k_x^2 + k_z^2 + 2\zeta k_0 k_x < m, \qquad (5.89)$$

where $k_0 = \hbar v_F C_1/(2\gamma\sqrt{\gamma^2 - C_1^2})$. Together Eqs. (5.88) and (5.89) define the surface states, which have an open parabolic-like form. The arcs are related by the transformation $k_x \leftrightarrow -k_x$ and are connected via bulk states in the presence of a magnetic field.

By using the Kubo formalism, the authors of [639] obtained the following expression for the Fermi arc QHE conductivity in the clean limit:

$$\sigma_{xz}^{\text{arcs}} = \frac{e^2}{2\pi\hbar} \operatorname{sgn}(C_-) \operatorname{sgn}(eB) \left[\frac{c\hbar S_{\text{arcs}}}{2\pi eB} + \frac{1}{2} \right]. \qquad (5.90)$$

Here $[\dots]$ denotes the integer part rounded down and

$$S_{\text{arcs}} = 2m \left(1 + \frac{\hbar^2 v_0^2}{4C_-^2 m} \right) \arctan\left(\frac{2|C_-|\sqrt{m}}{\hbar v_0} \right) - \frac{\hbar|v_0|\sqrt{m}}{|C_-|} \qquad (5.91)$$

is the area swept by the Fermi arc in momentum space. As is clear from Eq. (5.90), the conductivity due to the Fermi arcs is quantized. The situation is more complicated in Dirac semimetals, where, nevertheless, the quantized conductivity could be still observed under certain conditions. Note also that the quantized Fermi arc conductivity might occur also in thin wires, where an interplay of magnetic field and finite-size effects could be important [611].

5.6.3 *Magnetotransport and optical transmission*

Before finalizing this chapter, let us consider also how Fermi arcs are manifested in nonlocal transport and the alternating current (AC) conductivity. The knowledge of the AC surface conductivity is valuable because it determines the optical response of a material. The presence of a magnetic field can also allow for various nonlocal effects when a perturbation on one surface can lead to a signal on the other. In essence, this is due to coupling between the Fermi arcs on the opposite surfaces of a material caused by the field. In this subsection, we will briefly discuss such phenomena associated with the Fermi arcs transport, concentrating primarily on those that can be easily understood qualitatively and do not require a detailed investigation of the bulk transport properties.

In the previous sections, the effects of the surface states on the magneto-transport were already demonstrated, but the emphasis was made primarily on the case of the magnetic field perpendicular to the surface. Unfortunately, analytical investigations for the setup where magnetic field is parallel to the surface are not as simple.

By using a tight-binding approach, the effect of a long- and short-range disorder in the quantum limit was studied in the geometry with parallel electric and magnetic fields, $\mathbf{E} \parallel \mathbf{B}$ [637, 640]. It was shown that the internode scattering can be partially mediated by the surface states leading to the increase of the conductivity with the magnetic field. In the case of a long-range disorder and a strong magnetic field, the backscattering of the LLL quasiparticles could proceed through the Fermi arcs leading to the increase of the conductivity with the slab's width as L_y^2. Note also that the numerical results in [640] validate the semiclassical model of the coupled surface–bulk states discussed in Sec. 5.6.1. Indeed, they confirm that the counterpropagating surface modes become short-circuited by bulk states when the magnetic field is normal to the surface.

Let us now briefly discuss the AC surface conductivity. By building up on the idea of the surface–bulk orbits in a magnetic field developed in [632], the authors of [641] predicted two other interesting effects that could probe directly the physics of the Fermi arcs in experiment: (i) a nonlocal DC voltage and (ii) observation of additional sharp resonances in the optical transmission.

The setup for the nonlocal voltage probe is schematically depicted in Fig. 5.10(a), where a pair of electrodes is applied to each of the surfaces of a slab. Similarly to a usual metal, when the magnetic field is absent and the slab is sufficiently thick, a positive voltage applied to the top surface induces a small positive voltage on the bottom surface. However, as was discussed in Sec. 5.6.1, the presence of a magnetic field normal to the Weyl semimetal surface creates closed surface–bulk orbits that induce nonlocal effects in the DC conductivity. In such a regime, a positive voltage on the top surface induces a substantial negative voltage on the bottom surface, which scales linearly with a magnetic field [641]. In view of the closed surface–bulk orbits, the reversal of the voltage sign is easy to understand

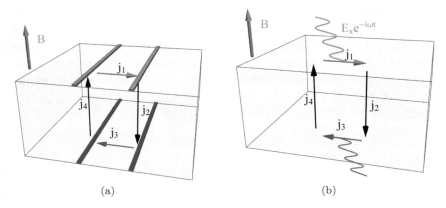

Fig. 5.10 (a) Experimental setup for a DC nonlocal current. Due to the surface–bulk orbits in a magnetic field, current j_1 injected at the top surface induces current j_3 at the bottom surface. (b) The cycling current j_1 generated by a pulse of electromagnetic waves on the top surface can lead to the electromagnetic emission from the lower surface.

because the velocities of the Fermi arc quasiparticles and, consequently, the currents j_1 and j_3 are opposite on the top and bottom surfaces of the slab.

Next, let us discuss the second proposal involving the AC conductivity, where the Weyl semimetal slab is irradiated with electromagnetic waves. The corresponding setup is shown in Fig. 5.10(b). As expected, in the case when the skin depth is much smaller than the slab thickness, the electromagnetic waves are mostly reflected from the slab. However, the nonlocal conductivity originating from the surface–bulk orbits may drastically change the conventional picture and lead to a resonance transmission. Let us sketch the physics behind this process. When an electromagnetic pulse hits the top surface at $t = 0$ it excites electrons on the Fermi arc and produces nonzero current j_1. Such a current will be going to the bulk LLs until all electrons excited by the electric field pulse are transferred, which takes about $t = t_{\mathrm{arcs}}/2$.[m] This induces the bulk current j_2. At time $t = (t_{\mathrm{arcs}} + t_{\mathrm{bulk}})/2$, the surface states at the bottom surface will start receiving the excited electrons, which then will return to the top surface via the bulk, therefore, closing the cycle. By making use of such arguments, one estimates that the transmission frequency due to the surface–bulk orbits is given by $\omega_{\mathrm{sb}} = 2\pi/(t_{\mathrm{arcs}} + t_{\mathrm{bulk}})$.

---

[m]Similarly to Sec. 5.6.1 $t_{\mathrm{arcs}} = 4\hbar b_z/(v_F eB \cos\theta)$ and $t_{\mathrm{bulk}} = 2L/(v_F \cos\theta)$.

As in mechanical systems, the resonance regime of the optical transmission takes place when the frequency of the incident wave is a multiple of ω_{sb}. When this condition is met, the resonant absorption of the radiation on one of the surfaces is converted into an electromagnetic emission on the other side. Because of some unavoidable reflection, of course, the transmission is not complete, but it can be enhanced significantly by applying a strong magnetic field and by choosing materials with a large value of the chiral shift.

Chapter 6

Anomalous Transport Properties

> *The truth may be puzzling. It may take some work to grapple with. It may be counterintuitive. It may contradict deeply held prejudices. It may not be consonant with what we desperately want to be true. But our preferences do not determine what's true.*
>
> Carl Sagan

Without much exaggeration, the electron transport properties are among the most important characteristics of solids. Not only they carry a trove of information about the fundamental physics of electron quasiparticles but also are of immense practical value. As is well known, the electron transport is determined largely by the band structure of materials. For example, while the absence of a band gap at the Fermi level in metals makes them good conductors, a sufficiently large gap between the valence and conductance bands is one of the definitive properties of insulators. In applied physics, the understanding of widely tunable transport properties in semiconductors opened the possibility to manipulate their conductivity and led to the invention of transistors, which nowadays are the basic elements of nearly all electronics. Of similar significance was also the discovery of superconductivity, which provides an example of an unusual electron transport in solids. Superconductivity in Weyl and Dirac semimetals will be discussed in Chapter 9. It is not surprising, therefore, that the search for novel materials with unconventional transport properties is one of the key driving forces of solid-state physics.

The experimental discovery of graphene, which has a 2D relativistic-like spectrum, ignited the interest of the condensed matter physics community to materials with low-energy Dirac quasiparticles. While a Dirac-like dispersion relation is unconventional for solids, its realizations come with a variety of interesting transport properties, including a remarkably high electron mobility, the integer quantum Hall effect, the Klein paradox, etc. The transport phenomena in 3D Weyl and Dirac semimetals are, perhaps, even more diverse and promising for applications than those in graphene. In particular, the low-energy description of electron quasiparticles in these 3D materials includes an additional degree of freedom known as chirality, which exactly coincides with the relativistic notion. As will be shown below, chirality plays a very important role in electron transport leading to a range of unusual or "anomalous" transport properties that are often related to the celebrated chiral anomaly [39, 40]. The examples include the so-called negative longitudinal magnetoresistivity and the generation of electric and/or chiral currents along an external magnetic or strain-induced pseudomagnetic field. From a conceptual viewpoint, the transport in Weyl and certain Dirac semimetals is also strongly affected by the nontrivial topology of their band structure, which is quantified by a nonvanishing Berry curvature (see Sec. 3.3). The hallmark topological effects are the anomalous Hall effect (AHE), whose intrinsic part is determined completely by the Berry curvature, and the electron transport due to the surface Fermi arcs states discussed in Sec. 5.5.

Since the in-depth description of all aspects of the transport properties in Dirac and Weyl semimetals is a formidable task, we will concentrate primarily on the anomalous features in this chapter. Some other electronic transport properties including the effects of disorder, thermoelectric, optical, and hydrodynamic phenomena will be discussed in Chapter 7.

6.1　Chiral anomaly

One of the most remarkable electronic properties of Dirac and Weyl semimetals is their ability to realize the celebrated *chiral anomaly* [39, 40], which was discovered by John Bell, Roman Jackiw, and Stephen Adler in elementary particle physics in 1969. Given its significance in various properties of Dirac and Weyl semimetals, it is instructive to briefly review the basic notions of this phenomenon.

Let us first describe what the chiral symmetry is. For this, it is convenient to use the model Hamiltonian of a Weyl semimetal

$$H_0 = \hbar v_F \left(\boldsymbol{\alpha} \cdot \mathbf{k} \right) - \left(\boldsymbol{\alpha} \cdot \mathbf{b} \right) \gamma_5 + b_0 \gamma_5. \tag{6.1}$$

This Hamiltonian is invariant under the global $U(1)$ transformations.[a] However, in view of the block-diagonal structure of Hamiltonian (3.5), there is a larger symmetry, which is generated by two independent global phase transformations of the upper and lower blocks. This $U_+(1) \times U_-(1)$ symmetry, or $U_R(1) \times U_L(1)$ in high energy physics notations, is known as the *chiral symmetry*. In the condensed matter setting, the same type of symmetry is usually called the valley symmetry, as the upper and lower blocks describe two valleys in the electron band structure. Note that the symmetry group $U_+(1) \times U_-(1)$ could be spanned by a combination of two transformations with the phase factors $e^{i\varphi}$ and $e^{i\theta\gamma_5}$.

According to Noether's theorem, the invariance of the action with respect to continuous phase transformations $e^{i\varphi}$ and $e^{i\theta\gamma_5}$ implies the existence of conserved electric and chiral charges. Mathematically, this is expressed in the form of two continuity relations for the corresponding current densities: $\partial_\mu j^\mu = 0$ and $\partial_\mu j_5^\mu = 0$, where

$$j^\mu(x) = \bar{\psi}(x)\gamma^\mu\psi(x) \tag{6.2}$$

is the conventional electric current density and

$$j_5^\mu(x) = \bar{\psi}(x)\gamma^\mu\gamma_5\psi(x) \tag{6.3}$$

is the chiral current density, $\psi(x)$ is the wave function, and $\bar{\psi} = \psi^\dagger\gamma^0$ is the Dirac conjugate bispinor. Note that here we use a relativistic-like notation for the four-vectors, although the Lorentz symmetry is absent. As usual, x^μ is the contravariant space–time four-vector, where $\mu = 0, \ldots, 3$. The components of spatial vector $\mathbf{x}$ are defined by x^i with $i = 1, 2, 3$. In a classical description, both electric and chiral charges are indeed conserved. However, both charges cannot be simultaneously conserved in quantum theory when fermions interact with electromagnetic fields. This is the essence of the chiral anomaly. The absolute status of the electric charge conservation, which is never violated in experiments, implies that the conservation of the chiral charge must be violated in the presence of background electromagnetic fields.

Let us briefly remind why such a violation of the classical conservation laws takes place (for a more detailed consideration, see [41, 42, 44, 357]).

---

[a]Gauging this symmetry and introducing interactions with the electromagnetic vector potential leads to the standard gradient invariance with respect to local phase transformations.

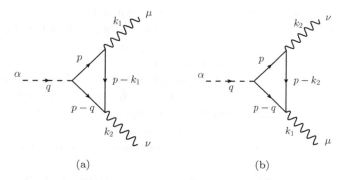

Fig. 6.1 The triangle diagrams describing the decay of the neutral pion π^0 into two γ quanta. Dashed lines correspond to the neutral pion π^0, solid lines denote fermions, and two wavy lines represent external photon lines.

The current algebra was a popular approach in the studies of hadron interactions in the 1960s, before the development of quantum chromodynamics [642]. One of its main ideas was to use the divergence of the chiral current as an interpolating field for the neutral pion field π^0, which was supplemented by the assumption of the partial conservation of the axial current (PCAC) [357, 642]. In such a framework, to the lowest order in perturbation theory, the amplitude of the neutral pion decay into two photons, $\pi^0 \to 2\gamma$, is determined by the triangle Feynman diagrams presented in Fig. 6.1. Note that the external photon lines are interchanged in the two diagrams.

In quantum field theory, the two Feynman diagrams in Fig. 6.1 represent one-loop contributions to the correlation function of the axial current j_5^α and two electromagnetic currents j^μ and j^ν, i.e.,

$$T^{\mu\nu\alpha}(k_1, k_2) = i \int d^4x_1 d^4x_2 \, \langle 0 | \, \mathcal{T} \left[j^\mu(x_1) j^\nu(x_2) j_5^\alpha(0) \right] | 0 \rangle \, e^{ik_1 x_1 + ik_2 x_2},$$

$$(6.4)$$

where k_1 and k_2 are momentum four-vectors and $\mathcal{T}$ is the time-ordering operator. In view of the PCAC relation, the decay amplitude for $\pi^0 \to 2\gamma$ is given by $q_\alpha T^{\mu\nu\alpha}$, where $q = k_1 + k_2$. Therefore, if one ignores the small contributions due to nonzero masses of the u and d quarks, which constitute the neutral pion, i.e., $\pi^0 \sim (u\bar{u} - d\bar{d})/\sqrt{2}$, then the conservation of the chiral current would mean that the amplitude of the π^0 decay vanishes. By making use of the diagrams in Fig. 6.1, we obtain the following analytic expression for the correlation function $T^{\mu\nu\alpha}$:

$$T^{\mu\nu\alpha} = -i \int \frac{d^4 p}{(2\pi)^4} \text{tr} \left[\frac{i}{p^\delta \gamma_\delta} \gamma^\alpha \gamma_5 \frac{i}{(p^\delta - q^\delta)\gamma_\delta} \gamma^\nu \frac{i}{(p^\delta - k_1^\delta)\gamma_\delta} \gamma^\mu - (\ldots) \right].$$

$$(6.5)$$

Here the ellipsis represents a term, where momenta k_1 and k_2, as well as indices μ and ν are interchanged. By making use of the identity $q_\alpha \gamma^\alpha \gamma_5 = \gamma_5 (p^\delta - q^\delta)\gamma_\delta + p^\delta \gamma_\delta \gamma_5$, we derive

$$q_\alpha T^{\mu\nu\alpha} = \Delta_1^{\mu\nu} + \Delta_2^{\mu\nu}, \qquad (6.6)$$

where

$$\Delta_1^{\mu\nu} = \int \frac{d^4 p}{(2\pi)^4} \text{tr} \left[\frac{i}{p^\delta \gamma_\delta} \gamma_5 \gamma^\nu \frac{i}{(p^\delta - k_1^\delta)\gamma_\delta} \gamma^\mu \right.$$

$$\left. - \frac{i}{(p^\delta - k_2^\delta)\gamma_\delta} \gamma_5 \gamma^\nu \frac{i}{(p^\delta - q^\delta)\gamma_\delta} \gamma^\mu \right]$$

$$(6.7)$$

is the contribution of the diagram in Fig. 6.1(a) and $\Delta_2^{\mu\nu}$ is given by an expression similar to $\Delta_1^{\mu\nu}$, but with the replacements $k_1 \leftrightarrow k_2$ and $\mu \leftrightarrow \nu$. Obviously, the chiral current would be conserved if the sum $\Delta_1^{\mu\nu} + \Delta_2^{\mu\nu}$ vanished. At first sight, this appears to be indeed the case. By shifting the integration variable $p \to p + k_2$ in the second term in Eq. (6.7), one finds that the two terms differ only by a sign and, consequently, $\Delta_1^{\mu\nu} = \Delta_2^{\mu\nu} = 0$. A careful examination reveals, however, that each integral is linearly divergent and, therefore, the shift of the loop momentum $p \to p + k_2$ should be performed with caution. Indeed, the shift produces an additional surface terms in amplitude (6.7). Let us note in passing that the problem is present only for linearly (and higher-order) divergent diagrams, while logarithmically diverging diagrams are not affected so much by a shift of the integration variable. If the loop integral is regularized, of course, no ambiguity arises in the calculation at all. One might argue, therefore, that the question of the chiral charge conservation reduces down to the issue of regularization in Feynman diagrams.

By parameterizing a general momentum shift in terms of the two independent momenta of the triangle diagrams

$$p \to p + \kappa k_1 + (\kappa - \beta) k_2 \qquad (6.8)$$

and using

$$\int_{-\infty}^{\infty} d^4 x \left[f(x + a) - f(x) \right] = 2i\pi^2 a^\mu \lim_{R \to \infty} R^2 R_\mu f(R) \qquad (6.9)$$

for a four-dimensional hypersphere of radius R, one can find that the divergence of the axial current is determined by the following relation (for details see, e.g., [357]):

$$q^\alpha T_{\mu\nu\alpha} = -\frac{1-\beta}{4\pi^2}\epsilon_{\mu\nu\sigma\rho}k_1^\sigma k_2^\rho. \tag{6.10}$$

Similarly, by contracting $T_{\mu\nu\alpha}$ with the photon momenta k_1 and k_2, one derives the following two relations:

$$k_1^\mu T_{\mu\nu\alpha} = \frac{(1+\beta)}{8\pi^2}\epsilon_{\nu\alpha\sigma\rho}k_1^\sigma k_2^\rho, \quad k_2^\nu T_{\mu\nu\alpha} = \frac{(1+\beta)}{8\pi^2}\epsilon_{\mu\alpha\rho\sigma}k_1^\sigma k_2^\rho, \tag{6.11}$$

which are directly connected with the electric current continuity relation $\partial_\mu j^\mu = 0$.

One might notice right away that there exists a preferred choice of parameter β that is consistent with the electric charge conservation, i.e., $\beta = -1$. Indeed, in this case, Eq. (6.11) reproduces the conventional Ward identities for the electric (vector) currents [42–45]. This choice for β enforces the gauge invariance of the theory and ensures the internal consistency of the theoretical framework at each order in perturbation theory. On the other hand, Eq. (6.10) gives

$$q^\alpha T_{\mu\nu\alpha} = -\frac{1}{2\pi^2}\epsilon_{\mu\nu\sigma\rho}k_1^\sigma k_2^\rho. \tag{6.12}$$

This result means that the naive chiral current conservation law is modified in quantum field theory as follows:

$$\partial_\mu j_5^\mu = \frac{1}{8\pi^2}\epsilon_{\mu\nu\sigma\rho}F^{\mu\nu}F^{\sigma\rho}. \tag{6.13}$$

The anomalous term on the right-hand side of Eq. (6.13) is known as the *Adler–Bell–Jackiw anomaly*, or, equivalently, the *chiral anomaly*. It is the paradigmatic and, perhaps, the best known example of quantum anomalies. Recall that a quantum anomaly corresponds to a situation when a symmetry of the classical action is absent in the quantum theory.[b] It is notable that the chiral anomaly is exact at one-loop order and does not receive any higher-order corrections.

---

[b]In the functional integral formalism, the chiral anomaly is connected with a noninvariance of the integration measure under the chiral phase transformation $\psi \to \psi e^{i\theta\gamma_5}$ [643].

It is truly amazing that such a subtlety connected with the regularization of divergent diagrams has a profound physical importance in quantum theory and is the crucial ingredient for the theoretical description of the pion decay time in the framework of the PCAC. Moreover, as will be shown below, the effects of the chiral anomaly are manifested in the physical properties of Dirac and Weyl semimetals whose low-energy quasiparticles have relativistic-like dispersions and well-defined chiralities. In particular, the chiral anomaly affects collective excitations, transport properties, etc.

Before concluding this section, it might be important to note that the issue of the chiral symmetry is very delicate in Dirac and Weyl semimetals. First of all, the underlying microscopic theory for electrons, which are described by the Schrödinger equation with a periodic lattice potential, does not have an inherent notion of chirality. Instead, chirality arises as an emergent concept that is well defined only for the low-energy excitations in the vicinity of Dirac points and Weyl nodes. In addition, while the formulation of the microscopic theory on a lattice provides a natural regularization, unfortunately, it obscures the status of the chiral anomaly and its implications. These issues are also well known and discussed extensively in the context of lattice calculations in high energy physics [46, 50, 54].

6.2 Negative magnetoresistivity

The negative magnetoresistivity is widely accepted as one of the hallmark properties of Dirac and Weyl semimetals. The term itself is an archetypal misnomer and means, in fact, that the resistivity decreases with a magnetic field rather than literary becomes negative. In either case, such a phenomenon is rare in solids because a magnetic field usually tends to hinder the electric current. In fact, the growth of resistivity with a magnetic field can be easily explained semiclassically. Since the magnetic field bends trajectories of charge carriers, the particle mean free path becomes shorter and, as a result, the resistivity increases.[c] In metals, the corresponding correction to the resistivity reads [288]

$$\delta\rho \propto (\Omega_{\mathrm{L}}\tau)^2, \qquad (6.14)$$

---

[c]Formally, the case of parallel electric and magnetic fields, $\mathbf{E} \parallel \mathbf{B}$, is an exception because the transport in the conventional semiclassical approach is not affected in such a case.

where $\Omega_L \sim eB/(mc)$ is the Larmor frequency, τ is the collision time, and m is the effective mass of electron quasiparticles.[d]

Unlike ordinarily metals, Dirac and Weyl semimetals with the relativistic-like energy spectrum of quasiparticles have an additional mechanism of the electron transport that relies on the quantum chiral anomaly discussed in Sec. 6.1. The realization of the quantum anomaly in solids is quite remarkable. Not only it allows one to test subtle effects of quantum physics, but also provides one of many bridges between condensed matter and high energy physics. In the rest of this section, we will first discuss the negative magnetoresistivity and then present a more rigorous treatment in terms of Green's function in a magnetic field by using the Kubo linear response theory.

6.2.1 *Qualitative description*

Historically, Holger Nielsen and Masao Ninomiya were the first who predicted [55] that the chiral anomaly could lead to a qualitatively new transport regime characterized by a negative longitudinal magnetoresistivity. The corresponding analysis assumed the limit of a strong magnetic field where the Landau levels are formed.

For the sake of simplicity, here we will start the discussion of anomalous transport by first using a one-dimensional (1D) chiral model with the following energy spectrum:

$$\epsilon_{\mathbf{p}} = -\lambda v_F p_z, \tag{6.15}$$

where $\lambda = \pm$ corresponds to the right- and left-handed quasiparticles, v_F is the Fermi velocity, and p_z is the 1D momentum.[e] Classically, when an external electric field E_z is applied to the system, charged quasiparticles accelerate according to $\dot{p}_z = -eE_z$, where $-e$ is the electron charge. Such an acceleration is equivalent to a spectral flow of chiral quasiparticles, which drives the chirality flipping at $p_z = 0$ when the zero energy is crossed.

---

[d]There are known effects, however, that could lead to a decrease of resistivity in nontopological materials too [288]. One of them is realized in the presence of magnetic impurities when the external magnetic field polarizes the spins of impurities and prevents electron scattering with spin flipping. Because of the net decrease of the scattering amplitude, the conductivity rises. This phenomenon is known as the weak antilocalization and will be discussed in more detail at the end of Sec. 7.1.4.

[e]Note that the momentum is defined as $\mathbf{p} = \hbar\mathbf{k}$, where $\mathbf{k}$ is the crystal wave vector.

The corresponding rate of the chiral charge production is given by

$$\dot{n}_\lambda = \frac{\dot{\epsilon}_{\mathbf{p}}}{2\pi\hbar v_F} = \lambda\frac{eE_z}{2\pi\hbar}, \tag{6.16}$$

where $(2\pi\hbar)^{-1}$ is the 1D density of states. Since the rate is proportional to the chirality index λ, it is clear that Eq. (6.16) describes the production of right-handed quasiparticles and the annihilation of the left-handed ones. Therefore, the net chiral (or axial) charge density $\rho_5 = -e(n_+ - n_-)$ is not conserved, but produced at the rate

$$\dot{\rho}_5 = -e\left(\dot{n}_+ - \dot{n}_-\right) = -\frac{e^2 E_z}{\pi\hbar}. \tag{6.17}$$

It is very important to emphasize that, unlike the chiral charge, the electric charge is conserved exactly (i.e., $\dot{n}_+ + \dot{n}_- = 0$), as expected. While very simple, the above 1D model is quite instructive as it closely resembles the realization of the chiral anomaly in 3D models in the limit of a very strong magnetic field. Indeed, in such a case, the effective dimensional reduction is achieved because a transverse motion of quasiparticles is quenched by the field.

Let us now proceed to the three-dimensional case. By applying an external magnetic field $\mathbf{B}$, one can show that the energy spectrum becomes discrete and the Landau levels are formed.[f] The corresponding band structure is schematically shown in Fig. 6.2. The key feature of the relativistic-like energy spectrum in a magnetic field is the gapless *lowest Landau level* (LLL) with the following chiral dispersion relation:

$$\epsilon_{n=0} = -s_{\mathrm{B}}\lambda v_F p_z, \tag{6.18}$$

where, by assumption, $\mathbf{B} \parallel \hat{\mathbf{z}}$ and $s_{\mathrm{B}} = \mathrm{sgn}\,(eB_z)$. Note that each Landau level, including the LLL, is highly degenerate with the number of states per unit area proportional to $|eB_z|$.[g]

As is clear from Eq. (6.18), the longitudinal component of the quasiparticle velocity $v_z = \partial\epsilon_{n=0}/\partial p_z$ has opposite signs for left- and right-handed Weyl nodes. This closely resembles the situation in the 1D model in Eq. (6.15). In fact, the whole low-energy dynamics in the 3D model is dominated by the LLL, which is effectively one-dimensional. One finds that the LLL not only plays the key role in realizing the quantum anomaly in

---

[f]The energy spectrum and the wave functions for Dirac quasiparticles are derived in Sec. 6.4.1. For the case of particles with nonrelativistic spectrum, see, for example, [285].
[g]For an intuitive derivation of the density of states in a magnetic field, see [285].

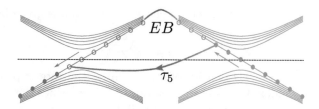

Fig. 6.2   Schematic illustration of the chiral anomaly in a Weyl semimetal with two Weyl nodes. A strong magnetic field leads to the formation of Landau levels, while an applied electric field tends to create an imbalance between the left- and right-handed electrons. This imbalance is washed out by chirality flipping processes.

the 3D case, but also saturates it [54]. As for the higher Landau levels, they do not contribute to the chiral anomaly.

In order to estimate the rate of chiral charge production in a 3D model in a magnetic field, one needs to take into account the density of the LLL states, which is given by $|eB_z|/(4\pi^2\hbar c)$ (see footnote g on p. 205). Then one derives the following production rate of the chiral charge density:

$$\dot{n}_\lambda = \frac{\dot{\epsilon}_{n=0}|eB_z|}{(2\pi\hbar)^2 c v_F} = \lambda\frac{e^2\,(\mathbf{E}\cdot\mathbf{B})}{(2\pi\hbar)^2 c}. \tag{6.19}$$

Consequently, the rate of change of the chiral charge density is given by

$$\dot{\rho}_5 = -e\,(\dot{n}_+ - \dot{n}_-) = -\frac{e^3\,(\mathbf{E}\cdot\mathbf{B})}{2\pi^2\hbar^2 c}. \tag{6.20}$$

As in the 1D case, the nonconservation of the chiral charge density is caused by the spectral flow of chiral states that cross the zero energy and, therefore, produce quasiparticles of one chirality and holes of the opposite chirality. In such a process, the chemical potential for the right-handed states μ_+ increases, while the chemical potential for the left-handed ones μ_- decreases. Therefore, in order to reach a steady state, some relaxation process is needed.

Because of the chiral nature of quasiparticles, their backscattering within a single Weyl node is inhibited in the limit of low temperature and strong magnetic field. The other scattering channel for the chiral quasiparticles of a given Weyl node is related to the transition to the other node. While this scattering is also suppressed partially by kinematic constraints, especially when the separation between the Weyl nodes is large, such a mechanism might be the dominant relaxation process at low temperature and strong magnetic fields.

To estimate the chiral chemical potential in the steady state, one can modify Eq. (6.20) by adding a phenomenological chirality relaxation term, i.e.,

$$\dot{\rho}_5 = -\frac{e^3\,(\mathbf{E}\cdot\mathbf{B})}{2\pi^2\hbar^2 c} - \frac{\rho_5}{\tau_5^{(2)}}. \tag{6.21}$$

Since $\dot{\rho}_5 = 0$ in the steady state, one obtains

$$\rho_5 = -\frac{e^3\,(\mathbf{E}\cdot\mathbf{B})\,\tau_5^{(2)}}{2\pi^2\hbar^2 c}. \tag{6.22}$$

In the limit of a strong magnetic field, $\rho_5 \propto |eB|\mu_5$ and the last estimate becomes equivalent to having a nonzero chiral chemical potential $\mu_5 = (\mu_+ - \mu_-)/2 \propto v_F|eE|\tau_5^{(2)}$ [55]. However, relation (6.22) can also be useful in the limit of high temperatures and weak magnetic fields. The schematic illustration of the chiral anomaly and the internode relaxation process is presented in Fig. 6.2.

One can use simple energy arguments in order to calculate the chiral anomaly contribution to the electrical conductivity. In particular, because of the difference between the chemical potentials μ_+ and μ_-, where $\mu_+ > \mu_-$ is assumed, the transfer of quasiparticles from the left-handed to the right-handed Weyl node costs energy, which is estimated as $\mu_5(\dot{n}_+ - \dot{n}_-)$ per unit time and unit volume. This should be compensated by the work done by the external electric field which is given by $(\mathbf{J}\cdot\mathbf{E})$. The energy balance leads to the following relation:

$$(\mathbf{J}\cdot\mathbf{E}) = \frac{e^2\,(\mathbf{E}\cdot\mathbf{B})}{2\pi^2\hbar^2 c}\mu_5. \tag{6.23}$$

By making use of the expression for the chiral chemical potential in the steady-state regime $\mu_5 = |e|v_F(\mathbf{E}\cdot\hat{\mathbf{B}})\tau_5^{(1)}$, one derives the following anomalous contribution to the conductivity tensor:

$$\sigma_{ij}^{(\mathrm{anom})} = \frac{|e|^3 v_F \tau_5^{(1)} B_i B_j}{2\pi^2\hbar^2 cB}. \tag{6.24}$$

Note that the corresponding anomalous part of the current is defined by $J_i^{(\mathrm{anom})} = \sigma_{ij}^{(\mathrm{anom})} E_j$. As we see from Eq. (6.24), the anomaly-related contribution to the conductivity grows *linearly* with the magnetic field strength B. In the special case of parallel fields, $\mathbf{E} \parallel \mathbf{B}$, we see that the above anomalous conductivity is positive, which implies that the *longitudinal* magnetoresistance receives a negative contribution. As for the *transverse* Ohmic conductivity for Dirac quasiparticles in the ultraquantum regime, it was calculated by Abrikosov in [644] and was shown to give a positive magnetoresistance with a linear dependence on the field. Note that the latter has no connection with the chiral anomaly.

Let us consider the regime of weak magnetic fields. In this case, the steady-state value of the chiral chemical potential can be obtained from the relation in Eq. (6.21). Indeed, as will be shown explicitly later, see Eq. (6.178), a small chiral charge density ρ_5 can be identified with the following value of the chiral chemical potential:

$$\mu_5 \simeq -\frac{3\pi^2 v_F^3 \hbar^3 \rho_5}{e\left(3\mu^2 + \pi^2 T^2\right)}, \tag{6.25}$$

where the electric chemical potential $\mu = \mu_+ + \mu_-$ and/or temperature T are assumed to be much larger than μ_5. Then, by combining Eqs. (6.22), (6.23), and (6.25), one obtains the following anomalous contribution to the conductivity tensor:

$$\sigma_{ij}^{(\text{anom})} = \frac{3e^4 v_F^3 \tau_5^{(2)} B_i B_j}{4\pi^2 \hbar c^2 \left(3\mu^2 + \pi^2 T^2\right)}. \tag{6.26}$$

In the special case when electric and magnetic fields are aligned, the chiral anomaly produces a positive contribution to the longitudinal conductivity that scales *quadratically* with the field, $\sigma = \sigma_0 + \sigma_1 B^2$. Here σ_0 denotes all other (e.g., Ohmic) nonanomalous contributions. This result translates into a negative magnetoresistivity, i.e., $\rho \sim 1/\sigma \approx 1/\sigma_0 - \sigma_1 B^2/\sigma_0^2$.

Before concluding this section, let us emphasize that the derivation of the anomalous conductivity in the regimes of strong and weak magnetic fields, with the results given by Eqs. (6.24) and (6.26), respectively, differ only by the dependence of the chirality flipping time τ_5 on magnetic field. The corresponding dependence will be further discussed in Sec. 7.1.4.

6.2.2 *Experimental results and nonlocal effects*

At present, a negative longitudinal magnetoresistivity has been observed in numerous experiments in Weyl [62–67] and Dirac [56–61] semimetals. By using the Weyl semimetal TaAs [64] as a representative example, the typical experimental results are presented in Fig. 6.3. The angular dependence of the magnetoresistivity is shown in Figs. 6.4(a) and 6.4(b), which was measured in the Dirac semimetal Na$_3$Bi. The experimental data clearly demonstrate a decrease (increase) of the magnetoresistivity (magnetoconductivity) with a magnetic field and a strong dependence on the field

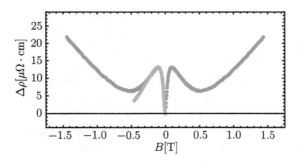

Fig. 6.3 Experimental data for the magnetoresistivity modification in the Weyl semimetal TaAs (red dotes) fitted with the semiclassical result (green solid line) at $T = 2$ K. (Adapted from [64].)

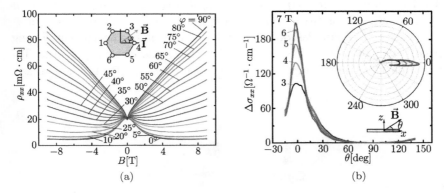

Fig. 6.4 (a) Resistivity ρ_{xx} in the Dirac semimetal Na$_3$Bi as a function of the magnetic field strength B at a few values of the in-plane angle φ between the current and the magnetic field. (b) The conductivity enhancement $\Delta\sigma_{xx} = \sigma_{xx}(\phi) - \sigma_{xx}(\pi/2)$ as a function of the out-of-plane angle θ at a few values of magnetic field. In both panels, temperature is $T = 4.5$ K. (Adapted from [57].)

direction. Such a dependence of the resistivity on a magnetic field obtained experimentally is quite nontrivial and signifies an interplay of several effects. As will be discussed briefly at the end of Sec. 7.1.4, the increase of the resistivity at very small fields could be related to the weak antilocalization effects. With the increase of a magnetic field, the chiral anomaly contribution overwhelms the weak antilocalization correction and the longitudinal magnetoresistivity becomes negative. It is worth noting, however, that the magnitude of the magnetoresistivity increase is quite large to be explained by the weak antilocalization effect alone. Further, the results in

Figs. 6.4(a) and 6.4(b) clearly demonstrate that the enhancement of conductivity is sensitive to the relative orientation of **B** and **E**.

While the observation of negative magnetoresistivity is widely accepted as a hallmark feature of the chiral anomaly, it is worth noting that, in principle, it might be also produced by other effects unrelated to the anomaly. Among them are the *current jetting* [645] due to an inhomogeneous distribution of the current and the electron scattering on long-range ionic impurities [646].[h] Therefore, the attribution of a negative magnetoresistance to the chiral anomaly should be interpreted with caution and the corresponding results should be always cross-verified via other observables.

To better understand subtleties of the observation of chiral anomaly signatures, it is instructive to consider the phenomenon of current jetting in more details. In essence, it is a nonuniform distribution of the current density that comes from (i) a strongly anisotropic magnetoresistivity tensor in materials and/or (ii) a point current injection (e.g., induced by point-like contacts). Examples of numerical simulations of the current jetting from [645] are shown in Fig. 6.5, which shows the electric potential distribution in a finite sample of the Weyl semimetal NbP. As one can see,

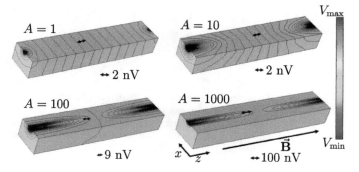

Fig. 6.5 Electric potential distribution calculated numerically for a few resistivity anisotropies $A = \rho_{xx}/\rho_{zz}$ in the Weyl semimetal NbP. Here ρ_{zz} and ρ_{xx} are the longitudinal and transverse resistivities with respect to the direction of magnetic field **B**, respectively. The electric current is injected through point-like electrodes at the upper left and right edges of samples. Black lines denote equipotential surfaces. The dimensions of samples are $0.4 \times 0.3 \times 2.0$ mm (width $\times$ thickness $\times$ length). (Adapted from [645].)

---

[h] As was shown in [647], the negative magnetoresistivity can appear under certain conditions in conventional centrosymmetric and time-reversal invariant materials without the chiral anomaly.

the potential distribution becomes highly distorted for a large magnetoresistivity anisotropy ρ_{xx}/ρ_{zz}, where ρ_{zz} and ρ_{xx} are the longitudinal and transverse resistivities with respect to the direction of the magnetic field, respectively. Therefore, the electric potential that is experimentally measured at points other than the current source and drain is not directly proportional to the intrinsic resistance of the material. Moreover, the corresponding data could even demonstrate a negative total longitudinal magnetoresistance, which was the case for small misalignments between the current and the magnetic field considered in [645].

It is reasonable to assume that the majority of existing experimental data is plagued to some degree by the current jetting effects. Fortunately, there exist a few ways to separate the latter from the effects of the chiral anomaly. The most obvious, although not the simplest in practice, is to ensure that the electric current is injected uniformly into the sample. Another approach is to use thin, needle-like crystals that allow for a relatively uniform distribution of the electric current across the whole sample. The corresponding optimal aspect ratio L/w, where L is the length and w is the width, is estimated to be $L/w \gtrsim \sqrt{\rho_{xx}/\rho_{zz}}$ (assuming that the width is larger than the thickness). When anisotropy is large, however, it is difficult to grow sufficiently long crystals. For example, a large anisotropy ratio ρ_{xx}/ρ_{zz} of the order of 2500 in NbP [645] suggests that the optimal aspect ratio should be larger than 50, while the crystals grown in laboratory usually have $L/w \approx 5$. Thus, the observation of the negative magnetoresistivity on its own should not be considered as a definitive proof of the chiral anomaly effects in Weyl and Dirac semimetals.

Besides the negative magnetoresistance, which is a local characteristic, it was proposed that the chiral charge imbalance induced by the chiral anomaly could be detected nonlocally [648]. The corresponding schematic setup is depicted in Fig. 6.6(a). Here, due to the long intervalley relaxation time,[i] the chiral chemical potential μ_5 locally induced via the chiral anomaly in the pair of electrodes on the left side could propagate over sufficiently long distances. Therefore, by applying a probe magnetic field far from the source and drain, one can convert μ_5 into a voltage drop measured in the pair of electrodes on the right side. Recently, the corresponding effect was tested experimentally in Cd_3As_2 [649] and $Bi_{0.97}Sb_{0.03}$ [650]. The nonlocal

---

[i]For example, according to [64], the chiral charge relaxation time in TaAs is $\tau_5 \approx 59.6$ ps at $T = 2$ K. The data for other materials are given in Appendix D.

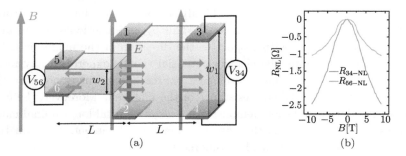

Fig. 6.6 (a) Schematic model setup for measuring nonlocal effects of the chiral anomaly. The voltage is applied to the electrodes 1–2 and the nonlocal voltage is measured via the pairs 3–4 and 5–6 separated by a distance L. (b) Nonlocal resistance R_{NL} obtained by extracting the diffusive Ohmic part as a function of magnetic field. Contacts 3–4 and 5–6 are shown in red and green lines. Temperature is $T = 20$ K. (Adapted from [649]).

part of resistance is shown in Fig. 6.6(b). The nonlocal part was obtained by subtracting the diffusive Ohmic part from the experimental data.

6.2.3 *Kubo linear response theory*

Let us discuss a more rigorous method for studying the transport in Dirac and Weyl semimetals that utilizes the Green's function technique and the Kubo linear response theory.

6.2.3.1 *Model*

The free low-energy Weyl Hamiltonian (3.7) in an external electromagnetic field takes the form

$$H_{\mathrm{W}} = \int d^3r\, \bar{\psi}(\mathbf{r}) \left[v_F\left(\boldsymbol{\gamma} \cdot \boldsymbol{\pi}\right) - \hbar v_F(\mathbf{b} \cdot \boldsymbol{\gamma})\gamma_5 - \mu\gamma^0 \right] \psi(\mathbf{r}), \qquad (6.27)$$

where $\psi(\mathbf{r})$ is the fermion field, $\boldsymbol{\pi} = -i\hbar\boldsymbol{\nabla} + e\mathbf{A}/c$ is the canonical momentum, $\mathbf{A}$ is the vector potential, which describes a constant magnetic field $\mathbf{B}$, $-e$ is the electron charge, c is the speed of light, μ is the electric chemical potential, and $\mathbf{b}$ is the chiral shift. Here we assumed that the parity-inversion ($\mathcal{P}$) symmetry is not broken and, therefore $b_0 = 0$. Note also that the same Hamiltonian H_{W} with the vanishing chiral shift, $\mathbf{b} = \mathbf{0}$, describes electron quasiparticles in a Dirac semimetal.

6.2.3.2 *Green's and spectral functions*

The inverse Green's function (fermion propagator) for Hamiltonian (6.27) reads (see also Appendices A.1 and A.2)

$$iG^{-1}(t, t'; r, r') = [(i\hbar\partial_t + \mu)\gamma^0 - v_F(\boldsymbol{\pi} \cdot \boldsymbol{\gamma}) + \hbar v_F(\boldsymbol{\gamma} \cdot \mathbf{b})\gamma_5]$$
$$\times \delta(t - t')\delta^3(\mathbf{r} - \mathbf{r}'). \tag{6.28}$$

Without loss of generality, let us choose an external magnetic field $\mathbf{B}$ to point in the $+z$-direction. In the Landau gauge, the corresponding the vector potential is given by $\mathbf{A} = (0, B_z x, 0)$. For simplicity, we assume that the chiral shift points in the same direction as the magnetic field $\mathbf{b} = b_z \hat{\mathbf{z}}$. The details of the derivation of Green's function $G(t, t'; r, r')$ from its inverse in Eq. (6.28) can be found in Appendix A.2. The final result takes the form

$$G(t, t'; r, r') = e^{i\Phi(\mathbf{r}_\perp, \mathbf{r}'_\perp)} \bar{G}(t - t'; \mathbf{r} - \mathbf{r}'), \tag{6.29}$$

where $\Phi(\mathbf{r}_\perp, \mathbf{r}'_\perp) = -eB_z(x+y')(x-y')/(2\hbar c)$ is the *Schwinger phase* [311], $\mathbf{r}_\perp = (x, y)$ is the position vector in the plane perpendicular to the field, and

$$\bar{G}(t - t'; \mathbf{r} - \mathbf{r}') = \int \frac{d\omega d^3 k}{(2\pi)^4} e^{-i\omega(t-t')+i\mathbf{k}\cdot(\mathbf{r}-\mathbf{r}')} \bar{G}(\omega; \mathbf{k}) \tag{6.30}$$

is the translation-invariant part of the quasiparticle propagator. Note that, in contrast, the Schwinger phase breaks translation invariance.[j]

Since Hamiltonian (6.27) is block-diagonal in the chirality space, it is convenient to represent Green's function in the following convenient form:

$$\bar{G}(\omega; \mathbf{k}) = \sum_{\lambda=\pm} \bar{G}^{(\lambda)}(\omega; \mathbf{k}) \mathcal{P}_5^{(\lambda)}, \tag{6.31}$$

where $\mathcal{P}_5^{(\lambda)} \equiv (1 + \lambda\gamma^5)/2$ are the chiral projectors with $\lambda = \pm$ and

$$\bar{G}^{(\lambda)}(\omega; \mathbf{k}) = ie^{-k_\perp^2 l_B^2} \sum_{\eta=\pm} \sum_{n=0}^{\infty} \frac{(-1)^n}{\epsilon_n^{(\lambda)}} \left\{ [\epsilon_n^{(\lambda)} \gamma^0 - \eta \hbar v_F(k_z - \lambda b_z)\gamma_z] \right.$$
$$\times [\mathcal{P}_- L_n(2k_\perp^2 l_B^2) - \mathcal{P}_+ L_{n-1}(2k_\perp^2 l_B^2)]$$
$$\left. + 2\eta \hbar v_F(\mathbf{k}_\perp \cdot \boldsymbol{\gamma}_\perp) L_{n-1}^1(2k_\perp^2 l_B^2) \right\} \frac{1}{\hbar\omega + \mu - \eta \epsilon_n^{(\lambda)}}. \tag{6.32}$$

---

[j]Several interesting phenomena are also predicted in the framework of the noncommutative theories [651–654], which are, however, yet to be realized in a condensed matter context.

Here $\mathbf{k}_\perp = (k_x, k_y)$ is the wave vector in the transverse plane, $\epsilon_n^{(\lambda)} = \sqrt{\hbar^2 v_F^2 (k_z - \lambda b_z)^2 + 2n\epsilon_L^2}$ is the energy of the nth Landau level, $\epsilon_L = \hbar v_F/l_B$ is the Landau energy scale, $l_B = \sqrt{\hbar c/|eB|}$ is the magnetic length, $L_n^\alpha(z)$ are the generalized Laguerre polynomials,[k] $\mathcal{P}_\pm \equiv (1 \pm i s_B \gamma_x \gamma_y)/2$ are (pseudo)spin projectors, and $s_B = \text{sgn}(eB)$.

It is also convenient to define the *spectral function* as the difference of the advanced and retarded propagators (see Appendix A)

$$A(\omega; \mathbf{k}) = \frac{1}{2\pi} \left[\bar{G}(\omega + i0; \mathbf{k}) - \bar{G}(\omega - i0; \mathbf{k}) \right]_{\mu=0} = \sum_{\lambda=\pm} A^{(\lambda)}(\omega; \mathbf{k}) \mathcal{P}_5^{(\lambda)}. \tag{6.33}$$

As is easy to verify, Green's function is related to the spectral function as follows:

$$\bar{G}(\Omega \pm i0; \mathbf{k}) = i \int_{-\infty}^{\infty} \hbar d\omega \frac{A(\omega; \mathbf{k})}{\hbar\Omega + \mu - \hbar\omega \pm i0}. \tag{6.34}$$

The explicit expression for $A^{(\lambda)}(\omega; \mathbf{k})$ is given by the same expression as $\bar{G}^{(\lambda)}(\omega; \mathbf{k})$ in Eq. (6.32), except for the last factor $(\hbar\omega + \mu - \eta\epsilon_n^{(\lambda)})^{-1}$ which is replaced with $\delta(\hbar\omega - \eta\epsilon_n^{(\lambda)})$.

In order to take into account scattering of quasiparticles on disorder, one needs to introduce a finite scattering time or quasiparticle width. Phenomenologically, this can be done by replacing the δ-function in the spectral function $A^{(\lambda)}(\omega; \mathbf{k})$ with a Lorentzian function, i.e.,

$$\delta(\omega - \eta E_n^{(\lambda)}) \to \frac{1}{\pi} \frac{\Gamma_n}{(\omega - \eta\epsilon_n^{(\lambda)})^2 + \Gamma_n^2}, \tag{6.35}$$

where Γ_n is the quasiparticle width in the nth Landau level. While such a simple model with a constant quasiparticle width simplifies the calculation of conductivity, it is clearly not very realistic. Nonetheless, it is still important and often captures the essence of the underlying physics. To go beyond the constant width approximation, one needs to use more realistic energy-dependent expressions for Γ_n. Analytical expressions for the quasiparticle width in Weyl and Dirac semimetals were obtained in the literature for several different types of scatterers [385, 547, 627, 630, 655–661] (see Sec. 7.1).

---

[k]By definition, $L_{-1}^\alpha \equiv 0$.

According to the *Kubo linear response theory*, the real part of the direct current (DC) conductivity tensor is defined in terms of the retarded current-current correlation function

$$\sigma_{nm} = \lim_{\Omega \to 0} \frac{\hbar}{\Omega} \mathrm{Im}\, \Pi_{nm}(\Omega + i0; \mathbf{0}), \tag{6.36}$$

where

$$\Pi_{nm}(\Omega; \mathbf{0}) = e^2 v_F^2 \frac{T}{\hbar} \sum_{l=-\infty}^{\infty} \int \frac{d^3 k}{(2\pi)^3} \mathrm{tr}\left[\gamma_n \bar{G}(i\omega_l; \mathbf{k}) \gamma_m \bar{G}(i\omega_l - \Omega; \mathbf{k})\right]. \tag{6.37}$$

Here $\omega_l = (2l+1)\pi T/\hbar$ is the fermion Matsubara frequency, l is an integer, and T is temperature in energy units.

By using the spectral function representation (6.34), it is straightforward to obtain

$$\Pi_{nm}(\Omega + i0; \mathbf{0}) = -e^2 v_F^2 \int\int d\omega d\omega' \frac{f^{\mathrm{eq}}(\hbar\omega) - f^{\mathrm{eq}}(\hbar\omega')}{\omega - \omega' - \Omega - i0}$$
$$\times \int \frac{d^3 k}{(2\pi)^3} \mathrm{tr}\left[\gamma_n A(\omega; \mathbf{k}) \gamma_m A(\omega'; \mathbf{k})\right], \tag{6.38}$$

where $f^{\mathrm{eq}}(\epsilon) = 1/[e^{(\epsilon-\mu)/T} + 1]$ is the Fermi–Dirac distribution function.

It is worth noting that the calculation of the diagonal components of the conductivity tensor can be significantly simplified because the trace in Eq. (6.38) is real. Then, one can use the identity

$$\frac{1}{\omega - \omega' - \Omega \mp i0} = \mathrm{p.v.} \frac{1}{\omega - \omega' - \Omega} \pm i\pi\delta(\omega - \omega' - \Omega) \tag{6.39}$$

to extract the imaginary part of the polarization tensor $\Pi_{ij}(\omega, \mathbf{k})$. Here p.v. stands for the principal value. The final result for σ_{nn} reads

$$\sigma_{nn} = -\pi e^2 v_F^2 \hbar^2 \sum_{\lambda=\pm} \int \frac{d\omega}{4T \cosh^2\left(\frac{\hbar\omega - \mu}{2T}\right)}$$
$$\times \int \frac{d^3 k}{(2\pi)^3} \mathrm{tr}\left[\gamma_n A^{(\lambda)}(\omega; \mathbf{k}) \gamma_n A^{(\lambda)}(\omega; \mathbf{k}) \mathcal{P}_5^{(\lambda)}\right]. \tag{6.40}$$

6.2.3.3 *Longitudinal conductivity*

As suggested by the results in Sec. 6.2.1, the chiral anomaly should dramatically affect the electron transport properties in Weyl and Dirac semimetals.

In particular, the anomaly should lead to a negative longitudinal magnetoresistivity. To verify such a qualitative feature of the anomalous transport, let us consider the longitudinal conductivity σ_{zz}, which is expected to be an increasing function of the magnetic field.

Since the LLL plays a special role in the chiral anomaly, it is instructive to separate the LLL and higher Landau-level (HLL) contributions to the conductivity tensor $\sigma_{zz} = \sigma_{zz}^{(\text{LLL})} + \sigma_{zz}^{(\text{HLL})}$. The LLL contribution $\sigma_{zz}^{(\text{LLL})}$ is given by

$$
\sigma_{zz}^{(\text{LLL})} = \frac{e^2 v_F^2 \hbar^2}{16\pi^3 l_B^2 T} \sum_{\lambda=\pm} \int \frac{d\omega dk_z}{\cosh^2\left(\frac{\hbar\omega-\mu}{2T}\right)} \frac{\Gamma_0^2}{\{\hbar^2 \left[\omega + s_B \lambda v_F(k_z - \lambda b_z)\right]^2 + \Gamma_0^2\}^2}
$$

$$
= \frac{e^2 v_F}{4\pi^2 l_B^2 \Gamma_0} = \frac{e^2 v_F |eB|\tau}{4\pi^2 \hbar^2 c}. \tag{6.41}
$$

Note that in the last part we used $\Gamma_0 = \hbar/\tau$ where τ is the relaxation time. Interestingly, this LLL contribution does not depend on temperature and the chemical potential. In the limit of strong magnetic field, this is expected to be the dominant contribution. In such a regime, as we see, the Kubo formalism with the exact Green's function predicts a linear growth of the longitudinal magnetoconductivity as a function of a magnetic field. This is in agreement with the result in Eq. (6.24) that was also obtained in the strong field limit.

The contribution of higher Landau levels to the longitudinal conductivity is given by the following expression:

$$
\sigma_{zz}^{(\text{HLL})} = \frac{e^2 v_F^2 \hbar^2}{4\pi^3 l_B^2 T} \sum_{n=1}^{\infty} \int \frac{d\omega dk_z}{\cosh^2\left(\frac{\hbar\omega-\mu}{2T}\right)}
$$

$$
\times \frac{\Gamma_n^2 \left[\left(\hbar^2\omega^2 + \epsilon_n^2 + \Gamma_n^2\right)^2 + 4\hbar^4 v_F^2 \omega^2 k_z^2 - 8n\epsilon_L^2 \hbar^2 \omega^2\right]}{\left[(\hbar\omega - \epsilon_n)^2 + \Gamma_n^2\right]^2 \left[(\hbar\omega + \epsilon_n)^2 + \Gamma_n^2\right]^2}. \tag{6.42}
$$

While the HLL contribution should be small in a strong magnetic field, its role is expected to grow with decreasing the field. This is confirmed by the numerical results in Fig. 6.7(a), where the LLL and HLL contributions to the longitudinal conductivity, as well as the total conductivity $\sigma_{zz} = \sigma_{zz}^{(\text{LLL})} + \sigma_{zz}^{(\text{HLL})}$ are presented. In agreement with Eq. (6.41), the LLL contribution linearly increases with B. The HLL part reveals the characteristic Shubnikov–de Haas oscillations (see also Sec. 3.9.1) and, on average,

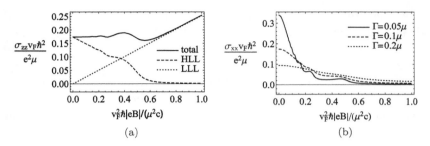

(a) (b)

Fig. 6.7 (a) The longitudinal conductivity σ_{zz} at $\Gamma = 0.1\mu$ as a function of a magnetic field. The solid line corresponds to the complete result $\sigma_{zz} = \sigma_{zz}^{(LLL)} + \sigma_{zz}^{(HLL)}$, as well as dashed and dotted lines represent the contributions of the higher $\sigma_{zz}^{(HLL)}$ and lowest Landau $\sigma_{zz}^{(LLL)}$ levels. (b) The transverse conductivity σ_{xx} as a function of a magnetic field for a few values of $\Gamma_n = \Gamma$. Here $\Gamma = 0.05\mu$ (solid line), $\Gamma = 0.1\mu$ (dashed line), and $\Gamma = 0.2\mu$ (dotted line).

decreases with the field. Since the LLL contribution dominates at sufficiently large values of B, the total magnetoconductivity grows with the magnetic field, implying that the resulting magnetoresistivity is negative.

Interestingly, in the low-energy model at hand, the chiral shift does not affect the longitudinal magnetoconductivity directly. From a technical viewpoint, it was removed from Eq. (6.42) by changing the integration variable $k_z \to k_z - \lambda b_z$. However, the chiral shift may still affect the conductivity indirectly via the quasiparticle width.

6.2.3.4 *Transverse conductivity*

Let us now discuss the transverse conductivity described by the diagonal components $\sigma_{xx} = \sigma_{yy}$. By making use of the definition in Eq. (6.40), one derives [548]

$$\sigma_{xx} = \frac{e^2 v_F^2 \hbar^2}{4\pi^3 l_B^2 T} \sum_{n=0}^{\infty} \int \frac{d\omega dk_z}{\cosh^2\left(\frac{\hbar\omega - \mu}{2T}\right)} \frac{\Gamma_{n+1}\Gamma_n}{\left[(\epsilon_n^2 + \Gamma_n^2 - \hbar^2\omega^2)^2 + 4\hbar^2\omega^2\Gamma_n^2\right]}$$

$$\times \frac{\left[(\hbar^2\omega^2 + \epsilon_n^2 + \Gamma_n^2)\left(\hbar^2\omega^2 + \epsilon_{n+1}^2 + \Gamma_{n+1}^2\right) - 4\hbar^4(v_F k_z)^2\omega^2\right]}{\left[(\epsilon_{n+1}^2 + \Gamma_{n+1}^2 - \hbar^2\omega^2)^2 + 4\hbar^2\omega^2\Gamma_{n+1}^2\right]}.$$

(6.43)

The corresponding numerical results for the transverse conductivity are presented in Fig. 6.7(b) for a few values of the quasiparticle width $\Gamma_n = \Gamma$. At intermediate values of the magnetic field, traces of the Shubnikov–de Haas oscillations are visible only at a sufficiently small quasiparticle width and fade away gradually with increasing Γ. Unlike the longitudinal component σ_{zz}, the transverse conductivity σ_{xx} decreases with a magnetic field on average.

The off-diagonal components of the transverse conductivity tensor, i.e., $\sigma_{xy} = -\sigma_{yx}$, are responsible for the Hall effect. Since the presence of disorder is not crucial for the Hall response [286], it is convenient to present the corresponding results in the limit $\Gamma_n \to 0$,[1] i.e.,

$$
\sigma_{xy} = -\frac{e^2 s_B}{4\pi^2 \hbar} \sum_{n=0}^{\infty} (2 - \delta_{n,0}) \int dk_z \frac{\sinh\left(\frac{\mu}{T}\right)}{\cosh\left(\frac{\epsilon_n}{T}\right) + \cosh\left(\frac{\mu}{T}\right)}
$$
$$
- \frac{e^2}{8\pi^2 \hbar} \sum_{\lambda=\pm} \lambda \int dk_z \frac{\sinh\left[\frac{\hbar v_F (k_z - \lambda b)}{T}\right]}{\cosh\left[\frac{\hbar v_F (k_z - \lambda b)}{T}\right] + \cosh\left(\frac{\mu}{T}\right)}. \qquad (6.44)
$$

The first term corresponds to the contribution of the filled states and vanishes as $\mu \to 0$. The second term, which stems from the LLL, is of topological origin and survives even at $\mu = 0$.

One should note that the integrals for the individual LLL contributions from each Weyl node ($\lambda = \pm$) in Eq. (6.44) are divergent and should be treated with care. This is an artifact of using the relativistic-like low-energy Hamiltonian (6.27) without imposing a cutoff at large momenta where its validity breaks down. Nevertheless, while the separate left- and right-handed contributions are divergent, their total result is finite and given by

$$
\sigma_{xy}^{\text{AHE}} = -\lim_{\Lambda \to \infty} \frac{e^2}{8\pi^2 v_F \hbar^2} T \left[\ln\left(\frac{\cosh\left[\frac{\hbar v_F (\Lambda - b_z)}{T}\right] + \cosh\left(\frac{\mu}{T}\right)}{\cosh\left[\frac{\hbar v_F (\Lambda + b_z)}{T}\right] + \cosh\left(\frac{\mu}{T}\right)} \right) \right.
$$
$$
\left. - \ln\left(\frac{\cosh\left[\frac{\hbar v_F (\Lambda + b_z)}{T}\right] + \cosh\left(\frac{\mu}{T}\right)}{\cosh\left[\frac{\hbar v_F (\Lambda - b_z)}{T}\right] + \cosh\left(\frac{\mu}{T}\right)} \right) \right] = \frac{e^2 b_z}{2\pi^2 \hbar}, \qquad (6.45)
$$

where Λ is a wave vector cutoff which is taken to infinity at the end of calculation. As we will show in Sec. 6.3, a more realistic two-band model, which has a natural momentum cutoff, gives the same answer.

---

[1]This also helps to avoid a problematic feature coming from modeling disorder via a constant-width Lorentzian distribution in Eq. (6.35), which leads to a divergent sum over the Landau levels.

The off-diagonal component of the conductivity in Eq. (6.45) describes the *anomalous Hall effect* (AHE), which is one of the hallmark features of $\mathcal{T}$ symmetry broken Weyl semimetals associated with a nonzero value of the chiral shift **b**. Since AHE vanishes at $b_z \to 0$, one might argue that such an effect should be absent in Dirac semimetals. This is not necessarily true, however, in the presence of a magnetic field because a nonzero chiral shift can be generated dynamically [548]. Only at $\mathbf{B} = \mathbf{0}$, Dirac semimetals can be distinguished unambiguously from Weyl semimetals with broken $\mathcal{T}$ symmetry by a nonvanishing AHE.

At nonzero chemical potential, the complete expression for the Hall conductivity (6.44) contains not only the *topological* contribution, but also an extra *matter* contribution due to filled states. In the limit of zero temperature, in particular, the full result is given by

$$
\sigma_{xy} = \frac{e^2 b_z}{2\pi^2 \hbar} - \frac{e^2 s_{\mathrm{B}} \, \mathrm{sgn}\,(\mu)}{2\pi^2 \hbar^2 v_F} \sum_{n=0}^{N_\mu} (2 - \delta_{n,0}) \sqrt{\mu^2 - 2n\epsilon_{\mathrm{L}}^2}, \tag{6.46}
$$

where $N_\mu = [\mu^2/(2\epsilon_{\mathrm{L}}^2)]$ is the highest occupied Landau level and $[\dots]$ denotes an integer part. As expected, the matter part of the Hall conductivity, described by the second term in Eq. (6.46), depends on the details of the low-energy quasiparticle spectrum and vanishes at the charge neutrality point $\mu = 0$.

Historically, one of the first studies of the transverse magnetoconductivity in systems with a relativistic-like dispersion relation was performed by Alexei Abrikosov [644]. Long before the experimental discovery of Dirac semimetals, he conjectured that a linear Dirac-like spectrum is realized in doped silver chalcogenides $Ag_{2+\delta}Se$ and $Ag_{2+\delta}Te$ with $\delta \simeq 0.01$. By assuming that the ultraquantum regime (with only the LLL partially occupied) is realized in these narrow-gap semiconductors for moderately strong magnetic fields, he was able to explain a positive transverse magnetoresistance growing linearly with the field, as observed in experiment. Unlike the simplified model of the quasiparticle width above, more realistic scattering on charged impurities with a screened Coulomb potential was used in his study. In the end, the following transverse components of the conductivity tensor were obtained:

$$
\sigma_{xx} = \alpha_{\mathrm{eff}}^2 \frac{ec n_{\mathrm{imp}}}{2\pi B} \ln(\varepsilon_e), \tag{6.47}
$$

$$
\sigma_{xy} = \frac{ec n_{\mathrm{el}}}{B}, \tag{6.48}
$$

where n_{imp} is the concentration of impurities, $n_{el} = \mu|eB|/(2\pi^2\hbar^2 cv_F)$ is the electron number density in the ultraquantum limit, $\alpha_{eff} = e^2/(\hbar v_F \varepsilon_e)$ is the effective coupling constant of quasiparticles in the material, and ε_e is the dielectric permittivity. Since the effects of disorder contribute to the expression for the Hall conductivity σ_{xy} only at the subleading order, they were ignored in Eq. (6.48). As is easy to check, the latter agrees with the matter contribution in Eq. (6.46) in the ultraquantum limit.

6.3 Anomalous Hall effect

As we saw in Sec. 6.2, a rigorous calculation of the Hall conductivity in Weyl semimetals produces an extra anomalous part, which is proportional to the chiral shift. This additional contribution corresponds to the AHE, which has a long and very interesting history in condensed matter physics.

As is well known, the standard Drude-like semiclassical approach offers a reliable treatment of the electron transport in solids. It is not omnipotent, however. For example, it does not provide any insight into the Hall conductivity in magnetic materials, which is much larger than the ordinary Hall effect in nonmagnetic conductors and exists even when an external magnetic field vanishes. While such an unusual Hall transport in magnetic materials was observed by Edwin Hall already in the 19th century [370], the first viable theoretical explanation was given almost 70 years later in the work by Robert Karplus and Joaquin Luttinger [372]. As they argued, the anomalous Hall effect can be understood if there is an *anomalous velocity* perpendicular to the electric field. The current consensus is that the AHE is, in fact, a rather complicated phenomenon [371]. It includes not only the intrinsic contributions originating from the Berry curvature, but also some extrinsic ones, for example, related to the disorder effects due to spin-dependent scattering of charge carriers.[m]

As in the case of conventional ferromagnetic materials, the AHE in Weyl semimetals [389, 391, 657, 662–665] could appear even in the absence of an applied magnetic field. As is easy to show by using the same arguments as in Sec. 5.2, the AHE has a topological origin and is triggered by a nontrivial Berry curvature. To demonstrate this, let us consider the simplest case of a Weyl semimetal with broken $\mathcal{T}$ symmetry that has a single pair of Weyl nodes separated by $2b_z\hat{\mathbf{z}}$ in momentum space. At any fixed value of k_z

---

[m]The spin-dependent scattering of charge carriers contributing to the AHE is also known as the side jumps or the skew scattering [371].

in the interval $|k_z| < |b_z|$, the $k_x - k_y$ plane can be considered as a 2D Chern insulator. One should recall that, in a Chern insulator phase, there is a chiral edge state contributing to the Hall conductivity as $\sim e^2/\hbar$ (see also Fig. 5.2(b)). Thus, by stacking all planes together, one can obtain the following total Hall conductivity proportional to the chiral shift:

$$\sigma_{xy}^{\text{AHE}} \propto \frac{e^2}{\hbar} \int_{BZ} \frac{dk_z}{2\pi} C(k_z) \propto \frac{e^2 b_z}{\hbar}. \tag{6.49}$$

Here

$$C(k_z) = \int \frac{dk_x dk_y}{(2\pi)^2} \Omega_z(\mathbf{k}) \propto \theta(b_z^2 - k_z^2) \tag{6.50}$$

is the momentum-dependent Chern number and $\mathbf{\Omega}(\mathbf{k})$ is the Berry curvature (see Sec. 3.3). Note that σ_{xy}^{AHE} vanishes when $b_z \to 0$. As is clear, in Weyl semimetals with multiple Weyl nodes, the sum over all pairs of Weyl nodes should be performed in Eq. (6.49). Therefore, for Weyl semimetals with broken $\mathcal{P}$ symmetry but preserved $\mathcal{T}$ symmetry, the symmetrical configuration of Weyl nodes gives $\sum_{n_W} \mathbf{b}_{n_W} = \mathbf{0}$. In other words, as expected from symmetry arguments, the AHE vanishes for such systems.

The anomalous Hall conductivity defined in Eq. (6.49) is the so-called intrinsic conductivity, which originates exclusively from the nontrivial topology of the material's electronic band structure. However, there is another contribution to the AHE conductivity related to the scattering on impurities. While in conventional ferromagnets it is hard to disentangle these two contributions [371], the AHE in Weyl semimetals is primarily of intrinsic origin [657]. At nonzero electric chemical potential μ, there is an extra matter contribution to the AHE, but it plays a minor role at small values of μ.

As will be seen in Sec. 6.6.2 below, the AHE is usually absent in a semiclassical treatment of the chiral kinetic theory (CKT), which uses a simple linear dispersion relation for Weyl quasiparticles, but misses some important information about the band topology. The correct description of the AHE in such a kinetic approach requires the inclusion of extra Chern–Simons contributions, which is done in the framework of the so-called *consistent CKT*. The need for extra topological terms in the CKT can be recognized by matching the CKT analysis with the study in a more realistic two-band model of a Weyl semimetal with a nonzero separation between the Weyl nodes. This way, one finds that the Chern–Simons terms originate from the filled states lying deeply below the Fermi surface. Note that this is also consistent with the analysis of the anomalous Hall conductivity within the Kubo linear response theory in Sec. 6.2.3.4.

In order to get a better understanding of the underlying physics of the AHE conductivity in Weyl semimetals, let use a two-band model of a Weyl semimetal defined in Sec. 3.5.1. The corresponding Hamiltonian reads

$$\mathcal{H}(\mathbf{k}) = -\mu + (\mathbf{d} \cdot \boldsymbol{\sigma}). \tag{6.51}$$

Here vector $\boldsymbol{\sigma} = (\sigma_x, \sigma_y, \sigma_z)$ has the three Pauli matrices as components and vector $\mathbf{d}$ is a periodic function in momentum space defined by the following components:

$$d_1 = \Lambda \sin(ak_x), \tag{6.52}$$

$$d_2 = \Lambda \sin(ak_y), \tag{6.53}$$

$$d_3 = t_0 + t_1 \cos(ak_z) + t_2 [\cos(ak_x) + \cos(ak_y)], \tag{6.54}$$

where a is the lattice spacing of a cubic lattice and Λ, t_0, t_1, and t_2 are material-dependent parameters. Recall that the model Hamiltonian in Eq. (6.51) with $|t_0 + 2t_2| \leq |t_1|$ describes a $\mathcal{T}$-symmetry broken Weyl semimetal with a single pair of Weyl nodes separated in the z-direction. The chiral shift b_z is given by

$$b_z = -\frac{1}{a} \arccos\left(\frac{-t_0 - 2t_2}{t_1}\right). \tag{6.55}$$

In order to visualize the energy spectrum and calculate the AHE conductivity, it is convenient to match the model parameters with those of a realistic phenomenological model of a Dirac semimetal introduced in Sec. 3.6.1, i.e.,

$$t_0 = M_0 - t_1 - 2t_2 \approx -1.0281 \text{ eV}, \quad \Lambda = \frac{A}{a} \approx 0.328 \text{ eV},$$

$$t_1 = -\frac{2M_1}{a^2} \approx 0.3783 \text{ eV}, \quad t_2 = -\frac{2M_2}{a^2} \approx 0.3684 \text{ eV}. \tag{6.56}$$

The energy spectrum of this model together with its linearized approximation (see Eq. (3.83)) is shown in Fig. 6.8(a). As one can see from this figure, the energy spectrum has a characteristic dome of the height $\epsilon_0 = |t_0 + t_1 + 2t_2|$. In addition, the difference between the two-band and linearized dispersion relation becomes noticeable at this energy scale.

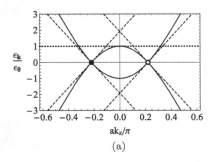
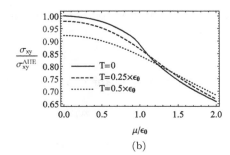

(a)

(b)

Fig. 6.8 (a) The energy spectrum of the two-band periodic model (6.51). The filled and empty points denote the right- and left-handed Weyl nodes. The linearized dispersion relation (see Eq. (3.83)) is shown by dashed lines. The horizontal dotted line corresponds to the energy at which the two separate Fermi surfaces merge, i.e., $\epsilon_0 = \lim_{k \to 0} \epsilon_k$. (b) The anomalous Hall conductivity σ_{xy} defined in Eq. (6.64) as a function of the electric chemical potential μ at $T = 0$ (solid line), $T = 0.25\epsilon_0$ (dashed line), and $T = 0.5\epsilon_0$ (dotted line).

By using the same approach as in Sec. 3.3.2, one can derive the following expression for the Berry curvature in the two-band model:

$$\Omega_j = \sum_{n,m=x,y,z} \frac{\epsilon_{jnm}\mathcal{F}_{nm}}{2}, \tag{6.57}$$

where ϵ_{jnm} is the Levi-Civita tensor and the Berry curvature tensor equals

$$\mathcal{F}_{nm}(\mathbf{k}) = \frac{1}{2\hbar|\mathbf{d}|^3} (\mathbf{d} \cdot [(\partial_{k_n}\mathbf{d}) \times (\partial_{k_m}\mathbf{d})]). \tag{6.58}$$

In order to calculate the AHE conductivity, it is convenient to use the Kubo linear response theory. Although the method was already introduced in Sec. 6.2.3, several key steps of the derivation are worth repeating here in order to highlight the modifications stemming from the different form of the Hamiltonian.

The conductivity tensor σ_{nm} is related to the polarization tensor $\Pi_{nm}(\Omega+i0; \mathbf{0})$ via the standard relation (6.36). In the two-band model, the polarization tensor can be expressed in terms of the quasiparticle Green's function as follows:

$$\Pi_{nm}(\Omega + i0; \mathbf{0}) = \frac{T}{\hbar} \sum_{l=-\infty}^{\infty} \int \frac{d^3k}{(2\pi)^3} \text{tr}\left[j_n(\mathbf{k})S(i\omega_l; \mathbf{k})j_m(\mathbf{k})S(i\omega_l - \Omega; \mathbf{k})\right], \tag{6.59}$$

where $j_n(\mathbf{k})$ are the components of the electric current density operator in momentum space, defined by

$$\mathbf{j}(\mathbf{k}) = -e\partial_{\mathbf{k}}H(\mathbf{k}). \tag{6.60}$$

The explicit form of the quasiparticle Green's function is given by

$$S(i\omega_l; \mathbf{k}) = i\frac{i\hbar\omega_l + \mu + (\mathbf{d}\cdot\boldsymbol{\sigma})}{(i\hbar\omega_l + \mu)^2 - |\mathbf{d}|^2}, \tag{6.61}$$

which can be conveniently rewritten by using the spectral representation in Eq. (6.34). Note that the spectral function $A(\omega; \mathbf{k})$ is defined by

$$A(\omega; \mathbf{k}) = \sum_{\eta=\pm} \frac{|\mathbf{d}| + \eta(\mathbf{d}\cdot\boldsymbol{\sigma})}{2|\mathbf{d}|}\delta\left(\hbar\omega - \eta|\mathbf{d}|\right). \tag{6.62}$$

As is clear from the sharp δ-function peaks, this spectral function describes noninteracting quasiparticles with the vanishing decay width. In realistic models, of course, the quasiparticle width should be nonzero. However, since the presence of disorder is not crucial for the AHE conductivity, the analysis below will be presented only for the clean case.

By making use of the spectral representation for Green's function, Eq. (6.36) can be recast in the following form:

$$\sigma_{nm} = \lim_{\Omega\to 0}\frac{i}{\Omega}T\sum_{l=-\infty}^{\infty}\int\frac{d^3k}{(2\pi)^3}\int\int d\omega d\omega'$$

$$\times \frac{\text{tr}\left[j_n(\mathbf{k})A(\omega; \mathbf{k})j_m(\mathbf{k})A(\omega'; \mathbf{k})\right]}{(i\hbar\omega_l + \mu - \hbar\omega)\left(i\hbar\omega_l - \hbar\Omega - i0 + \mu - \hbar\omega'\right)}. \tag{6.63}$$

Furthermore, the off-diagonal components of the conductivity tensor with $n \neq m$ are given by [432]

$$\sigma_{nm} = e^2\int\frac{d^3k}{(2\pi)^3}\left[f^{\text{eq}}(|\mathbf{d}|) - f^{\text{eq}}(-|\mathbf{d}|)\right]\mathcal{F}_{nm}(\mathbf{k})$$

$$\stackrel{T\to 0}{=} -e^2\int\frac{d^3k}{(2\pi)^3}\mathcal{F}_{nm}(\mathbf{k})\left[1 - \theta(|\mu| - |\mathbf{d}|)\right]. \tag{6.64}$$

The topological origin of this contribution to the Hall conductivity becomes obvious in the limit $T \to 0$ and $\mu \to 0$, when the Berry curvature tensor remains the only function in the integrand. After performing the integration, we derive the following result for the anomalous Hall conductivity:

$$\sigma_{xy}^{\text{AHE}} = -\sigma_{yx}^{\text{AHE}} \stackrel{\mu=T=0}{=} \frac{e^2 b_z}{2\pi^2\hbar}. \tag{6.65}$$

As expected, this result perfectly agrees with those in Eqs. (6.45) and (6.49) obtained by carefully regularizing the linearized model and using the qualitative topological considerations, respectively.

The AHE conductivity given by Eq. (6.64) is shown in Fig. 6.8(b) as a function of the electric chemical potential μ for a few values of temperature. As we see, even when disorder effects are neglected, the conductivity gradually deteriorates with increasing the electric chemical potential at all temperatures. At $T = 0$, in particular, the decrease of σ_{xy} as a function of μ is also evident from the last expression in Eq. (6.64). Another notable feature of the AHE conductivity is a partially smeared step-like feature at $\mu \approx \epsilon_0$. Although the change of the conductivity slope is rather small, it is not accidental but connected with the onset of merging of the two Fermi surfaces around the Weyl nodes of opposite chiralities. Overall, the AHE conductivity decreases gradually as a function of the chemical potential because, as is evident from Fig. 6.8(a), the deviations of the energy dispersion relations from the perfect relativistic-like form grow progressively away from the low-energy region.

6.4 Chiral separation and chiral magnetic effects

Let us now turn our attention to other anomalous transport effects in topological semimetals. Since the low-energy properties of Dirac and Weyl semimetals are described by relativistic-like Hamiltonians, it is natural to expect that some relativistic properties can be manifested in these materials as well. Two of such celebrated phenomena, introduced originally in high energy physics, are the *chiral separation effect* (CSE) [144, 147, 148] and *chiral magnetic effect* (CME) [144–146, 156].

Conceptually, the CSE is associated with a nondissipative chiral (axial) current density induced in a chiral relativistic plasma when an external magnetic field **B** is applied. In the case of noninteracting fermions, the expression for the chiral current density can be easily derived and is given by

$$\mathbf{j}_5 = \mathbf{j}_+ - \mathbf{j}_- = \frac{e^2 \mu}{2\pi^2 \hbar^2 c} \mathbf{B}, \tag{6.66}$$

where μ is the chemical potential. Here $\mathbf{j}_+$ and $\mathbf{j}_-$ are the current densities of the right- and left-handed fermions, respectively. Since the spin is aligned either along or against the direction of the particle momentum for the right- and left-handed fermions, respectively, the CSE current can be interpreted alternatively as a spin or pseudospin polarization of magnetized chiral matter.

Unlike the CSE, the CME requires the presence of a nonzero chiral imbalance, which can be formally quantified by the chiral chemical potential μ_5 [145].[n] The chiral chemical potential couples to the difference between number densities of the right- and left-handed fermions and enters the low-energy Hamiltonian (6.27) via the term $-\mu_5\bar{\psi}\gamma^0\gamma_5\psi$. A chiral asymmetry in magnetized relativistic matter leads to a nondissipative electric current

$$\mathbf{j} = \frac{e^2\mu_5}{2\pi^2\hbar^2c}\mathbf{B},\tag{6.67}$$

which is parallel to the external magnetic field $\mathbf{B}$. In regard to the experimental verification of the CME current in high energy physics, it is fair to say that a definite confirmation is still lacking. Indeed, while the charge-dependent particle correlations observed in heavy-ion collisions at RHIC [666–671] and LHC [672] appear to be in qualitative agreement with the predictions of the CME [673–675], the results in [676, 677] suggest that similar charge-dependent azimuthal correlations could be also attributed to background effects.

It is natural to expect that the CME can also be realized in Weyl and Dirac semimetals. However, this is not as straightforward as it may seem at first sight. Indeed, a naive version of the CME phenomenon seems to imply that a persistent electric current can be induced in the ground state of a solid when a magnetic field is applied. This would be at odds with the well-known fact that a nonequilibrium distribution of electrons is needed to drive a nonvanishing current. As expected, a careful analysis demonstrates that the static CME current vanishes in Weyl semimetals in equilibrium [390, 678–681]. The same conclusion was reached also in relativistic quantum field theoretical models with a lattice regularization [682]. In the case of a two-band model, the details of derivation will be outlined in Sec. 6.4.4.

Similarly, as will be demonstrated below using the framework of a linearized low-energy model [683], a nonzero chiral chemical potential induces no steady-state CME currents in systems of finite volume. With that being said, the CME is, however, expected to play an important role in

---

[n]In view of the chiral anomaly, the chiral charge is not conserved and, therefore, the chiral chemical potential μ_5 is not a chemical potential of conventional type and should be interpreted with caution.

out-of-equilibrium processes. In particular, the celebrated negative magnetoresistance (see also Sec. 6.2) can be easily explained as the consequence of
the CME with a chiral chemical potential generated by the chiral anomaly.

By taking into account that the CME is essentially an out-of-equilibrium
phenomenon, the calculation of the corresponding conductivity crucially
depends on the order of the limits $\mathbf{q} \to \mathbf{0}$ and $\Omega \to 0$ [7, 684] in the polarization tensor $\Pi_{nm}(\Omega; \mathbf{q})$ defined by Eq. (6.37). One finds, in particular,
that the result for the current vanishes identically in the *static limit*, when
$\Omega = 0$ is set before the limit $\mathbf{q} \to \mathbf{0}$ is taken, but could be nonzero in the
uniform limit, when the order of the limits is reversed. The uniform limit
determines the response to a slowly varying field, which might indeed allow
for the CME current in an out-of-equilibrium state. On the other hand, the
static limit describes a system in the ground state without any currents.

6.4.1 *Model and wave functions in magnetic field*

In order to understand the origin of the chiral separation and chiral magnetic effects, let us start by considering a low-energy theory with relativistic-
like Dirac fermions confined to a slab of finite thickness in the z-direction,
i.e., $-L_z \leq z \leq L_z$. For technical reasons, it will be convenient to assume
that the fermions are massive. The model Hamiltonian reads

$$H^{(\mathrm{D})} = \int d^3 r \bar{\psi}(\mathbf{r}) \left[v_F \left(\boldsymbol{\gamma} \cdot \boldsymbol{\pi} \right) - m(z) \right] \psi(\mathbf{r}). \tag{6.68}$$

Here $\boldsymbol{\pi} = -i\hbar\boldsymbol{\nabla} + e\mathbf{A}/c$ is the canonical momentum, $\mathbf{A} = (0, Bx, 0)$ is
the vector potential, which describes a magnetic field $\mathbf{B}$ pointing in the
$\hat{\mathbf{z}}$-direction, and $m(z) = M\theta(z^2 - L_z^2) + m\theta(L_z^2 - z^2)$ is a coordinate-
dependent gap function (or a mass expressed in the energy units), where
$\theta(x)$ is the unit step function. In the limit $m \to 0$, the model Hamiltonian in Eq. (6.68) describes a Dirac semimetal with a gapless spectrum in
the bulk. In general, however, it is assumed that the band gap could be
nonzero. This might be useful, for example, in applications to such materials as bismuth alloy $\mathrm{Bi}_{1-x}\mathrm{Sb}_x$ at small concentrations of antimony, where
the low-energy spectrum is described by the Dirac equation with a nonzero
Dirac gap [89, 513, 685].

The model parameter M is introduced in order to describe the boundary
effects. Formally, it defines an effective energy gap in vacuum, which will be
assumed much larger than all characteristic energy scales inside the slab.
In studies of graphene, similar boundary conditions with an infinite gap

outside the material are known as the infinite mass boundary conditions [81, 686, 687]. In high energy physics, a similar model is known as the Bogolyubov bag model [688] generalized to the case of a nonzero mass.°
In such a model, hadrons are described as bags of finite size with massless fermions (quarks) confined inside. In order to prevent the fermions from escaping the bag, it is required that the component of the electric current normal to the surface vanishes. The boundary conditions that impose such a requirement necessarily break the chiral symmetry (which is present in an infinite system at $m = 0$) because they allow for the chirality (or helicity) flip when particles scatter on the boundary.

To proceed with the analysis, let us start from deriving the Landau-level wave functions for the Dirac Hamiltonian (6.68) (see also [693, 694]). The eigenstates, which satisfy the equation $H\psi(\mathbf{r}) = \epsilon\psi(\mathbf{r})$, can be sought in the form $\psi(\mathbf{r}) = e^{ik_z z + is_\text{B} k_y y}\phi(x)$, where $\mathbf{k}$ is the wave vector and $s_\text{B} = \text{sgn}\,(eB)$. The function $\phi(x)$ satisfies the following equation:

$$\gamma^0 \left[\hbar v_F \left(k_z \gamma_z + s_\text{B} k_y \gamma_y - i\gamma_x \partial_x \right) + \frac{v_F eBx}{c}\gamma_y - m \right] \phi(x) = \epsilon\phi(x).$$
$$(6.69)$$

It is convenient to replace the x coordinate with a new dimensionless variable

$$\xi = \frac{1}{l_\text{B}} \left(l_\text{B}^2 k_y + x \right),$$
$$(6.70)$$

where $l_\text{B} = \sqrt{\hbar c / |eB|}$ is the magnetic length, and rewrite Eq. (6.69) as follows:

$$\left[\partial_\xi + is_\text{B}\xi\gamma_y\gamma_x - il_\text{B}\gamma_x \left(k_z\gamma_z + \frac{m}{\hbar v_F} - \frac{\epsilon}{\hbar v_F}\gamma^0 \right) \right] \phi(\xi) = 0. \qquad (6.71)$$

By introducing a set of linearly independent bispinors $u_s^{\pm}$ with $s = \pm$ that satisfy the following relations:

$$i\gamma_y\gamma_x u_s^{\pm} = \mp s_\text{B} u_s^{\pm}, \qquad (6.72)$$

$$\gamma_y\gamma_x \left(\epsilon\gamma^0 - \hbar v_F k_z\gamma_z \right) u_s^{\pm} = s\sqrt{\hbar^2 v_F^2 k_z^2 - \epsilon^2}\, u_s^{\pm}, \qquad (6.73)$$

$$\frac{\gamma_x \left(\epsilon\gamma^0 - \hbar v_F k_z\gamma_z - m \right)}{\sqrt{\epsilon^2 - \hbar^2 v_F^2 k_z^2 - m^2}} u_s^- = u_s^+, \qquad (6.74)$$

°For a review of other bag models in hadron physics, see [689]. The general form of the boundary conditions for relativistic matter in a finite volume was also discussed in [690–692].

one can express the solution for function $\phi_s(\xi)$ as a linear combination

$$\phi_s(\xi) = \Phi_+^s(\xi)u_s^+ + \Phi_-^s(\xi)u_s^-. \tag{6.75}$$

Then, by using the relations in Eqs. (6.72)–(6.74), one can rewrite Eq. (6.71) in the following form:

$$\left[\partial_\xi \Phi_\pm^s(\xi) \mp \xi \Phi_\pm^s(\xi) + i\kappa \Phi_\mp^s(\xi)\right] = 0, \tag{6.76}$$

$$\left[\partial_\xi^2 \mp 1 - \xi^2 + \kappa^2\right] \Phi_\pm^s(\xi) = 0, \tag{6.77}$$

where $\kappa = \sqrt{\epsilon^2 - \hbar^2 v_F^2 k_z^2 - m^2}/\epsilon_L$ and $\epsilon_L = v_F \hbar/l_B$ is the Landau energy scale. For definiteness, let us choose $s_B = 1$. Then solutions to Eqs. (6.76) and (6.77) are given in terms of the parabolic cylinder functions [695]

$$\Phi_-^s(\xi) = D_{\kappa^2/2}(\sqrt{2}\xi), \quad \Phi_+^s(\xi) = \frac{i\kappa}{\sqrt{2}}D_{\kappa^2/2-1}(\sqrt{2}\xi). \tag{6.78}$$

The normalizability requires that the wave functions are finite as $|\xi| \to \infty$. This is the case only if $\kappa^2/2 = n$, where $n = 0, 1, 2 \ldots$ are nonnegative integers. Then the parabolic cylinder functions $D_n\left(\sqrt{2}\xi\right)$ are expressed in terms of the Hermite polynomials $H_n(\xi)$ as follows [695]:

$$D_n(\sqrt{2}\xi) = \frac{1}{\sqrt{2^n}}e^{-\xi^2/2}H_n(\xi). \tag{6.79}$$

By using the definition of κ, one also derives the explicit expressions for the Landau-level energies

$$\epsilon_n = \pm\sqrt{\hbar^2 v_F^2 k_z^2 + m^2 + 2n\epsilon_L^2}. \tag{6.80}$$

It is important to note that the function $\Phi_+^s(\xi)$ is not finite as $|\xi| \to \infty$ for the $n = 0$ Landau level. Therefore, the corresponding bispinor u_s^+ should vanish. In order to see the implications of this requirement, one can use Eq. (6.74) to express u_s^+ in terms of u_s^-, i.e.,

$$u_s^+ = -\gamma_x \frac{m + s\sqrt{\epsilon_n^2 - \hbar^2 v_F^2 k_z^2}}{\sqrt{2n\epsilon_L^2}} u_s^-. \tag{6.81}$$

As is obvious, bispinor u_s^+ in the LLL vanishes identically if and only if $s = -1$. This confirms the well-known fact that the LLL is spin polarized because only one spin polarization is allowed. For higher Landau levels, no such constraint appears and, therefore, both spin polarizations $s = \pm$ are possible.

The explicit expression for bispinor $\phi_{n,s}(\xi)$ in the nth Landau level reads

$$\phi_{n,s}(\xi) = \left[\frac{e^{-\xi^2/2}}{\sqrt{2^n}} H_n(\xi) - i\gamma_x \frac{m + s\sqrt{m^2 + 2n\epsilon_L^2}}{\sqrt{2\epsilon_L^2}} \frac{e^{-\xi^2/2}}{\sqrt{2^{n-1}}} H_{n-1}(\xi) \right] u_{n,s}^-, \tag{6.82}$$

where, as follows from Eqs. (6.72) through (6.74),

$$u_{n,s}^- = \begin{pmatrix} 0 \\ \chi_2 \\ 0 \\ \chi_4 \end{pmatrix}, \quad \chi_4 = \frac{s\sqrt{\epsilon_n^2 - \hbar^2 v_F^2 k_z^2}}{\epsilon_n - \hbar v_F k_z} \chi_2. \tag{6.83}$$

Note that we took into account Eq. (6.81) in order to rewrite $u_{n,s}^+$ in terms of $u_{n,s}^-$.

The final Landau-level wave functions in an infinite sample are given by

$$\psi_{0,s=-}(\mathbf{r}) = C_0 \, e^{ik_z z + ik_y y} \, \phi_{0,s=-}(\xi), \tag{6.84}$$

$$\psi_{n,s}(\mathbf{r}) = e^{ik_z z + ik_y y} [C_1 \phi_{n,s=+}(\xi) + C_2 \phi_{n,s=-}(\xi)], \tag{6.85}$$

where

$$\phi_{0,s=-}(\xi) = \begin{pmatrix} 0 \\ Y_0(\xi) \\ 0 \\ -\frac{m}{\epsilon_0 - \hbar v_F k_z} Y_0(\xi) \end{pmatrix}, \tag{6.86}$$

$$\phi_{n>0,s}(\xi) = \begin{pmatrix} -is\frac{\sqrt{m^2 + 2n\epsilon_L^2}}{\epsilon_n - \hbar v_F k_z} \frac{m + s\sqrt{m^2 + 2n\epsilon_L^2}}{\sqrt{2n\epsilon_L^2}} Y_{n-1}(\xi) \\ Y_n(\xi) \\ i\frac{m + s\sqrt{m^2 + 2n\epsilon_L^2}}{\sqrt{2n\epsilon_L^2}} Y_{n-1}(\xi) \\ s\frac{\sqrt{m^2 + 2n\epsilon_L^2}}{\epsilon_n - \hbar v_F k_z} Y_n(\xi) \end{pmatrix}. \tag{6.87}$$

Here $Y_n(\xi) = e^{-\xi^2/2} H_n(\xi)/(\sqrt{2^n n! \sqrt{\pi}})$ are the harmonic oscillator functions.

Let us now discuss what happens in finite samples with a slab geometry. The corresponding wave functions can be obtained as linear combinations of the wave functions (6.84) and (6.85) with pairs of opposite momenta k_z. For example, for the LLL wave function inside the slab, one has

$$\Psi(\mathbf{r})_{0,|z|<L_z} = C_0 e^{ik_z z + ik_y y} \phi_{0,s=-}(\xi)$$
$$+ \tilde{C}_0 e^{-ik_z z + ik_y y} \phi_{0,s=-}(\xi)|_{k_z \to -k_z}. \tag{6.88}$$

The wave functions outside the slab can be obtained from Eqs. (6.84) and (6.85) by replacing m with M and k_z with $\pm i k_z'$, where $+$ $(-)$ corresponds to the vacuum region at $z > L_z$ ($z < -L_z$). As is clear, the energy dispersion relation on the vacuum side is given by $\epsilon_n = \pm\sqrt{M^2 + 2n\epsilon_L^2 - \hbar^2 v_F^2 (k_z')^2}$. For simplicity, one can assume that the vacuum mass is infinite, i.e., $M \to \infty$. Then since the energies on both sides of the slab boundary should match, one must require that $\hbar v_F k_z' \approx M$. By taking this into account, one derives the following LLL wave functions in vacuum:

$$\psi(\mathbf{r})_{0,z>L_z} = C_0' Y_0(\xi) e^{ik_y y} e^{-M(z-L_z)/(\hbar v_F)} \begin{pmatrix} 0 \\ 1 \\ 0 \\ -i \end{pmatrix}, \tag{6.89}$$

$$\psi(\mathbf{r})_{0,z<-L_z} = C_0'' Y_0(\xi) e^{ik_y y} e^{M(z+L_z)/(\hbar v_F)} \begin{pmatrix} 0 \\ 1 \\ 0 \\ i \end{pmatrix}. \tag{6.90}$$

By matching the semimetal and vacuum wave functions at the surfaces,

$$\Psi(z = L_z)_{0,|z|<L_z} = \psi(z = L_z)_{0,z>L_z}, \tag{6.91}$$

$$\Psi(z = -L_z)_{0,|z|<L_z} = \psi(z = -L_z)_{0,z<-L_z}, \tag{6.92}$$

it is easy to show that the z component of the wave vector should satisfy the following characteristic relation:

$$\hbar v_F k_z \cos(2L_z k_z) + m \sin(2L_z k_z) = 0. \tag{6.93}$$

Since $k_z = 0$ corresponds to a trivial solution, it should be excluded. The coefficients C_0, $\tilde{C}_0$, C_0', and C_0'' can be straightforwardly determined from the matching of the wave functions at $z = \pm L_z$ and the normalization conditions. The same type of calculations can be also repeated for the HLLs, where two independent solutions $\Psi_{n>0,|z|<L_z}^{(1,2)}$ appear for each Landau level. Their explicit form can be found in [683]. It should be noted that the

characteristic relation in Eq. (6.93) remains valid also for the HLLs inside the slab.

6.4.2 *Chiral separation effect in relativistic-like model*

By using the results from Sec. 6.4.1, let us calculate the chiral (axial) current in the slab. In terms of the Landau-level wave functions, the chiral current density is given by

$$j_{5,z} = -\sum_{e,h} ev_F \int \frac{dk_y}{2\pi} \sum_{k_z} \Big[f^{eq}(\epsilon_{n=0}) \Psi^\dagger(\mathbf{r})_{n=0,|z|<L_z} \gamma^0 \gamma_z \gamma_5 \Psi(\mathbf{r})_{n=0,|z|<L_z}$$

$$+ \sum_{i=1}^{2} \sum_{n=1}^{\infty} f^{eq}(\epsilon_n) \Psi^{(i)\,\dagger}(\mathbf{r})_{n,|z|<L_z} \gamma^0 \gamma_z \gamma_5 \Psi^{(i)}(\mathbf{r})_{n,|z|<L_z} \Big], \qquad (6.94)$$

Here the LLL and HLL contributions are given by the first and the second terms in the square brackets, respectively. The electric chemical potential μ and temperature T are taken into account via the standard Fermi–Dirac distribution function

$$f^{eq}(\epsilon_n) = \frac{1}{e^{(\epsilon_n - \mu)/T} + 1}, \qquad (6.95)$$

where ϵ_n is defined in Eq. (6.80) and $\sum_{e,h}$ denotes a summation over the electron and hole contributions (or the particle and antiparticle contributions in relativistic physics). Note that for holes one should replace $\mu \to -\mu$ and $e \to -e$. In the zero-temperature limit, which will be discussed below, the distribution function reduces to a unit step-function and the calculation simplifies considerably. Note that the following relation is valid at $T \to 0$:

$$\lim_{T \to 0} \sum_{e,h} e f^{eq}(\epsilon_n) = e\, \text{sgn}\,(\mu)\, \theta\left(\mu^2 - \epsilon_n^2\right). \qquad (6.96)$$

The most interesting contribution to the chiral current density comes from the spin-polarized LLL. It reads

$$j_{5,z}^{(LLL)} = \frac{e^2 B v_F \, \text{sgn}\,(\mu)}{2\pi \hbar c L_z} \sum_{k_z} \theta\left(\mu^2 - \hbar^2 v_F^2 k_z^2 - m^2\right)$$

$$\times \frac{L_z \left(m^2 + \hbar^2 v_F^2 k_z^2\right)\left[1 - \cos\left(2k_z z\right)\cos\left(2k_z L_z\right)\right]}{2L_z(m^2 + \hbar^2 v_F^2 k_z^2) + m v_F}, \qquad (6.97)$$

where the characteristic relation in Eq. (6.93) was used. As one sees from Eqs. (6.93) and (6.97), for a Dirac material with a nonzero gap, $m \neq 0$, the

LLL contribution to the chiral current density has a nontrivial dependence on the z coordinate. On the other hand, in the gapless (chiral) limit, the chiral current density is uniform. The corresponding explicit expression reads

$$\lim_{m \to 0} j_{5,z}^{(LLL)} = \frac{e^2 B v_F \, \text{sgn}\,(\mu)}{4\pi \hbar c L_z} \sum_{k_z} \theta\left(\mu^2 - \hbar^2 v_F^2 k_z^2\right) = \frac{e^2 B v_F \, \text{sgn}\,(\mu)}{4\pi \hbar c L_z} k_{\text{max}},$$

(6.98)

where it was taken into account that Eq. (6.93) simplifies down to $\cos\left(2 L_z k_z\right) = 0$ in the gapless case. This gives the following allowed values of the wave vector: $k_{z,j} = (2l - 1)\pi/(4 L_z)$, where l is a positive integer. Because of the unit step function in Eq. (6.98), the sum returns the integer result $k_{\text{max}} = [2 L_z |\mu|/(\pi \hbar v_F) + 1/2]$, where $[\ldots]$ represents the integer part. As expected, in the limit $L_z \to \infty$, the expression for the current density reduces to the conventional relation for the CSE in an infinite system, i.e.,

$$\lim_{L_z \to \infty} \lim_{m \to 0} j_{5,z}^{(LLL)} = \frac{e^2 B \mu}{2\pi^2 \hbar^2 c}.$$

(6.99)

Note that the result is saturated by the LLL contribution.

It can be verified by direct calculations that the contribution to $j_{5,z}$ from each HLL $(n > 0)$ vanishes. From a technical viewpoint, this comes due to the exact cancelation of the individual contributions from the two types of states, $\Psi_{|z|<L_z}^{(1)}$ and $\Psi_{|z|<L_z}^{(2)}$, in HLLs. Thus, as in the case of an infinite system [147], only the LLL contributes to the CSE, $j_{5,z} = j_{5,z}^{(LLL)}$.

As is evident from Eq. (6.98), the chiral current density is quantized in Dirac materials of a finite thickness. The height of quantization steps $\Delta j_{5,z}$ is proportional to the magnetic field and inversely proportional to the slab thickness, i.e., $\Delta j_{5,z} = e^2 B v_F/(4\pi \hbar c L_z)$. Interestingly, this step size is sensitive to the Fermi velocity.

Numerical results for the dimensionless chiral current density $\tilde{j}_{5,z} = 2\pi^2 c \hbar L_z j_{5,z}/(e^2 v_F B)$ are shown in Figs. 6.9(a) and 6.9(b) for the cases of vanishing as well as nonzero mass, respectively. As one can see, the presence of a nonzero energy gap m leads to spatial oscillations of $j_{5,z}$ inside the slab. Additionally, at $m \neq 0$, the height of steps varies with the position inside the slab. This effect is most pronounced at small values of μ. As one can check, this behavior is dominated by modes with small wave vectors k_z, which are the most susceptible to the effects of the gap. One should also observe that the amplitude of the oscillations changes across the sample and has the tendency to increase near the slab boundaries. The physical origin of these

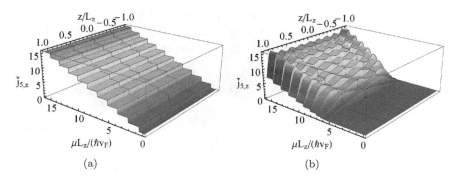

Fig. 6.9   The dimensionless chiral current density $\tilde{j}_{5,z}$ as a function of the electric chemical potential μ and coordinate z for (a) $m = 0$ and (b) $m = 6\,\hbar v_F/L_z$.

oscillations is related to the interference of the counter-propagating plane waves carrying opposite chiralities that produces a chiral standing wave.

Conceptually, it is important to emphasize that the chiral charge density

$$\rho_5 = -\sum_{e,h} ev_F \int \frac{dk_y}{2\pi} \sum_{k_z} \Big[f^{eq}(\epsilon_{n=0}) \Psi^\dagger(\mathbf{r})_{n=0,|z|<L_z} \gamma_5 \Psi(\mathbf{r})_{n=0,|z|<L_z}$$

$$+ \sum_{i=1}^{2} \sum_{n=1}^{\infty} f^{eq}(\epsilon_n) \Psi^{(i)\,\dagger}(\mathbf{r})_{n,|z|<L_z} \gamma_5 \Psi^{(i)}(\mathbf{r})_{n,|z|<L_z} \Big] = 0 \qquad (6.100)$$

vanishes everywhere inside the slab. Therefore, contrary to naive expectation, there is no chiral charge accumulation near the surfaces of the slab or anywhere in the bulk.

6.4.3   *Chiral magnetic effect in relativistic-like model*

Let us now discuss the chiral magnetic effect. Since it requires the presence of a nonzero chiral chemical potential μ_5, which is meaningful only in the massless case, it is convenient to set $m \to 0$ in the rest of this subsection. By definition, the electric current density is given by

$$j_z = -\sum_{e,h} ev_F \int \frac{dk_y}{2\pi} \sum_{k_z} \Big[f^{eq}(\epsilon_{n=0}) \, \Psi^\dagger(\mathbf{r})_{n=0,|z|<L_z} \gamma^0 \gamma_z \Psi(\mathbf{r})_{n=0,|z|<L_z}$$

$$+ \sum_{i=1}^{2} \sum_{n=1}^{\infty} f^{eq}(\epsilon_n) \, \Psi^{(i)\,\dagger}(\mathbf{r})_{n,|z|<L_z} \gamma^0 \gamma_z \Psi^{(i)}(\mathbf{r})_{n,|z|<L_z} \Big], \qquad (6.101)$$

where the first and the second terms in the square brackets correspond to the LLL and HLL contributions, respectively.

Let us start by considering the case of an infinite system out of equilibrium. In order to calculate the current densities, we will retain only the LLL contribution, which saturates the CME. In such an approximation, the distribution function for the fermions of chirality λ is given by

$$f_\lambda^{\text{eq}}(\epsilon_{n=0}) = \frac{1}{e^{(\hbar v_F k_z - \mu_\lambda)/T} + 1}, \tag{6.102}$$

where the wave vector k_z takes continuous rather than discrete values. By substituting the LLL wave function at $m = 0$, i.e.,

$$\psi(\mathbf{r})_{n=0} = C_0 Y_0(\xi) e^{ik_z z + ik_y y} \begin{pmatrix} 0 \\ \frac{1+\lambda}{2} \\ 0 \\ \frac{1-\lambda}{2} \end{pmatrix} \tag{6.103}$$

into the expression for the current, which is similar to Eq. (6.101) but the sum over k_z is replaced with the integration, one derives the following chiral current density:

$$j_{z,\lambda} = \lambda \frac{e^2 B \mu_\lambda}{4\pi^2 \hbar^2 c}. \tag{6.104}$$

By adding and subtracting the contributions of opposite chiralities, we easily obtain the CME and CSE currents

$$j_z = \frac{e^2 B \mu_5}{2\pi^2 \hbar^2 c}, \tag{6.105}$$

$$j_{z,5} = \frac{e^2 B \mu}{2\pi^2 \hbar^2 c}. \tag{6.106}$$

It can be verified by direct calculations that the CME and the CSE currents are saturated by the LLL contribution. The higher Landau levels contribute to neither of them.

As was already discussed at the beginning of this section, the CME electric current (6.105) is unphysical if interpreted as the ground-state property of a solid but could describe an observable out-of-equilibrium effect if treated properly. This will be also confirmed by the consistent chiral kinetic theory in Sec. 6.6.4, where the inclusion of topological Chern–Simons terms resolves the interpretation problem and clarifies the difference between the electric currents in and out of equilibrium. Similarly, it can be demonstrated explicitly within a more realistic two-band model with the periodic dispersion relation that an external magnetic field does not induce any electric currents in equilibrium. While the corresponding detailed calculations will be given in Sec. 6.4.4, it is instructive to provide a few qualitative arguments that explain the absence of the CME current in the ground state.

By retaining only the z component of momentum, the electric current can be schematically written as an integral over the Brillouin zone (BZ)

$$j_z \propto e^2 B \sum_n \int_{\mathrm{BZ}} dk_z f^{\mathrm{eq}}\left(\epsilon_n(k_z)\right) \frac{\partial \epsilon_n(k_z)}{\partial k_z}. \qquad (6.107)$$

By changing the integration variable in Eq. (6.107) from k_z to energy ϵ_n, one can show that j_z should vanish for any continuous periodic energy dispersion in the Brillouin zone, provided the distribution function depends explicitly only on the energy ϵ_n [390]. Of course, this is a very special case, but the qualitative conclusion remains true even in a more general situation because of the well-known theorem in solid-state physics [356], which states that an electric current can exist only in states with nonequilibrium distributions of electrons.

In conclusion, while the CME is an interesting phenomenon associated with the chiral anomaly, the naive predictions regarding a nonzero current in the equilibrium ground-state of solid-state systems cannot be taken at face value. Within the linearized relativistic-like models such predictions are plagued with at least two serious problems: (i) the concept of well-defined chirality breaks down beyond the low-energy sector of the theory and (ii) some important contributions from the filled states deep below the Fermi surface might be missing. One should also remember that finite size and boundary effects drastically modify the naive CME predictions even in the case of a truly relativistic spectrum [683]. In particular, realistic boundary conditions (e.g., with no electric current across the boundary) break the chiral symmetry and profoundly change the CME for the nonanomalous electric current. In contrast, the chiral current density due to the CSE effect is much more robust and can be easily realized even in the ground state, e.g., as a usual magnetization or a standing wave with counter-propagating modes of opposite chiralities.

6.4.4 *Two-band model with periodic dispersion*

In order to study the CSE and CME currents in Dirac and Weyl semimetals and to show the relation to high energy physics, a low-energy model with a linear relativistic-like energy dispersion was employed for simplicity in the preceding two subsections. It is very important, however, to check whether the conclusions drawn there can be reconfirmed in more realistic two-band models with periodic dispersion relations. Generically, the effects of the band spectrum periodicity could be indeed essential for anomalous transport properties. For example, as was already shown in Sec. 6.3, the AHE is easily missed in the linearized models of Weyl semimetals but appears naturally in a two-band model with a periodic spectrum. Its importance will be also demonstrated in Sec. 6.6.3, where it will be used to justify topological contributions to the charge and current densities in the consistent chiral kinetic theory.

In order to investigate the anomalous transport responsible for the chiral magnetic and chiral separation effects, here we employ a two-band model Hamiltonian similar to that introduced in Eq. (6.51) in Sec. 6.3 (see also Sec. 3.5.1)

$$H(\mathbf{k}) = -\mu + d_0 + (\mathbf{d} \cdot \boldsymbol{\sigma}), \tag{6.108}$$

where the components of the vector function $\mathbf{d}$ are defined in Eqs. (6.52)–(6.54). In order to break $\mathcal{P}$ symmetry, we also added an extra function d_0, which takes the following explicit form:

$$d_0 = \frac{t_0}{2} \sin{(ak_z)}. \tag{6.109}$$

As is easy to see, this function determines the energy separation between Weyl nodes $2|b_0|$, where

$$b_0 = \frac{t_0}{2} \sin{(ab_z)} \tag{6.110}$$

and b_z is the chiral shift defined in Eq. (6.55). The energy spectrum of Hamiltonian (6.108) is given by

$$\epsilon_{\mathbf{k}} = -\mu + d_0 \pm |\mathbf{d}|. \tag{6.111}$$

The energy spectrum for $d_0 = 0$ and $d_0 \neq 0$ is visualized in Figs. 6.10(a) and 6.10(b), respectively. In order to plot these energy relations, we used the numerical values of model parameters in Eq. (6.56). As expected, the results in Figs. 6.10(a) and 6.10(b) confirm that the momentum-dependent parameter d_0 induces a nonzero energy separation between the Weyl nodes of opposite chirality. This term implies that the Weyl nodes have different energies *relative* to the Fermi level in the material. Naively, this can be interpreted as having a nonzero chiral chemical potential μ_5, which leads to the CME in a magnetic field.

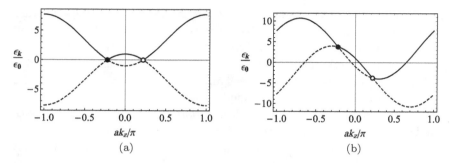

Fig. 6.10　The energy spectrum (6.111) of the two-band model in Eq. (6.108) for (a) $d_0 = 0$ and (b) $d_0 \neq 0$. The numerical constants are given in Eq. (6.56) as well as $\mu = 0$. The filled and empty points denote the right- and left-handed Weyl nodes. In addition, $\epsilon_0 = \lim_{\mathbf{k} \to 0} \epsilon_{\mathbf{k}}$.

In order to have a meaningful discussion of the CSE and CME currents, it is crucial to clarify the notion of chirality in the two-band model. Strictly speaking, the concept of chirality is well defined only for low-energy quasiparticles in the vicinity of Weyl nodes and its generalization to the whole Brillouin zone is dubious. This problem is well known and extensively discussed, for example, in the context of lattice formulations of relativistic field theories [46, 50, 54]. Additionally, according to the no-go theorem of Nielsen and Ninomiya [53], it is impossible to formulate a local, Hermitian, and chirally symmetric theory on a lattice without introducing an equal number of right- and left-handed fermions (and, therefore, doubling the number of fermion species). While the latter is a minor issue for solid-state physics, it creates serious problems for numerical lattice simulations of chiral gauge theories, such as the electroweak theory, where the use of chiral fermion representations is critical. Fortunately, this is not an insurmountable problem. By using the Ginsparg–Kaplan equation [696, 697] for the lattice Dirac operator, it is possible to define a modified chiral symmetry, which leaves the lattice action for massless fermions invariant. In essence, this is possible because of a relatively large freedom to choose the subleading terms in powers of the lattice spacing constant. They allow for modified definitions of lattice actions and the chiral symmetry generators. While such modifications often come with a substantial increase of the numerical cost of simulations, at least they allow to implement chiral representations in lattice formulations of relativistic quantum field theories. In the two-band model, we will use a heuristic approach instead.

To define a simplified version of chirality that extends to the whole Brillouin zone of the two-band model at hand, we will exploit its symmetry properties. As is clear, the definition of chirality needs to agree with the usual chirality of low-energy quasiparticles near the Weyl nodes and, at the same time, be well defined in the whole Brillouin zone. One of the reasonable definitions is given, for example, by the sign of a product of the three velocity components

$$\lambda(\mathbf{k}) = \text{sign}\left[v_x(\mathbf{k})v_y(\mathbf{k})v_z(\mathbf{k})\right], \tag{6.112}$$

where $v_i = \partial_{k_i} d_i$ with $i = x, y, z$ is the quasiparticle velocity. As is easy to verify, this indeed agrees with the conventional definition of chirality for the standard Weyl Hamiltonian $H(\mathbf{k}) = \lambda \hbar v_F (\boldsymbol{\sigma} \cdot \mathbf{k})$.[P]

Since both CME and CSE currents are linear in magnetic field, it is sufficient to perform a simple linear response analysis where the interaction of the gauge field $\mathbf{A} = (0, xB, 0)$ with the electric current is included via the standard minimal coupling term

$$H_{\text{int}} = \mathbf{j} \cdot \mathbf{A}(\mathbf{r}), \tag{6.113}$$

where the electric current density operator in momentum space is given by

$$\mathbf{j}(\mathbf{k}) = -e\partial_{\mathbf{k}}H(\mathbf{k}). \tag{6.114}$$

By making use of the quasiparticle Green's function $G(r, r')$, the electric current density can be expressed as follows:

$$\mathbf{J}(r) = -\lim_{r' \to r} \text{tr}\left[\mathbf{j}(-i\boldsymbol{\nabla})G(r, r')\right], \tag{6.115}$$

where $r = (t, \mathbf{r})$ and $r' = (t', \mathbf{r}')$. To linear order in the background electromagnetic fields, Green's function (or, equivalently, the Feynman propagator) takes the form

$$G(r, r') \approx G^{(0)}(r - r') + G^{(1)}(r, r'), \tag{6.116}$$

where the zeroth-order Green's function $G^{(0)}(r-r')$ is translation invariant. Its Fourier transform reads

$$G^{(0)}(\omega; \mathbf{k}) = i\frac{\omega + \mu - d_0 + (\mathbf{d} \cdot \boldsymbol{\sigma})}{[\omega + \mu - d_0 + i0\,\text{sgn}\,(\omega)]^2 - |\mathbf{d}|^2}. \tag{6.117}$$

[P]One can also use the reflection symmetry $k_z \to -k_z$ of Hamiltonian (6.108) and define the chirality as $\lambda(\mathbf{k}) = -\text{sgn}\,(k_z)$. This definition distinguishes electron quasiparticles from the two different valleys and, therefore, could also be viewed as the valley index. Such a definition of chirality is limited, however, only to Weyl semimetals with broken $\mathcal{T}$ symmetry.

The first-order correction to Green's function is given by

$$G^{(1)}(r, r') = -i \int d^4 r'' G^{(0)}(r - r'') H_{\text{int}}(r'') G^{(0)}(r'' - r'). \qquad (6.118)$$

After performing the Fourier transform, one derives the following expression for the electric current density along the magnetic field[q]:

$$J_z = -\frac{B}{2} \int \frac{d\omega d^3 k}{(2\pi)^4} \text{tr} \left\{ j_z(\mathbf{k}) \left[\partial_{k_z} G^{(0)}(\omega, \mathbf{k}) \right] j_y(\mathbf{k}) G^{(0)}(\omega; \mathbf{k}) \right.$$

$$\left. - j_z(\mathbf{k}) G^{(0)}(\omega; \mathbf{k}) j_y(\mathbf{k}) \left[\partial_{k_x} G^{(0)}(\omega; \mathbf{k}) \right] \right\}. \qquad (6.119)$$

The expression for the chiral current is similar, but the chirality operator $\lambda(\mathbf{k})$ should be inserted in the integrand. By integrating over the frequency (for details, see [432]), one obtains the following expression for the electric current density:

$$J_z = e^2 B \int \frac{d^3 k}{(2\pi)^3} \text{sgn} \left(\mu - d_0 \right) \left((\partial_{k_z} \mathbf{d}) \cdot \left[(\partial_{k_x} \mathbf{d}) \times (\partial_{k_y} \mathbf{d}) \right] \right)$$

$$\times \delta \left[(\mu - d_0)^2 - |\mathbf{d}|^2 \right] - e^2 B \int \frac{d^3 k}{(2\pi)^3} \frac{(\partial_{k_z} d_0)}{|\mathbf{d}|} \left(\mathbf{d} \cdot \left[(\partial_{k_x} \mathbf{d}) \times (\partial_{k_y} \mathbf{d}) \right] \right)$$

$$\times \left\{ \frac{1}{2|\mathbf{d}|^2} \left[1 - \theta \left(|\mu - d_0| - |\mathbf{d}| \right) \right] - \delta \left[(\mu - d_0)^2 - |\mathbf{d}|^2 \right] \right\}. \qquad (6.120)$$

After integrating over the whole Brillouin zone, it is straightforward to show that the electric current along the external magnetic field vanishes, $J_z = 0$. This result obtained in the two-band model provides a direct support for the absence of the CME current in the global equilibrium state and complements the arguments presented at the beginning of this section and in Sec. 6.4.3.

Unlike the CME, the CSE current is nontrivial in the two-band model. Since neither a chiral chemical potential μ_5 nor an energy separation between the Weyl nodes is required for inducing the CSE current, it suffices to set $d_0 = 0$ in the following discussion. It is convenient to use the generalized chirality defined by Eq. (6.112). For a sufficiently small electric chemical potential, $|\mu| < \epsilon_0$, direct calculations lead to the same standard

---

[q]It can be shown [432, 433] that other components of the electric and chiral current densities vanish if only a magnetic field is applied.

expression for the chiral current (6.99) as that obtained in relativistic models. On the other hand, some deviations in the CSE current appear for $|\mu| > \epsilon_0$, when the notion of chirality starts to deteriorate. In order to understand the physical reason for this, one should note that ϵ_0 defines the energy above which separate sheets of the Fermi surface, corresponding to the opposite chirality Weyl nodes, start to merge. This is evident, for example, from Fig. 6.8(a) or Fig. 6.10(a). It should be also clear that the definition of chirality of quasiparticles with energies larger than ϵ_0 becomes ambiguous. Nevertheless, as can be verified by utilizing a different definition of chirality, e.g., $\lambda(\mathbf{k}) = -\operatorname{sgn}(k_z)$, the final result for the CSE current is not very sensitive to a specific choice of the definition. This indicates that the CSE current is determined primarily by the states in the vicinity of Weyl nodes, where any reasonable definition works well and produces nearly the same chiral response.

In summary, while the key anomalous aspects of the electric and chiral charge transport in Weyl and Dirac semimetals can be captured within relativistic-like low-energy models, the periodicity of the band structure plays a significant role. This is particularly important for the electric current due to the CME, which is absent in an equilibrium ground state. This does not prevent, of course, the possibility that the CME could be realized in out-of-equilibrium processes. In fact, it is likely that the CME is manifested in a negative longitudinal magnetoresistance [7]. Furthermore, the CME driven by time-dependent perturbations could have nontrivial implications in finite samples of Weyl semimetals. Indeed, as shown in [608] by using numerical simulations, the Fermi arc surface states in a magnetic field (see also Chapter 5) can carry an electric current comparable to that of the bulk states even for large samples.

Unlike the CME, the CSE current can be manifested as a spin polarization even in the ground state of Dirac and Weyl semimetals. With that being said, the expression for the CSE current in Eq. (6.99) is a direct consequence of the chiral anomaly and the relativistic-like spectrum of low-energy quasiparticles. Indeed, when the quasiparticle chirality becomes ambiguous away from the Weyl nodes, the predictions for the CSE and other related effects start to deteriorate.

6.5 Planar Hall effect

Another interesting phenomenon observed in topological materials is the *planar Hall effect* (PHE), whose manifestation as well as relation to the chiral anomaly and the CME in Weyl semimetals were emphasized by Anton Burkov in [698].[r] In essence, the PHE is connected with the appearance of a transverse voltage when the in-plane magnetic field is neither aligned nor perpendicular to the current. The key feature of the PHE, which makes it distinct from the usual Hall effect is that the voltage, the magnetic field, and the current are coplanar. It should be mentioned that the name planar Hall effect is a misnomer because, in drastic contrast to the usual Hall effect, the relevant components of the resistivity tensor are even functions of a magnetic field.

The PHE can also be realized in usual ferromagnetic metals [699, 700], where it originates from the interplay of the magnetic order and the spin–orbit interaction. As will be explicitly shown below, the latter is not required in topological materials, where the PHE stems from the chiral anomaly. The PHE in Weyl and Dirac semimetals was studied theoretically in [698, 701–703].[s] Shortly after the theoretical predictions, the PHE was observed in GdPtBi [706, 707], Na_3Bi [707], Cd_3As_2 [708, 709], and $ZrTe_5$ [710].

Let us provide a phenomenological derivation of the PHE. The starting point is the system of diffusion equations for the electric μ and chiral μ_5 chemical potentials [698, 711]

$$\partial_t \mu = D\Delta\mu + \xi\left(\mathbf{B}\cdot\boldsymbol{\nabla}\right)\mu_5, \tag{6.121}$$

$$\partial_t \mu_5 = D\Delta\mu_5 - \frac{\mu_5}{\tau_5} + \xi\left(\mathbf{B}\cdot\boldsymbol{\nabla}\right)\mu, \tag{6.122}$$

where D is the diffusion constant, which, for simplicity, is assumed to be the same for the electric and chiral charges, $\xi = e/[2\pi^2\nu(\mu)\hbar^2 c]$ is the transport coefficient associated with the CME or CSE currents (see also Sec. 6.4), $\nu(\mu)$ is the density of states, whose precise form is not important now, and τ_5 is the chiral charge relaxation time. By assumption, μ on the right-hand side of the diffusion equations above is the electrochemical potential that includes the electric potential. It is also important that μ_5 vanishes in the equilibrium state.[t]

---

[r]Originally, this effect was dubbed "the giant planar Hall effect" in [698].
[s]An expression for the PHE current appeared in earlier studies, e.g., [704, 705], where μ_5 in the CME current is given by its expression $\mu_5 \propto (\mathbf{E}\cdot\mathbf{B})$ generated via the chiral anomaly.
[t]As is explained in Sec. 6.6.2, in the case of a nonzero energy separation $2b_0$ between the Weyl nodes of opposite chirality, the CME current $\propto \mu_5 + b_0$ vanishes in the equilibrium state, where the chiral chemical potential equals $\mu_5 = -b_0$.

The right-hand side of Eq. (6.121) is nothing else but $\nabla \cdot \mathbf{j}/[e\nu(\mu)]$, where the electric current density $\mathbf{j}$ is given by

$$\mathbf{j} = \frac{\sigma}{e}\nabla\mu + \frac{e^2\mu_5\mathbf{B}}{2\pi^2\hbar^2 c}. \tag{6.123}$$

Here $\sigma = e^2\nu(\mu)D$ is the Drude conductivity and the second term describes the CME current. The latter appears out of equilibrium when μ_5 is nonzero. Without loss of generality, let us assume that the steady-state current $\mathbf{I} = I\hat{\mathbf{x}}$ is driven along the x-direction through the wire with cross-section L_y^2 and length L_x. The corresponding schematic setup of the PHE is depicted in Fig. 6.11(a). Then Eq. (6.123) leads to

$$\partial_x\mu = \frac{eI}{\sigma L_y^2} - \frac{\mu_5}{L_5}\cos\theta, \tag{6.124}$$

$$\partial_y\mu = -\frac{\mu_5}{L_5}\sin\theta, \tag{6.125}$$

where θ is the angle between the magnetic field $\mathbf{B}$ and the current $\mathbf{I}$. In the last equation, it was convenient to introduce the chiral anomaly length scale $L_5 = 2\pi^2\hbar^2 c\sigma/(|e|^3 B) = D/(\xi B)$. Its inverse value $1/L_5$ quantifies the strength of coupling between the electric and chiral charges induced by the chiral anomaly. By taking into account the spatial symmetries in the problem, the dependence on the z-coordinate can be ignored.

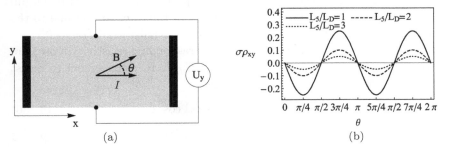

Fig. 6.11 (a) Schematic setup for the observation of the PHE. Here $\mathbf{I}$ is an injected electric current, $\mathbf{B}$ is an external magnetic field, θ is the angle between them, and U_y is the PHE voltage. (b) The dependence of the off-diagonal resistivity ρ_{xy} on the angle θ between $\mathbf{I}$ and $\mathbf{B}$. Here $L_5/L_\mathrm{D} = 1$ (solid line), $L_5/L_\mathrm{D} = 2$ (dashed line), and $L_5/L_\mathrm{D} = 3$ (dotted line).

By substituting Eqs. (6.124) and (6.125) into the steady-state equation (6.122) and ignoring the dependence of μ_5 on the transverse coordinate y, one obtains

$$\partial_x^2 \mu_5 - \frac{\mu_5}{\lambda^2} = -\frac{eI\cos\theta}{\sigma L_5 L_y^2}, \tag{6.126}$$

where $\lambda^2 = L_5^2 L_D^2 / \left(L_5^2 + L_D^2\right)$ and $L_D = \sqrt{D\tau_5}$ is the chiral charge diffusion length. In order to solve Eq. (6.126), it is reasonable to assume that $\mu_5(x = 0) = \mu_5(x = L_x) = 0$, which corresponds to the absence of the chiral charge at the contacts. Such boundary conditions are justified since the contacts are assumed to be made from nontopological materials. Note that, as will be shown explicitly below, the effect of the boundary conditions is not important at large L_x. The solution to Eq. (6.126) reads

$$\mu_5(x) = \frac{eI\lambda^2\cos\theta}{\sigma L_5 L_y^2}\left[1 - \frac{\cosh\left(\frac{L_x - 2x}{2\lambda}\right)}{\cosh\left(\frac{L_x}{2\lambda}\right)}\right]. \tag{6.127}$$

The voltage drops U_x and U_y can be easily obtained now by integrating the electrochemical potential gradients in Eqs. (6.124) and (6.125)

$$U_x = \frac{1}{e}\int_0^{L_x} dx\,\partial_x\mu = \frac{IL_x}{\sigma L_y^2} - \frac{I\lambda^2\cos^2\theta}{\sigma L_5^2 L_y}\left[L_x - 2\lambda\tanh\left(\frac{L_x}{2\lambda}\right)\right], \tag{6.128}$$

$$U_y = \frac{1}{eL_x}\int_0^{L_x} dx \int_0^{L_y} dy\,\partial_y\mu = -\frac{I\lambda^2\cos\theta\sin\theta}{\sigma L_5^2 L_y L_x}\left[L_x - 2\lambda\tanh\left(\frac{L_x}{2\lambda}\right)\right], \tag{6.129}$$

where averaging over the x coordinate was performed in the last equation. As is clear from Eq. (6.129), the voltage U_y in the direction normal to the current vanishes in the absence of the chiral anomaly, i.e., $1/L_5 \to 0$.

The independent components of the corresponding resistivity tensor are given by $\rho_{xx} = L_y^2 U_x/(IL_x)$ and $\rho_{yx} = L_y U_y/I$. It is convenient to recast these components in the following form:

$$\rho_{xx} = \rho_\perp - \left(\rho_\perp - \rho_\parallel\right)\cos^2\theta, \tag{6.130}$$

$$\rho_{yx} = -\left(\rho_\perp - \rho_\parallel\right)\sin\theta\cos\theta, \tag{6.131}$$

where

$$\rho_\parallel = \frac{1}{\sigma}\left[1 - \frac{\lambda^2}{L_5^2} + \frac{2\lambda^3}{L_5^2 L_x}\tanh\left(\frac{L_x}{2\lambda}\right)\right], \tag{6.132}$$

$$\rho_\perp = \frac{1}{\sigma} \tag{6.133}$$

are the components of the resistivity tensor along and perpendicular to the direction of the magnetic field, respectively. As is clear, Eq. (6.130) reproduces the celebrated negative magnetoresistivity for the longitudinal direction $\theta = 0$ (see also Sec. 6.2). The off-diagonal component ρ_{yx} describes the PHE but, unlike the case of the usual Hall effect, $\rho_{yx} \neq -\rho_{xy}$. One should note that the off-diagonal component is an even function of the magnetic field, $\rho_{yx}(B) = \rho_{yx}(-B)$, implying that the PHE remains unchanged when the direction of magnetic field is flipped.

The dependence of ρ_{yx} on θ for a few values of L_5/L_D is shown in Fig. 6.11(b). As expected, the PHE resistivity is a periodic function of angle θ with period π. The maximum absolute values ρ_{yx} are reached at $\theta = \pi/4 + n\pi/2$, where n are integers. As one might anticipate from Eqs. (6.131), the amplitude of the PHE resistivity oscillations decreases with $L_5 \propto 1/B$, which is equivalent to its growth with a magnetic field.

The representative data for the PHE in GdPtBi is shown in Figs. 6.12(a) and 6.12(b) for a few values of a magnetic field and a few values of temperature, respectively. Note that, at zero magnetic field, GdPtBi is a zero-gap semiconductor with quadratic bands, which split in the presence of a magnetic field and form Weyl nodes near the Fermi level. As is easy to see, the theoretical predictions in Fig. 6.11(b) are in good qualitative agreement with the experimental findings. The amplitude of the PHE resistivity increases linearly with a magnetic field and tends to saturate with temperature.

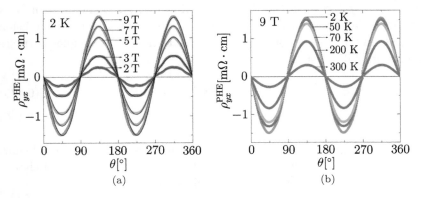

Fig. 6.12 The PHE resistivity ρ_{yx} as a function of the angle between the current and the magnetic field (see Fig. 6.11(a)) (a) at $T = 2$ K for several values of a magnetic field and at (b) $B = 9$ T for a few values of temperature. (Adapted from [706].)

In summary, the PHE provides another interesting manifestation of the chiral anomaly in solids. The combined observation of a negative magnetoresistivity and the PHE constitutes a strong evidence in favor of relativistic-like low-energy quasiparticles in Dirac and Weyl semimetals. When also supported by the spectroscopic measurements (see Sec. 4.3), the observation of such anomalous transport properties provides a very solid evidence for topological semimetals. In this connection, it is should be emphasized that the PHE alone cannot be considered as a definite proof of the chiral anomaly since it might originate from orbital effects [699].

6.6   Kinetic theory approach

To finalize this chapter, we discuss how anomalous properties of Weyl and Dirac semimetals can be described in such standard and time-honed approach as kinetic theory. This is a standard semiclassical framework for studying the transport properties and other nonequilibrium processes in media made of charged or neutral particles. In particular, it can be applied to electron quasiparticles in solids.[u] One should keep in mind, however, that the application of kinetic theory could be limited when the interaction between quasiparticle is not weak or when there are strong correlation effects. The key equation of kinetic theory is the celebrated Boltzmann equation

$$\partial_t f + (\dot{\mathbf{p}} \cdot \partial_{\mathbf{p}}) f + (\dot{\mathbf{r}} \cdot \boldsymbol{\nabla}) f = I_{\text{coll}}(f), \qquad (6.134)$$

where $f = f(t, \mathbf{p}, \mathbf{r})$ is the distribution function, $\mathbf{p}$ is momentum, $\mathbf{r}$ is coordinate, and the term $I_{\text{coll}}(f)$ on the right-hand side is the collision integral. In the presence of an external force $\mathbf{F}$, the motion of quasiparticles is conventionally described by the following equations:

$$\dot{\mathbf{r}} = \mathbf{v}, \qquad (6.135)$$

$$\dot{\mathbf{p}} = \mathbf{F}, \qquad (6.136)$$

where $\mathbf{v_p} = \partial_{\mathbf{p}} \epsilon_{\mathbf{p}}$ is the quasiparticle velocity. As we will discuss below, the equations of motion (6.135) and (6.136) are not universal and change for chiral quasiparticles with a nontrivial Berry curvature.

---

[u]It is worth noting that, with minor modifications, the results of this section will also apply to truly relativistic forms of chiral matter.

6.6.1 *Kinetic theory with Berry curvature*

As was first shown in [379, 380, 712] by using the wave-packet formalism, the kinetic equations should be modified in the presence of the Berry curvature (see also the discussion in Sec. 3.3). More recently, the interest to kinetic theory for systems with a nontrivial Berry curvature was also revived in the high-energy physics community [152, 713, 714]. This has led to a refined formulation of the *chiral kinetic theory* (CKT) applicable to generic chiral media composed of massless relativistic fermions. Despite being semiclassical, such a theory is able to capture the effects of chiral anomaly discussed in Sec. 6.1, as well as describe the chiral magnetic, chiral separation, and chiral vortical [149–152] effects.[v] In view of a close similarity between relativistic chiral matter and low-energy models of Weyl and Dirac semimetals, the CKT found immediate applications in studies of anomalous transport properties in topological materials [715]. In agreement with other more sophisticated approaches, the CKT is able to reproduce almost all hallmark effects discussed earlier in this chapter, including the negative magnetoresistivity, the chiral magnetic and the chiral separation effects, and much more. Starting from its success in describing transport phenomena, the CKT became a standard tool for other studies of Weyl and Dirac semimetals. It is instructive, therefore, to review it here.

Let us start with the derivation of kinetic theory in a generic model with nontrivial Berry curvature. As we will see, the CKT is just a special case of such a theory for chiral quasiparticles. In the literature, there are several approaches that allow one to account for the Berry curvature and/or chirality within the framework of kinetic theory. The oldest one is the wave-packet formalism considered in [379, 380, 712] and reviewed in detail in [377]. It should be noted, however, that the concept of chirality was not introduced explicitly in these papers. The chiral particles were discussed explicitly by using the Hamiltonian approach in [713, 716]. The CKT can be also derived directly from quantum field theory [714, 717–723]. The corresponding derivation relies on the quantum kinetic equation for the Wigner function [724–730]. In such a scheme, the kinetic theory is obtained by employing an expansion of the quantum kinetic equation in powers of gauge fields and derivatives [718, 719, 723, 731]. Unlike the CKT, the Wigner function approach could be applied even in a strong magnetic

---

[v]In essence, the chiral vortical effect (CVE) implies a current along the plasma's vorticity ω. For additional details, see also Sec. 7.4.2.

field [721, 732]. For the sake of simplicity, however, here we will follow [152] that uses a path-integral approach to derive the CKT.

In the path-integral formulation, the transition amplitude between two states $|\psi(t_f)\rangle$ and $|\psi(t_i)\rangle$ can be written as

$$\langle\psi(t_f)|\,\psi(t_i)\rangle = \int \mathcal{D}r\mathcal{D}k e^{iS/\hbar}, \qquad (6.137)$$

where S is the action, defined by

$$S = \int_{t_i}^{t_f} dt\, [(\mathbf{p}\cdot\dot{\mathbf{r}}) - H], \qquad (6.138)$$

and H is the Hamiltonian. In the case of the CKT, the latter is the Weyl Hamiltonian (3.3), i.e., $H = \lambda v_F\,(\boldsymbol{\sigma}\cdot\mathbf{p})$. By making use of a suitable unitary transformation, the Hamiltonian can be diagonalized, i.e.,

$$U_{\mathbf{p}}^{\dagger} H U_{\mathbf{p}} = H_{\text{diag}}. \qquad (6.139)$$

In the path integral formulation, one needs to perform such a diagonalization at each time step. By assuming that quasiparticle momenta are $\mathbf{k}_1$ and $\mathbf{k}_2$ at times t_1 and t_2, respectively, one has

$$\ldots U_{\mathbf{p}_2} U_{\mathbf{p}_2}^{\dagger} e^{-iH(\mathbf{p}_2)\Delta t/\hbar} U_{\mathbf{p}_2} U_{\mathbf{p}_2}^{\dagger} U_{\mathbf{p}_1} U_{\mathbf{p}_1}^{\dagger} e^{-iH(\mathbf{p}_1)\Delta t/\hbar} U_{\mathbf{p}_1} U_{\mathbf{p}_1}^{\dagger} \ldots$$

$$= \ldots U_{\mathbf{p}_2} e^{-iH_{\text{diag}}(\mathbf{p}_2)\Delta t/\hbar} U_{\mathbf{p}_2}^{\dagger} U_{\mathbf{p}_1} e^{-iH_{\text{diag}}(\mathbf{p}_1)\Delta t/\hbar} U_{\mathbf{p}_1}^{\dagger} \ldots$$

$$\approx \ldots U_{\mathbf{p}_2} e^{-iH_{\text{diag}}(\mathbf{p}_2)\Delta t/\hbar} e^{-i\hat{\mathbf{A}}_{\mathbf{p}_2}\cdot\Delta\mathbf{p}} e^{-iH_{\text{diag}}(\mathbf{p}_1)\Delta t/\hbar} U_{\mathbf{p}_1}^{\dagger} \ldots \qquad (6.140)$$

Here it was assumed that $\Delta\mathbf{p} = \mathbf{p}_2 - \mathbf{p}_1$ and $\Delta t = t_2 - t_1$ are small. To leading order in $\Delta\mathbf{p}$, one can approximate $U_{\mathbf{p}_2}^{\dagger} U_{\mathbf{p}_1} \approx 1 - U_{\mathbf{p}_2}^{\dagger}(\Delta\mathbf{p}\cdot\partial_{\mathbf{p}_2})U_{\mathbf{p}_1}$. It is also convenient to introduce the shorthand notation $\hat{\mathbf{A}}_{\mathbf{p}} = -i\hbar U_{\mathbf{p}}^{\dagger}\partial_{\mathbf{p}}U_{\mathbf{p}}$, which represents nothing else but the Berry connection matrix. Note that, in general, the corresponding matrix contains diagonal and nondiagonal elements. Therefore, the next step is to perform the Abelian projection [733] of the Berry connection. In the classical limit, the effects of the off-diagonal components of the latter are negligible and the Berry connection can be treated as a diagonal matrix. It should be noted, however, that this approximation breaks down at the degeneracy points, i.e., the Weyl nodes $\mathbf{p} = \mathbf{0}$. Moreover, the Berry connection (as well as the Berry curvature) acquires a non-Abelian structure [373] for multiple degenerate states. For example, this is the case for certain Dirac semimetals (see also Sec. 3.6 and [378]). Here, for simplicity, we concentrate only on a simple nondegenerate case and avoid the neutrality point, i.e., consider $\mathbf{p} \neq \mathbf{0}$.

The amplitude (6.137) can be rewritten as

$$U_{\mathbf{p}_f} \int \mathcal{D}r\mathcal{D}k \exp\left\{ i \int_{t_i}^{t_f} dt[(\mathbf{p}\cdot\dot{\mathbf{r}}) - H_{\text{diag}} - (\hat{\mathbf{A}}_{\mathbf{p}}\cdot\dot{\mathbf{p}})]/\hbar \right\}U_{\mathbf{p}_i}. \quad (6.141)$$

Since the matrix structure in the exponent is diagonal and the off-diagonal components of the Berry connection matrix are ignored in the classical limit, one extracts the following expression for the effective action:

$$S = \int_{t_i}^{t_f} dt\,[(\mathbf{p}\cdot\dot{\mathbf{r}}) - \epsilon_{\mathbf{p}} - (\mathbf{A}_{\mathbf{p}}\cdot\dot{\mathbf{p}}) - e\,(\mathbf{A}\cdot\dot{\mathbf{r}})\,/c + e\phi], \quad (6.142)$$

where $\epsilon_{\mathbf{p}}$ is the quasiparticle dispersion relation. By using the usual minimal coupling, we also included the effects of an external electromagnetic field via the vector $\mathbf{A}$ and scalar ϕ potentials in the action. By definition, the charge of electron quasiparticles is $-e$.

The extremum of this effective action defines the semiclassical equations of motion for quasiparticles

$$\dot{\mathbf{r}} = \mathbf{v}_{\mathbf{p}} + [\dot{\mathbf{p}}\times\mathbf{\Omega}], \quad (6.143)$$

$$\dot{\mathbf{p}} = -e\tilde{\mathbf{E}} - \frac{e}{c}\,[\dot{\mathbf{r}}\times\mathbf{B}], \quad (6.144)$$

where $\mathbf{v}_{\mathbf{p}} = \partial_{\mathbf{p}}\epsilon_{\mathbf{p}}$ is the quasiparticle velocity, $\mathbf{\Omega} = [\partial_{\mathbf{p}}\times\mathbf{A}_{\mathbf{p}}]$ is the Berry curvature, $\mathbf{B}$ is the magnetic field, and $\tilde{\mathbf{E}} = \mathbf{E} + (1/e)\mathbf{\nabla}\epsilon_{\mathbf{p}}$, where $\mathbf{E}$ is an electric field.[w] Comparing Eqs. (6.135) and (6.136) with Eqs. (6.143) and (6.144), it is important to note that the quasiparticle velocity obtains an anomalous term determined by the Berry curvature and the time derivative of momentum. Another important modification, which is not immediately evident from the above simplified derivation, is an additional Zeeman-like term in the quasiparticle energy dispersion relation $\epsilon_{\mathbf{p}} = \epsilon_{\mathbf{p}}(B = 0) - (\mathbf{B}\cdot\mathbf{m})$ [377, 380, 714], where $\mathbf{m}$ is the orbital magnetic moment.[x] By solving Eqs. (6.143) and (6.144) in terms of $\dot{\mathbf{r}}$ and $\dot{\mathbf{p}}$, one obtains

$$\Theta\dot{\mathbf{r}} = \mathbf{v}_{\mathbf{p}} - e[\tilde{\mathbf{E}}\times\mathbf{\Omega}] - \frac{e}{c}\mathbf{B}\,(\mathbf{v}\cdot\mathbf{\Omega}), \quad (6.145)$$

$$\Theta\dot{\mathbf{p}} = -e\tilde{\mathbf{E}} - \frac{e}{c}\,[\mathbf{v}\times\mathbf{B}] + \frac{e^2}{c}\mathbf{\Omega}(\tilde{\mathbf{E}}\cdot\mathbf{B}), \quad (6.146)$$

where $\Theta = [1 - e\,(\mathbf{B}\cdot\mathbf{\Omega})\,/c]$.

---

[w] As argued in [374, 377, 712, 734], additional terms related to Berry curvature appear in Eqs. (6.143) and (6.144) when parameters of the Hamiltonian are perturbed.
[x] The corresponding correction can be obtained in the wave-packet approach [380].

As is obvious from Eqs. (6.145) and (6.146), the semiclassical dynamics of quasiparticles is strongly modified by the Berry curvature effects in the presence of a magnetic field. As was already mentioned in Sec. 3.3.3, there is an anomalous contribution to the velocity determined by the Berry curvature (see the second term in Eq. (6.145)). Since the equations of motion are modified, the conventional form of Liouville's theorem, which implies the conservation of phase-space volume, is not valid any more [381, 716]. Indeed, let us consider the time evolution of a phase-space volume element $\Delta V = \Delta r \Delta p$. By differentiating it with respect to time, $d \ln \Delta V / dt = \boldsymbol{\nabla} \cdot \dot{\mathbf{r}} + \partial_{\mathbf{p}} \cdot \dot{\mathbf{p}}$, and then using the equations of motion (6.145) and (6.146), it is straightforward although somewhat tedious to show that the result equals to a total derivative, i.e., $\boldsymbol{\nabla} \cdot \dot{\mathbf{r}} + \partial_{\mathbf{p}} \cdot \dot{\mathbf{p}} = -d \ln \Theta / dt$. After integrating this expression, the time evolution of the phase-space volume element is given by

$$\Delta V(t) = \frac{\Delta V_0}{\Theta}, \tag{6.147}$$

where ΔV_0 is the phase-space volume at t_0. Since $\Theta = [1 - e(\mathbf{B} \cdot \boldsymbol{\Omega})/c]$ depends on the Berry curvature and the magnetic field, the phase-space volume ΔV changes with time.

Interestingly, the breakdown of Liouville's theorem can be easily cured if the density of states ν is also modified by the Berry curvature, i.e.,

$$\nu = \frac{1}{(2\pi\hbar)^3} \quad \rightarrow \quad \nu = \frac{\Theta}{(2\pi\hbar)^3}. \tag{6.148}$$

After such an adjustment, $\nu \Delta V = const$, i.e., the number of states in the phase-space volume remains constant. This implies that the mean value of an observable $O(\mathbf{r}, \mathbf{p})$ at a spatial point $\mathbf{r}$ can be calculated as

$$O(\mathbf{r}) = \int \frac{d^3 p}{(2\pi\hbar)^3} \Theta(\mathbf{p}) O(\mathbf{r}, \mathbf{p}) f(\mathbf{p}), \tag{6.149}$$

where $f(\mathbf{p})$ is a distribution function.

Before proceeding to the specific case of the CKT, it is important to discuss the limitations of the kinetic theory approach. Obviously, the semiclassical theory breaks down when a magnetic field is very strong and quantum effects cannot be ignored. Furthermore, kinetic theory becomes unreliable when interactions are strong and quasiparticles no longer move freely between collisions. Interestingly, the formalism could be extended to the second order in external fields. In such a case, the general structure of the kinetic equations remains the same, but the Berry curvature, the dispersion relation, and the velocity are modified [423, 735, 736]. In some applications, one should also bear in mind that the conventional formulation of the chiral kinetic theory is not Lorentz covariant. While the latter can be recovered in principle [718, 719, 723, 737, 738], the corresponding framework becomes significantly more complicated, especially when interactions included via the collision integral are important.

6.6.2   *Chiral kinetic theory*

In this subsection, the explicit formulation of the chiral kinetic theory is provided. The CKT describes the semiclassical motion of quasiparticles governed by the Weyl Hamiltonian $H = \lambda v_F (\boldsymbol{\sigma} \cdot \mathbf{p})$, where λ is the chirality. Since chirality is the key ingredient of the CKT, it is convenient to keep track of the chirality index λ. The Berry curvature for the Weyl Hamiltonian (3.3) has been already derived in Sec. 3.3. It has a simple monopole-like structure and is proportional to chirality λ, i.e.,

$$\boldsymbol{\Omega}_\lambda = \lambda \hbar \frac{\hat{\mathbf{P}}}{2p^2}. \tag{6.150}$$

Therefore, as is evident from Eqs. (6.145) and (6.146), the dynamics of quasiparticles is affected by their chirality when external electromagnetic fields are applied to the system.

The Boltzmann equation (6.134) in the case of CKT reads [152, 713–715]

$$\partial_t f_\lambda + \frac{1}{1 - \frac{e}{c}(\mathbf{B}_\lambda \cdot \boldsymbol{\Omega}_\lambda)} \left\{ \left(-e\tilde{\mathbf{E}}_\lambda - \frac{e}{c}[\mathbf{v_p} \times \mathbf{B}_\lambda] + \frac{e^2}{c}(\tilde{\mathbf{E}}_\lambda \cdot \mathbf{B}_\lambda)\boldsymbol{\Omega}_\lambda) \cdot \partial_\mathbf{p} f_\lambda \right.$$

$$\left. + (\mathbf{v_p} - e[\tilde{\mathbf{E}}_\lambda \times \boldsymbol{\Omega}_\lambda] - \frac{e}{c}(\mathbf{v_p} \cdot \boldsymbol{\Omega}_\lambda)\mathbf{B}_\lambda) \cdot \boldsymbol{\nabla} f_\lambda \right\} = I_{\text{coll}}(f_\lambda), \tag{6.151}$$

where f_λ is the distribution function for fermions of chirality λ. In equilibrium, for example, the distribution function takes the usual Fermi–Dirac form

$$f_\lambda^{\text{eq}} = \frac{1}{e^{(\epsilon_\mathbf{p} - \mu_\lambda)/T} + 1}. \tag{6.152}$$

Here $\mu_\lambda = \mu + \lambda \mu_5$ is the effective chemical potential for the right- ($\lambda = +$) and left-handed ($\lambda = -$) quasiparticles, μ is the electric chemical potential, μ_5 is the chiral chemical potential, T is temperature in the energy units, and $\epsilon_{\mathbf{p}}$ is the dispersion relation. The latter is given by [377, 380, 714]

$$\epsilon_{\mathbf{p}} = v_F p - (\mathbf{m}_\lambda \cdot \mathbf{B}_\lambda) = v_F p \left[1 + \frac{e}{c}(\mathbf{B}_\lambda \cdot \mathbf{\Omega}_\lambda)\right], \qquad (6.153)$$

where $\mathbf{m}_\lambda = -e v_F p \mathbf{\Omega}_\lambda / c$ is the magnetic moment. One should note that the equations of the CKT presented above may include both electromagnetic fields $\mathbf{E}$ and $\mathbf{B}$, as well as pseudoelectromagnetic fields $\mathbf{E}_5$ and $\mathbf{B}_5$. In Weyl and Dirac semimetals, the latter can be induced by strain as was discussed in Sec. 3.7. Therefore, it was convenient to introduce the chiral fields $\mathbf{E}_\lambda = \mathbf{E} + \lambda \mathbf{E}_5$ and $\mathbf{B}_\lambda = \mathbf{B} + \lambda \mathbf{B}_5$ that interact with fermions of a given chirality λ.

By making use of Eq. (6.153), one derives the following expression for the quasiparticle velocity

$$\mathbf{v}_{\mathbf{p}} = \partial_{\mathbf{p}} \epsilon_{\mathbf{p}} = v_F \hat{\mathbf{p}} \left[1 - 2\frac{e}{c}(\mathbf{B}_\lambda \cdot \mathbf{\Omega}_\lambda)\right] + \frac{e v_F}{c} \mathbf{B}_\lambda (\hat{\mathbf{p}} \cdot \mathbf{\Omega}_\lambda). \qquad (6.154)$$

It should be noted that the CKT expressions above are valid only for sufficiently weak magnetic fields. By using Eq. (6.150), one finds the following criterion $\hbar |e \mathbf{B}_\lambda| / (c p^2) \ll 1$, where the characteristic value of momentum can be estimated as the smaller of $p \simeq |\mu| / v_F$ and $p \simeq T / v_F$.

For completeness, let us also briefly discuss the collision integral on the right-hand side in Eq. (6.151). As in nonrelativistic kinetic theory [739, 740], the simplest approximation for the Boltzmann equation is the collisionless limit with $I_{\text{coll}} = 0$. Such an approximation was proposed long time ago by Anatoly Vlasov [741, 742] and can be justified under certain conditions in plasma when the rate of collisions is small compared to the rate of processes under consideration. In general, however, the collision integral is essential for the description of nonequilibrium dynamics and dissipative effects in matter. Since the inclusion of the collision integral leads an integro-differential kinetic equation that is difficult to solve, various approximations are often used to simplify the analysis. One of the simplest and quite popular models is the so-called relaxation-time approximation which uses $I_{\text{coll}} = -(f - f^{\text{eq}})/\tau$, where τ is the relaxation time.

In order to capture the specifics of Weyl semimetals, however, one needs to distinguish between the intra- and internode scattering processes. This can be achieved by introducing two different types of the collision integrals. Because of a separation between the Weyl nodes in energy and/or momentum, the internode scattering time τ_5 is usually much larger than τ. In some cases, it might be even justified to set $\tau_5 \to \infty$. For certain phenomena, however, the corresponding collision term plays a critical role and, in such cases, it cannot be neglected in the kinetic equation [648, 743, 744]. In general, the scattering time may depend on the energy and the momentum of quasiparticles, as well as on a magnetic field. A few different models of the relaxation processes due to disorder will be discussed in Sec. 7.1.

In the CKT, the charge and current densities are given by [713, 714]

$$\rho_\lambda = -\sum_{e,h} e \int \frac{d^3 p}{(2\pi\hbar)^3} \left[1 - \frac{e}{c}(\mathbf{B}_\lambda \cdot \mathbf{\Omega}_\lambda)\right] f_\lambda, \tag{6.155}$$

$$\mathbf{j}_\lambda = -\sum_{e,h} e \int \frac{d^3 p}{(2\pi\hbar)^3} \left\{\mathbf{v_p} - \frac{e}{c}(\mathbf{v_p} \cdot \mathbf{\Omega}_\lambda)\mathbf{B}_\lambda - e[\tilde{\mathbf{E}}_\lambda \times \mathbf{\Omega}_\lambda]\right\} f_\lambda$$

$$- \sum_{e,h} e\mathbf{\nabla} \times \int \frac{d^3 p}{(2\pi\hbar)^3} f_\lambda v_F p \mathbf{\Omega}_\lambda, \tag{6.156}$$

where $\sum_{e,h}$ denotes a summation over electrons and holes. Note that, in the case of holes, one should replace $\mathbf{\Omega}_\lambda \to -\mathbf{\Omega}_\lambda$, $\mu_\lambda \to -\mu_\lambda$, and flip sign of the electric charge $e \to -e$. As one can see, there are several types of contributions with the Berry curvature in Eq. (6.156). The last term, in particular, describes a magnetization current induced by the Berry curvature, $\mathbf{j}_\Omega = c\mathbf{\nabla} \times \mathbf{M}_\Omega$. As a total curl, the latter does not affect the continuity equations for the electric and chiral charges. The magnetization current does, however, play an important role in Maxwell's equations.

It is instructive to discuss the continuity relations for the electric and chiral charges in the CKT. By using Eqs. (6.151), (6.155), and (6.156), as well as taking into account that the collision term conserves the electric and chiral charges, it is straightforward to obtain

$$\frac{\partial \rho_5}{\partial t} + (\mathbf{\nabla} \cdot \mathbf{j}_5) = -\frac{e^3}{2\pi^2\hbar^2 c}[(\mathbf{E} \cdot \mathbf{B}) + (\mathbf{E}_5 \cdot \mathbf{B}_5)], \tag{6.157}$$

$$\frac{\partial \rho}{\partial t} + (\mathbf{\nabla} \cdot \mathbf{j}) = -\frac{e^3}{2\pi^2\hbar^2 c}[(\mathbf{E} \cdot \mathbf{B}_5) + (\mathbf{E}_5 \cdot \mathbf{B})]. \tag{6.158}$$

One may interpret the first equation as a generalization of the anomalous chiral charge continuity equation to the case with background electromagnetic and pseudoelectromagnetic fields.[y] In Weyl semimetals, this corresponds to a chiral charge pumping between the Weyl nodes of opposite chirality and does not pose any problem (see also Sec. 6.2). The situation

---

[y]The nonconservation of the chiral charge in strain-induced pseudoelectromagnetic fields was first discussed in [408, 413].

is drastically different from the second equation, which predicts a local nonconservation of the electric charge when both electromagnetic and pseudoelectromagnetic fields are applied to a system. Unlike the chiral charge nonconservation, such an effect is physically unacceptable and should be forbidden in the theory.

6.6.3 Consistent chiral kinetic theory

As was stated in Sec. 6.6.2, the local charge nonconservation described by Eq. (6.158) signifies an inconsistency of the conventional formulation of CKT. Below we formulate a consistent chiral kinetic theory that resolves the problem.

6.6.3.1 Consistent versus covariant currents

A simple solution to the puzzle of the local nonconservation of the electric charge is to reinterpret the right-hand side in Eq. (6.158) as additional topological contributions to the electric charge and current densities, which are defined by the following Chern–Simons terms [420]:

$$\rho_{\text{CS}} = -\frac{e^3}{2\pi^2\hbar^2 c^2}\left(\mathbf{A}^5 \cdot \mathbf{B}\right), \tag{6.159}$$

$$\mathbf{j}_{\text{CS}} = -\frac{e^3}{2\pi^2\hbar^2 c}A_0^5\mathbf{B} + \frac{e^3}{2\pi^2\hbar^2 c}\left[\mathbf{A}^5 \times \mathbf{E}\right]. \tag{6.160}$$

As is easy to verify, the sum of the time derivative of ρ_{CS} and the divergence of $\mathbf{j}_{\text{CS}}$ reproduce up to a sign the right-hand side in Eq. (6.158). Note that A_0^5 and $\mathbf{A}^5$ are the components of the axial vector potential A_μ^5 ($\mu = 0, 1, 2, 3$) that may depend on the time and spatial coordinates. The pseudoelectromagnetic fields are defined as $\mathbf{E}_5 = -\boldsymbol{\nabla}A_0^5 - \partial_t\mathbf{A}^5/c$ and $\mathbf{B}_5 = \boldsymbol{\nabla} \times \mathbf{A}^5$. After including the topological contributions given in Eqs. (6.159) and (6.160), the conservation of the electric charge is enforced locally. This is clear from the fact that the electric four-current $J^\mu = (c\rho + c\rho_{\text{CS}}, \mathbf{j} + \mathbf{j}_{\text{CS}})$ satisfies the usual continuity equation $\partial_\mu J^\mu = 0$. While this formally resolves the problem, one might wonder what is the physical origin of the topological terms. As we will argue below, they come from filled states deeply below the Fermi surface. The CKT, where the consistent currents with the Chern–Simons contributions are used instead the of covariant ones, is called the consistent chiral kinetic theory [420]. It should be noted that the role of consistent and covariant currents in application to Weyl semimetals was first uncovered by Karl Landsteiner [681].

As one can see from Eqs. (6.159) and (6.160), even a constant axial vector potential A_μ^5 could give rise to nonzero charge and current densities and, therefore, could have observable effects. At first sight, this might appear surprising since there are no pseudoelectromagnetic fields for a constant A_μ^5. One has to remember, however, that there is no usual gauge symmetry associated with the pseudoelectromagnetic fields and, therefore, an axial vector potential cannot be gauged away. Moreover, the constant components of the axial vector potential can be reinterpreted as the parameters that define the separations between the Weyl nodes in energy and momentum $b_0 = -eA_0^5$ and $\mathbf{b} = -e\mathbf{A}^5/(\hbar c)$, respectively. After taking this into account, one finds that the last term in Eq. (6.160) describes the anomalous Hall effect that was discussed in Sec. 6.3. As for the first term $\sim b_0\mathbf{B}$, it is critical for ensuring the absence of the CME current in the equilibrium ground state of Weyl semimetals (see also the discussion in Sec. 6.4).

Historically, a similar problem with the local charge conservation was first encountered in high energy physics in the presence of a background vector gauge field A_μ and its axial counterpart A_μ^5. Indeed, a symmetric treatment of the left- and right-handed contributions in the celebrated triangle diagrams produces an anomalous continuity equation for the so-called *covariant electric current* j^μ [41, 681, 745]. Of course, such a definition of the electric current is inconsistent with the gauge symmetry and contradicts particle phenomenology. By using a different regularization of ultraviolet divergences, one can also define the so-called *consistent electric current* J^μ, which is nonanomalous and consistent with the gauge invariance. The difference between the covariant and consistent currents is given by the Bardeen–Zumino polynomial [746, 747]. In the case under consideration, it takes the same form as the Chern–Simons four-current

$$j_{\text{CS}}^\mu = -\frac{e^3}{4\pi^2\hbar^2 c}\epsilon^{\mu\nu\rho\eta}A_\nu^5 F_{\rho\eta}. \tag{6.161}$$

Here $\epsilon^{\mu\nu\rho\eta}$ is the four-dimensional Levi-Civita symbol with $\epsilon^{0123} = -1$ and the signature of the Minkowski metric is assumed to be $(+, -, -, -)$. After separating the time-like and space-like components in Eq. (6.161), it is easy to verify that this result reproduces exactly the Chern–Simons terms in Eq. (6.159) and (6.160).

Similarly to the electric current, the Chern–Simons terms (Bardeen–Zumino polynomial) in the chiral current density could be defined as

$$j_{5,\text{CS}}^\mu = -\frac{e^3}{12\pi^2\hbar^2 c}\epsilon^{\mu\nu\rho\eta}A_\nu^5 F_{\rho\eta}^5. \tag{6.162}$$

This correction to axial current density can be derived rigorously in high-energy physics under some natural assumptions about the ultraviolet sector of the theory. Its role in Weyl semimetals, where the axial gauge fields are effective fields induced by strains and the concept of chirality breaks down beyond the low-energy region, is not clear. Nevertheless, if Eq. (6.162) is taken at face value, the continuity relation for the consistent chiral current reads

$$\partial_\mu J_5^\mu = -\frac{e^3}{2\pi^2\hbar^2c}\left[(\mathbf{E}\cdot\mathbf{B}) + \frac{1}{3}(\mathbf{E}_5\cdot\mathbf{B}_5)\right]. \qquad (6.163)$$

It is interesting to note that the anomalous term related to the strain-induced pseudomagnetic fields has a factor $1/3$.

6.6.3.2 *Microscopic origin of the Chern–Simons terms*

In order to understand the underlying physics better, it is important to discuss the *microscopic origin* of the Chern–Simons terms (6.159) and (6.160). For this purpose, it is convenient to employ the same two-band model that was used in Sec. 6.4.4. Recall that its Berry curvature is given by the following expression:

$$\Omega_j = \sum_{n,m=x,y,z} \frac{\epsilon_{jnm}\mathcal{F}_{nm}(\mathbf{k})}{2}, \qquad (6.164)$$

where

$$\mathcal{F}_{nm}(\mathbf{k}) = \frac{1}{2\hbar|\mathbf{d}|^3}\left(\mathbf{d}\cdot[(\partial_{k_n}\mathbf{d})\times(\partial_{k_m}\mathbf{d})]\right) \qquad (6.165)$$

is the Berry curvature tensor and $\mathbf{d}$ is a periodic function of the wave vector defined explicitly in Eqs. (6.52) through (6.54). Unlike the approximate monopole-like Berry curvature of linearized models, which is defined only in the vicinity of Weyl nodes (see also Sec. 3.3), the expression in Eq. (6.164) is valid in the whole Brillouin zone. Among other things, it captures exactly all nontrivial effects due the chiral shift b_z defined by Eq. (6.55) and the energy separation b_0 defined by Eq. (6.110).

In order to extract the topological contributions in the kinetic theory framework, it suffices to take into account only the filled states deeply below the Fermi surface. In this case, the explicit expressions for the electric charge and current densities are given by [432]

$$\rho_{\text{CS}} = -\frac{e^2}{c} \int \frac{d^3p}{(2\pi\hbar)^3} (\mathbf{B} \cdot \mathbf{\Omega}) = \frac{e^2}{2\pi^2\hbar c} (\mathbf{b} \cdot \mathbf{B}), \qquad (6.166)$$

$$\mathbf{j}_{\text{cs}} = -e^2 \int \frac{d^3p}{(2\pi\hbar)^3} \left\{ \frac{(\mathbf{v_k} \cdot \mathbf{\Omega})\mathbf{B}}{c} + [\mathbf{E} \times \mathbf{\Omega}] \right\}$$

$$= \frac{e^2}{2\pi^2\hbar^2 c} b_0 \mathbf{B} - \frac{e^2}{2\pi^2\hbar} [\mathbf{b} \times \mathbf{E}], \qquad (6.167)$$

where, by definition, $\mathbf{v_k} = \hbar^{-1} [at_0 \cos(ak_z)\hat{\mathbf{z}} + \partial_{\mathbf{k}}|\mathbf{d}|]$. For simplicity, it was assumed that only electric $\mathbf{E}$ and magnetic $\mathbf{B}$ fields are nonzero. A more general case is discussed in detail in [432]. Note that the above results capture the topology of the whole Brillouin zone via the Berry curvature in Eq. (6.164).

In the absence of strain-induced pseudoelectromagnetic fields, the topological contributions in Eqs. (6.166) and (6.167) are completely equivalent to the Chern–Simons terms in Eqs. (6.159) and (6.160). Note that they originate primarily from the filled electron states deeply below the Fermi surface. Since information about such states is absent in linearized semiclassical low-energy models, it should not be surprising why the conventional covariant formulation of the CKT lacks the corresponding topological terms.

The Chern–Simons current (6.161) can also be viewed as a boundary condition for the spectral flow at the cut-off scale in relativistic-like models. As is clear from Eqs. (6.166) and (6.167), such a cutoff appears naturally in proper lattice theories and nothing should be imposed by hand. In addition, ρ_{cs} and $\mathbf{j}_{\text{cs}}$ are proportional to the winding number of the mapping of a two-dimensional section of the Brillouin zone onto the unit sphere [432]. The situation is somewhat different for the Chern–Simons terms in strain-induced pseudoelectromagnetic fields that contribute to the chiral charge and chiral current densities. They appear to be nonuniversal and depend on model details connected with strain effects [432, 433].

An alternative way to explain the result in Eq. (6.158) is to properly take into account the spatial boundary of a Weyl semimetal while still maintaining the covariant form of currents. Then Eq. (6.158) can be interpreted as a charge pumping between the bulk and the surface of a material. As argued in [408, 413, 745], there is an effective accumulation of electric charges on the surfaces of Weyl semimetals because of the anomaly in the covariant current. The presence of the term $\sim \mathbf{B}_5 \cdot \mathbf{E}$ in the divergence of electric current is explained by a transfer of electrons from the bulk to the surface via a semiclassical evolution of electron states in the Brillouin zone. In turn, the term $\sim \mathbf{E}_5 \cdot \mathbf{B}$ is responsible for a nonequilibrium distribution that can be also relaxed by transferring charge to the surface. The apparent drawback of these proposals is the nonlocality of the transfer process that is inconsistent with the local charge conservation. A step toward the unification of the seemingly different consistent and covariant descriptions was made in [601]. By using a lattice model, the Fermi arcs were identified as the source of the covariant anomaly terms in Eqs. (6.157) and (6.158). Further, the explicit connection of the charge growth (loss) near the Fermi surface and loss (growth) at the band bottom was demonstrated. The latter appears to be in perfect agreement with the consistent current framework discussed above.

6.6.3.3 *Axion electrodynamics*

Finally, let us discuss an alternative way to describe the Chern–Simons terms in Weyl semimetals that is known in the literature as the *axion electrodynamics* [748–752]. Its core is the axion term in the electromagnetic sector of the theory described by the following additional term in the Lagrangian density:

$$\mathcal{L}_\theta = -\frac{e^2}{4\pi^2 \hbar} \theta(t, \mathbf{r}) F_{\mu\nu} \epsilon^{\mu\nu\rho\lambda} F_{\rho\lambda} = -\frac{e^2}{4\pi^2 \hbar} \theta(t, \mathbf{r}) \left(\mathbf{E} \cdot \mathbf{B} \right). \quad (6.168)$$

In the high-energy physics literature, θ is a pseudoscalar field, whose dependence on time and coordinates is crucial. Otherwise, this is a full derivative, which is irrelevant for the equation of motion. Its interaction with electromagnetic fields solves the so-called strong $\mathcal{CP}$ problem in QCD [357]. This axion field was actively searched in experiments. In the condensed matter context, the axion electrodynamics was used for investigating the properties of ^{3}He [73, 753], topological insulators [97–100, 754], and Weyl semimetals [390, 556, 663–665]. Moreover, it is important to emphasize that the use of axion electrodynamics provides another example of successful cross-pollination between high energy and condensed matter physics.

Because of the additional θ-dependent terms in the Lagrangian density (6.168), the equation for Gauss's and Ampere's laws are different in axion electrodynamics, namely [748]

$$\boldsymbol{\nabla} \cdot \mathbf{E} = 4\pi\rho + \frac{e^2}{\pi\hbar} \mathbf{B} \cdot \boldsymbol{\nabla}\theta, \quad (6.169)$$

$$\boldsymbol{\nabla} \times \mathbf{B} = \frac{4\pi}{c}\mathbf{J} - \frac{e^2}{\pi c \hbar}\left(\frac{\partial\theta}{\partial t}\mathbf{B} + [\boldsymbol{\nabla}\theta \times \mathbf{E}]\right) + \frac{1}{c}\frac{\partial\mathbf{E}}{\partial t}. \tag{6.170}$$

As is easy to check, the θ-dependent contributions to the charge and current densities are the same as the Chern–Simons terms in Eqs. (6.166) and (6.167), provided the following identifications are made: $\boldsymbol{\nabla}\theta = 2\mathbf{b}$ and $\partial_t\theta = -2b_0$.

6.6.4 *Chiral magnetic and chiral separation effects*

In this subsection, we demonstrate how the consistent CKT works by applying it for the description of the chiral magnetic and chiral separation effects (for a different approach, see Sec. 6.4). For this purpose, it is sufficient to set $\mathbf{E}_\lambda = \mathbf{0}$ and assume that $\mathbf{B}_\lambda$ is uniform and time independent. Furthermore, in order to calculate the currents, it is convenient to expand the distribution function (6.152) in powers of the background field, i.e.,

$$f_\lambda^{\mathrm{eq}} \approx f_\lambda^{(0)} + \frac{e}{c}v_F p\,(\mathbf{B}_\lambda \cdot \boldsymbol{\Omega}_\lambda)\,\frac{\partial f_\lambda^{(0)}}{\partial\epsilon_{\mathbf{p}}}, \tag{6.171}$$

where $\epsilon_{\mathbf{p}} = v_F p$ and the unperturbed distribution is given by

$$f_\lambda^{(0)} = \frac{1}{e^{(\epsilon_{\mathbf{p}} - \mu_\lambda)/T} + 1}. \tag{6.172}$$

Then, to linear order in $\mathbf{B}_\lambda$, the "matter" (i.e., nontopological) contributions to the charge and current densities are given by

$$\rho_\lambda \approx -\sum_{e,h} e \int \frac{d^3 p}{(2\pi\hbar)^3}\left\{\left[1 - \frac{e}{c}(\mathbf{B}_\lambda \cdot \boldsymbol{\Omega}_\lambda)\right]f_\lambda^{(0)} + v_F p\frac{e}{c}\,(\mathbf{B}_\lambda \cdot \boldsymbol{\Omega}_\lambda)\,\frac{\partial f_\lambda^{(0)}}{\partial\epsilon_{\mathbf{p}}}\right\}, \tag{6.173}$$

$$\mathbf{j}_\lambda \approx -\sum_{e,h} e \int \frac{d^3 p}{(2\pi\hbar)^3}v_F \hat{\mathbf{p}}\left\{\left[1 - 2\frac{e}{c}(\mathbf{B}_\lambda \cdot \boldsymbol{\Omega}_\lambda)\right]f^{(0)}\right.$$
$$\left. + v_F p\frac{e}{c}\,(\mathbf{B}_\lambda \cdot \boldsymbol{\Omega}_\lambda)\,\frac{\partial f_\lambda^{(0)}}{\partial\epsilon_{\mathbf{p}}}\right\}. \tag{6.174}$$

The terms odd in momentum including those with the Berry curvature vanish after the integration over the angles. The integrations in the remaining terms, as well as the summation over electron and hole states can be performed analytically with the help of formulas in Appendix B. The final results read

$$\rho_\lambda \approx \sum_{e,h} \frac{eT^3}{\pi^2 v_F^3 \hbar^3} \mathrm{Li}_3(-e^{\frac{\mu_\lambda}{T}}) = -e\frac{\mu_\lambda}{6\pi^2 v_F^3 \hbar^3}\left(\mu_\lambda^2 + \frac{\pi^2 T^2}{3}\right), \quad (6.175)$$

$$\mathbf{j}_\lambda \approx \sum_{e,h} \frac{\lambda e^2 \mathbf{B}_\lambda T}{4\pi^2 \hbar^2 c} \ln\left(1 + e^{\frac{\mu_\lambda}{T}}\right) = \lambda e^2 \frac{\mu_\lambda \mathbf{B}_\lambda}{4\pi^2 \hbar^2 c}, \quad (6.176)$$

where $\mathrm{Li}_m(x)$ is the polylogarithm function [443]. By using Eq. (6.175), one can easily derive the electron number and chiral number densities

$$n = \sum_{\lambda=\pm} \rho_\lambda/(-e) = \frac{\mu\left(\mu^2 + 3\mu_5^2 + \pi^2 T^2\right)}{3\pi^2 v_F^3 \hbar^3}, \quad (6.177)$$

$$n_5 = \sum_{\lambda=\pm} \lambda\rho_\lambda/(-e) = \frac{\mu_5\left(\mu_5^2 + 3\mu^2 + \pi^2 T^2\right)}{3\pi^2 v_F^3 \hbar^3}. \quad (6.178)$$

Note that n_5 quantifies an imbalance between the number densities of right- and left-handed quasiparticles, which is induced by a nonzero chiral chemical potential μ_5.

By using Eq. (6.176), we also derive the electric and chiral currents densities

$$\mathbf{j} = \sum_{\lambda=\pm} \mathbf{j}_\lambda = e^2\frac{\mu_5\mathbf{B} + \mu\mathbf{B}_5}{2\pi^2 \hbar^2 c}, \quad (6.179)$$

$$\mathbf{j}_5 = \sum_{\lambda=\pm} \lambda\mathbf{j}_\lambda = e^2\frac{\mu\mathbf{B} + \mu_5\mathbf{B}_5}{2\pi^2 \hbar^2 c}. \quad (6.180)$$

These currents describe the celebrated chiral magnetic and chiral separation effects, respectively, generalized to the case of nonzero strain-induced pseudomagnetic fields. It is worth emphasizing that these results were obtained in the weak field limit and without the explicit use of the Landau levels.

As is clear, both CME and CSE currents should be nondissipative since neither magnetic nor pseudomagnetic fields can do any work in the system. The presence of the CSE current in the ground state of the matter could be easily reconciled with the general principles of solid-state physics. Indeed, such a current can be viewed as a simple spin polarization. It also helps that there are multiple ways for breaking the exact conservation of chiral charge. These include, for example, the chiral anomaly, the chiral symmetry breaking on the surface of materials, as well as the deterioration of the chirality itself beyond the low-energy regime. The situation with the CME is more puzzling since it seems to imply that there is a persistent electric current in the ground state of a Weyl semimetal. This is resolved by noting that the covariant current in Eq. (6.179) should be amended by the Chern–Simons contribution in Eq. (6.167). Thus, the total physical (consistent) current reads

$$\mathbf{J} = \mathbf{j} + \mathbf{j}_{cs} = e^2 \frac{(\mu_5 + b_0)\mathbf{B} + \mu\mathbf{B}_5}{2\pi^2\hbar^2 c}. \tag{6.181}$$

In equilibrium, as is clear from the schematic band structure of a Weyl semimetal with broken $\mathcal{T}$ and $\mathcal{P}$ symmetries in Fig. 6.13(a), the chiral chemical potential is given by $\mu_5 = -b_0$. Largely, this is due to the fact that the Fermi level must be the same for both Weyl nodes. By substituting $\mu_5 = -b_0$ into the last expression for the electric current, one finds that the part proportional to $\mathbf{B}$ vanishes. This result also agrees with the calculations in a two-band model of Weyl semimetals with periodic dispersion presented in Sec. 6.4.4.

As is easy to check, there is another term in the CME current (6.181), which is proportional to the pseudomagnetic field $\mathbf{B}_5$. Such a current describes the so-called chiral pseudomagnetic effect [413]. Physically, it can be identified with a (bound) magnetization current, which is given by a curl of magnetization $\mathbf{M} \propto \mu\mathbf{A}^5$ (recall that $\mathbf{A}^5$ is observable). As such, it does not contribute to the electron transport. This interpretation was supported by numerical calculations in [413], where several strain profiles were employed. As an example, let us consider a strain that is modeled by

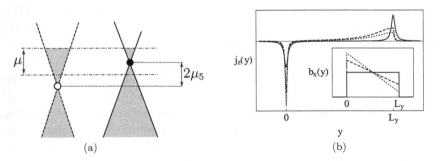

(a) (b)

Fig. 6.13 (a) Schematic band structure of a Weyl semimetal with broken $\mathcal{T}$ and $\mathcal{P}$ symmetries. In the equilibrium state, the chiral chemical potential μ_5 is determined by the energy separation between the Weyl nodes of opposite chirality (filled and empty dots), i.e., $\mu_5 = -b_0$. (b) Schematic representation of the bound current density in a slab of Weyl semimetal for a few profiles of the chiral shift given in the inset. The semimetal is at $0 < y < L_y$ and vacuum is at $y > L_y$ and $y < 0$.

the following dependence of the chiral shift on the y coordinate inside the slab:

$$b_x(y) = \left(\frac{b_f - b_i}{L_y}y + b_i\right)[\theta(y) - \theta(y - L_y)], \qquad (6.182)$$

where the boundaries of the slab are assumed to be at $y = 0$ and $y = L_y$. A few profiles of $b_x(y)$ with different slopes and the corresponding currents are schematically shown in Fig. 6.13(b). While a linear dependence of the chiral shift on the coordinate inside the slab causes an asymmetry and a nonzero magnetization current in the bulk, the total electric current vanishes. Experimentally, one could probe the local bound currents with the scanning superconducting quantum interference devices (SQUIDs) and torque experiments.

Since mechanical deformations by themselves cannot break $\mathcal{T}$ symmetry, the effects linear in pseudomagnetic field should vanish after summation over all nodes in Weyl semimetals invariant under time inversion. In practice, this means that the chiral pseudomagnetic effect is absent in such Weyl semimetals.

6.6.5 *Pseudomagnetic lens*

As we argued before, the semiclassical formalism of kinetic theory provides a natural and simple way to study the motion of electron quasiparticles in solids under weak perturbations. In this subsection, in order to illustrate the use of the CKT, let us have a closer look at the semiclassical motion of quasiparticles in external magnetic and pseudomagnetic fields in Weyl semimetals. As will become clear from the analysis, such fields could be used for focusing and manipulating charged carriers depending on their chirality. In essence, the underlying physics is similar to that in conventional magnetic lenses used for deflecting and focusing electron beams in the cathode ray tubes and the electron microscopes [755].

In connection to magnetic lensing, it is instructive to recall a few facts on the history of geometric electron optics, which started from the pioneering work by Hans Busch in 1926 [756], where it was suggested that the magnetic field of a short coil can act as a converging lens for electrons. This proposal was instrumental for the construction of the first electron microscope by Max Knoll and Ernst Ruska [757–760]. Because the wavelength of electrons can easily be a few orders of magnitude smaller than the wavelength of the visible light, the resolution of the electron microscope quickly surpassed that of the conventional optical instruments.

Having the conventional magnetic lens setup in mind, a lens able to separate quasiparticles of different chiralities in both magnetic and strain-induced pseudomagnetic fields was proposed in [424]. The lensing of Weyl quasiparticles was also discussed in the framework of Weyl metamaterials in [425], where an inhomogeneous magnetization was employed to focus quasiparticles. In this connection, one should recall that electron lensing could be also achieved in 2D Dirac semimetals such as graphene, where the Veselago lens [214] with a negative refraction index is realized by utilizing p–n junctions (see Sec. 2.2.2).

In geometrical particle optics, the motion of charged quasiparticles in external electromagnetic fields is described analogously to the propagation of light rays. Mathematically, such a description relies on the *eikonal* or Wentzel–Kramers–Brillouin (WKB) approximation [285, 755, 761]. The eikonal approximation is an appropriate and efficient means to solve wave equations when the de Broglie wavelength of particles are small compared to the characteristic inhomogeneities of the system or external fields. In mathematical physics, the WKB approximation[z] is also a well-established technique for finding approximate solutions to linear differential equations with coefficients varying slowly in space [763].

The starting point of the eikonal approximation for Weyl quasiparticles is the abbreviated action $S_0 \equiv S_0(\mathbf{r})$,[aa] which is introduced by replacing

$$\mathbf{p} \to \boldsymbol{\nabla} S_0 + \frac{e}{c} \mathbf{A}_\lambda \tag{6.183}$$

in the dispersion relation of quasiparticles $\epsilon(\mathbf{p})$. Here the vector potential takes the form $\mathbf{A}_\lambda = (-yB_\lambda/2, xB_\lambda/2, 0)$ and describes a constant background field $\mathbf{B}_\lambda$ pointing in the $+z$-direction. As will become evident below, the energy dispersion should include the terms up to the second order in (pseudo)magnetic fields and is given by [423]

$$\epsilon = v_F p + \lambda \frac{e \hbar v_F}{2cp^2} (\mathbf{B}_\lambda \cdot \mathbf{p}) + \frac{e^2 \hbar^2 v_F}{16c^2 p^3} \left[2\mathbf{B}_\lambda^2 - \frac{(\mathbf{B}_\lambda \cdot \mathbf{p})^2}{p^2} \right] + O\left(B_\lambda^3\right). \tag{6.184}$$

This is to be compared with the first-order expression for the energy in Eq. (6.153).

---

[z]In mathematical physics, it is also known as the JWKB approximation. The additional letter J stands for Harold Jeffreys, who developed a general method of approximating solutions to linear second-order differential equations in 1924 [762].
[aa]Note that the abbreviated action S_0 is related to the full action as $S = -\epsilon t + S_0$ [761].

To simplify the qualitative study, it is convenient to consider only the case where the electrons move along the (pseudo)magnetic field, i.e., the z-axis (which will be identified with the optical axis of the lens), and remain sufficiently close to it (the paraxial approximation). This allows one to consider the distance from the optical axis $r_\perp = \sqrt{x^2 + y^2}$ as a small parameter. By keeping only the leading order terms in $r_\perp^2$, one can obtain the following abbreviated action:

$$S_0 = \frac{\epsilon}{v_F}\left[Cz + \frac{r_\perp^2}{2}A(z) + O\left(r_\perp^4\right)\right], \tag{6.185}$$

where

$$C = 1 - \lambda\frac{B_\lambda}{B_\epsilon} - \frac{5}{4}\left(\frac{B_\lambda}{B_\epsilon}\right)^2 + O\left(\frac{B_\lambda^3}{B_\epsilon^3}\right) \tag{6.186}$$

and the function $A(z)$ is a solution of

$$a_1\partial_z A(z) + [A(z)]^2 + a_2^2 = 0. \tag{6.187}$$

Here the following shorthand notation was used:

$$a_1 \approx 1 + \frac{5}{4}\left(\frac{B_\lambda}{B_\epsilon}\right)^2, \quad a_2^2 \approx \frac{eB_\lambda^2}{2c\hbar B_\epsilon}\left(1 + 2\lambda\frac{B_\lambda}{B_\epsilon}\right), \tag{6.188}$$

where $B_\epsilon = 2c\epsilon^2/(e\hbar v_F^2)$ sets a characteristic scale, determined by the quasiparticle energy ϵ.

Let us analyze the simplest case of uniform (pseudo)magnetic fields that are nonzero only for $0 < z < L$, i.e., $B_\lambda(z) = B_\lambda\theta(z)\theta(L - z)$. Then, by solving Eq. (6.187) in the three regions, i.e., at $z < 0$, $0 < z < L$, and $z > L$, as well as matching the abbreviated action at the edges $z = 0$ and $z = L$, the following lens equation is obtained:

$$(z_1 + g_\lambda)(z_2 - h_\lambda) = -F_\lambda^2. \tag{6.189}$$

Here z_1 and z_2 are the coordinates of the quasiparticle source and its image, respectively. As is clear from Eq. (6.189), $z = -g_\lambda$ and $z = h_\lambda$ are the locations of the principal foci, and F_λ is the principal focal length. Their explicit form reads

$$g_\lambda \simeq \frac{l_\epsilon B_\epsilon}{B_\lambda\left(1 + \lambda B_\lambda/B_\epsilon\right)}\cot\left[\frac{LB_\lambda\left(1 + \lambda B_\lambda/B_\epsilon\right)}{l_\epsilon B_\epsilon}\right], \tag{6.190}$$

$$F_\lambda \simeq \frac{l_\epsilon B^*}{B_\lambda\left(1 + \lambda B_\lambda/B_\epsilon\right)\sin\left[\frac{LB_\lambda(1 + \lambda B_\lambda/B_\epsilon)}{l_\epsilon B_\epsilon}\right]}, \tag{6.191}$$

where $h_\lambda = L + g_\lambda$ and the de Broglie wavelength of Weyl quasiparticles is $l_\epsilon = \hbar v_F/\epsilon$. As expected, when the source is placed at the left focal point, i.e., $z_1 \to -g_\lambda$, the image is located at infinity, $z_2 \to \infty$. Similarly, when the source is at infinity, $z_1 \to -\infty$, its image appears at the right focal point, $z_2 \to h_\lambda$.

By using Eq. (6.191), one finds that a combination of magnetic and pseudomagnetic fields leads to different focal lengths for quasiparticles of opposite chirality. In the special case with $B_5 \neq 0$ but $B = 0$, it is trivial to show that $F_+ = F_-$. This is not the case already when $B \neq 0$ but $B_5 = 0$ because a small difference between the focal lengths F_+ and F_- is induced by the Berry curvature effects quantified by the terms $\propto \lambda B_\lambda/B_\epsilon$. While such effects are of the second order in weak (pseudo)magnetic fields, they are still important in a (pseudo)magnetic lens that is able not only to focus quasiparticles but also to spatially separate chiral charges. As is clear, additional possibilities for manipulating quasiparticles appear when both magnetic and pseudomagnetic fields are used [424].

The numerical estimates for the principal focal lengths in Cd_3As_2, where $v_F \approx 1.5 \times 10^8$ cm/s [107] (see also Appendix D), are shown in Fig. 6.14(a) as functions of B_5. As expected, in the presence of both magnetic and pseudomagnetic fields, the focal lengths are different for quasiparticles of

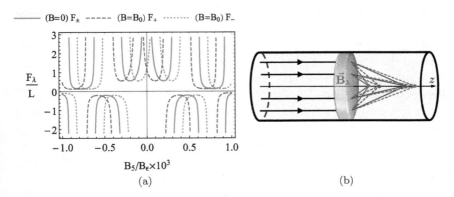

(a)   (b)

Fig. 6.14 (a) The focal length F_λ for quasiparticles of chirality λ as a function of the pseudomagnetic field B_5. While red solid lines correspond to quasiparticles of both chiralities at $B = 0$, blue dashed and green dotted lines represent right- and left-handed quasiparticles at $B = B_0 = 10^{-4}B_\epsilon$, respectively. In addition, $\epsilon = 100$ meV and $L = 10^{-2}$ cm. (b) A schematic illustration of the quasiparticles focusing inside a Dirac semimetal crystal that has the $\mathbb{Z}_2$ Weyl structure. Red and blue lines correspond to right- and left-handed quasiparticles, respectively. Solid and dashed lines correspond to different pairs of Weyl nodes.

opposite chiralities. It should be noted that the dependence of focal lengths F_λ on a (pseudo)magnetic field is quasiperiodic. Also, formally, the results for F_λ diverge at $B_\lambda/B_\epsilon \approx l_\epsilon \pi n/L$, where n is an integer. The presence of such divergences for certain values of background fields implies a breakdown of the paraxial approximation and the need for a more accurate analysis.

It is interesting to note that a (pseudo)magnetic lens can be used for splitting beams of chiral quasiparticles in certain Dirac materials too. As was discussed in Sec. 3.6, A_3Bi (A = Na, K, Rb) and some phases of Cd_3As_2 are, in fact, $\mathbb{Z}_2$ Weyl semimetals that contain two pairs of Weyl nodes separated in momentum. The pairs are, however, superimposed leading to a Dirac semimetal phase. Since a strain-induced pseudomagnetic field $\mathbf{B}_5$ is determined by the chiral shift vector (see the discussion in Sec. 3.7), the field is the same in magnitude but opposite in direction for the two superimposed pairs of Weyl nodes. Therefore, if the Berry curvature effects were neglected in Eq. (6.191), the focal length of right-handed (left-handed) quasiparticles from the first pair of Weyl nodes would coincide exactly with the focal length of left-handed (right-handed) quasiparticles of the second pair of Weyl nodes. In other words, after passing through the lens, a generic beam of quasiparticles would split into two nonchiral Dirac beams each composed of quasiparticles from one of the two Dirac cones. The Berry curvature gives additional corrections and makes all focal lengths different. This is illustrated schematically in Fig. 6.14(b). Therefore, after passing through the lens, a beam of quasiparticles is split into four chiral beams, realizing the complete chirality and nodal splitting. It is worth noting that even in the absence of a (pseudo)magnetic lens, the motion of the charge carriers in Dirac semimetals might be quite nontrivial due to non-Abelian Berry curvature effects [378].

Before finalizing this subsection, let us note that strains are not the only way to create effective gauge fields. For example, textures with nontrivial magnetization can also be used to produce axial gauge fields [425, 429] in Weyl and Dirac semimetals. According to [425], a magnetization with a radial gradient in the plane perpendicular to the initial velocities of quasiparticles gives rise to an effective axial gauge field. Similarly to the (pseudo)magnetic lens above, such fields can cause the trajectories of incoming quasiparticles to converge at a focal point.

Chapter 7

Electronic Transport

Everything flows, nothing stands still.

Heraclitus

In addition to various anomalous features connected with the condensed matter manifestation of the chiral anomaly and nontrivial topology, Dirac and Weyl semimetals have other interesting electronic transport properties. The electrical conductivity is one of the most fundamental characteristics of any material. In order to correctly describe the dissipative electron transport in Weyl and Dirac semimetals, it is crucial to know how different types of disorder affect the conductivity and how a steady state in an applied electric field is established.

Another important class of electronic transport properties is related to a thermoelectric response, where a temperature gradient plays the role of a driving force for charge carriers. As it turns out, thermoelectric transport could also have anomalous features similar to those discussed in Chapter 6. However, since the thermoelectric properties are connected with thermally excited quasiparticles, the corresponding currents are not truly topological.

In addition to the conventional direct current (DC) transport, the investigation of optical properties and the alternating current (AC) characteristics could also illuminate the properties of novel phases of matter. The linear energy spectrum and the nontrivial topology of Weyl and Dirac quasiparticles affect many optical properties, such as the generation of higher harmonics, the photovoltaic effect, the Kerr rotation, etc.

Another interesting aspect of the transport properties of Dirac and Weyl semimetals is related to the possibility of hydrodynamic regime for electrons. In such a regime, the collective behavior of electrons resembles a viscous liquid rather than a motion of independent carriers. Historically, this possibility was first proposed by Radii Gurzhi in the 1960s [240, 241] and was first tested experimentally in ultrapure 2D materials in the 1990s [764, 765]. Later electron hydrodynamics was extensively studied in graphene (see Sec. 2.5.2). While relativistic hydrodynamics is very important in high energy physics, heavy-ion collisions, and astrophysics, such systems are usually hard to control experimentally. Thus, the hydrodynamic regime in Dirac and Weyl semimetals provides a very convenient alternative way for studying the properties of relativistic-like fluids.

It should be emphasized that the investigation of transport properties and the prediction of novel phenomena are crucial not only from the fundamental viewpoint but also are pivotal for future practical applications of Dirac and Weyl semimetals. In particular, as will be discussed in Sec. 7.3, Weyl semimetals could be suitable for sensitive infrared detectors and ultrafast control of chiral currents.

7.1   Effects of disorder

The correct description of disorder is crucial for determining the electron transport properties in solids. In this connection, it suffices to recall the paradigmatic Drude model, where one of its key ingredients, namely, the mean free scattering time, is determined at low temperature largely by disorder. The same applies to the electron transport in Dirac and Weyl semimetals, where the effects of disorder were investigated extensively [385, 547, 627, 630, 655–661, 766]. Similarly to the study of the Fermi arc transport properties in Sec. 5.5, one could employ the Green's functions method. As will be shown below, the key effects of disorder could be included via perturbative corrections to the self-energy and the vertex function. As a simpler alternative, one can also use a semiclassical Boltzmann description with a suitable collision integral to calculate the transport scattering time.

In what follows, we will start by presenting the main methods used for calculating the quasiparticle width. Then a few simple examples of disorder will be discussed and the quasiparticle width will be calculated. At the end of this section, the results for more advanced disorder models and the effects of external fields will be briefly discussed.

7.1.1 Model and disorder potentials

For the sake of simplicity, let us consider a Hamiltonian for a single Weyl node given by

$$H_\lambda = \int d^3 r \psi^\dagger(\mathbf{r})[-\mu \mathbb{1}_2 - i\lambda\hbar v_F (\boldsymbol{\sigma} \cdot \boldsymbol{\nabla}) + \hat{U}(\mathbf{r})]\psi(\mathbf{r}), \qquad (7.1)$$

where $\hat{U}(\mathbf{r})$ is the disorder potential that does not mix Weyl fermions of different chiralities. Since only the intranode contribution to the conductivity is considered in this section, the chiral shift is omitted in Eq. (7.1). For simplicity, the disorder potential is assumed to be diagonal in the (pseudo)spin space, i.e., $\hat{U}(\mathbf{r}) = \mathbb{1}_2 U(\mathbf{r})$.[a] While the results below will be given for $\lambda = +$, the generalization to the case of multiple Weyl nodes is straightforward. One of the simplest forms of disorder is the short-range disorder, which could be modeled by the δ-function potential

$$U_{\text{loc}}(\mathbf{r}) = \sum_j u(\mathbf{r} - \mathbf{r}_j) = \sum_j u_0 \delta(\mathbf{r} - \mathbf{r}_j), \qquad (7.2)$$

where u_0 is the strength of disorder potential and $\mathbf{r}_j$ denotes the position of impurities. Clearly, the use of a local potential is not applicable for electrically charged impurities, which are described by a long-range, partially screened Coulomb potential, i.e.,

$$U_{\text{C}}(\mathbf{r}) = \sum_j \frac{e^2}{\varepsilon_e |\mathbf{r} - \mathbf{r}_j|} e^{-4\pi e^2 \nu(\mu) |\mathbf{r} - \mathbf{r}_j|/\varepsilon_e}. \qquad (7.3)$$

Here, by assuming the Thomas–Fermi approximation, the screening length is determined by the density of states $\nu(\mu)$ at the Fermi level, see Eq. (3.136) in Sec. 3.9. The electric permittivity of the material is denoted by ε_e.

The translation invariance, which is broken by disorder, is effectively restored by averaging over the random positions of impurities, i.e.,

$$\langle A \rangle_{\text{dis}} = \frac{1}{V} \sum_j \int d^3 r_j \, A(\mathbf{r}_j), \qquad (7.4)$$

where $A(\mathbf{r}_j)$ is a function of $\mathbf{r}_j$ and V is the volume of the system. This is often well justified since the transport properties are usually measured

---

[a] For the case of a vector disorder with $\hat{U}(\mathbf{r}) = \boldsymbol{\sigma} U(\mathbf{r})$ see, for example, [658].

in macroscopically large samples and the coherence length of electrons is typically much smaller than the system size. As is clear, the presence of disorder further reduces the coherence length of electrons. Then, the sample can be viewed as a collection of several phase-coherent subsystems. Therefore, most transport observables come from measuring the average over all such subsystems. Since this averaging is imposed by the system itself, it is called the *self-averaging*. In real materials, it is almost impossible to control precisely the positions of all defects. Therefore, the self-averaging is equivalent to averaging over all impurity positions.

In the case of the local disorder potential (7.2), after averaging over the random positions of impurities, one can obtain the following useful results:

$$\langle u(\mathbf{r} - \mathbf{r}_j) \rangle_{\text{dis}} = u_0 n_{\text{imp}}, \tag{7.5}$$

$$\langle u(\mathbf{r} - \mathbf{r}_j) u(\mathbf{r}' - \mathbf{r}_j) \rangle_{\text{dis}} = u_0^2 n_{\text{imp}} \delta(\mathbf{r} - \mathbf{r}'), \tag{7.6}$$

where $n_{\text{imp}} = N_{\text{imp}}/V$ is the concentration, N_{imp} is the total number of impurities, and V is the volume of the system. As we will see below (see also the discussion in Sec. 5.5), the only effect of nonzero $\langle u(\mathbf{r} - \mathbf{r}_j) \rangle_{\text{dis}}$ is a renormalization of the electric chemical potential. Therefore, it is often justified to neglect the average value of the disorder potential, i.e., to set $\langle u(\mathbf{r} - \mathbf{r}_j) \rangle_{\text{dis}} = 0$. In terms of the crossed diagram technique [136, 138], the correlator in Eq. (7.6) defines a new element. Although such a treatment of disorder might be viewed as old-fashioned, compared to some advanced modern techniques (e.g., the supersymmetry approach [767] and the replica trick [768]), it is certainly one of the most direct and transparent methods.

For the future use in calculations, it is convenient to have an explicit expression for the Fourier transform of the disorder potential. In the case of a short-range disorder, one finds

$$u_{\mathbf{k}} = u_0. \tag{7.7}$$

For the long-range Coulomb disorder defined in Eq. (7.3), the corresponding result reads

$$u_{\mathbf{k}} = \frac{4\pi e^2}{\varepsilon_e k^2 + 4\pi e^2 \nu(\epsilon_{\mathbf{k}})}. \tag{7.8}$$

7.1.2   Green's function approach

As was explicitly demonstrated before (see, for example, Secs. 5.5 and 6.2.3.2), the Green's function approach is a powerful tool for investigating various properties of condensed matter systems, including the effects of disorder. Diagrammatically, Green's function in the presence of disorder is given by the following infinite series of diagrams:

$$G = \longrightarrow + \longrightarrow\!\!\bullet\!\!\longrightarrow + \longrightarrow\!\!\bullet\!\!\longrightarrow\!\!\bullet\!\!\longrightarrow + \longrightarrow\!\!\bullet\!\!\longrightarrow\!\!\bullet\!\!\longrightarrow\!\!\bullet\!\!\longrightarrow + \cdots ,$$
(7.9)

where the solid lines correspond to the bare Green's function calculated without any impurities and the dashed lines with crosses at the ends represent the insertions of impurity potential. By averaging over disorder and omitting the diagrams with crossed impurity lines,[b] one finds diagrammatically

$$\langle G \rangle_{\text{dis}} = \longrightarrow = \longrightarrow + \longrightarrow\!\!\bullet\!\!\longrightarrow + \longrightarrow\!\!\bullet\!\!\longrightarrow\!\!\bullet\!\!\longrightarrow ,$$
(7.10)

where the right-hand side contains the bare Green's function contribution and two diagrams connected with disorder.

By using Hamiltonian (7.1) in the clean limit $\hat{U}(\mathbf{r}) \to 0$, it is straightforward to find the following (free) retarded and advanced Green's functions:

$$S^{\text{R/A}}(\omega, \mathbf{k}) = i \frac{\hbar\omega + \mu + \lambda\hbar v_F (\boldsymbol{\sigma} \cdot \mathbf{k})}{(\hbar\omega + \mu)^2 - \epsilon_{\mathbf{k}}^2 \pm i0 \, \text{sgn}(\hbar\omega + \mu)} ,$$
(7.11)

where $\epsilon_{\mathbf{k}} = \pm\hbar v_F k$. In a general case, the Fourier transform of the disorder propagator is given by

$$D(\omega, \mathbf{k}) = 2\pi n_{\text{imp}} |u_{\mathbf{k}}|^2 \delta(\omega),$$
(7.12)

where $u_{\mathbf{k}}$ denotes the Fourier transform of the impurity potential. The latter is given by Eqs. (7.7) and (7.8) for the local and Coulomb potentials, respectively.

The full Green's function averaged over disorder takes the following form:

$$G^{\text{R/A}}(\omega, \mathbf{k}) = \frac{i}{\hbar\omega + \mu - \lambda\hbar v_F (\boldsymbol{\sigma} \cdot \mathbf{k}) - \Sigma(\omega, \mathbf{k})} ,$$
(7.13)

where $\Sigma(\omega, \mathbf{k})$ is the electron self-energy function.

---

[b]The contribution of the diagrams with crossed impurity lines is usually small, see, for example, [138, 139].

In perturbation theory, $G^{R/A}$ is determined by replacing the full Green's functions (double lines) with the free propagators (thin lines). Then, in the case of short-range disorder, the leading one-loop order approximation for the self-energy is given by

$$\Sigma(\Omega, \mathbf{k}) = \quad + \quad \simeq \quad + $$

$$= n_{\text{imp}} u_0 - i \int \frac{d\omega d^3 k'}{(2\pi)^4} S^{R/A}(\omega, \mathbf{k}') D(\Omega - \omega, \mathbf{k} - \mathbf{k}')$$

$$\simeq n_{\text{imp}} u_0 \mp i n_{\text{imp}} u_0^2 \frac{(\hbar\Omega + \mu)^2}{4\pi\hbar^3 v_F^3}. \tag{7.14}$$

Note that, in the last expression, a divergent real part of the self-energy, stemming from the second-order correction, was omitted. Such a divergence comes from an unbounded linear dispersion relation and is unphysical. Moreover, as one can see from definition (7.13), a constant real part of the self-energy, including the first-order term $n_{\text{imp}} u_0$, plays a minor role since it simply redefines the electric chemical potential $\mu \to \mu - \text{Re}\Sigma$ due to the presence of disorder. Such an effect might be of interest when disorder is added dynamically to the system, but can be ignored otherwise. As for the imaginary part, it is important since it defines the quasiparticle width

$$\Gamma(\epsilon) = \mp \text{Im}\,\Sigma = \frac{\pi n_{\text{imp}} u_0^2 \nu(\epsilon)}{2}. \tag{7.15}$$

The corresponding scattering time or, more precisely, the *quasiparticle lifetime*, is usually defined as $\tau(\epsilon) = \hbar / [2\Gamma(\epsilon)]$.

Having calculated the quasiparticle width, one can derive the conductivity by using the Kubo linear response theory discussed in Sec. 6.2.3. The corresponding result for the DC conductivity reads [769]

$$\sigma_{xx} = \frac{e^2}{12\pi^2 \hbar^2 v_F} \int_{-\infty}^{\infty} d\epsilon \frac{1}{4T \cosh\left(\frac{\epsilon - \mu}{2T}\right)} \frac{\epsilon^2 + 3\Gamma^2(\epsilon)}{\Gamma(\epsilon)}$$

$$\overset{T \to 0}{=} \frac{e^2}{12\pi^2 \hbar^2 v_F} \frac{\mu^2 + 3\Gamma^2(\mu)}{\Gamma(\mu)}. \tag{7.16}$$

It is instructive to evaluate the conductivity for three different models of disorder: (i) the simplest model with $\Gamma(\epsilon) = const$, (ii) local disorder with $\Gamma(\epsilon) \propto \epsilon^2$, which is consistent with the result in Eq. (7.15), and

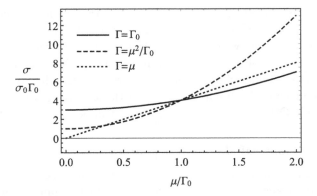

Fig. 7.1 The electrical conductivity (7.16) at $\Gamma = \Gamma_0$ (solid line), $\Gamma = \mu^2/\Gamma_0$ (dashed line), and $\Gamma = \mu$ (dotted line). Here $\sigma_0 = e^2/(12\pi^2\hbar^2 v_F)$ and the precise value of Γ_0 is not important.

(iii) long-range disorder with $\Gamma(\epsilon) \propto \mu^3/\epsilon^2$, where the impurity concentration is proportional to the electron density, see Sec. 7.1.3. As will be shown below, the conductivity in Eq. (7.16) agrees with a similar result in the semiclassical approach, provided the quadratic term $\sim \Gamma^2(\mu)$ in the numerator is omitted.

Schematically, the results for the three different types of disorder are shown in Fig. 7.1. As one sees, a phenomenological model with a constant quasiparticle width $\Gamma(\epsilon) = const$ gives the same qualitative behavior of the conductivity as the model with short-range disorder, i.e., $\Gamma(\epsilon) \propto \epsilon^2$. In particular, both conductivities are quadratic functions of μ and attain a finite value at $\mu = 0$. The case of long-range disorder, on the other hand, is quite different, since the conductivity is linear in μ.

Before discussing the semiclassical Boltzmann approach, it is worth noting that the perturbative expansion (7.14) works well only for sufficiently weak and dilute disorder. A more reliable and complicated method is the self-consistent Born approximation, where one needs to solve the integral equation for the self-energy function

$$\Sigma(\Omega, \mathbf{k}) = -i \int \frac{d\omega d^3 k'}{(2\pi)^4} G(\omega, \mathbf{k}') D(\Omega - \omega, \mathbf{k} - \mathbf{k}'). \qquad (7.17)$$

Note that, for the self-energy, it suffices to consider only the most important second diagram in Eq. (7.14), where the thin line (free propagator) should be replaced with the double line (full Green's function). The corresponding calculations for neutral as well as charged impurities are presented in [655, 656]. By using the model with a neutral Gaussian disorder $u(\mathbf{r}) \propto \exp\left(-r^2/L^2\right)$, the existence of a critical disorder strength, separating the regions with $\Gamma(0) = 0$ from $\Gamma(0) \neq 0$, was demonstrated in [655]. In the case of strong disorder, it was shown that, unlike the metallic regime, the conductivity grows with increasing the disorder potential. The critical behavior is absent, however, for the long-range Coulomb potential [655, 656]. Other nonperturbative results will be discussed in Sec. 7.1.5.

7.1.3 *Semiclassical Boltzmann approach*

As a simpler alternative to a more powerful but cumbersome Green's function method, one can employ the semiclassical kinetic approach formulated in Sec. 6.6. In such a treatment, one needs to solve the Boltzmann equation (6.151) with an appropriate collision term. In the case of scattering on impurities, the collision has the following general form[c]:

$$I_{\text{coll}}\left[f_\lambda(\mathbf{p})\right] = \sum_{\lambda'} \int d^3 p' W(\mathbf{p}, \lambda; \mathbf{p}', \lambda') \left[f_{\lambda'}(\mathbf{p}') - f_\lambda(\mathbf{p})\right], \quad (7.18)$$

where $W(\mathbf{p}, \lambda; \mathbf{p}', \lambda')$ is the transition probability between the states $(\mathbf{p}, \lambda)$ and $(\mathbf{p}', \lambda')$ that depends on the details of the scattering potential. Since this collision integral involves integration over momentum $\mathbf{p}'$, the corresponding kinetic equation turns into an integro-differential equation that is usually very hard to solve. A much simpler version of the collision term is given by the relaxation-time approximation, which is defined by

$$I_{\text{coll}}\left[f_\lambda(\mathbf{p})\right] = -\frac{f_\lambda(\mathbf{p}) - f_\lambda^{\text{eq}}(\mathbf{p})}{\tau(\mathbf{p})}, \quad (7.19)$$

where $f_\lambda^{\text{eq}}(\mathbf{p}) = [e^{(\epsilon_\mathbf{p} - \mu_\lambda)/T} + 1]^{-1}$ is the equilibrium (Fermi–Dirac) distribution function.[d] By comparing the approximate expression (7.19) with the collision integral (7.18), the relaxation time $\tau(\mathbf{p})$ can be estimated as follows:

$$\frac{1}{\tau(\mathbf{p})} = \sum_{\lambda'} \int d^3 p' W(\mathbf{p}, \lambda; \mathbf{p}', \lambda'). \quad (7.20)$$

---

[c]Note that $f_{\lambda'}(\mathbf{p}')[1 - f_\lambda(\mathbf{p})] - f_\lambda(\mathbf{p})[1 - f_{\lambda'}(\mathbf{p}')] = f_{\lambda'}(\mathbf{p}') - f_\lambda(\mathbf{p})$.
[d]In general, this is a local distribution function with spatially dependent μ_λ and T. However, for homogeneous external fields, these parameters do not depend on coordinates.

Note that $\tau(\mathbf{p})$ can have a nontrivial dependence on momentum. This is natural indeed since even an isotropic contribution due to the intranode scattering is expected to depend on the absolute value of $\mathbf{p}$ (or, equivalently, on energy $\epsilon_\mathbf{p}$). The internode scattering is nonisotropic and, therefore, can produce a directional dependence. In Eq. (7.20), it might be tempting to identify the two terms with $\lambda' = \pm$ as partial contributions of the intranode (i.e., chirality preserving) and the internode (i.e., chirality flipping) processes, with the relaxation times defined by

$$\frac{1}{\tau_0} = \int d^3p' W(\mathbf{p}, \lambda; \mathbf{p}', \lambda), \qquad (7.21)$$

$$\frac{1}{\tau_5} = \int d^3p' W(\mathbf{p}, \lambda; \mathbf{p}', -\lambda), \qquad (7.22)$$

respectively. Then, as is clear, $1/\tau = 1/\tau_0 + 1/\tau_5$.[e] In practice, τ_5 is typically much longer than τ_0 (see the numerical values in Appendix D). Then, for most transport properties, it is justified to ignore the chirality-flipping processes by setting $\tau_5 \to \infty$ or, equivalently, $\tau \approx \tau_0$.[f]

When deviations of the system from equilibrium are small, which is assumed here, one can solve the kinetic equation iteratively. To linear order in the electric field, the corresponding approximate solution reads

$$f_\lambda = f_\lambda^{\text{eq}} + \frac{e\tau}{1 - \frac{e}{c}(\mathbf{B} \cdot \boldsymbol{\Omega}_\lambda)} \frac{\partial f_\lambda^{\text{eq}}}{\partial \epsilon_\mathbf{p}} [(\mathbf{v}_\mathbf{p} \cdot \mathbf{E}) - \frac{e}{c}(\tilde{\mathbf{E}} \cdot \mathbf{B})(\mathbf{v}_\mathbf{p} \cdot \boldsymbol{\Omega}_\lambda)], \qquad (7.23)$$

where, for simplicity, we treated $\mathbf{E}$ and $\mathbf{B}$ fields as constant. By recalling the key definitions for relativistic-like Weyl quasiparticles from Sec. 6.6, the velocity is defined as $\mathbf{v}_\mathbf{p} = \partial_\mathbf{p} \epsilon_\mathbf{p}$ and the energy is given by

$$\epsilon_\mathbf{p} = v_F p \left[1 + \frac{e}{c}(\mathbf{B} \cdot \boldsymbol{\Omega}_\lambda) \right]. \qquad (7.24)$$

By making use of the distribution function in Eq. (7.23), the current density of chiral quasiparticles can be calculated from its definition

$$\mathbf{j}_\lambda = -\sum_{e,h} e \int \frac{d^3p}{(2\pi\hbar)^3} \left\{ \mathbf{v}_\mathbf{p} - \frac{e}{c}(\mathbf{v}_\mathbf{p} \cdot \boldsymbol{\Omega}_\lambda)\mathbf{B} - e[\tilde{\mathbf{E}} \times \boldsymbol{\Omega}_\lambda] \right\} f_\lambda$$

$$- \sum_{e,h} e \boldsymbol{\nabla} \times \int \frac{d^3p}{(2\pi\hbar)^3} f_\lambda v_F p \boldsymbol{\Omega}_\lambda, \qquad (7.25)$$

---

[e] For several alternative implementations for the collision integrals of intra- and internode processes see, for example, [647, 705, 743, 744, 770].

[f] As discussed in Sec. 6.2.1, the internode relaxation time τ_5 is essential for determining the out-of-equilibrium chiral chemical potential μ_5 induced by the chiral anomaly.

where $\sum_{e,h}$ represents the sum over the electron (particle) and hole (antiparticle) contributions. Recall that for holes (antiparticles) one should replace $\Omega_\lambda \to -\Omega_\lambda$, $\mu_\lambda \to -\mu_\lambda$, and $e \to -e$. In addition, since a linearized quasiparticle dispersion is used, the Chern–Simons terms discussed in Sec. 6.6.3 should be added in the electric and chiral current densities.

After substituting the distribution function (7.23) into Eq. (7.25) and calculating the integrals over the momenta,[g] it is straightforward to derive an explicit expression for the current densities. Since the calculation details are tedious and the intermediate expressions are bulky, they will be omitted here. In addition, since the purely anomalous parts of the currents were already discussed in Chapter 6, they will not be repeated here.

To start with, let us present the resulting dissipative part for the longitudinal current density (i.e., in the direction of magnetic field $\mathbf{B} \parallel \hat{\mathbf{z}}$),

$$j_{z,\lambda}^{(\mathrm{diss})} = \frac{\tau e^2 \mu_\lambda^2}{6\pi^2 v_F \hbar^3} \left(1 + \frac{v_F^4 e^2 B^2}{5\mu_\lambda^4} \right) E_z, \tag{7.26}$$

which was obtained in the limit $T \to 0$ and under the assumption that $\tau = const$. Since $b_0 = 0$ in $\mathcal{P}$-symmetric Weyl semimetals, one can also set $\mu_5 = -b_0 = 0$. Consequently, the following longitudinal electrical conductivity is obtained after the summation over λ:

$$\sigma_{zz} = \frac{\tau e^2 \mu^2}{3\pi^2 v_F \hbar^3} \left(1 + \frac{v_F^4 e^2 B^2}{5\mu^4} \right). \tag{7.27}$$

Here the quadratic positive dependence on a magnetic field indicates the celebrated "negative" magnetoresistivity that was first predicted in [55] (see also Sec. 6.2). Similar results were also reconfirmed within the kinetic theory framework more recently in [770–773].

As we argued earlier, the relaxation time τ may depend on the quasiparticle energy and the applied magnetic field. Furthermore, the corresponding functional dependence could strongly vary with the type of disorder. Therefore, the model results in this subsection should be viewed only as qualitative estimates. Nevertheless, even such a simple consideration is able to capture the anomalous enhancement of conductivity by the magnetic field

---

[g]It is helpful to use the expressions in Appendix B for calculating the momentum integrals.

via the term proportional to B^2. The origin of the latter could be traced back to the chiral anomaly.

In a more refined Boltzmann approach, one replaces the relaxation time with the *transport scattering time*, which is defined in terms of the quasiparticle scattering amplitude as follows:

$$\frac{1}{\tau_{\mathrm{tr}}} = \int \frac{d^3 k'}{(2\pi)^3} \left[1 - \cos\left(\theta_{\mathbf{kk'}}\right)\right] W_{\mathbf{k'},\mathbf{k}}. \tag{7.28}$$

Here the factor $\left[1 - \cos\left(\theta_{\mathbf{kk'}}\right)\right]$ accounts for the fact that the forward scattering $\theta_{\mathbf{kk'}} = 0$ should not contribute to the transport scattering time,[h]

$$W_{\mathbf{k'},\mathbf{k}} = \frac{2\pi}{\hbar} n_{\mathrm{imp}} \left|\langle \mathbf{k'}| U |\mathbf{k}\rangle\right|^2 \delta\left(\epsilon_{\mathbf{k'}} - \epsilon_{\mathbf{k}}\right) \tag{7.29}$$

is the probability of elastic collisions, $\langle \mathbf{k'}| U |\mathbf{k}\rangle$ is the matrix element of the disorder potential, and $\theta_{\mathbf{kk'}}$ is the angle between wave vectors $\mathbf{k}$ and $\mathbf{k'}$. For Weyl quasiparticles with a relativistic-like dispersion, one then obtains [547, 630, 655, 774]

$$\frac{\hbar}{\tau_{\mathrm{tr}}(\epsilon)} = \pi n_{\mathrm{imp}} \nu(\epsilon) \int_0^\pi d\theta \, \sin\left(\theta\right) |u_q|^2 \left[1 - \cos\left(\theta\right)\right] \frac{1 + \cos\left(\theta\right)}{2}. \tag{7.30}$$

Note that the extra factor $\left[1 + \cos\left(\theta\right)\right]/2$ arises from the matrix element of the impurity potential and suppresses the large angle scattering. As is easy to check, the absolute value of a momentum transfer that appears in the Fourier transform of disorder potential u_q is given by $q = |\mathbf{k'} - \mathbf{k}|_{k=k'} = 2(\epsilon/v_F)\sin\left(\theta/2\right)$. In the case of local disorder (7.2), it is straightforward to obtain

$$\tau_{\mathrm{tr}}(\epsilon) = \frac{3}{2}\tau(\epsilon) = \frac{3\hbar}{2\pi n_{\mathrm{imp}} u_0^2 \nu(\epsilon)}, \tag{7.31}$$

where $\tau(\epsilon) = \hbar/\left[2\Gamma(\epsilon)\right]$ and $\Gamma(\epsilon)$ is given in Eq. (7.15). Note that the transport time has an extra factor of $3/2$ compared to the relaxation time. Since both timescales are comparable, however, they lead to similar qualitative estimates for the conductivity.

---

[h]In the Green's function approach, this factor arises from the vertex correction. The corresponding study was done, for example, in [630, 658].

By making use of the transport scattering time, one obtains the following expression for the conductivity:

$$\operatorname{Re}\sigma_{xx}(\Omega) = \frac{e^2 v_F^2}{3}\int_{-\infty}^{\infty} d\epsilon \frac{\nu(\epsilon)}{4T\cosh^2\left(\frac{\epsilon-\mu}{2T}\right)}\frac{1/\tau_{\mathrm{tr}}(\epsilon)}{\Omega^2 + 1/\tau_{\mathrm{tr}}^2(\epsilon)}$$

$$\overset{\Omega\to 0}{=} \frac{e^2 v_F^2 \hbar}{2\pi n_{\mathrm{imp}}u_0^2}, \tag{7.32}$$

which agrees with the conventional Drude expression for the DC conductivity, i.e.,

$$\sigma_{xx}^{(\mathrm{Drude})} = \frac{e^2 v_F^2}{3}\nu(\mu)\tau_{\mathrm{tr}}(\mu) = \frac{e^2 v_F^2 \hbar}{2\pi n_{\mathrm{imp}}u_0^2}. \tag{7.33}$$

It is interesting to point out that, for a short-range disorder, the transport scattering time is inversely proportional to the density of states $\nu(\mu)$ and, as a result, the DC conductivity is independent of the electric chemical potential. In fact, it is determined exclusively by impurity concentration n_{imp} and disorder strength u_0.

It is also instructive to demonstrate the use of the kinetic theory description for the electron transport in semimetals with electrically charged impurities. In such a case, impurities interact with electrons via a partially screened Coulomb potential, see Eq. (7.3). The Fourier transform of the corresponding disorder potential is given in Eq. (7.8). For such disorder, the transport scattering time (7.30) equals [547]

$$\frac{1}{\tau_{\mathrm{tr}}(\epsilon)} = \frac{\pi\alpha^2 n_{\mathrm{imp}}v_F^3 \hbar^2}{4\epsilon^2}\int_0^{\pi} d\theta \frac{\sin^3(\theta)}{\left[\sin^2(\theta/2)+\alpha/(2\pi)\right]^2}$$

$$= \frac{4\pi^3 n_{\mathrm{imp}}v_F^3 \hbar^2}{3\epsilon^2}g(\alpha), \tag{7.34}$$

where $\alpha = e^2/(\varepsilon_e \hbar v_F)$ is an effective coupling constant and

$$g(\alpha) = \frac{3\alpha^2}{\pi^2}\left[\frac{1}{2}\left(1+\frac{\alpha}{\pi}\right)\ln\left(1+\frac{2\pi}{\alpha}\right)-1\right]. \tag{7.35}$$

The function $g(\alpha)$ approaches zero as $3\alpha^2/(2\pi^2)\ln(2\pi/\alpha)$ for small values of α and is approximately $1-2\pi/\alpha$ for large α.

By substituting the transport scattering time (7.34) into the Drude expression for the DC conductivity in Eq. (7.33), one finds that $\sigma \sim \mu^4/n_{\mathrm{imp}}$ for materials with electrically charged impurities. In the case of donor impurities, the functional dependence is also different because the number of impurities is proportional to the number of electrons, i.e., $n_{\text{imp}} \propto \mu^3$. Then the transport scattering time $\tau_{\text{tr}} \sim 1/\left[\mu g(\alpha)\right]$ and the DC conductivity becomes a linear function of the electric chemical potential $\sigma \sim \mu/g(\alpha)$.

As the analysis in this subsection demonstrates, the semiclassical Boltzmann framework provides a simple and straightforward method for studying transport properties. Of course, such an approach is subject to the inherent limitations of kinetic theory and lacks the sophistication of the Green's function formalism. In particular, there could be some important discrepancies due to its inability to capture all quantum effects. Nevertheless, the semiclassical approach gives reasonable qualitative results, sheds light on the underlying processes, and provides an intuitive understanding of transport properties. For example, the scattering times in the case of short-range disorder, $\tau(\epsilon) \sim 1/\left[n_{\text{imp}} u_0^2 \nu(\epsilon)\right]$, and long-range disorder, $\tau(\epsilon) \sim \epsilon^2/n_{\text{imp}}$, appear to be a good agreement with those obtained in the Green's function formalism.

7.1.4 *Magnetic field effects*

Another important topic that deserves a separate discussion is the interplay of disorder and a magnetic field. Indeed, in addition to the anomalous effects discussed in Sec. 6.2, a background magnetic field could have other interesting effects on the transport properties in disordered Weyl semimetals.

Let us recall that the Landau-level spectrum implies that the DOS is an oscillating function of a magnetic field. This affects various observable properties in a qualitative manner. For example, it is responsible for the Shubnikov–de Haas oscillations of the electrical conductivity and the de Haas–van Alphen oscillations of the magnetization (see Sec. 3.9). Such transport and thermodynamic properties could be studied easily by using either the Kubo formalism or the chiral kinetic theory (see Chapter 6). This requires the knowledge of the quasiparticle width or the transport scattering time. The latter depends not only on the type of disorder but also can be modified by the presence of a magnetic field [8, 659, 775]. The transverse magnetoresistivity in the presence of short-range disorder was addressed in [659] and the case of long-range charged impurities was investigated in [659, 775].

As was discussed in Sec. 7.1.2, the quasiparticle width due to scattering is captured by the imaginary part of the self-energy. Therefore, in order to investigate the effect of disorder on the transport properties in the presence of a magnetic field, let us start with calculating the self-energy in the simplest Born approximation, i.e.,

$$\Sigma(\Omega, \mathbf{k}) = -i \int \frac{d\omega d^3 k}{(2\pi)^4} S(\omega, \mathbf{k}') D(\Omega - \omega, \mathbf{k} - \mathbf{k}'). \tag{7.36}$$

Here $D(\omega, \mathbf{k})$ is the disorder propagator, which is nothing else but the Fourier transform of the impurity correlation function, see Eq. (7.12). The translation invariant Green's function $S(\omega, \mathbf{k})$ can be obtained from Eq. (6.32) and equals

$$S(\omega; \mathbf{k}) = i \sum_{\eta=\pm} \sum_{n=0}^{\infty} \frac{(-1)^n e^{-k_\perp^2 l_B^2}}{\epsilon_n (\hbar\omega + \mu - \eta\epsilon_n)} \{2\eta\hbar v_F (\mathbf{k}_\perp \cdot \boldsymbol{\sigma}_\perp) L_{n-1}^1 \left(2k_\perp^2 l_B^2\right)$$

$$+ (\epsilon_n - \eta\hbar v_F k_z \sigma_z) \left[\mathcal{P}_- L_n \left(2k_\perp^2 l_B^2\right) - \mathcal{P}_+ L_{n-1} \left(2k_\perp^2 l_B^2\right)\right]\}, \tag{7.37}$$

where, for the sake of concreteness, quasiparticles are assumed to be right-handed ($\lambda = +$). This is not a limitation since the internode effects are neglected and the broadening should be independent of the chirality because of $\mathcal{P}$ symmetry. Concerning the other notation, $L_n^\alpha(z)$ are the generalized Laguerre polynomials[i] and $\mathcal{P}_\pm = (1 \pm s_B \sigma_z)/2$ are the (pseudo)spin projectors.

Let us consider the case of short-range disorder where the potential is given in Eq. (7.2). In this case, the disorder propagator takes a very simple form given in Eq. (7.12). The integration over $\mathbf{k}_\perp$ in the expression for self-energy (7.36) can be performed exactly by using the following table integral [281]:

$$\int_0^\infty e^{-st} t^\alpha L_n^\alpha(t) dt = \frac{\Gamma(\alpha + n + 1)(s - 1)^n}{n! s^{\alpha + n + 1}}, \tag{7.38}$$

which is valid for $\mathrm{Re}\, s > 0$ and $\mathrm{Re}\, \alpha > -1$. Then, in the limit $\mu \to 0$, one finds that the self-energy has a diagonal form in the (pseudo)spin space, i.e., $\Sigma(\epsilon) = \mathrm{diag}\,[\Sigma_+(\epsilon), \Sigma_-(\epsilon)]$, where $\epsilon = \hbar\Omega$ and

$$\Sigma_+(\epsilon) = \frac{n_{\mathrm{imp}} u_0^2}{2\pi l_B^2} \sum_{n \geq 1}^{\infty} \int \frac{dk_z}{2\pi} \frac{\epsilon - \hbar v_F k_z}{\epsilon^2 + i0\,\mathrm{sgn}\,(\epsilon) - 2n\epsilon_L^2 - \hbar^2 v_F^2 k_z^2}, \tag{7.39}$$

$$\Sigma_-(\epsilon) = \frac{n_{\mathrm{imp}} u_0^2}{2\pi l_B^2} \sum_{n \geq 0}^{\infty} \int \frac{dk_z}{2\pi} \frac{\epsilon + \hbar v_F k_z}{\epsilon^2 + i0\,\mathrm{sgn}\,(\epsilon) - 2n\epsilon_L^2 - \hbar^2 v_F^2 k_z^2}. \tag{7.40}$$

---

[i]By definition, $L_{-1}^\alpha \equiv 0$.

Since the real part of the self-energy can be absorbed into the definition of the reference point for the energy or a shifted value of the chemical potential, we will ignore it in the following. Instead, we will concentrate on the imaginary part, which determines the quasiparticle width. As we see from the explicit expressions, the $n = 0$ Landau level contributes to Σ_- but not to Σ_+. Moreover, as is easy to see, the corresponding components of the imaginary parts differ only by a constant, i.e.,

$$\operatorname{Im}\Sigma_-(\epsilon) = \operatorname{Im}\Sigma_+(\epsilon) - \Gamma_0, \tag{7.41}$$

where

$$\Gamma_0 = \frac{n_{\text{imp}}u_0^2\epsilon_{\text{L}}^2}{4\pi\hbar^3 v_F^3} \tag{7.42}$$

is the partial contribution to the quasiparticle width from the lowest Landau level. After integrating over k_z, one obtains the following explicit expression for the imaginary part:

$$\operatorname{Im}\Sigma_-(\epsilon) = -\Gamma_0 \sum_{n=0}^{N_\epsilon} \frac{|\epsilon|}{\sqrt{\epsilon^2 - 2n\epsilon_{\text{L}}^2}}. \tag{7.43}$$

Here $N_\epsilon = \left[\epsilon^2/(2\epsilon_{\text{L}}^2)\right]$ and $[\ldots]$ denotes the integer part. For small energies $|\epsilon| < \epsilon_{\text{L}}$, only the lowest Landau level is filled and the result for the imaginary part of the self-energy reads $\operatorname{Im}\Sigma_-(\epsilon) = -\Gamma_0$, which also implies that $\operatorname{Im}\Sigma_+(\epsilon) = 0$. At larger energies, the higher Landau levels start to fill up and the corresponding approximate result can be obtained by using the Euler–Maclaurin summation formula [659], i.e.,

$$\operatorname{Im}\Sigma_-(\epsilon) \simeq -\Gamma_0 \left(\frac{|\epsilon|}{\sqrt{\epsilon^2 - 2N_\epsilon\epsilon_{\text{L}}^2}} + \frac{\epsilon^2}{\epsilon_{\text{L}}^2} - \frac{|\epsilon|}{\epsilon_{\text{L}}^2}\sqrt{\epsilon^2 - 2(N_\epsilon - 1)\epsilon_{\text{L}}^2} \right.$$
$$\left. + \frac{1}{2} + \frac{|\epsilon|}{2\sqrt{\epsilon^2 - 2(N_\epsilon - 1)\epsilon_{\text{L}}^2}} \right). \tag{7.44}$$

By noting that self-energy (7.44) contains square root singularities, one might call the validity of the perturbative Born approximation in question. One of the ways to resolve the problem with singularities is to use a self-consistent treatment, where instead of expressions (7.39) and (7.40), the following set of equations is used:

$$\Sigma_+(\epsilon) = \frac{\Gamma_0}{\pi} \sum_{n\geq 1}^{N_{\max}} \int d\tilde{k}_z \frac{\epsilon - \tilde{k}_z - \Sigma_-(\epsilon)}{\left[\epsilon - \tilde{k}_z - \Sigma_-(\epsilon)\right]\left[\epsilon - \tilde{k}_z - \Sigma_+(\epsilon)\right] - 2n\epsilon_{\text{L}}^2}, \tag{7.45}$$

$$\Sigma_-(\epsilon) = \frac{\Gamma_0}{\pi} \sum_{n \geq 0}^{N_{\max}} \int d\tilde{k}_z \frac{\epsilon + \tilde{k}_z - \Sigma_+(\epsilon)}{\left[\epsilon - \tilde{k}_z - \Sigma_-(\epsilon)\right]\left[\epsilon - \tilde{k}_z - \Sigma_+(\epsilon)\right] - 2n\epsilon_L^2},$$

$$(7.46)$$

where $\tilde{k}_z = \hbar v_F k_z$. One can check that the results of the Born approximation are reproduced in the limit of weak disorder and at small energies $|\epsilon| < \epsilon_L$.

In the regime where several Landau levels are occupied ($\epsilon \gg \epsilon_L$), the contribution of the lowest Landau level is relatively small and the difference between the two (pseudo)spin components of the self-energy can be neglected, $\Sigma_+(\epsilon) \approx \Sigma_-(\epsilon) = \Sigma(\epsilon)$. As in the simple Born approximation, one can ignore the real part of the self-energy and concentrate on the imaginary part, which describes the total broadening $\Gamma = -\operatorname{Im}\Sigma(\epsilon)$ and can be conveniently represented as a sum of individual Landau-level contributions, i.e., $\Gamma = \sum_{n=0}^{N_{\max}} \Gamma_n$, where $N_{\max}$ is the index of the highest Landau level within the bandwidth. Note that the dependence of Γ on energy is suppressed for the brevity of notation. This leads to a self-consistent approximation with the value of total broadening Γ determined by [659]

$$\Gamma = \sum_{n=0}^{N_{\max}} \Gamma_n = \frac{\Gamma_0 \Gamma}{\pi} \sum_{n=0}^{N_{\max}} \int d\tilde{k}_z \frac{\epsilon^2 + 2n\epsilon_L^2 + \Gamma^2}{(\epsilon^2 - 2n\epsilon_L^2 - \Gamma^2 - \tilde{k}_z^2)^2 + 4\epsilon^2\Gamma^2}. \quad (7.47)$$

After performing the integration on the right-hand side and assuming that disorder is weak (i.e., $\Gamma \ll \epsilon$), we arrive at the following equation for Γ:

$$\Gamma = \frac{\Gamma_0 \epsilon}{\sqrt{2}} \sum_{n=0}^{N_{\max}} \sqrt{\frac{\epsilon^2 - 2n\epsilon_L^2 - \Gamma^2 + \sqrt{(\epsilon^2 - 2n\epsilon_L^2 - \Gamma^2)^2 + 4\epsilon^2\Gamma^2}}{(\epsilon^2 - 2n\epsilon_L^2 - \Gamma^2)^2 + 4\epsilon^2\Gamma^2}}. \quad (7.48)$$

One should note that disorder shifts the Landau-level energies, $2n\epsilon_L^2 \to 2n\epsilon_L^2 + \Gamma^2$. In what follows, it is instructive to limit the study to the values of ϵ that are close to the shifted energies of Landau levels. This also simplifies the self-consistent analysis of broadening.

When the energy is sufficiently small, $\epsilon_L \ll \epsilon \ll \epsilon_L(\epsilon_L/\Gamma_0)^{1/5}$, one finds that Landau levels are well-separated and the broadening is dominated by the same Landau level. The corresponding parametric energy dependence is given by

$$\Gamma \simeq \left(\frac{\Gamma_0}{2}\right)^{2/3} \epsilon^{1/3}. \tag{7.49}$$

At large energies, $\epsilon \gg \epsilon_{\mathrm{L}}(\epsilon_{\mathrm{L}}/\Gamma_0)^{1/5}$, the levels overlap and their broadening is given by

$$\Gamma \simeq \Gamma_0 \left(\frac{\epsilon}{\epsilon_{\mathrm{L}}}\right)^2. \tag{7.50}$$

Note that this result comes from many lower Landau levels and coincides with the zero magnetic field limit in Eq. (7.15).

In Fig. 7.2, the energy dependence of the quasiparticle width, obtained self-consistently from Eq. (7.48), is compared with the result for the width in the Born approximation, $\Gamma_- = -\mathrm{Im}\,\Sigma_-$, which is defined by Eq. (7.43). As expected, with decreasing the disorder strength u_0 and/or the impurity concentration n_{imp}, the results of the self-consistent analysis gradually approach the Born approximation. In addition, outside the low-energy regime $\epsilon \lesssim \epsilon_{\mathrm{L}}$, the quasiparticle width in both approaches has a general tendency to grow with energy.

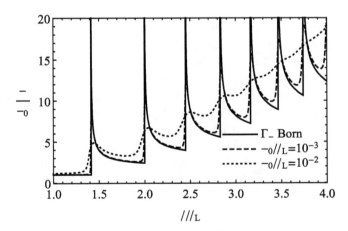

Fig. 7.2 The quasiparticle width in the presence of a magnetic field calculated by using the Born approximation result, $\Gamma_- = -\mathrm{Im}\,\Sigma_-$, given in Eq. (7.43) (solid line) and self-consistent results obtained from Eq. (7.48) at $\Gamma_0/\epsilon_{\mathrm{L}} = 10^{-3}$ (dashed line) and $\Gamma_0/\epsilon_{\mathrm{L}} = 10^{-2}$ (dotted line). The cutoff for the Landau-level summation is $N_{\mathrm{max}} = 100$.

Having obtained the quasiparticle width, one can now straightforwardly calculate the transverse electrical conductivity.[j] The corresponding analysis can be done along the same lines as in Sec. 6.2.3. Here we will discuss only a few qualitative results.

In the low-temperature and weak disorder regime ($T \lesssim \epsilon_L$ and $\Gamma_0 \lesssim \epsilon_L$) or, equivalently, in the strong magnetic field limit, the transport is dominated by the lowest Landau level. In this case, the conductivity is linear in magnetic field and proportional to disorder concentration, $\sigma_{xx} \propto n_{\text{imp}} u_0^2 |eB|$. At high temperature, $T \gtrsim \epsilon_L$, or in the weak magnetic field limit, the transport is dominated by higher Landau levels and the role of the lowest Landau level is negligible. In such a case, the conductivity can be described by a Drude-like formula [659, 775]

$$\sigma_{xx}^D = \frac{e^2 v_F^2}{6\pi} \int \frac{d\epsilon}{4T \cosh\left(\frac{\epsilon}{2T}\right)} \frac{\nu(\epsilon)\tau_{\text{tr}}(\epsilon)}{1 + \omega_c^2 \left[\tau_{\text{tr}}(\epsilon)\right]^2}, \qquad (7.51)$$

where the cyclotron frequency is $\omega_c = \hbar v_F^2/(l_B^2 \epsilon)$. At moderate temperatures when the Landau levels do not overlap, the conductivity scales as $\sigma_{xx} \propto n_{\text{imp}} u_0^2 T^6 B^{-2}$. At higher temperature, when the Landau levels strongly overlap, the magnetic field effects appear only in the form of a subleading correction $\propto -B^{1/3} T^{-1} n_{\text{imp}}^{4/3} u_0^{8/3}$ to the well-known result at $B = 0$, i.e., $\sigma_{xx} \propto e^2 v_F^2/(n_{\text{imp}} u_0^2)$.

Here it might be also instructive to briefly comment on the case of charged impurities. The key difference from short-range impurities is the dependence of the effective disorder strength on a magnetic field [659]. From physics viewpoint, this is the result of an enhanced screening by the field.[k] In the strong magnetic field limit ($\epsilon_L \gg T$), for example, the effective disorder strength behaves as $u_0 \propto 1/|eB|$. Therefore, by utilizing the lowest Landau-level approximation mentioned above, i.e., $\sigma_{xx} \propto n_{\text{imp}} u_0^2 |eB|$, one derives $\sigma_{xx} \propto 1/|eB|$. This agrees with Abrikosov's result for the magnetoconductivity in Eq. (6.47). In the weak field limit, on the other hand, the magnetic field correction to conductivity is quadratic, $\propto n_{\text{imp}}^{4/3} B^2$ [775].

---

[j]It is worth noting that disorder can also lead to a vertex renormalization. Since the latter often has only a quantitative rather than qualitative effect, it could be omitted in a simplified analysis. It might be also instructive to recall that, in the absence of a magnetic field, the vertex correction changes the quantum scattering time $\tau_q = \hbar/(2\Gamma)$ to the transport scattering time $\tau_{\text{tr}} = 3\tau_q/2$ (see also the discussion in Sec. 7.1.3).

[k]This is easiest to understand in the strong magnetic field limit, in which the density of zero energy states is proportional to $|eB|$. This implies that the effective Debye screening length is inversely proportional to $|eB|$.

Another interesting phenomenon related to the interplay of magnetic field and disorder in Weyl and Dirac semimetals is the *weak antilocalization* effect that tends to increase sample's conductivity. In the absence of magnetic fields, it stems from a strong spin–orbit coupling that causes a destructive interference between pairs of time-reversed quasiparticle paths around the impurity. By noting that the magnetic field breaks $\mathcal{T}$ symmetry, the weak antilocalization effect is expected to be destroyed gradually with increasing the field strength.[1] The effect was studied rigorously in a two-band model of Weyl semimetals by using the Bethe–Salpeter equation for the scattering amplitude [661]. It was shown that the corresponding correction to conductivity weakly depends on magnetic field $\propto -B^2$ and dominates at small B. In stronger magnetic fields, the effects of the chiral anomaly $\propto B^2$ overwhelm the weak antilocalization correction and the longitudinal magnetoconductivity becomes positive. This qualitative scenario is in good agreement with the experimental data in [56, 57, 59–61] and [62–66] for Dirac and Weyl semimetals, respectively. It is worth noting, however, that the magnitude of the magnetoresistivity increase observed experimentally at weak fields may not be explained by the weak antilocalization alone.

7.1.5 *Nonperturbative effects*

While transport properties of Dirac and Weyl semimetals can be studied by perturbative techniques in certain regimes, sometimes a truly nonperturbative analysis is required. In particular, this is the case when the electric chemical potential is very small $\mu \ll T$ and electrically charged impurities are present. In such a regime, the quasiparticle scattering is determined almost exclusively by a weakly screened Coulomb interaction. The corresponding effects can be accounted for properly by performing a resummation of infinite sets of relevant Feynman diagrams (e.g., as in the self-consistent Born approximation) or by other advanced methods (e.g., the Schwinger–Dyson equation). By using the self-consistent Born approximation, the conductivity in Weyl and Dirac semimetals was studied in [630, 655]. Unlike a perturbative study, the self-consistent Born approach usually does not allow for simple analytical solutions. Therefore, by omitting the technical details, let us mention only a few qualitative results

---

[1]For the same reasons, the weak antilocalization is likely to be suppressed or even absent in Weyl semimetals with broken $\mathcal{T}$ symmetry.

here. In the case of a long-range Gaussian potential discussed in [655], the conductivity agrees well with the Boltzmann approach for small disorder strengths. However, when the strength exceeds a certain critical value, the conductivity starts to increase with disorder.

A nonperturbative treatment is often essential not only for producing more reliable results, but also for providing an insight into qualitatively new features in transport. For example, a nonzero DOS originating from disorder [627, 629, 630] is one of such nonperturbative effects in Weyl and Dirac semimetals at vanishing μ. Moreover, even a single impurity could affect the local DOS leading to impurity resonances. In this subsection, a few nonperturbative effects will be briefly discussed.

7.1.5.1 *Impurity resonance*

For illustrative purposes, let us start with a simple case of a single spinless, short-range impurity in a Weyl or Dirac semimetal. The corresponding Hamiltonian for a Weyl node is given in Eq. (7.1), where an impurity is described by the potential $U(\mathbf{r}) = U\delta(\mathbf{r})$ and the electric chemical potential is set to zero. A convenient approach to investigate impurity effects is to use the T-matrix formalism [776]. The T-matrix for a local disorder is defined as $T_{\mathbf{k},\mathbf{k}'} = \delta_{\mathbf{k},\mathbf{k}'}T$, where

$$T = U + U \int \frac{d^3 k}{(2\pi)^3} (-i) S^{\mathrm{R}}(\omega, \mathbf{k}) T. \tag{7.52}$$

This gives

$$T = \left[1 - U \int \frac{d^3 k}{(2\pi)^3} (-i) S^{\mathrm{R}}(\omega, \mathbf{k}) \right]^{-1} U. \tag{7.53}$$

Here the retarded Green's function for the free Hamiltonian in the frequency–momentum space reads

$$S^{\mathrm{R}}(\omega, \mathbf{k}) = i \frac{\hbar\omega + \lambda\hbar v_F (\boldsymbol{\sigma} \cdot \mathbf{k})}{(\hbar\omega)^2 - (\hbar v_F k)^2 + i0 \, \mathrm{sgn}\,(\omega)}. \tag{7.54}$$

The full Green's function in the T-matrix approach is defined by

$$G^{\mathrm{R}}(\omega, \mathbf{k}, \mathbf{k}') = \delta_{\mathbf{k},\mathbf{k}'} S^{\mathrm{R}}(\omega, \mathbf{k}) + (-i) S^{\mathrm{R}}(\omega, \mathbf{k}) T_{\mathbf{k},\mathbf{k}'} S^{\mathrm{R}}(\omega, \mathbf{k}'). \tag{7.55}$$

The poles of $G^{\mathrm{R}}(\omega, \mathbf{k}, \mathbf{k}')$ define bound states and *impurity resonances* [213, 777, 778]. The latter are manifested as sharp features in the frequency and coordinate dependence of the local DOS, which is defined as

$$\nu(\omega, \mathbf{r}) = -\frac{1}{\pi}\text{Im tr}\left[(-i)G^{\text{R}}(\omega, \mathbf{r}, \mathbf{r})\right], \qquad (7.56)$$

where the full Green's function in the mixed frequency-coordinate representation is

$$G^{\text{R}}(\omega, \mathbf{r}, \mathbf{r}') = S^{\text{R}}(\omega, \mathbf{r} - \mathbf{r}') + (-i)S^{\text{R}}(\omega, \mathbf{r})T S^{\text{R}}(\omega, -\mathbf{r}'). \qquad (7.57)$$

Here, without the loss of generality, we assumed that the defect is located at the origin $\mathbf{r} = \mathbf{0}$.

As follows from Eq. (7.55) or (7.57), the poles of the full Green's function include also the poles of the T-matrix. Therefore, in order to determine the positions of impurity resonances and bound states, one needs to solve the following characteristic equation:

$$\frac{1}{U} = \text{Re}\int\frac{d^3k}{(2\pi)^3}S^{\text{R}}(\omega, \mathbf{k}). \qquad (7.58)$$

After straightforward calculations, its explicit form reads

$$\frac{1}{U} = \frac{1}{2\pi^2\hbar v_F^3}\left[-v_F\Lambda\omega + \frac{\omega^2}{2}\ln\left|\frac{\omega + v_F\Lambda}{\omega - v_F\Lambda}\right|\right], \qquad (7.59)$$

where Λ is the wave vector cutoff. The results for the frequency dependence of the local density of states (7.56) are shown in Fig. 7.3 for a few disorder strengths. As one can see, an impurity resonance appears as a peak in the local density of states. With the rise of impurity strength, the peak gradually moves to smaller frequencies. Notice also that a frequency asymmetry of the DOS at $U \neq 0$ is related to the fact that the impurity potential breaks the particle–hole symmetry. The impurity resonances for other possible types of impurity potential in Weyl and Dirac semimetals are analyzed in [779].

7.1.5.2 *Rare-region states and charge puddles*

Let us consider now the case when several impurities are present. In particular, we explain how electrically charged impurities could lead to the formation of electron and hole puddles [629, 630] and, therefore, a nonzero DOS in Weyl and Dirac semimetals at vanishing μ. Note that the same effect also appears in narrow gap semiconductors [780]. To understand the origin of charged puddles, one can study the problem in the self-consistent Thomas–Fermi approximation [629]. In this approximation, the local Fermi

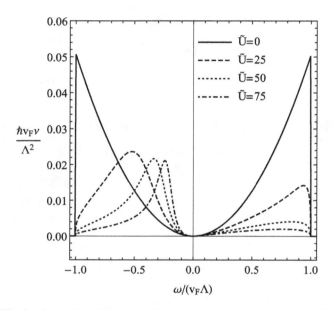

Fig. 7.3 The local density of states at the impurity site as a function of frequency ω. Disorder strengths are $\tilde{U} = 0$ (solid line), $\tilde{U} = 25$ (dashed line), $\tilde{U} = 50$ (dotted line), and $\tilde{U} = 75$ (dot-dashed line). Here, $\tilde{U} = U\Lambda^2/(\hbar v_F)$.

energy is determined by chemical potential μ and the local value of the Coulomb potential energy, i.e., $\mu + eu(r)$, where

$$u(r) = \frac{e^2}{\varepsilon_e r} e^{-r/r_{\mathrm{sc}}}. \tag{7.60}$$

In the last expression, $r_{\mathrm{sc}} = \sqrt{\varepsilon_e/(4\pi e^2 \langle \nu(\mu) \rangle)}$ is the screening length which can be determined self-consistently in terms of the local density of states $\langle \nu(\mu) \rangle$. Note that, for small values of the chemical potential, r_{sc} is expected to be much larger than the average spatial separation between the impurities. In such a case, the local Coulomb potential energy is determined by a superposition of many individual contributions from random and uncorrelated impurities. The corresponding values have a Gaussian distribution with variance U^2/e^2, where U^2 is the mean-squared value of the electron potential energy defined by

$$U^2 = n_{\mathrm{imp}} \int d^3r\, e^2 u^2(r) = \frac{2\pi e^4 n_{\mathrm{imp}}}{\varepsilon_e^2} r_{\mathrm{sc}}. \tag{7.61}$$

Then the spatially averaged DOS can be determined as follows:

$$\langle \nu(\mu) \rangle = \int_{-\infty}^{\infty} du\nu(\mu + eu) \frac{\exp\left[-e^2 u^2/(2U^2)\right]}{\sqrt{2\pi U^2/e^2}} = \frac{U^2 + \mu^2}{2\pi^2 \hbar^3 v_F^3}, \qquad (7.62)$$

Interestingly, this spatially averaged DOS is nonvanishing in the limit $\mu \to 0$. Indeed, by supplementing the results in Eqs. (7.61) and (7.62) at $\mu = 0$ with the definition of the screening length r_{sc} in terms of the local density of states $\langle \nu(0) \rangle$, one obtains

$$U = 2^{1/6} \sqrt{\pi v_F \hbar} \frac{e \left(n_{\text{imp}}\right)^{1/3}}{\sqrt{\varepsilon_e}}. \qquad (7.63)$$

The corresponding average DOS from Eq. (7.62) is given by $\langle \nu(0) \rangle \propto e^2 n_{\text{imp}}^{2/3}$ [629]. The physical interpretation of this result is the following. A superposition of the Coulomb potentials from many randomly distributed charged impurities produces an effective electric potential with fluctuations of typical size r_{sc} and amplitude U. The corresponding regions with positive and negative values of the Coulomb energy are nothing else but electron and hole puddles, respectively. Note that, at $\mu = 0$, the screening length (which is also the typical size of puddles) is given by

$$r_{sc} = \frac{\varepsilon_e \hbar v_F}{2^{2/3} e^2 \left(n_{\text{imp}}\right)^{1/3}}, \qquad (7.64)$$

As is easy to see, in the weakly coupled regime with $e^2/(\varepsilon_e \hbar v_F) \ll 1$, this length scale is indeed much larger than the average separation between impurities $(n_{\text{imp}})^{-1/3}$. The formation of electron and hole puddles is the formal reason for having a nonzero DOS at $\mu = 0$. This finding demonstrates that Dirac and Weyl semimetal phases with the vanishing DOS are very elusive in reality.

By making use of the local density of states, the electrical conductivity can be estimated by using the Drude formula $\sigma = e^2 v_F^2 \langle \nu(\mu) \rangle \tau_{\text{tr}}/3$, where the transport time is discussed in Sec. 7.1.3 (see also Eqs. (7.30) and (7.34)). One finds, in particular, that the minimum conductivity is achieved at $\mu = 0$ due to a small nonzero density of states induced by electron and hole puddles. The corresponding estimate reads $\sigma_{\min} \propto e^2 n_{\text{imp}}^{1/3}$ [629].

In passing, let us also address the case of short-range disorder. Usually, the Dirac point is stable with respect to randomly-distributed local disorder [385, 629, 655, 781–783]. Indeed, this can be understood from a simple-dimensional analysis [629]. An electron wave packet of size ξ experiences disorder from approximately ξ^3 impurities. While the average value of potential can be set to zero, there is a statistical deviation of the number of impurities $\propto \sqrt{\xi^3}$. Therefore, the averaged disorder potential experienced by an electron is of the order of $1/\xi^{3/2}$. This should be compared with the electron kinetic energy, which scales as $k \propto 1/\xi$. As we see, the short-range disorder is irrelevant in the long-wavelength case (i.e., small k). Obviously, this argument does not apply to the case of the Coulomb potential, where the potential energy scales as the statistical deviation of the number of impurities divided by the range of potential estimated by the size of wave packet, i.e., $\xi^{1/2}$.

The low-energy behavior of Weyl and Dirac semimetals with short-range disorder is also unusual. It is dominated by the *rare-region* effects [627]. In this connection, it is worth noting that there is a similarity between the rare-region effects and another classical phenomenon known as the Lifshitz tails [784]. While the latter correspond to the exponential tails in energy bands due to impurity-induced states inside the band gap, the rare region effects change the low-energy DOS in the gapless case. Usually, it is assumed that weak disorder is perturbatively irrelevant in 3D Dirac and Weyl semimetals or, in other words, that such disorder does not affect the vanishing DOS at the Dirac point. Qualitatively, this follows from the fact that the quasiparticle width decreases with energy more rapidly than the energy itself, see for example, Eq. (7.15). A nonzero DOS is induced by exponentially rare regions that host power-law bound resonances induced by disorder [627]. Note that the effect would be missed if the conventional disorder averaging is performed by using perturbative calculations.

The presence of the rare region states was also confirmed by using numerical lattice simulations in [785–787].[m] It is worth noting, however, that the stability of the rare regions states as well as their ability to generate a finite DOS at Weyl nodes was questioned in [788, 789]. In particular, it was suggested that a disorder potential in a Weyl semimetal leads to the impurity resonance states that contribute to the spectral density in the vicinity of zero energy, but not exactly at zero. Therefore, the DOS should still vanish at a Weyl node.[n] This criticism, however, does not apply in the case of a Dirac point, where two Weyl nodes of opposite chirality overlap.

---

[m] As the numerical study in [626] showed, the rare region states could hybridize with the Fermi arcs leading to dissipative surface transport. For a discussion of the Fermi arc transport, see also Sec. 5.5.

[n] In the case of a single Weyl node and a spherical impurity, for example, the impurity-induced average DOS vanishes as $\nu(\epsilon) \propto |\epsilon| \ln |\epsilon|$ near the node.

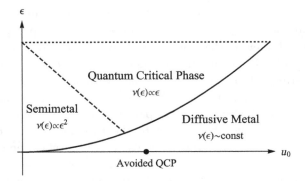

Fig. 7.4 Schematic crossover diagram for Weyl and Dirac semimetals in the plane of energy ϵ and disorder strength u_0. The quantum critical point (QCP) is avoided due to disorder effects. Here the dotted line corresponds to an energy cutoff in the low-energy relativistic-like description.

It also ignores the internode scattering processes in real Weyl semimetals with an even number of Weyl nodes.

Nonperturbative effects of rare region states might be important for the quantum critical transitions. In particular, these states could lead to an avoided or hidden quantum critical point (QCP) [786, 787] between semimetal and metal phases. Indeed, this is supported by the numerical calculations showing that the DOS is an analytic function of energy near a Dirac point or a Weyl node. The corresponding crossover diagram is schematically shown in Fig. 7.4. Depending on the energy scale and the disorder strength, a semimetal (with $\nu \propto \epsilon^2$), a quantum critical phase (with $\nu \propto \epsilon$), and a diffusive metal (with $\nu \sim const$) can be realized. Note that the validity of the phase diagram extends only up to cutoff energy denoted by the black dashed line, where the low-energy relativistic-like description breaks down.

Before concluding this subsection, let us note that another powerful method for investigating disorder effects is the renormalization group technique. While it will not be considered in this book, an interested reader may consult [790–792].

7.2 Thermoelectric and thermal transport

Thermoelectrics is an important subfield of condensed matter physics. It is not only of significant theoretical interest, but also has a wide range of practical applications that include various cooling devices, thermometers, and thermoelectric generators. Generally, thermoelectric devices are thermodynamic engines that do not contain any moving parts. They convert heat into electricity or, vice versa, use electricity to pump heat. It is not surprising, therefore, that there is a plethora of research papers as well as a number of books on various aspects of thermoelectrics [793, 794].

The discovery of topological materials with a linear relativistic-like energy spectrum brought new possibilities in thermoelectrics. Like electric transport, its thermoelectric counterpart in Dirac and Weyl semimetals is expected to be very efficient due to the high mobility of quasiparticles. Besides that, such materials may also reveal novel exotic phenomena absent in conventional metals and semimetals. Indeed, since properties of topological semimetals are strongly affected by the Berry curvature and the chiral anomaly (see Sec. 6.2), it is natural to expect that anomalous effects will be manifested in the thermoelectric transport and other thermal effects [770–773, 795–801]. In particular, the latter include the anomalous Nernst, Ettingshausen, and thermal Hall effects.

In this section, the key aspects of the thermoelectric and thermal transport in Dirac and Weyl semimetals will be presented. We will start by introducing several key ideas of the thermal transport theory, including the general formalism for calculating thermoelectric coefficients. Then we will discuss the thermal conductivity and related properties in Dirac and Weyl semimetals by using the chiral kinetic theory and the Kubo formalism. As in the rest of the book, the main attention will be paid to the electron contributions and, therefore, all phonon effects will be ignored. At the end, a few experimental results on the thermoelectric and thermal transport will be discussed.

7.2.1 *General notions*

Let us start by discussing the description of thermoelectric phenomena in the linear response regime [235, 793, 794, 802]. By assuming $\mu_5 = 0$ for simplicity, one can express the electric and heat currents as follows:

$$J_n = eL_{nm}^{11}(eE_m - \nabla_m \mu) - eL_{nm}^{12} \nabla_m \left(\frac{1}{T}\right), \qquad (7.65)$$

$$J_n^Q = -L_{nm}^{21} \frac{eE_m - \nabla_m \mu}{T} + L_{nm}^{22} \nabla_m \left(\frac{1}{T}\right), \qquad (7.66)$$

where $\mathbf{E}$ is the electric field, $\nabla \mu$ is the gradient of the electric chemical potential, J_n is the electric current, and J_n^Q is the *heat current*. Note that the latter is related to the energy current J_n^E as $J_n^Q = J_n^E + \mu J_n/e$. This relation has a clear physical meaning, namely, only the electrons with energies exceeding the Fermi level contribute to the heat transport. Note also that the Onsager reciprocal relation demands that $L_{nm}^{12} = L_{nm}^{21}$.

In experiment, the relevant observables are the electrical conductivity σ_{nm}, Seebeck S_{nm}, and thermal conductivity κ_{nm} tensors. The electrical conductivity tensor σ_{nm} is defined through the electric current when $\nabla T = \nabla \mu = 0$, i.e.,

$$\sigma_{nm} = e^2 L_{nm}^{11}. \tag{7.67}$$

The *Seebeck coefficient* or *thermopower* S_{nm} can be extracted by measuring an electric potential generated by a thermal gradient at $\mathbf{J} = \nabla \mu = 0$. In this case, one has

$$S_{nm} = -\frac{1}{eT^2}(L^{11})_{nl}^{-1} L_{lm}^{12}. \tag{7.68}$$

Finally, the *thermal conductivity* κ_{nm} is defined through the heat current in the absence of an electric current, i.e., $J_n^Q = -\kappa_{nm}\nabla_m T$. Then

$$\kappa_{nm} = \frac{1}{T^2}\left[L_{nm}^{22} - \frac{1}{T}L_{nl}^{21}(L^{11})_{lj}^{-1}L_{jm}^{12}\right]. \tag{7.69}$$

As in the case of electric transport, one of the simplest approaches to study thermal and thermoelectric properties in Weyl and Dirac semimetals is based on the chiral kinetic theory (see Sec. 6.6.2). It is critical, of course, that the latter captures the key effects of the Berry curvature. An even more powerful approach is the Kubo linear response theory (see Sec. 6.2.3), although its application to thermoelectric and thermal transport is not as straightforward. Indeed, additional complications come from the special types of driving forces associated with temperature gradients. Unlike external electromagnetic fields that are easy to include in the Hamiltonian formulation, such forces are thermodynamic or statistical.

As was first shown by Joaquin Luttinger [803], a nonuniform temperature distribution can be described by using a fictitious gravitational potential that couples to the energy density of the system. Such a potential is naturally incorporated in the Hamiltonian and allows to use the standard Kubo theory to calculate the energy or heat flow. In the end, the response to a temperature gradient is obtained by using arguments similar to those employed in the Einstein relations.

The Luttinger approach to the thermoelectric response was extended to electronic systems in quantizing magnetic fields [804]. It was argued that the local electric and heat currents could be separated into *transport* and *magnetization* parts. The latter is a divergence-free current that can significantly modify the local current density but produces no net current passing through the sample. It was also argued that the Onsager relations apply to transport currents observed experimentally, but not to local currents. From a technical viewpoint, there are two new types of terms in the expressions for electric and heat currents related to the electromagnetic orbital magnetization **M** and its heat counterpart $\mathbf{M}^Q$.°

The effects of the Berry curvature on the anomalous thermoelectric transport were addressed for the first time in [805]. It was shown, in particular, that the Berry curvature gives an extra correction to the orbital magnetization [381, 806]. The latter contributes to the transport current due to a statistical force in a way similar to the anomalous velocity term produced by a mechanical force. While the contribution from the carrier magnetic moments in the local current will be subtracted out in the transport current, the Berry-phase correction to the magnetization will remain. After accounting for the orbital magnetization effects, the modified expressions for thermoelectric currents, including the anomalous Nernst and Hall ones, satisfy the Onsager relations.

The separation of the transport and magnetization terms is also crucial in the Kubo approach [804, 807], which was applied in thermoelectric transport studies of Weyl semimetals in [795]. In the rest of this section, however, we will concentrate primarily on the chiral kinetic theory approach.

Let us derive an explicit expression for the transport currents. According to the formalism in Sec. 6.6.2, the local electric current density for the electrons with chirality λ is given by

$$\mathbf{j}_\lambda = -\sum_{e,h} e \int \frac{d^3p}{(2\pi\hbar)^3} \left\{ \mathbf{v_p} - \frac{e}{c}(\mathbf{v_p} \cdot \mathbf{\Omega}_\lambda)\mathbf{B}_\lambda - e[\tilde{\mathbf{E}}_\lambda \times \mathbf{\Omega}_\lambda] \right\} f_\lambda$$

$$+ \sum_{e,h} \int \frac{d^3p}{(2\pi\hbar)^3} c \left[\mathbf{\nabla} \times \mathbf{m}_\lambda \right] f_\lambda. \tag{7.70}$$

Note that the last term in Eq. (7.70) is the magnetization current associated with the magnetic moment $\mathbf{m}_\lambda = -ev_F p \mathbf{\Omega}_\lambda / c$ of carriers with

---

°The heat magnetization is a combination of the gravitomagnetic energy and the orbital magnetization $\mathbf{M}^Q = \mathbf{M}_\epsilon + \mu\mathbf{M}/e$.

a relativistic-like energy dispersion relation. The *transport* current $\mathbf{J}_\lambda$ is defined by subtracting the magnetization current from $\mathbf{j}_\lambda$, i.e.,

$$\mathbf{J}_\lambda = \mathbf{j}_\lambda - c\,[\boldsymbol{\nabla} \times \mathbf{M}_\lambda], \tag{7.71}$$

where $\mathbf{M}_\lambda$ is the magnetization density. In equilibrium, it can be obtained by differentiating the grand canonical potential Ω with respect to the magnetic field. In the case of a nonzero Berry curvature, the definition of the grand canonical potential reads

$$\Omega = \sum_{e,h} T \int \frac{d^3p}{(2\pi\hbar)^3} \left[1 - \frac{e}{c}(\mathbf{B}_\lambda \cdot \boldsymbol{\Omega}_\lambda)\right] \ln\left[1 - f_\lambda^{\text{eq}}(\epsilon_\mathbf{p})\right], \tag{7.72}$$

where the quasiparticle energy $\epsilon_\mathbf{p}$ inside the equilibrium distribution function includes the interaction energy with the magnetic moment, i.e., $\epsilon_\mathbf{p} = v_F p - (\mathbf{m}_\lambda \cdot \mathbf{B}_\lambda)$, as given by Eq. (6.153). Then, for the magnetization density, one obtains

$$\mathbf{M}_\lambda = -\left(\frac{\partial \Omega}{\partial \mathbf{B}_\lambda}\right)_{\mu,T} = \sum_{e,h} \int \frac{d^3p}{(2\pi\hbar)^3} \left[\mathbf{m}_\lambda f_\lambda^{\text{eq}} + T\frac{e}{c}\boldsymbol{\Omega}_\lambda \ln\left(1 - f_\lambda^{\text{eq}}\right)\right]. \tag{7.73}$$

Note that $\mathbf{M}_\lambda$ contains the orbital magnetization term $\propto \mathbf{m}_\lambda f_\lambda^{\text{eq}}$, which gives a correction to the current that cancels exactly the last term in Eq. (7.70), as well as an extra term proportional to the Berry curvature. This term plays an important role in restoring the Mott relation (see Eq. (7.85) below) between the electrical and thermoelectric conductivities. Note also that, in kinetic theory, the definition of the magnetization in Eq. (7.73) could be extended to a nonequilibrium case by replacing $f_\lambda^{\text{eq}} \to f_\lambda$.

By making use of the magnetization in Eq. (7.73), one derives the following expression for the transport electric current (7.71)

$$\mathbf{J}_\lambda = -\sum_{e,h} e \int \frac{d^3p}{(2\pi\hbar)^3} \left\{\mathbf{v}_\mathbf{p} - \frac{e}{c}(\mathbf{v}_\mathbf{p} \cdot \boldsymbol{\Omega}_\lambda)\mathbf{B}_\lambda - e[\tilde{\mathbf{E}}_\lambda \times \boldsymbol{\Omega}_\lambda]\right\} f_\lambda$$
$$+ \sum_{e,h} \frac{e}{c} \int \frac{d^3p}{(2\pi\hbar)^3} \left[\frac{\boldsymbol{\nabla}T}{T} \times \boldsymbol{\Omega}_\lambda\right] \left[(v_F p - \mu)f_\lambda - T\ln(1 - f_\lambda)\right]. \tag{7.74}$$

It is worth pointing out that, in local equilibrium, the last term can be rewritten via the entropy density $s_\mathbf{p}$ as follows:

$$\frac{e}{c} \int \frac{d^3p}{(2\pi\hbar)^3} \left[\boldsymbol{\nabla}T \times \boldsymbol{\Omega}_\lambda\right] s_\mathbf{p}, \tag{7.75}$$

where

$$s_{\mathbf{p}} = -f_\lambda^{\mathrm{eq}} \ln\left(f_\lambda^{\mathrm{eq}}\right) - \left(1 - f_\lambda^{\mathrm{eq}}\right) \ln\left(1 - f_\lambda^{\mathrm{eq}}\right). \tag{7.76}$$

The case of the heat current is more complicated [807, 808]. By omitting the details of the derivation, the final expression for the transport heat current is given by

$$\mathbf{J}_\lambda^Q = \sum_{e,h} \int \frac{d^3p}{(2\pi\hbar)^3} (\epsilon_{\mathbf{p}} - \mu) \left\{ \mathbf{v}_{\mathbf{p}} - \frac{e}{c}(\mathbf{v}_{\mathbf{p}} \cdot \mathbf{\Omega}_\lambda)\mathbf{B}_\lambda - e[\tilde{\mathbf{E}}_\lambda \times \mathbf{\Omega}_\lambda] \right\} f_\lambda$$

$$+ \sum_{e,h} eT \int \frac{d^3p}{(2\pi\hbar)^3} \left[\mathbf{E}_\lambda \times \mathbf{\Omega}_\lambda\right] \ln\left(1 - f_\lambda\right)$$

$$- \sum_{e,h} T \int \frac{d^3p}{(2\pi\hbar)^3} \left[\mathbf{\nabla}T \times \mathbf{\Omega}_\lambda\right] \left[\frac{\pi^2}{3} + f_\lambda \ln^2\left(1/f_\lambda - 1\right) - \ln^2\left(1 - f_\lambda\right)\right.$$

$$\left. - 2\mathrm{Li}_2\left(1 - f_\lambda\right)\right]. \tag{7.77}$$

In addition to the usual energy flow term $\propto (\epsilon_{\mathbf{p}} - \mu)\mathbf{v}_{\mathbf{p}}$, this expression also contains a number of terms associated with a nonzero Berry curvature. The latter include contributions $\propto [\mathbf{E}_\lambda \times \mathbf{\Omega}_\lambda]$ due to the anomalous velocity, as well as those induced by a temperature gradient, i.e., $\propto [\mathbf{\nabla}T \times \mathbf{\Omega}_\lambda]$.

7.2.2 *Longitudinal transport*

It is convenient to start our discussion of thermoelectric properties in Dirac and Weyl semimetals from the longitudinal transport along the electric field, assuming that a temperature gradient points in the same direction. This case is not only the easiest to understand but it is also instructive since the effects of the chiral anomaly might be manifested here.

As is clear from the general relations in Eqs. (7.65) and (7.66), the longitudinal transport coefficients are defined by the diagonal terms with $n = m$. Note that, without loss of generality, one can assume a homogeneous chemical potential $\mu = const$. Further, for qualitative estimates, it is sufficient to use the kinetic equation in Eq. (6.151) with the collision integral in the relaxation-time approximation, i.e.,

$$I_{\mathrm{coll}}(f_\lambda) = -\frac{f_\lambda - f_\lambda^{\mathrm{eq}}}{\tau}. \tag{7.78}$$

We use this approximation in the rest of this section to calculate the electron transport properties in Dirac and Weyl semimetals. It should be noted, however, that more sophisticated approaches can be also utilized [647, 705, 743, 744, 770].

7.2.2.1 *Thermoelectric transport in absence of (pseudo)magnetic fields*

In the absence of (pseudo)magnetic fields (i.e., $B_\lambda = 0$), it is straightforward to find a steady-state solution to the kinetic equation (6.151) by using the same general method as in Sec. 7.1.3 in the case of a constant temperature. To linear order in an applied (pseudo)electric field and a temperature gradient, the solution is given by

$$f_\lambda = f_\lambda^{\text{eq}} + \tau \frac{\partial f_\lambda^{\text{eq}}}{\partial \epsilon_{\mathbf{p}}} \left[e \left(\mathbf{v_p} \cdot \mathbf{E}_\lambda \right) + \left(\mathbf{v_p} \cdot \nabla T \right) \frac{\epsilon_{\mathbf{p}} - \mu}{T} \right]. \qquad (7.79)$$

Note that here $\epsilon_{\mathbf{p}} = v_F p$ and $\mathbf{v_p} = v_F \hat{\mathbf{p}}$. By making use of the definitions in Eqs. (7.74) and (7.77) and assuming that the relaxation time τ is independent of the quasiparticle momentum, one derives the following expressions for the electric and heat currents:

$$\mathbf{J}_\lambda = \frac{e^2 \tau}{6\pi^2 v_F \hbar^3} \left(\mu^2 + \frac{\pi^2 T^2}{3} \right) \mathbf{E}_\lambda - \frac{e\mu T^3 \tau}{9 v_F \hbar^3} \nabla \left(\frac{1}{T} \right), \qquad (7.80)$$

$$\mathbf{J}_\lambda^Q = -\frac{e\mu T^2 \tau}{9 v_F \hbar^3} \mathbf{E}_\lambda + \frac{\tau T^3}{18 v_F \hbar^3} \left(\mu^2 + \frac{7\pi^2 T^2}{5} \right) \nabla \left(\frac{1}{T} \right), \qquad (7.81)$$

where formulas from Appendix B were used and the summation over electrons and holes was performed according to the prescription in Sec. 6.6.2.

By comparing the results in Eqs. (7.80)–(7.81) with the general relations (7.65)–(7.66), one can easily extract transport coefficients L_{nn}^{11}, L_{nn}^{12}, and L_{nn}^{22}. Then, by making use of the definitions in Eq. (7.67)–(7.69), one can finally obtain the longitudinal components of electrical conductivity σ_{nn}, Seebeck coefficient S_{nn}, and thermal conductivity κ_{nn}. The corresponding explicit results are given by

$$\sigma_{nn} = \frac{e^2 \tau}{3\pi^2 v_F \hbar^3} \left(\mu^2 + \frac{\pi^2 T^2}{3} \right), \qquad (7.82)$$

$$S_{nn} = -\frac{2\pi^2 T}{e} \frac{\mu}{3\mu^2 + \pi^2 T^2}, \qquad (7.83)$$

$$\kappa_{nn} = \frac{\tau T}{9 v_F \hbar^3} \left(\mu^2 + \frac{7}{5}\pi^2 T^2 - \frac{4\mu^2 \pi^2 T^4}{3\mu^2 + \pi^2 T^2} \right). \qquad (7.84)$$

Note that the *Mott relation*

$$L_{nm}^{12} = \frac{\pi^2 T^3}{3} \frac{dL_{nm}^{11}}{d\mu} \qquad (7.85)$$

holds even for the simplest model with constant τ. On the other hand, as is easy to verify, the *Wiedemann–Franz law*

$$\kappa_{nm} = \frac{\pi^2 T}{3e^2} \sigma_{nm} \qquad (7.86)$$

is broken.

In general, the relaxation time τ is a nontrivial function of the quasiparticle energy. As a result, the dependence of transport coefficients on temperature and chemical potential is modified compared to the model with constant τ [769, 772]. Below we will summarize the corresponding results and highlight their physical meaning.

In the case of short-range disorder described by Eq. (7.2), the transport scattering time equals

$$\tau_{\text{sr}} = \frac{3}{2\pi n_{\text{imp}} u_0^2 \nu(\epsilon_{\mathbf{p}})} \left(1 + \frac{5 n_{\text{imp}}^2 u_0^4 \epsilon_{\mathbf{p}}^2}{16\pi^2 \hbar^6 v_F^6} \right), \qquad (7.87)$$

where n_{imp} is the concentration of impurities, u_0 is the disorder strength, and $\nu(\epsilon_{\mathbf{p}})$ is the DOS given by Eq. (3.136). Note that here we included an additional factor $3/2$ in order to obtain the transport scattering time, rather than the conventional quasiparticle lifetime, see the discussion in Sec. 7.1.3. The second term in parentheses of Eq. (7.87) stems from the first-order Born approximation [772, 809] and is important for describing L_{nn}^{12} and the Seebeck coefficient.

For comparison, let us also mention the case of long-range disorder in Weyl and Dirac semimetals in the limit of a small chemical potential. In such a case, the Coulomb potential of charged impurities is poorly screened. This is due to the fact that the perturbative DOS vanishes when $\mu \to 0$, and a nonperturbative contribution associated with electron and hole puddles is rather small. Then, according to Eq. (7.34), the scattering time equals

$$\tau_{\text{lr}} = \frac{3\epsilon_{\mathbf{p}}^2}{4\pi^3 \hbar^2 v_F^3 n_{\text{imp}} g(\alpha)}, \qquad (7.88)$$

where the functional dependence $g(\alpha)$ on the effective coupling constant α is given in Eq. (7.35).

Table 7.1 The transport coefficients for three models with different dependence of the relaxation time on energy: $\tau = const$, $\tau = \tau_{\rm sr}$, and $\tau = \tau_{\rm lr}$. The explicit expressions for $\tau_{\rm sr}$ and $\tau_{\rm lr}$ are given in Eqs. (7.87) and (7.88), respectively. A small μ expansion is used for the case $\tau = const$ and a small T/μ expansion is used for $\tau = \tau_{\rm lr}$. Note that the electrical conductivity is given by $\sigma_{nn} = e^2 L^{11}_{nn}$. The subleading order corrections to the transport coefficients can be found in [772].

τ	L^{11}_{nn}	L^{12}_{nn}	L^{22}_{nn}	S_{nn}	κ_{nn}
$\tau = const$	$\dfrac{T^2 \tau}{9\hbar^3 v_F}$	$\dfrac{2\mu T^3 \tau}{9\hbar^3 v_F}$	$\dfrac{7\pi^2 T^5 \tau}{90\hbar^3 v_F}$	$-\dfrac{2\mu}{eT}$	$\dfrac{7\pi^2 T^3 \tau}{45\hbar^3 v_F}$
$\tau = \tau_{\rm sr}$	$\dfrac{v_F^2 \hbar}{2\pi n_{\rm imp} u_0^2}$	$\dfrac{5 n_{\rm imp} u_0^2 T^3 \mu}{48\pi v_F^4 \hbar^5}$	$\dfrac{\pi \hbar v_F^2 T^3}{6 n_{\rm imp} u_0^2}$	$-\dfrac{5 n_{\rm imp}^2 u_0^4 T \mu}{24 e v_F^6 \hbar^6}$	$\dfrac{\pi \hbar v_F^2 T}{6 n_{\rm imp} u_0^2}$
$\tau = \tau_{\rm lr}$	$\dfrac{\mu}{8\pi^5 v_F \hbar^2 g(\alpha)}$	$\dfrac{T^3}{6\pi^3 \hbar^2 v_F g(\alpha)}$	$\dfrac{T^3 \mu}{24\pi^3 \hbar^2 v_F g(\alpha)}$	$-\dfrac{4\pi^2 T}{3e\mu}$	$\dfrac{T\mu}{24\pi^3 \hbar^2 v_F g(\alpha)}$

The results for thermoelectric transport coefficients in the three models with different energy dependence of the scattering time τ on energy are summarized in Table 7.1. As is easy to check, the Wiedemann–Franz law holds at sufficiently small temperatures for realistic models with long-range and weak short-range disorder. This is in stark contrast to the case of a naive model with $\tau = const$. Furthermore, the thermal conductivity is linear in T for the models with nontrivial dependence of τ on energy. It should be emphasized, however, that such a dependence is not a distinctive feature of only Dirac and Weyl semimetals but observed for other materials too [288, 356, 793, 794].

It is unclear if any specific features of the transport coefficients in Table 7.1 could be used as unambiguous signatures of a relativistic-like dispersion relation or nontrivial topology in Weyl and Dirac semimetals. One might try to argue, for example, that the coefficients have a distinctive dependence on chemical potential μ. The latter, however, is very difficult to control or verify in experiment. A more promising approach is to look for qualitatively new effects associated with a nonzero Berry curvature, which can be turned on by an external magnetic field.[p] In essence, this is the same approach that helped to reveal the celebrated negative magnetoresistivity (see Sec. 6.2). In the following subsection, therefore, we will briefly discuss the longitudinal thermoelectric transport in the presence of a magnetic field.

---

[p]Strictly speaking, there is one effect induced by the Berry curvature in the transverse transport, namely, the anomalous Hall effect discussed in Sec. 6.3, that can be realized even in the absence of magnetic fields. For the corresponding discussion, see Sec. 7.2.3.

7.2.2.2 *Thermoelectric transport in magnetic fields*

As we already discussed in Sec. 3.3 and reiterated at the end of Sec. 7.2.2.1, the nontrivial topological properties of Weyl semimetals are encoded in the Berry curvature. The numerous effects due to the Berry curvature could be naturally activated by applying background magnetic fields. Several such effects were considered in Chapter 6, including the negative magnetoresistivity, the chiral separation effect, the out-of-equilibrium chiral magnetic effect (CME), etc.

The longitudinal transport coefficients in the presence of magnetic fields can be obtained by a straightforward generalization of the approach considered in Sec. 7.2.2.1. By using the relaxation-time approximation and solving the Boltzmann equation (6.151) to linear order in external fields and gradients, one can find the expression for f_λ. The corresponding solution is similar to that in Eq. (7.79) albeit becomes more complicated because of its dependence on a magnetic field. The distribution function can then be used in the expressions for electric (7.74) and heat (7.77) current densities. Since the corresponding expressions are bulky and contain a lot of information, only the key qualitative features will be discussed here. Moreover, the emphasis will be made primarily on those thermoelectric transport properties that distinguish Weyl semimetals from nontopological materials.

Let us start by considering the case with a temperature gradient ∇T parallel to an external magnetic field $\mathbf{B} \parallel \hat{\mathbf{z}}$. As was discussed in Sec. 6.2, the longitudinal electrical conductivity σ_{zz} has a nontrivial correction due to the chiral anomaly. In particular, σ_{zz} receives a positive contribution proportional to the square of the magnetic field, i.e.,

$$\sigma_{zz} \approx \sigma_{zz}(B = 0) \left(1 + \sigma_1 B^2\right). \tag{7.89}$$

The direct calculation of the heat current shows that the thermal conductivity has a similar contribution with a quadratic dependence on the field, i.e.,

$$\kappa_{zz} \approx \kappa_{zz}(B = 0) \left(1 + \kappa_1 B^2\right), \tag{7.90}$$

where both κ_1 and σ_1 are positive parameters. Since, in general, $\kappa_1 \neq \sigma_1$, a deviation from the Wiedemann–Franz law (7.86) should be expected. Such a deviation was predicted in [771, 773] and later also reconfirmed in the hydrodynamic studies in [810]. It is worth noting, however, that the Wiedemann–Franz law holds (for the same configuration with $\nabla T \parallel \mathbf{B}$) in the studies of [770, 772], which neglected the internode scattering. Indeed, it was found that the parameters determining quadratic terms are equal, i.e., $\sigma_1 = \kappa_1 = \left[\hbar v_F^2 e/(c\mu^2)\right]^2$. It is interesting to mention that a term quadratic in a magnetic field could also appear in the thermoelectric coefficient L_{zz}^{12}. Its origin is usually connected with the mixed axial-gravitation anomaly [681, 811, 812]. The corresponding effect was observed experimentally in the Weyl semimetal NbP [813].

It is also interesting to discuss transport in the case with a temperature gradient perpendicular to a magnetic field, e.g., $\nabla T \parallel \hat{\mathbf{x}}$ and $\mathbf{B} \parallel \hat{\mathbf{z}}$. In this case, the expression for thermal conductivity is modified by magnetic field as follows [771, 772]:

$$\kappa_{xx} \approx \kappa_{xx}(B=0)\left(1 - \kappa_2 B^2\right), \tag{7.91}$$

where κ_2 is a positive coefficient that is determined by the scattering time, $\kappa_2 \propto \tau^2$. For example, $\kappa_2 = e^2 v_F^4 \tau^2/(c\mu)^2$ [772]. Note that the Wiedemann–Franz law holds in this case. As it turns out, a decreasing dependence of κ_{xx} with a magnetic field in Eq. (7.91) is not a unique feature of Dirac and Weyl materials. Indeed, for $\nabla T \perp \mathbf{B}$, the decrease of thermal conductivity with magnetic field is the usual response also in topologically trivial metals [288, 289, 814].

7.2.3 *Transverse Hall-like response*

Let us now discuss the transverse thermoelectric transport in Dirac and Weyl semimetals, which is captured by the off-diagonal components of transport coefficients $L_{nm}^{\alpha\beta}$. As we will see, this type of transport is truly anomalous. As is clear already from the expressions for the currents in Eqs. (7.74) and (7.77), the off-diagonal transport coefficients are strongly affected by the Berry curvature and a magnetic field. In the following analysis, we will separate (whenever possible) the effects due to magnetic fields from those arising only from a nontrivial topology of the band structure in Dirac or Weyl semimetals.

7.2.3.1 *Effects induced by magnetic field*

Let us first discuss the effects of an external magnetic field ($\mathbf{B} \parallel \hat{\mathbf{z}}$) on the transverse (Hall-like) thermoelectric coefficients in a linearized model of Weyl and Dirac semimetals. In the case of electrical conductivity, magnetic field $\mathbf{B}$ leads to the usual Hall effect. At low temperatures, the relevant off-diagonal component of the conductivity tensor reads [772, 773]

$$\sigma_{xy} \approx -\frac{e^3 B \tau^2 v_F \mu}{12\pi^2 \hbar^3 c}. \tag{7.92}$$

In the derivation of this result, a constant relaxation time $\tau = const$ and a small μ is assumed. By using the Mott relation (7.85), one can also obtain

$$L_{xy}^{12} \approx -\frac{eB\tau^2 v_F T^3}{36\hbar^3 c}. \tag{7.93}$$

Finally, by using the definition in Eq. (7.68), one can calculate the off-diagonal Seebeck coefficient, i.e.,

$$S_{xy} \approx \frac{v_F^2 \tau B}{4cT}. \tag{7.94}$$

This nonzero off-diagonal component of the Seebeck tensor describes the *Nernst* or *Nernst–Ettingshausen effect* in Dirac and Weyl semimetals. From a physics viewpoint, the Nernst effect is an electric potential induced in the direction perpendicular to both temperature gradient and magnetic field. The inverse effect is the *Ettingshausen effect*. It is connected with the generation of a temperature gradient in the presence of a perpendicular electric current and a magnetic field, i.e., $\partial_y T \propto B_z j_x$.

The off-diagonal component of thermal conductivity is also nonzero. At low temperatures, it is given by the following expression:

$$\kappa_{xy} \approx -\frac{7\pi^2 eB v_F \tau^2 T^3}{45\mu c \hbar^3}. \tag{7.95}$$

This transport coefficient describes the thermal Hall effect, which is also known as the *Leduc–Righi effect*. In essence, it corresponds to the generation of a transverse heat flow (or secondary temperature gradient) when an external magnetic field is applied to the system, i.e., $\partial_x T \propto B_z \partial_y T$.

It should be noted that all of the above transport coefficients can be affected by the energy dependence of the relaxation time τ. Nevertheless, the form of the qualitative results should remain the same at sufficiently low temperature, provided $\tau \approx \tau(\mu)$.

Let us also discuss briefly other thermoelectric effects. One of them is the *Seebeck effect* related to the generation of a voltage when a temperature gradient is applied to a junction made of two different materials. This effect is quantified by the diagonal components of the Seebeck tensor. The Peltier effect is, in essence, an inverse to the Seebeck effect. It describes the appearance of a temperature gradient when a current flows through a junction of two conductors. Since the Seebeck coefficient depends, in general, on temperature, there will be a difference in heat production between conductors with constant and inhomogeneous temperature profiles. This phenomenon is known as the *Thomson effect* and is determined by a derivative of the diagonal components of the Seebeck tensor with respect to temperature.

Since the validity of the kinetic theory approach is limited only to weak magnetic fields, it is also instructive to discuss briefly the case of strong fields where the Landau levels are formed. As was argued in [798, 815], the drift term $\propto \mathbf{E} \times \mathbf{B}$, which has the same effect on both electrons and holes, should cause a strong dependence of the DOS on B. As a result, the thermopower should become proportional to the strength of magnetic field. In gapped semiconductors, the thermopower saturates at sufficiently large B when the chemical potential falls below the band edge. This effect is absent in gapless Dirac and Weyl semimetals, where the transverse thermopower continues to grow as $S_{xx} \propto TB$. The longitudinal thermopower S_{zz}, on the other hand, should saturate. For the Hall thermoelectric conductivity quantified by L_{xy}^{12}, one finds that a relativistic-like energy spectrum implies a constant value $L_{xy}^{12} \propto T^3$ at large B. This is in contrast to L_{xy}^{12} decaying at large B in materials with the usual quadratic energy spectrum. It was also shown by using the Wigner function approach in [721] that the heat current in a strong magnetic field acquires a dissipative term $J_z^Q = -e v_F \tau |eB| T \partial_z T / (6 c \hbar^2)$, which is proportional to the strength of magnetic field B.

7.2.3.2 *Anomalous thermoelectric transport*

By recalling the discussion of anomalous electric transport in Sec. 6.3, it is reasonable to expect that nontrivial topological properties of Weyl semimetals could also give rise to interesting thermoelectric characteristics. In particular, in Weyl semimetals with broken $\mathcal{T}$ symmetry, there is an intriguing possibility to realize anomalous thermoelectric effects even in the absence of a magnetic field. As we will argue in this subsection, there are indeed anomalous contributions to the thermoelectric response and thermal conductivity similar to the anomalous Hall effect (AHE), predicted in [389, 391, 657, 662–665]. For reference, recall that the latter is described by the following off-diagonal conductivity tensor:

$$\sigma_{nm}^{\text{AHE}} = \epsilon_{nml} \frac{e^2 b_l}{2\pi^2 \hbar} \tag{7.96}$$

at $\mu = 0$ and $T \to 0$. As in the case of the AHE [432, 657], the origin of the anomalous thermal and thermoelectric response is related to the filled states deeply below the Fermi surface [795]. The AHE cannot be obtained reliably in models with a linearized spectrum using the chiral kinetic theory.[q] However, the anomalous contributions to the thermoelectric coefficients can be easily reproduced by using a two-band model described in Secs. 3.5.1 and 6.3, which captures well the band structure topology of Weyl semimetals.

By utilizing this two-band model below, we show that the anomalous (nondissipative) parts of the thermoelectric and thermal transport coefficients can be derived within the framework of Kubo theory. In order to calculate thermoelectric coefficients, we will follow the same general method as in Sec. 6.3 but generalized to the case of nonvanishing temperature and chemical potential gradients [803, 816]. By making use of the expression for the off-diagonal components of the electrical conductivity tensor in Eq. (6.64), we derive

$$L_{nm}^{11} = \frac{1}{e^2}\sigma_{nm} = \int \frac{d^3k}{(2\pi)^3}\left[f^{\text{eq}}(|\mathbf{d}|) - f^{\text{eq}}(-|\mathbf{d}|)\right]\mathcal{F}_{nm}(\mathbf{k}), \qquad (7.97)$$

where $\mathcal{F}_{nm}(\mathbf{k})$ is the Berry curvature tensor defined in Eq. (6.58). Note that the anomalous Hall conductivity σ_{xy}^{AHE} in Eq. (7.96) is reproduced from Eq. (7.97) in the limit of vanishing temperature ($T \to 0$) and zero chemical potential ($\mu = 0$). One also finds that, in general, the effects of nonzero T and μ tend to suppress the anomalous Hall effect.

As mentioned earlier, in order to correctly describe the flow of thermoelectric and heat currents in Weyl semimetals within the linear response Kubo theory, it is also important to correctly separate transport and magnetization contributions [804, 807]. By omitting the details of derivation, the final result for the nondissipative off-diagonal part of the transport coefficient L_{nm}^{21} reads [795]

$$L_{nm}^{21} = -T\mu \int \frac{d^3k}{(2\pi)^3}\left[f^{\text{eq}}(|\mathbf{d}|) - f^{\text{eq}}(-|\mathbf{d}|)\right]\mathcal{F}_{nm}(\mathbf{k}) + \frac{T}{e}\sum_{l=x,y,z}\epsilon_{nml}M_l,$$
$$(7.98)$$

---

[q]The anomalous Hall effect is reproduced in the consistent chiral kinetic theory, which includes the additional topological Chern–Simons terms in the electric charge and current densities [420], see Sec. 6.6.2.

where the magnetization is given by

$$M_l = e \sum_{n,m=x,y,z} \frac{\epsilon_{nml}}{2} \int \frac{d^3k}{(2\pi)^3} \mathcal{F}_{nm}(\mathbf{k}) \left\{ T \ln \left(\frac{1 + e^{(\mu - |\mathbf{d}|)/T}}{1 + e^{(\mu + |\mathbf{d}|)/T}} \right) \right.$$

$$\left. + |\mathbf{d}| \left[f^{\text{eq}}(|\mathbf{d}|) + f^{\text{eq}}(-|\mathbf{d}|) \right] \right\}. \tag{7.99}$$

The presence of the Berry curvature in the last expression suggests that the magnetization in Weyl semimetals has a topological origin. This becomes obvious in the limit $T \to 0$ and small μ, where

$$M_l \simeq -e\mu \sum_{n,m=x,y,z} \frac{\epsilon_{nml}}{2} \int \frac{d^3k}{(2\pi)^3} \mathcal{F}_{nm}(\mathbf{k}) = \frac{e\mu b_l}{2\pi^2 \hbar}. \tag{7.100}$$

Similarly to topological contributions in electric current, this magnetization is determined by the winding number of the mapping of a two-dimensional section of the Brillouin zone onto a unit sphere [73, 100]. Note, however, that unlike a truly topological response in Weyl semimetals, the thermoelectric one quantified by L_{nm}^{21} is proportional to the electric chemical potential.

From the physics viewpoint, the nondissipative coefficient L_{nm}^{21} in Eq. (7.98) describes an anomalous heat current in response to an external electric field, i.e.,

$$\mathbf{J}_{\text{Ett}}^Q = \frac{e}{T} L_{xy}^{21} [\mathbf{E} \times \hat{\mathbf{b}}], \tag{7.101}$$

where $\hat{\mathbf{b}} = \mathbf{b}/b$. This relation determines the anomalous Ettingshausen effect. Unlike the conventional Ettingshausen effect, where the presence of an external magnetic field is crucial, its anomalous counterpart is allowed by the chiral shift. In addition, by using the Onsager reciprocal relation $L_{nm}^{21} = L_{nm}^{12}$, one finds that there should exist an anomalous Nernst current, i.e.,

$$\mathbf{J}_{\text{Ner}} = e L_{xy}^{21} \left[\mathbf{\nabla} \left(\frac{1}{T} \right) \times \hat{\mathbf{b}} \right]. \tag{7.102}$$

This anomalous current can be also obtained directly by using the kinetic theory approach in a two-band model with a periodic dispersion relation [773]. While heat (7.101) and electric (7.102) currents are clearly connected with a nontrivial topology, they also depend on an electric chemical potential and temperature. Moreover, as one can see from Eqs. (7.98) and (7.100), L_{xy}^{21} vanishes when $\mu = 0$. Therefore, the anomalous Nernst and

Ettingshausen effects are induced only at a nonzero density and, as such, cannot be viewed as purely topological.

Finally, let us discuss anomalous contributions to thermal conductivity. The nondissipative off-diagonal part of the corresponding transport coefficient reads [795]

$$L_{nm}^{22} = T \int \frac{d^3k}{(2\pi)^3} \left[f^{\text{eq}}(|\mathbf{d}|) - f^{\text{eq}}(-|\mathbf{d}|) \right] \left(\mu^2 - |\mathbf{d}|^2 \right) \mathcal{F}_{nm}(\mathbf{k})$$
$$- 2T \sum_{l=x,y,z} \epsilon_{nml} M_l^Q, \tag{7.103}$$

where the heat magnetization is defined by

$$M_l^Q = \sum_{n,m=x,y,z} \frac{\epsilon_{nml}}{2} \int \frac{d^3k}{(2\pi)^3} \mathcal{F}_{nm}(\mathbf{k}) \{ T(\mu - |\mathbf{d}|) \ln \left(1 + e^{(\mu - |\mathbf{d}|)/T} \right)$$
$$- T(\mu + |\mathbf{d}|) \ln \left(1 + e^{(\mu + |\mathbf{d}|)/T} \right) + T^2 \text{Li}_2(-e^{(\mu - |\mathbf{d}|)/T})$$
$$- T^2 \text{Li}_2(-e^{(\mu + |\mathbf{d}|)/T}) - |\mathbf{d}| \left[(|\mathbf{d}| - \mu) f^{\text{eq}}(|\mathbf{d}|) - (|\mathbf{d}| + \mu) f^{\text{eq}}(-|\mathbf{d}|) \right] \}. \tag{7.104}$$

Here $\text{Li}_2(x)$ is the dilogarithm function [443]. The results simplify significantly in the limit $\mu \to 0$ and small T, where $L_{xy}^{22} = T^2 \kappa_{xy}^{\text{ATHE}}$ and the heat conductivity is given by

$$\kappa_{xy}^{\text{ATHE}} \simeq -\frac{\pi^2 T}{3} \int \frac{d^3k}{(2\pi)^3} \mathcal{F}_{xy}(\mathbf{k}) = \frac{Tb_z}{6\hbar}. \tag{7.105}$$

This transport coefficient represents the *anomalous thermal Hall effect* (ATHE) in Weyl semimetals. Physically, it describes a heat current perpendicular to both temperature gradient and chiral shift, i.e.,

$$\mathbf{J}_{\text{ATHE}}^Q = \frac{T^3}{6\hbar} \left[\mathbf{\nabla} \left(\frac{1}{T} \right) \times \mathbf{b} \right]. \tag{7.106}$$

Unlike the conventional Leduc–Righi effect, an external magnetic field is not needed for generating the ATHE current.

It is worth emphasizing that, after taking orbital magnetization effects into account, both Mott relation (7.85) and Wiedemann–Franz law (7.86) hold [807] for small T and μ. It is worth noting also that the ATHE conductivity was first obtained in [665, 772] by using the Wiedemann–Franz law and the anomalous Hall conductivity (7.96).

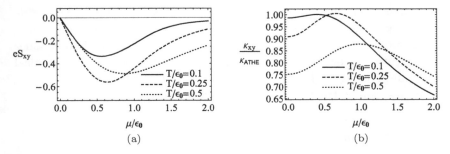

Fig. 7.5 (a) The off-diagonal component of the Seebeck tensor S_{xy} for $T/\epsilon_0 = 0.1$ (solid line), $T/\epsilon_0 = 0.25$ (dashed line), and $T/\epsilon_0 = 0.5$ (dotted line). (b) The off-diagonal component of the thermal conductivity tensor κ_{xy} for $T/\epsilon_0 = 0.1$ (solid line), $T/\epsilon_0 = 0.25$ (dashed line), and $T/\epsilon_0 = 0.5$ (dotted line). In both panels $\epsilon_0 = |\mathbf{d}(0)|$ is the quasiparticle energy at $\mathbf{k} = 0$.

The anomalous off-diagonal components of the Seebeck tensor S_{xy} and the thermal conductivity κ_{xy} are shown in Figs. 7.5(a) and 7.5(b), respectively, for a few values of temperature. To plot the results, we used the numerical values of parameters in Eq. (6.56).

As one can see from Fig. 7.5(a), the off-diagonal component of the Seebeck tensor S_{xy} is linear in μ for small values of the chemical potential. At large values, however, the dependence on μ becomes nonmonotonic. In the two-band model at hand, in particular, it shows a maximum (in absolute value) at $\mu/\epsilon_0 \approx 0.5$. With increasing temperature, the width of the corresponding peak broadens. One finds that the temperature dependence of S_{xy} is also nonmonotonic. Indeed, as a function of T, its absolute value grows at small temperatures but decreases at large ones. The off-diagonal component of thermal conductivity κ_{xy}, which is shown in Fig. 7.5(b), is also a nonmonotonic function of μ, which flattens gradually with temperature. Interestingly, with increasing temperature, the location of maximum tends to move to larger values of μ, although the width of the peak becomes broader.

By noting that a sizable separation of Weyl nodes and a well-defined concept of chirality are limited only to the low-energy sector of the theory, the deterioration of the Seebeck tensor S_{xy} and the thermal conductivity κ_{xy} should not be surprising at sufficiently high values of temperature or chemical potential. Indeed, both high T and large μ cause the states far away from the Weyl nodes to contribute to transport and, therefore, to blur significantly the topological signal.

In conclusion, similarly to the electrical conductivity, the thermoelectric and thermal conductivities provide another way of probing nontrivial topological properties of Weyl semimetals related to a separation of Weyl nodes in momentum. However, since nonzero temperature and electric chemical potential play a crucial role in the thermoelectric and thermal coefficients, the corresponding transport properties are not truly topological.

7.2.4 *Experimental signatures and figure of merit*

In this subsection, we review briefly the experimental studies and applications of thermoelectric transport in Dirac and Weyl semimetals. Because of the rapid development in the field, this can be only illustrative with emphasis on developments that seem most closely related to the theoretical ideas discussed earlier.

7.2.4.1 *Experimental results*

The thermoelectric properties of the Dirac semimetal Cd_3As_2 in a magnetic field were investigated in [817–821]. The experimental results for the diagonal S_{xx} and off-diagonal S_{xy} components of the Seebeck tensor are presented in Figs. 7.6(a) and 7.6(b), respectively. It is assumed that $\mathbf{B} \parallel \hat{\mathbf{z}}$ and $-\nabla T \parallel \hat{\mathbf{x}}$.

As one can see from Fig. 7.6(a), the thermopower S_{xx} increases with a magnetic field and saturates at large values of the field strength. The off-diagonal component S_{xy}, which describes the Nernst effect, is shown in Fig. 7.6(b). As expected in the case of a Dirac semimetal with unbroken $\mathcal{T}$ symmetry, it does not vanish only in the presence of a nonzero magnetic field. Similarly to S_{xx}, the off-diagonal component saturates at high magnetic fields but also shows quantum oscillations at sufficiently low temperatures. It is likely that the Nernst signal contains both conventional contribution due to the Hall effect, see Eq. (7.94), and anomalous one that is related to the Berry curvature, see Eqs. (7.98), (7.101), and (7.102). Indeed, as argued in [548], a nontrivial anomalous contribution should also appear because an external magnetic field splits Dirac nodes of Cd_3As_2 into pairs of opposite chirality Weyl nodes. Such a conclusion is indirectly supported by an observation of beating in the quantum oscillations of S_{xy} [820]. Physically, it arises from the interference of closely spaced oscillations related to bulk states with split Weyl nodes. Finally, we note that as in the case

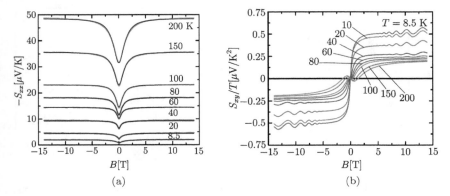

Fig. 7.6 The dependence of the (a) diagonal S_{xx} and (b) off-diagonal S_{xy} components of the Seebeck tensor in Cd_3As_2 on magnetic field for several values of temperature. (Adapted from [820].)

with parallel electric and magnetic fields when the chiral anomaly leads to a negative magnetoresistivity, the thermopower also becomes "negative" for $\nabla T \parallel \mathbf{B}$ and decreases as B^2 [817].

Let us now turn to the case of Weyl semimetals. The magnetic field dependence of thermopower and thermal conductivity was measured in Weyl semimetals with broken $\mathcal{P}$ symmetry such as NbP [822, 823] as well as in TaAs and TaP [824]. The representative experimental data for the off-diagonal component of the Seebeck tensor S_{xy} in TaAs [824] are shown in Fig. 7.7(a) for magnetic fields ranging from 0.5 T to 14 T. Similarly to the model results in Fig. 7.5(a), the temperature dependence of thermopower measured in experiment is nonmonotonic and shows a temperature-broadened peak. In addition, the position of the peak shifts to higher temperatures as a magnetic field increases.

It should be noted that the thermopower can reach relatively large values in Weyl semimetals. In particular, the peak values are about 150 $\mu V/K$ in TaAs [824], 200 $\mu V/K$ in TaP [824], and as large as 800 $\mu V/K$ in NbP [822, 823]. The latter clearly surpasses the typical values of thermopower in many metals ($1-100~\mu V/K$) and is comparable to those in some semiconductors (e.g., about $200-300~\mu V/K$ in BiTe [794] and 1000 $\mu V/K$ in InSb [825]).

An evidence of chiral anomaly effects in the diagonal components of the Seebeck tensor was reported in the magnetic-field-induced Weyl semimetal

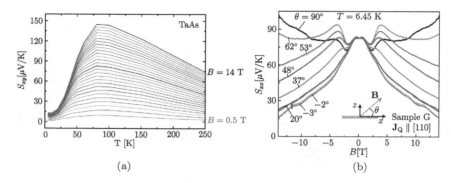

(a)   (b)

Fig. 7.7   (a) The dependence of the off-diagonal component of the Seebeck tensor S_{xy} in TaAs on temperature at different values of a magnetic field. (Adapted from [824].) (b) The diagonal component of the Seebeck tensor S_{xx} in GdPtBi as a function of magnetic field at different orientations of a magnetic field. (Adapted from [826].)

GdPtBi [826].[r] As one can see from Fig. 7.7, the longitudinal thermopower response, which is characterized by the diagonal components of the See-beck tensor S_{xx}, is very different from the transverse one. In particular, it gets suppressed when $\mathbf{B} \parallel \boldsymbol{\nabla} T$. A similar decreasing ("negative") dependence of the thermopower on a magnetic field was also observed in Cd_3As_2 [817].

In the extreme quantum limit, in which only the lowest Landau level is occupied, the thermoelectric properties of the Dirac semimetal $ZrTe_5$ and the Weyl semimetal TaP were measured in [827, 828]. In agreement with the theoretical predictions of [815], the thermoelectric Hall conductivity $-eL_{xy}^{12}/T^2$ acquires a universal value independent of magnetic field and carrier concentration.

7.2.4.2   *Thermoelectric figure of merit*

It is important to note that a high thermoelectric conductivity does not necessary make Dirac and Weyl semimetals good thermoelectric devices. The ability of a material to produce a large thermoelectric power is quantified by the dimensionless *thermoelectric figure of merit*. In the SI system

---

[r]Note that GdPtBi is a half-Heusler compound, which is a gapless semiconductor with quadratic bands when no magnetic field is applied. A nonzero magnetic field splits its bands and creates Weyl nodes.

of units, it is defined as [793, 794]

$$ZT = \frac{\sigma S^2 T}{\kappa}. \tag{7.107}$$

The physical meaning of the figure of merit is simple. It formalizes the fact that a good thermoelectric material should have high electrical σ and thermoelectric S conductivities, but its thermal conductivity κ should be small. Note that κ consists of the electron as well as phonon contributions. Since the former is usually related to the electrical conductivity σ by the Wiedemann–Franz law, the ratio σ/κ cannot be arbitrary large even if the phonon contribution were neglected. Taking phonons into account will only increase the net thermal conductivity and, thus, suppress the figure of merit. Therefore, the standard approach to increase the efficiency is to seek for materials where the propagation of phonons is hindered.

Another important characteristic is the so-called *power factor*, defined by

$$P = \sigma S^2. \tag{7.108}$$

It quantifies the maximal electrical power of a thermoelectric generator that can be extracted for a given temperature difference. Note, however, that it does not quantify the efficiency as the figure of merit does.

If the off-diagonal components of the response coefficients such as S_{xy} and κ_{xy} cannot be neglected, the following relations should be used for the figure of merit and the power factor [798]:

$$ZT = \frac{T \left(S_{xx} \kappa_{xx} - S_{xy} \kappa_{xy} \right)^2}{\left(\kappa_{xx}^2 + \kappa_{xy}^2 \right) \left(\kappa_{xx} \rho_{xx} - T S_{xy}^2 \right)}, \tag{7.109}$$

$$P = \frac{\left(S_{xx} \kappa_{xx} - S_{xy} \kappa_{xy} \right)^2}{\kappa_{xx} \left(\kappa_{xx} \rho_{xx} - T S_{xy}^2 \right)}, \tag{7.110}$$

where $\rho_{nm} = (\sigma)_{nm}^{-1}$ are the components of the resistivity tensor.

It is instructive to mention characteristic numerical values of the figure of merit and the power factor in Weyl and Dirac semimetals. Because of high thermal conductivity, their figure of merit is usually relatively small. For example, $ZT \approx 0.1 - 0.2$ in the Dirac semimetal Cd_3As_2 [818, 819]. It was argued in [829], however, that applying a strong magnetic field could increase the figure of merit in Cd_3As_2 by almost an order of magnitude, reaching $ZT \approx 1$ at room temperature. The enhancement of the power factor is more modest and reaches about 2. The corresponding values are comparable to those of commercially utilized bismuth alloys [794] but lower than the figure of merit predicted, for example, for PbTe [794, 830, 831]. As was discussed at the end of Sec. 7.2.2.2, the thermopower S_{xx} can be dramatically enhanced by applying a magnetic field since S_{xx} grows linearly with B in the quantum limit. This was indeed observed in $Pb_{1-x}Sn_xSe$ [832], whose low-energy quasiparticles are massive Dirac fermions, in strong fields up to about 30 T. According to estimates in [798], the figure of merit in such an ultraquantum regime can reach very high values. For example, it is possible to get $ZT \approx 3$ for the parameters considered in [832].

7.3   Optical response

While the direct current (DC) conductivity is arguably one of the simplest transport properties of solids, the alternating current (AC) or optical response driven by an oscillating electric field is also very interesting and provides important information on various material properties. The AC response is quantified by a frequency-dependent optical conductivity. As we will discuss below, like the DC transport, the AC one in Weyl and Dirac semimetals has unique features too.

Theoretically, the optical conductivity of Dirac and Weyl semimetals was studied in many research papers [385, 769, 833–842]. Related optical effects, including unconventional dichroism as well as the Faraday and Kerr effects in Weyl semimetals, were investigated theoretically in [843–845]. Experimentally, a wide range of optical properties in Dirac and Weyl semimetals was studied in [846–850, 850–852] and [853–864], respectively.

Curiously, it was suggested that the optical response of Weyl and Dirac semimetals might be invaluable in a direct detection of sub-MeV dark matter particles, assuming that the latter scatter on electrons via kinetically mixed dark photons [865].

In the following few subsections, we will review various theoretical aspects of optical and magneto-optical conductivity, optical activity and other related effects in Dirac and Weyl semimetals. The corresponding experimental findings will be discussed in Sec. 7.3.5.

7.3.1   *Optical conductivity*

Let us start our discussion of the optical response in Dirac and Weyl semimetals from a simple case without an external magnetic field. The AC conductivity can be calculated by using the Kubo linear response theory outlined in Secs. 6.2.3 and 7.1. One might notice that several frequency-dependent results have been already presented there. In principle, going beyond the DC regime is straightforward. The main difference is that the calculation of the AC linear response does not require taking the limit $\Omega \to 0$.

It is instructive to start our discussion of AC conductivity from the simplest case with no disorder. Generically, the conductivity will receive contributions from two types of processes: the interband and intraband ones. In the case of a simple relativistic-like Weyl Hamiltonian (7.1) with vanishing disorder potential, the corresponding absorptive parts of the conductivity are given by the following explicit expressions [769]:

$$\operatorname{Re}\sigma_{xx}^{\text{inter}}(\Omega) = \frac{e^2}{24\pi\hbar v_F}\Omega\frac{\sinh\left(\frac{\hbar\Omega}{2T}\right)}{\cosh\left(\frac{\mu}{T}\right) + \cosh\left(\frac{\hbar\Omega}{2T}\right)}$$

$$\stackrel{T\to 0}{=} \frac{e^2\Omega}{24\pi\hbar v_F}\theta\left(\hbar\Omega - 2\mu\right), \tag{7.111}$$

$$\operatorname{Re}\sigma_{xx}^{\text{intra}}(\Omega) = \frac{e^2}{6\pi\hbar^3 v_F}\delta(\Omega)\left(\mu^2 + \frac{\pi^2 T^2}{3}\right). \tag{7.112}$$

Note that, in the limit of vanishing temperature and zero chemical potential, only the interband contribution does not vanish and is linear in frequency Ω.[s] This is in agreement with the original prediction in [385]. As for the intraband contribution (7.112), it represents the conventional Drude conductivity that is proportional to the δ-function and diverges when $\Omega \to 0$. The Drude weight does not vanish only when either $\mu \neq 0$ or $T \not\to 0$. This is expected, of course, in the clean limit.

The real part of the interband optical conductivity has a characteristic step-like feature at $\hbar\Omega = 2\mu$. Physically, this is a manifestation of the Pauli blocking principle, which forbids transitions between fully-occupied states. For a relativistic-like spectrum of Dirac materials, the corresponding transitions induced by an external oscillating field are schematically shown in Fig. 7.8.

Generically, the optical conductivity $\sigma(\Omega)$ may also have a nonvanishing imaginary part. From physics viewpoint, the imaginary part of $\sigma(\Omega)$ modifies the dielectric constant, $\varepsilon_e(\Omega) = 1 + 4\pi i \sigma(\Omega)/\Omega$, turning it into a frequency-dependent function. The expression for the imaginary part of

---

[s]For comparison, in multi-Weyl semimetals, the real part of optical conductivity is strongly anisotropic [840]. Indeed, depending on the direction, the conductivity can show either the conventional $\sigma \sim \Omega$ (along the direction of linear dispersion) or a modified $\sigma \sim \Omega^{2/n_w - 1}$ (along the direction of nonlinear dispersion) dependence, where n_w is the topological charge of a Weyl node.

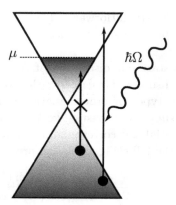

Fig. 7.8   Schematic description of optical transitions in a Weyl or Dirac semimetal. At vanishing temperature, electrons from the valence band can be excited only if $\hbar\Omega > 2\mu$. For smaller frequencies, the Pauli principle blocks the transition.

$\sigma(\Omega)$ can be obtained either by a direct calculation within the Kubo formalism or by using the Kramers–Kronig relation. The latter reads

$$\operatorname{Im}\sigma_{ij}(\Omega) = -\frac{2\Omega}{\pi}\,\text{p.v.}\int_0^\Lambda d\omega \frac{\operatorname{Re}\sigma_{ij}(\omega)}{\omega^2 - \Omega^2}. \tag{7.113}$$

Note that the upper limit in the integration is given by an ultraviolet cutoff Λ that can be interpreted either as (i) the band width $\Lambda \simeq v_F\pi/a$ where a is a lattice spacing or (ii) as the energy scale where a relativistic-like dispersion of quasiparticles becomes inapplicable. By making use of the Kramers–Kronig relation, one reveals that $\operatorname{Im}\sigma_{xx}^{\text{inter}}(\Omega)$ has a logarithmic divergence [834, 837], i.e.,

$$\operatorname{Im}\sigma_{xx}^{\text{inter}}(\Omega) \simeq -\frac{e^2\Omega}{24\pi^2\hbar v_F}\ln\frac{\Lambda^2}{\Omega^2}. \tag{7.114}$$

In the limit $T \to 0$ and $\mu \to 0$, this is the only contribution that survives. It leads to an enhancement of the dielectric constant at small frequencies. At nonzero T or μ, one finds that the intraband part of the conductivity has a nonzero imaginary part too, i.e.,

$$\operatorname{Im}\sigma_{xx}^{\text{intra}}(\Omega) = \frac{e^2}{3\pi^2\hbar^3 v_F\Omega}\left(\mu^2 + \frac{\pi^2 T^2}{3}\right), \tag{7.115}$$

which is inversely proportional to the frequency Ω. This is a characteristic dependence for a metal phase, which is formally realized when either $\mu \neq 0$ or $T \not\to 0$. In such a regime, the imaginary part of optical conductivity determines the low-frequency skin depth given by $\delta = c/\sqrt{2\pi\Omega \operatorname{Im} \sigma_{xx}^{\text{intra}}(\Omega)} = \sqrt{2}c/\Omega_e$, where Ω_e is the plasmon frequency.

As was already shown in Sec. 7.1, the presence of disorder significantly modifies the conductivity. Therefore, it is instructive to analyze the AC conductivity in the model of a Weyl semimetal defined by Hamiltonian (7.1). For simplicity, the effects of disorder will be captured by a quasiparticle width $\Gamma(\omega)$ that depends only on frequency. Then, the interband and intraband contributions are given by [835]

$$\operatorname{Re} \sigma_{xx}^{\text{inter}}(\Omega) = \frac{e^2}{6\pi^2 \hbar v_F} \int_{-\infty}^{\infty} d\omega \frac{f^{\text{eq}}(\hbar\omega) - f^{\text{eq}}(\hbar\omega + \hbar\Omega)}{\Omega}$$

$$\times \frac{\Gamma^2(\omega)\Gamma(\omega+\Omega) + \hbar^2\omega^2\Gamma(\omega+\Omega) + \Gamma(\omega)\Gamma^2(\omega+\Omega) + \hbar^2(\omega+\Omega)^2\Gamma(\omega)}{[\Gamma(\omega) + \Gamma(\omega+\Omega)]^2 + \hbar^2\Omega^2},$$

$$(7.116)$$

$$\operatorname{Re} \sigma_{xx}^{\text{intra}}(\Omega) = \frac{e^2}{6\pi^2 \hbar v_F} \int_{-\infty}^{\infty} d\omega \frac{f^{\text{eq}}(\hbar\omega) - f^{\text{eq}}(\hbar\omega + \hbar\Omega)}{\Omega}$$

$$\times \frac{\Gamma^2(\omega)\Gamma(\omega+\Omega) + \hbar^2\omega^2\Gamma(\omega+\Omega) + \Gamma(\omega)\Gamma^2(\omega+\Omega) + \hbar^2(\omega+\Omega)^2\Gamma(\omega)}{[\Gamma(\omega) + \Gamma(\omega+\Omega)]^2 + \hbar^2(2\omega+\Omega)^2}.$$

$$(7.117)$$

In a model with a constant quasiparticle width $\Gamma = const$ and in the limit of vanishing temperature, it is straightforward to obtain the following analytical expressions:

$$\operatorname{Re} \sigma_{xx}^{\text{inter}}(\Omega) = \frac{e^2}{6\pi^2 \hbar^2 v_F} \Gamma \left\{ 1 + \frac{\hbar\Omega}{4\Gamma} \left[\arctan\left(\frac{2\mu + \hbar\Omega}{2\Gamma}\right) \right. \right.$$

$$\left. \left. - \arctan\left(\frac{2\mu - \hbar\Omega}{2\Gamma}\right) \right] \right\}, \tag{7.118}$$

$$\operatorname{Re} \sigma_{xx}^{\text{intra}}(\Omega) = \frac{e^2}{12\pi^2 \hbar^2 v_F} \frac{\Gamma}{1 + [\hbar\Omega/(2\Gamma)]^2} \left(1 + \frac{3\mu^2 + \Omega^2}{3\Gamma^2}\right). \tag{7.119}$$

Unlike the Drude conductivity in usual metals $\propto \Gamma/(4\Gamma^2 + \hbar^2\Omega^2)$, the result in Eq. (7.119) does not vanish at large Ω but approaches a constant value $\propto \Gamma$.

In the cases of local disorder where $\Gamma(\omega) \propto \omega^2$ and screened charged disorder where $\Gamma(\omega) \propto \mu^3/\omega^2$, which were discussed in Sec. 7.1.2, the frequency dependence of the real part of the total optical conductivity $\operatorname{Re} \sigma_{xx}(\Omega) = \operatorname{Re} \sigma_{xx}^{\text{inter}}(\Omega) + \operatorname{Re} \sigma_{xx}^{\text{intra}}(\Omega)$ is shown in Figs. 7.9(a) and 7.9(b), respectively.

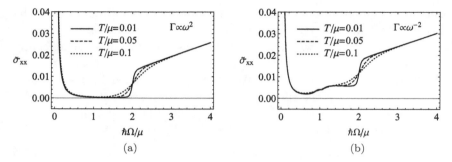

Fig. 7.9 The real part of the total optical conductivity $e^2 \mu \tilde{\sigma}_{xx}/(\hbar^2 v_F) = \text{Re}\,\sigma_{xx}(\Omega) = \text{Re}\,\sigma_{xx}^{\text{inter}}(\Omega) + \text{Re}\,\sigma_{xx}^{\text{intra}}(\Omega)$, where the interband and interband contributions are given in Eqs. (7.116) and (7.117). Panel (a) corresponds to $\Gamma(\omega) = 0.01\hbar^2\omega^2/\mu$ and panel (b) describes the case $\Gamma(\omega) = 0.01\mu^3/(\hbar\omega)^2$. In both panels $T/\mu = 0.01$ (solid lines), $T/\mu = 0.05$ (dashed lines), and $T/\mu = 0.1$ (dotted lines).

It is interesting that, in the case of local disorder with $\Gamma(\omega) \propto \omega^2$, the behavior of $\sigma(\Omega)$ is qualitatively the same as in the simplest model with $\Gamma(\omega) = const.$ Irrespective of the type of disorder, the optical conductivity has a characteristic step-like feature at $\hbar\Omega = 2\mu$. As was already mentioned earlier and illustrated in Fig. 7.8, the origin of this step is connected with the Pauli blocking of occupied states. Unsurprisingly, with increasing temperature and the strength of disorder potential, the step gets gradually washed away.

In the model with short-range disorder, one finds that the low-frequency ($\Omega \ll T$) optical conductivity has a square root dependence on Ω [547], i.e.,

$$\text{Re}\,[\sigma_{xx}(\Omega) - \sigma_{xx}(0)] \approx -\frac{e^2\hbar^3 v_F^3 \sqrt{\pi v_F}}{12\left(n_{\text{imp}}u_0^2\right)^{3/2} T} \sqrt{\Omega}. \tag{7.120}$$

In the opposite limit $\Omega \gg T$, one obtains [385]

$$\text{Re}\,\sigma_{xx}(\Omega) \approx \frac{e^2\Omega}{24\pi\hbar v_F}\left(1 - \frac{8n_{\text{imp}}u_0^2\Omega}{15\pi^2 v_F^3}\right). \tag{7.121}$$

Note that, for simplicity, in both cases the vertex corrections leading to a renormalization of the relaxation time $\tau \to \tau_{\text{tr}} = 3\tau/2$ were neglected. As is clear from Eq. (7.121), to leading order in a small disorder strength, the high-frequency optical conductivity is independent of disorder. By performing a renormalization group analysis [837], one could argue that the leading-order linear dependence of optical conductivity in Eq. (7.121) is a universal property, which is independent of the actual nature of point-like impurity scatterers.

Before concluding this subsection, it should be noted that the real part of optical conductivity satisfies the so-called f-sum rule.[t] If the width of the energy band (or the cutoff that defines the range of validity of a relativistic-like dispersion relation) $\hbar\Lambda$ is much larger than the chemical potential and temperature, then the sum rule can be written as follows:

$$\int_{+0}^{\Lambda} d\Omega \, \mathrm{Re} \left[\sigma_{xx}^{\mathrm{intra}}(\Omega) + \sigma_{xx}^{\mathrm{inter}}(\Omega) \right] = \frac{e^2 \Lambda^2}{48\pi v_F \hbar}. \tag{7.122}$$

Note that the right-hand side is independent of μ and T. This is a very powerful relation implying that the total spectral weight due to both intra- and interband processes is conserved. The utility of the sum rule can be already seen in the simplest case with the intraband optical conductivity given by the Drude expression (7.112). The corresponding intraband part of the spectral weight is

$$\int_{+0}^{\infty} d\Omega \, \mathrm{Re} \, \sigma_{xx}^{\mathrm{intra}}(\Omega) = \frac{e^2}{12\pi v_F \hbar^3} \left(\mu^2 + \frac{\pi^2 T^2}{3} \right). \tag{7.123}$$

Since this contribution grows with μ and T, the conservation of the total spectral weight implies that the interband contribution should decrease accordingly.

7.3.2 *Magneto-optical conductivity*

Similarly to the electron DC transport, the AC transport can be easily modified by an external magnetic field. It is natural, therefore, to discuss the key effects of magnetic field on the optical properties in Dirac and Weyl semimetals.

The magneto-optical conductivity of Dirac and Weyl semimetals was studied theoretically in [833, 835, 838, 841]. Here it is instructive to start a discussion with the diagonal components of magneto-optical conductivity. This will be a natural generalization of the results for the optical conductivity in Sec. 7.3.1. The off-diagonal components, which are crucial for the interaction of Weyl and Dirac semimetals with light, will be discussed in Sec. 7.3.4. Within the Kubo linear response theory, the magneto-optical

---

[t]The term "f-sum rule" is borrowed from atomic physics.

conductivity is given by [833]

$$\mathrm{Re}\,\sigma_{xx}(\Omega) = -\frac{e^2 \hbar v_F^2}{8l_B^2} \sum_{n=0}^{\infty} \int \frac{dk_z}{2\pi} \left\{ \left(1 - \frac{\hbar^2 v_F^2 k_z^2}{\epsilon_n \epsilon_{n+1}}\right) \delta\left(\hbar\Omega + \epsilon_n - \epsilon_{n+1}\right) \right.$$

$$\times \frac{f^{\mathrm{eq}}(\epsilon_n) - f^{\mathrm{eq}}(-\epsilon_n) - f^{\mathrm{eq}}(\epsilon_{n+1}) + f^{\mathrm{eq}}(-\epsilon_{n+1})}{\epsilon_n - \epsilon_{n+1}} + \left(1 + \frac{\hbar^2 v_F^2 k_z^2}{\epsilon_n \epsilon_{n+1}}\right)$$

$$\left. \times \delta\left(\hbar\Omega - \epsilon_n - \epsilon_{n+1}\right) \frac{f^{\mathrm{eq}}(\epsilon_n) - f^{\mathrm{eq}}(-\epsilon_n) + f^{\mathrm{eq}}(\epsilon_{n+1}) - f^{\mathrm{eq}}(-\epsilon_{n+1})}{\epsilon_n + \epsilon_{n+1}} \right\},$$

$$(7.124)$$

where ϵ_n are the Landau-level energies defined in Eqs. (3.131) and (3.132).

Note that the presence of the δ-functions in the optical conductivity (7.124) enforces the selection rules for the optical transitions, which can occur only between Landau levels with indices differing by ± 1. This applies to both intra- and interband transitions. As usual, in order to account for disorder effects, one can simply replace the δ-functions with the Lorentzian distributions of nonzero width $\Gamma(\epsilon)$, i.e.,

$$\delta(\epsilon) \to \frac{1}{\pi} \frac{\Gamma(\epsilon)}{\Gamma^2(\epsilon) + \epsilon^2}, \qquad (7.125)$$

where the dependence of quasiparticle width $\Gamma(\epsilon)$ on energy is determined by the type of disorder (see Sec. 7.1).

The frequency dependence of the magneto-optical conductivity given by Eq. (7.124) is shown in Fig. 7.10 for a few values of the electric chemical potential. As one can see, the real part of conductivity contains a series of sharp peaks added on top of a linearly growing function given in Eq. (7.111). At $\mu = 0$, for example, the peaks occur at $\Omega l_B/v_F = \sqrt{2(n+1)} + \sqrt{2n}$, which correspond to the onset of new interband transitions. As is clear, the peak spacing scales as $\sqrt{B}$.[u] Because of the 3D nature of Weyl and Dirac

---

[u]As one might expect, in multi-Weyl semimetals, the corresponding magneto-optical conductivity becomes anisotropic and acquires a different scaling relation with an external magnetic field. For example, when the magnetic field is along the direction of the linear dispersion relation, the peak separation scales as $B^{n_w/2}$ [839], where n_w is the topological charge of a Weyl node. In addition, while the longitudinal magneto-optical conductivity for the directions of the nonlinear dispersion relation shows peaks on top of a linear growth, the peaks for the conductivity in the direction of the linear dispersion occur on top of a constant background.

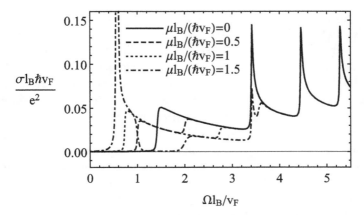

Fig. 7.10 The real part of the total optical conductivity $\mathrm{Re}\,\sigma_{xx}(\Omega)$ given in Eq. (7.124), where electric chemical potential is $\mu l_{\mathrm{B}}/(\hbar v_F) = 0$ (solid lines), $\mu l_{\mathrm{B}}/(\hbar v_F) = 0.5$ (dashed lines), $\mu l_{\mathrm{B}}/(\hbar v_F) = 1$ (dotted lines), and $\mu l_{\mathrm{B}}/(\hbar v_F) = 1.5$ (dot-dashed lines). In addition, $\Gamma l_{\mathrm{B}}/(\hbar v_F) = T l_{\mathrm{B}}/(\hbar v_F) = 0.01$.

semimetals, the Landau levels are not discrete like in graphene but dispersive. This is also reflected in a rather rich structure of magneto-optical conductivity. As a chemical potential is increased, the peaks at low energies disappear and the optical spectral weight is transferred into an intraband peak at low frequency $\Omega l_{\mathrm{B}}/v_F \propto \sqrt{2(n+1)} - \sqrt{2n}$. For example, the peak in Fig. 7.10 originates from transitions between the first and second Landau levels of the same band. The redistribution is particularly strong when μ coincides with the energies of Landau levels (e.g., the dot-dashed line in Fig. 7.10). It is worth noting that an additional step-like feature for $\mu l_{\mathrm{B}}/(\hbar v_F) < \sqrt{2}$ (see dashed and dotted lines in Fig. 7.10) is related to the transition between the zero and first Landau levels. It disappears above $\mu l_{\mathrm{B}}/(\hbar v_F) = \sqrt{2}$, where the optical spectral weight is completely transferred into the intraband peak.

Magneto-optical conductivity can be also used as a tool for detecting qualitative effects due to the chiral anomaly [835]. As we discussed in Sec. 6.2, the latter can be activated by applying parallel electric and magnetic fields that induce a steady state with a nonzero chiral chemical potential $\mu_5 \propto (\mathbf{E} \cdot \mathbf{B})$. For a minimal model with a single pair of Weyl nodes, this means that the nodes of opposite chirality should have different effective chemical potentials $\mu_{\pm} = \mu \pm \mu_5$. Such a splitting, in turn, should affect the characteristic step-like feature of optical conductivity $\mathrm{Re}\,\sigma_{xx}(\Omega)$

at $\hbar\Omega = 2\mu$, see Eq. (7.111) and Fig. 7.9, that originates from the Pauli blockade. Indeed, since the optical conductivity is an additive sum of contributions from all Weyl nodes, the corresponding feature should split into two smaller steps at $\hbar\Omega_{\pm} = 2(\mu \pm \mu_5)$. In principle, such a two-step behavior can be resolved in the regime of sufficiently low temperatures and clean samples. Its experimental observation would provide a direct evidence of the chiral anomaly manifestation in optical processes.

Concerning the general features of longitudinal magneto-optical conductivity, one can argue that a magnetic field should redistribute the spectral weight from higher to lower frequencies [841]. This conclusion follows from a combination of two effects. On the one hand, because of the "positive" magnetoconductivity discussed in Sec. 6.2, the conductivity is enhanced at small Ω due to the chiral anomaly. On the other hand, according to the sum rule in Eq. (7.122), the total spectral weight must remain constant. Then, using the same notation as in Sec. 6.5, the overall frequency dependence of the real part of optical conductivity should take the following form [841]:

$$\operatorname{Re}\sigma_{xx}(\Omega) = \frac{\sigma(0)}{1 + \Omega^2\tau^2}\left[1 + \left(\frac{L_{\mathrm{D}}}{L_5}\right)^2\frac{1 - \Omega^2\tau\tau_5}{1 + \Omega^2\tau_5^2}\right], \qquad (7.126)$$

where τ and τ_5 are the electric and chiral charge relaxation times as well as $L_{\mathrm{D}} = \sqrt{D\tau_5}$ is the chiral charge diffusion length. The extra length scale L_5 is determined by the density of states at the Fermi surface, the mean free path ℓ_{mfp}, and is inversely proportional to the magnetic field, i.e., $L_5 \propto \mu^2\ell_{\mathrm{mfp}}/|eB|$. While the prefactor in Eq. (7.126) is nothing else as the usual Drude conductivity, the second term in the square brackets describes the chiral anomaly effects. As expected, this term is positive at small and negative at large frequencies.

7.3.3   *Optical activity*

The phenomenon of natural optical activity was discovered by François Arago in 1811, who observed a rotation of the plane of polarization of light in quartz crystals. This is, however, only one example of a large family of optical activity phenomena, which are characterized by different interaction of right- and left-handed polarized light with matter. Among them, of particular interest are the effects related to magnetic fields, e.g., the magneto-optical activity discovered by Michael Faraday in 1846, as well as various magnetochiral effects that occur in chiral media[v] in the presence of a magnetic field. The classical examples of the optical and magnetic optical activity include the circular dichroism and the Faraday effect. The former is a rotation of the plane of polarization caused by different absorption and refraction of left- and right-handed polarized light on chiral molecules, i.e., molecules that break mirror symmetry. On the other hand, the Faraday effect is a rotation of the plane of polarization caused by the magnetic field (which breaks $\mathcal{T}$ symmetry). The magnetochiral effect is more subtle as it requires breaking of both mirror and $\mathcal{T}$ symmetries.

From a theoretical point of view, optically active media are characterized [866–868] by an additional gyrotropic contribution to the dielectric tensor $\varepsilon_{ij}(\Omega, \mathbf{k}) = \varepsilon_{ij}(\Omega, \mathbf{0}) + i\gamma_{ijl}k_l$, where γ_{ijl} is the third-rank gyrotropic tensor and $\mathbf{k}$ is the wave vector. The gyrotropic tensor is antisymmetric in indices i and j, i.e., $\gamma_{ijl} = -\gamma_{jil}$. In the absence of dissipation, it is also real: $\gamma_{ijl}^* = \gamma_{ijl}$. Additional restrictions on the tensor structure could come from the crystal symmetries of solids. In particular, γ_{ijl} is nontrivial only in noncentrosymmetric materials. Isotropy implies a particularly simple expression for the gyrotropic tensor: $\gamma_{ijl} = c\gamma\epsilon_{ijl}/\Omega$. Then the corresponding refractive indices for the right- and left-polarized light are equal to $n_{\pm}^2 = n_0^2 \pm \gamma n_0$, where $\varepsilon_{ij}(\Omega, \mathbf{0}) = \varepsilon_e\delta_{ij}$ and $n_0 = \sqrt{\varepsilon_e}$. Since the linearly polarized light can be represented as a combination of right- and left-handed circular modes with equal amplitudes, the difference in the two refractive indices gives rise to a rotation of the polarization plane. As is easy to check, the resulting rotation angle per unit distance is $\Omega\gamma/(2c)$.

By using various approaches, including the Kubo formalism and the semiclassical kinetic theory, the gyrotropic tensor in Weyl semimetals was calculated in [679, 680, 843, 869, 870]. Among other effects, it was found that the dynamic CME can play an important role in optical activity. The chiral anomaly, however, is not a necessary ingredient for producing a nontrivial gyrotropic tensor. The latter could originate from the orbital magnetization effects (see Sec. 6.6.2) [679]. An effective medium theory for electromagnetic wave propagation through materials with a dynamic chiral

---

[v]Here the chiral media include not only Dirac and Weyl semimetals but also other materials with a given handedness. The latter could be manifested by chiral molecular (or macroscopic) structures that do not possess mirror symmetry. For example, dextrose and levulose, which are isomers of glucose, are "chiral" materials that cause the plane of polarization of light to rotate either clockwise (dextrose) or counterclockwise (levulose).

magnetic effect was developed in [871]. In addition, disorder-induced corrections were also discussed in the latter study.

The chiral anomaly effect that becomes significant in the presence of an external magnetic field is only one of several possible sources of optical activity in Weyl semimetals. Another interesting possibility can be realized in Weyl semimetals with broken $\mathcal{T}$ symmetry. Recall that such semimetals are characterized by a nonzero separation between the Weyl nodes, determined by the chiral shift **b**. Since the chiral shift is odd under $\mathcal{T}$ symmetry, it can produce an unusual optical response, which is somewhat similar to the magnetochiral effect, albeit exists in the absence of any external magnetic field.

A dynamical version of the anomalous Hall effect results in an optical activity that can be observed as the birefringence of transmitted light [663]. Similarly to the anomalous Hall effect in the DC regime (see Sec. 6.3) this effect is triggered by the chiral shift that causes anomalous currents and transforms a linearly polarized light entering Weyl semimetal into an elliptically polarized light.[w]

The anomalous Hall effect and the Fermi arc surface states play an important role in the Kerr and Faraday effects, as well as in the linear dichroism [845].[x] A comprehensive analysis of the optical response in Weyl semimetals accounting for both surface and bulk contributions was performed in [872]. It was suggested, in particular, that a lot of information about the electronic structure of Weyl semimetals, including the electric chemical potential, the separation between Weyl nodes, the properties of surface states, etc. can be extracted from optical measurements. The corresponding calculations were done by using a two-band model similar to that in Sec. 5.3, where the chiral shift and the Fermi arcs naturally appear. The Faraday and Kerr rotations are present only for surfaces where the Fermi arcs are absent when electromagnetic waves propagate along the gyrotropy axis defined by the chiral shift [845, 873]. Otherwise, there is a magnetic linear dichroism signal.

Let us start our discussion of the optical activity effects in Weyl semimetals in the absence of an external magnetic field. Recall in this

---

[w] Note that the presence of a nonzero electric chemical potential can hinder most effects related to the propagation of light through Weyl semimetals. Therefore, the effects discussed in this subsection are valid only at $\mu \to 0$. Some predictions may also survive in sufficiently thin samples.

[x] The linear dichroism of light is related to the dependence of absorption on the orientation of polarization plane.

connection that, similarly to a magnetic field, a nonzero chiral shift **b** breaks $\mathcal{T}$ symmetry and causes the anomalous Hall effect. The latter can be taken into account via the following corrections to the off-diagonal conductivity [840, 845] (cf. with Eq. (6.65)):

$$\mathrm{Re}\,\sigma_{xy}(\Omega) = \frac{e^2 b}{2\pi^2 \hbar} + \frac{e^2 b}{24\pi^2 \hbar v_F^2} \frac{\Omega^2}{\Lambda^2 - b^2}, \qquad (7.127)$$

where $\mathbf{b} = b\hat{\mathbf{z}}$ and Λ is a momentum cutoff of the linearized model. The real and imaginary interband parts of σ_{xx} are given by Eqs. (7.111) and (7.114), respectively.

To describe the underlying physics of the *anomalous* Kerr and Faraday effects,[y] let us consider a Weyl semimetal in the form of a slab of thickness L. Even for such a simple geometry, qualitatively different regimes can be realized depending on the thickness of slab.

One limiting case is realized when the slab is an ultra-thin film with the thickness less than the wavelength of electromagnetic radiation, i.e., $L \lesssim 2\pi c/\Omega$. In this case, the interaction of an electromagnetic wave with the film can be captured simply by using an appropriate boundary condition between the two topologically trivial (vacuum) regions on the opposite sides of the slab (labeled 1 and 2): $(\mathbf{B}_1 - \mathbf{B}_2) \times \hat{\mathbf{z}} = \mathbf{J}_{\mathrm{film}}$. When there are no Fermi arcs on the semimetal surface (i.e., in the $x-y$ plane), the main contribution to current $\mathbf{J}_{\mathrm{film}}$ comes from the anomalous Hall effect represented by the terms $\propto b$ in conductivity (7.127). Then, by matching the incident, reflected, and transmitted electromagnetic waves, the *anomalous* Kerr and Faraday effects can be derived. By omitting the details, one obtains the following explicit expressions for the anomalous Kerr and Faraday angles [845]:

$$\tan\theta_K = -\frac{36\pi\alpha_0 v_F^2 bL}{\pi^2 \alpha_0^2 \Omega^2 L^2 + 6\pi^2 \alpha v_F \Omega L + 36\alpha_0^2 v_F^2 b^2 L^2}, \qquad (7.128)$$

$$\tan\theta_F = \frac{6\alpha_0 v_F bL}{6\pi v_F + \alpha_0 \Omega L}, \qquad (7.129)$$

where $\alpha_0 = e^2/(\hbar c)$ is the fine structure constant. By assuming that $\Omega \ll v_F \Lambda$, here we took only the first term in Eq. (7.127) into account. Since the results in Eqs. (7.128) and (7.129) are proportional to the chiral

---

[y]Recall that the usual Kerr and Faraday effects describe a rotation of the plane of polarization of reflected and transmitted light, respectively, in the presence of a magnetic field.

shift b, it is clear that both effects are examples of anomalous optical activity caused by a nontrivial band topology of Weyl semimetals. It is also interesting to note that the value of effects could be rather large. Indeed, an order of magnitude estimate reveals that the rotation angle can easily reach values of about $\pi/2$ for a micrometer-thick film [845, 872].

For a sufficiently thick Weyl semimetal, which could be modeled as a semi-infinite slab, the chiral shift gives rise to the birefringence [845, 872, 873]

$$n_{\pm}^2 = \varepsilon(\Omega) \pm \frac{2e^2 b}{\pi \hbar \Omega}, \tag{7.130}$$

where $\varepsilon(\Omega) = \varepsilon_{\mathrm{e}} + 4\pi i \sigma_{xx}(\Omega)/\Omega$. In this case, the anomalous Kerr angle is given by

$$\theta_{\mathrm{K}} \simeq \frac{\alpha_0^2 c^2 b}{3\pi v_F \Omega \left\{ \mathrm{Re}[\varepsilon(\Omega)] \right\}^{3/2} \left\{ \mathrm{Re}[\varepsilon(\Omega)] - 1 \right\}}. \tag{7.131}$$

This result could be easily understood by noting a similarity of the Chern–Simons or, equivalently, the anomalous Hall effect current $\propto [\mathbf{b} \times \mathbf{E}]$ to the gyrotropic term in ferromagnetic materials [866, 867].

Let us now briefly discuss the role of the Fermi arcs on the Weyl semimetal surfaces. Because of the anisotropy associated with the Fermi arc states, the absorption of light depends on the orientation of its plane of polarization with respect to the direction of the chiral shift $\mathbf{b}$. Indeed, the refractive indices for the light polarized parallel and perpendicular to $\mathbf{b}$ are given by $n_{\parallel}^2 = \epsilon(\Omega)$ and $n_{\perp}^2 \simeq \epsilon(\Omega) - [4\pi \mathrm{Re}\, \sigma_{xy}(\Omega)]^2 / [\Omega^2 \epsilon(\Omega)]$, respectively. Since the electric and magnetic fields in the incident electromagnetic wave lie in the $y - z$ plane (assuming that $\mathbf{b} \parallel \hat{\mathbf{z}}$), the off-diagonal components of the dielectric permittivity ε_{xy} can sustain neither Kerr nor Faraday effects. For $\mathbf{E} \parallel \hat{\mathbf{z}}$, there will be no rotation of $\mathbf{E}$ because the Chern–Simons terms vanish in this case. When $\mathbf{E} \parallel \hat{\mathbf{y}}$, the longitudinal, i.e., normal to the surface, component of the electric field may develop due to the nontrivial topology of Weyl semimetals. The longitudinal component can be probed via the anomalous magnetic linear dichroism signal quantified by $\mathrm{Im}(n_{\perp} - n_{\parallel}) \propto b^2/\Omega^2$. An interplay of the bulk and surface states in the transmission, reflection, and polarization of the corresponding electromagnetic waves is discussed in more detail in [872].

In connection to the optical properties caused by the nontrivial topology in Weyl semimetals, there is also an interesting possibility of an unconventional "negative" refraction in a narrow window of frequencies [873]. In such a regime, refracted light bends in the direction opposite to that in conventional materials. In addition, Weyl semimetals might be suitable for the creation of nonreciprocal waveguides, where light propagates only in one direction and the anomalous Hall effect plays the key role [874].

Optical activity of Dirac and Weyl semimetals is also nontrivial in the presence of external magnetic fields. The corresponding analysis in a two-band model was performed by using the kinetic theory approach in [875]. It was found that the simultaneous presence of a magnetic field and the chiral shift might produce magnetochiral terms in both longitudinal and transverse electrical conductivities. The effect is proportional to $\tau \mu \, (\mathbf{b} \cdot \mathbf{B})$ and resembles the electric magnetochiral effect in generic chiral conductors of specific handedness [876].

The chiral anomaly allows for another nontrivial effect induced by parallel electric and magnetic fields [679, 680, 843] and described by the gyrotropic tensor $\gamma_{ijl} = c\gamma(\Omega)\epsilon_{ijl}/\Omega$, where $\gamma(\Omega) \propto i\tau\tau_5^{1/3} |(\mathbf{E} \cdot \mathbf{B})|^{1/3}$. As discussed in Sec. 6.2, the effect of $(\mathbf{E} \cdot \mathbf{B})$ is to produce an out-of-equilibrium steady state with a nonzero chiral chemical potential μ_5. The value of μ_5 is determined by the combined effect of chiral charge pumping via the chiral anomaly and internode relaxation processes with a characteristic timescale τ_5. The imaginary part of γ gives rise to circular dichroism, i.e., the conversion of linearly polarized light into an elliptically polarized one. Indeed, since refractive indices of the right- and left-handed modes are different, $n_\pm^2 = \varepsilon(\Omega) \pm ck\gamma(\Omega)/\Omega$, the ellipticity of transmitted light is given by

$$\tan\theta \simeq \frac{L}{2}\mathrm{Im}\,\gamma(\Omega), \qquad (7.132)$$

where L is the thickness of a Weyl semimetal slab and Ω is the frequency of light. However, a nontrivial gyrotropic tensor is not a hallmark feature of Weyl semimetals or the chiral anomaly [679, 680, 870]. The corresponding dynamical chiral or gyrotropic magnetic effect originates primarily from the intrinsic orbital magnetic moment and may exist in other materials.

In a strong magnetic field, a semiclassical analysis becomes unreliable. Instead, one can utilize the Kubo formalism to calculate the off-diagonal components of optical conductivity. By following the same steps as in the calculation of longitudinal conductivity in Sec. 7.3.2, one can also derive the expression for $\sigma_{xy}(\Omega)$. The explicit form of the imaginary part $\mathrm{Im}\,\sigma_{xy}(\Omega)$, in particular, is similar to the right-hand side of Eq. (7.124) up to the following replacements: $f^{\mathrm{eq}}(\epsilon_n) \to -f^{\mathrm{eq}}(\epsilon_n)$ and $f^{\mathrm{eq}}(\epsilon_{n+1}) \to -f^{\mathrm{eq}}(\epsilon_{n+1})$, as well as change the overall sign in front of the second term in the curly brackets [833], i.e.,

$$
\text{Im}\,\sigma_{xy}(\Omega) = -\frac{e^2 \hbar v_F^2}{8 l_B^2} \sum_{n=0}^{\infty} \int \frac{dk_z}{2\pi} \left\{ \left(1 - \frac{\hbar^2 v_F^2 k_z^2}{\epsilon_n \epsilon_{n+1}}\right) \delta\left(\hbar\Omega + \epsilon_n - \epsilon_{n+1}\right) \right.
$$
$$
\times \frac{f^{\text{eq}}(\epsilon_{n+1}) + f^{\text{eq}}(-\epsilon_{n+1}) - f^{\text{eq}}(\epsilon_n) - f^{\text{eq}}(-\epsilon_n)}{\epsilon_n - \epsilon_{n+1}} - \left(1 + \frac{\hbar^2 v_F^2 k_z^2}{\epsilon_n \epsilon_{n+1}}\right)
$$
$$
\left. \times \delta\left(\hbar\Omega - \epsilon_n - \epsilon_{n+1}\right) \frac{f^{\text{eq}}(\epsilon_{n+1}) + f^{\text{eq}}(-\epsilon_{n+1}) - f^{\text{eq}}(\epsilon_n) - f^{\text{eq}}(-\epsilon_n)}{\epsilon_n + \epsilon_{n+1}} \right\}.
$$

$$(7.133)$$

In order to determine the rotation of the polarization plane for the transmitted (Faraday effect) or reflected (Kerr effect) light, it is convenient to use the following combinations of conductivities: $\sigma_{\pm}(\Omega) = \sigma_{xx}(\Omega) \pm i\sigma_{xy}(\Omega)$. The frequency dependence of the real parts of $\sigma_{\pm}(\Omega)$ is illustrated in Fig. 7.11 for $\mu l_B/(\hbar v_F) = 1$, which corresponds to a strong magnetic field limit. As one can see, the behavior of σ_+ and σ_- is very different in the region of small Ω. Indeed, the lowest peak in σ_- is enhanced, while the corresponding peak in σ_+ is absent. This translates into different propagation of electromagnetic light with left- and right-handed polarizations.

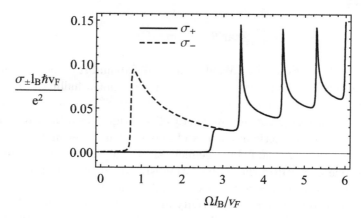

Fig. 7.11   The real part of $\sigma_{\pm}(\Omega) = \sigma_{xx}(\Omega) \pm i\sigma_{xy}(\Omega)$ for $\mu l_B/(\hbar v_F) = 1$, where $\sigma_{xx}(\Omega)$ and $\sigma_{xy}(\Omega)$ is given in Eqs. (7.124) and (7.133), respectively. The solid line corresponds to σ_+ and the dashed one to σ_-. In addition, $\Gamma l_B/(\hbar v_F) = T l_B/(\hbar v_F) = 0.01$.

7.3.4 *Second-order optical effects*

Nontrivial properties of Weyl and Dirac semimetals could be revealed also in the second-order optical effects. The latter include the *photovoltaic effect* (PVE), the *photogalvanic effect* (PGE), and the *second harmonic generation* (SHG). In essence, the PVE is the generation of an electric current in a material irradiated with electromagnetic waves. It was discovered by Edmond Becquerel in 1839, at the age of 19, when he was experimenting in his father's laboratory. Nowadays, the PVE is the cornerstone of modern photodiodes and solar cells. The PGE is one of the photovoltaic effects where the generation of electric current is sensitive to the polarization of electromagnetic waves [868, 877]. The general expression for the photocurrent can be written as follows:

$$\partial_t J_j = \sum_{l=x,y,z} \beta_{jl}\, i\, [\mathbf{E} \times \mathbf{E}^*]_l + \sum_{l,m=x,y,z} \chi_{jlm} \frac{E_l E_m^* + E_m E_l^*}{2}$$
$$+ \sum_{l,m,n=x,y,z} \zeta_{jlmn} k_l E_m E_n^*. \tag{7.134}$$

A nonzero tensor β_{jl} appears in gyrotropic materials with broken inversion symmetry and describes the *circular* PGE. Such an effect is realized when a material is illuminated with a circularly polarized light. The direction of the corresponding photocurrent depends on the polarization of light. The second term in Eq. (7.134) describes the *linear* PGE.[z] Finally, the third term captures the photon drag effect, which is associated with a momentum transfer from photons to electrons.

The SHG, as suggested by its name, is the generation of oscillating electric current with the double frequency of a driving force. The intraband contribution to the SHG is similar to the linear PGE described by the second term in Eq. (7.134), but $E_l E_m^*$ should be replaced with $E_l E_m$.

The generation of photocurrents in Dirac and Weyl semimetals in the absence of external fields was extensively discussed both theoretically [838, 878–890] and experimentally [855, 859–861, 863, 864]. Since Dirac and Weyl semimetals possess gapless energy spectrum, this means that, in principle, photons of an arbitrary long wavelength could be used to induce a photocurrent. As will be also discussed in Sec. 7.3.5, this

---

[z]The part of the linear PGE current originating exclusively from intrinsic properties of the band structure is often called the *shift current*.

makes Dirac and Weyl semimetals potentially good candidates for infrared detectors.

Let us consider the underlying physics of the circular PGE in Weyl semimetals with broken $\mathcal{P}$ symmetry [882, 883]. In such a case, the tensor β_{ij} equals

$$\beta_{ij} = -\frac{\pi e^3}{\hbar} \int \frac{d^3 k}{(2\pi)^3} \left(\partial_{k_i} \epsilon_{\mathbf{k}} \right) \Omega_j \delta \left(\hbar\omega - \epsilon_{\mathbf{k}} \right), \tag{7.135}$$

where $\mathbf{\Omega}$ is the Berry curvature. By using an isotropic model of a Weyl semimetal with a monopole-like Berry curvature $\Omega_j = \lambda \hat{\mathbf{k}}/(2\hbar k^2)$, it is straightforward to obtain the following result for the trace of β_{ij}:

$$\mathrm{tr}\,(\beta) = -\frac{e^3}{4\pi\hbar^2} n_{\mathrm{w}}, \tag{7.136}$$

where n_{w} is the topological charge of a Weyl node.[aa] It is important to emphasize that this expression is valid only when one of the Weyl nodes is Pauli blocked, i.e., $-2(b_0 - \mu) < \hbar\omega < 2(b_0 + \mu)$, where $2b_0$ is the energy separation between the Weyl nodes. In the case of circularly polarized light, $[\mathbf{E} \times \mathbf{E}^*] = iE^2 \hat{\mathbf{n}}$, where $\hat{\mathbf{n}}$ is the unit vector normal to the polarization plane. Therefore, the final result for the PGE current of a given handedness $(\lambda = \pm)$ reads

$$\partial_t \mathbf{J}_\lambda = -\lambda i \frac{e^3}{12\pi\hbar^2} [\mathbf{E} \times \mathbf{E}^*], \tag{7.137}$$

where we assumed an isotropic case with $\beta_{jj} = \mathrm{tr}(\beta)/3$. One can rewrite the amplitude of the electromagnetic wave in terms of the light intensity $I = cE^2/(8\pi)$ as follows:

$$\partial_t J_{\mathrm{R,L}} = \pm \frac{2e^3 I}{c\hbar^2}, \tag{7.138}$$

where, by definition, $J_{R,L} = \pm (J_{\pm,x} + J_{\pm,y} + J_{\pm,z})$. This result is quite remarkable since the prefactor is composed of only fundamental constants e, c, and $\hbar$. It turns out, however, that this property could be modified, especially at higher frequencies, by other effects, e.g., due to higher-order corrections to the low-energy Hamiltonian and interaction effects.

[aa] Recall that $n_{\mathrm{w}} = \lambda$ for usual Weyl semimetals with the chirality of Weyl nodes λ and $n_{\mathrm{w}} = 2\lambda, 3\lambda$ for double- and triple-Weyl semimetals, respectively.

It is natural to expect that an external magnetic field affects the nonlinear optical response. Theoretically, the generation of a photocurrent in Dirac and Weyl semimetals in a magnetic field can be studied by using a kinetic theory approach [838, 844, 875, 884, 886, 887] or the Floquet formalism [838]. In an isotropic model of a Weyl semimetal, it was found that the SHG is linear in $\mathbf{B}$, i.e., $J^{\mathrm{SHG}}(\Omega) \propto B|\mathbf{E}|^2/(\mu\Omega^2)$. Since the corresponding current has roots in the chiral anomaly, the overall prefactor depends on the orientation of a magnetic field. Experimentally, it was suggested that the SHG in a magnetic field could be manifested in a nonlinear Kerr rotation.

The PVE was investigated theoretically in [886, 887] in the limit of weak and strong magnetic fields.[bb] It was found that important ingredients allowing for a nontrivial PVE current are broken spatial inversion and particle–hole symmetries. When a magnetic field is weak, the qualitative analysis could be easily done in the chiral kinetic theory. The corresponding PVE current is given by an expression resembling that of the SHG, i.e., $J^{\mathrm{PVE}}(\Omega) \propto B|\mathbf{E}|^2/\Omega^2$. In a strong field [886], where the Landau quantization is important, the current comes primarily from transitions between the zeroth and first Landau levels. The overall form of the expression remains similar to that in the weak field case.

Before concluding the discussion of the second-order optical effects, it is instructing to address an interplay between the nontrivial topology of Weyl semimetals and the effects of a magnetic field. By using the kinetic theory approach, it was argued in [844, 875] that the second-order nonlinear optical response should include the PGE and the SHG currents proportional to both magnetic field and chiral shift. Moreover, these currents can be generated even in Dirac semimetals such as Cd_3As_2 and Na_3Bi with a pair of Dirac nodes, described by a model similar to that in Sec. 3.6.1. In order to properly take into account the separation between the Weyl nodes in the kinetic approach, a two-band model similar to that in Sec. 5.3 was employed in [844]. It was found that the final result for the photogalvanic current contains the following terms:

$$J_j^{\mathrm{PGE}} \overset{\mathbf{b}\|\mathbf{B}}{\propto} (\mathbf{B}\cdot\mathbf{b}) \sum_{l,m,n=x,y,z} \epsilon_{jlm} M_{ln} E_m^* E_n \frac{\tau}{\mu b_z^2} + h.c., \qquad (7.139)$$

$$J_j^{\mathrm{PGE}} \overset{\mathbf{b}\perp\mathbf{B}}{\propto} (\mathbf{B}\cdot\mathbf{E}) [\mathbf{b}\times\mathbf{E}^*]_j \frac{\tau}{\mu b_z^2} + h.c., \qquad (7.140)$$

[bb]In [886, 887] the PVE was called the helical magnetic and magneto-gyrotropic photogalvanic effects, respectively.

where $M_{ln} = \text{diag}(-1, -1, 2)$, $\mathbf{B} \parallel \hat{\mathbf{z}}$, and the chiral shift is either parallel or perpendicular to the magnetic field. The same expressions with the replacement $\mathbf{E}^* \to \mathbf{E}$ are also valid for the SHG. It can be easily verified that when the separation between Weyl nodes is large, the deviations from the linearized dispersion relation and the ideal model of a Weyl semimetal with two independent nodes are weak. Therefore, the PGE currents in Eqs. (7.139) and (7.140) vanish. It is clear, therefore, that a simple linearized model of a Weyl semimetal might not be sufficient for the description of topology effects in optical response.

7.3.5 *Experimental results*

Let us now turn to the experimental results concerning the optical properties in Dirac and Weyl semimetals. Since the optical response is one of the most direct ways of testing electron properties of Dirac and Weyl semimetals, it should not be surprising that there is already a trove of data collected in numerous experiments [846–850, 850–864, 891].

We start by reviewing the case of Dirac semimetals. In agreement with the theoretical result in Eq. (7.111), a linear dependence of optical conductivity on frequency was observed experimentally in the Dirac semimetal ZrTe$_5$ [846].[cc] The corresponding optical conductivity derived from the reflectivity is shown in Fig. 7.12(a) for a few values of temperature. It was proposed that the large peak around 4000 cm^{-1} stems from a Van Hove singularity in the joint DOS. In addition, the characteristic relativistic-like dependence of the plasma frequency $\Omega_e \propto \sqrt{\mu^2 + \pi^2 T^2/3}$ (see Eq. (8.59)) on temperature T was observed. It is interesting that the optical conductivity in the Dirac semimetal Cd$_3$As$_2$ might have a nonlinear dependence on frequency [848], which could probably stem from interband effects. At sufficiently large energies, the Dirac semimetal Cd$_3$As$_2$ might also support massless Kane electrons rather than symmetry-protected 3D Dirac quasiparticles [848, 893].

In the presence of a background magnetic field, the formation of relativistic-like Landau levels in the Dirac semimetal ZrTe$_5$ was experimentally reported in [847]. The experimental data for the reflectivity R presented in Fig. 7.12(b) is in good agreement with the theoretical

---

[cc]It is fair to mention, however, that the Dirac nature of quasiparticles in ZrTe$_5$ is debated. For example, the optical conductivity results in [892] suggest that the energy spectrum can be linear in two directions and quadratic in the third one.

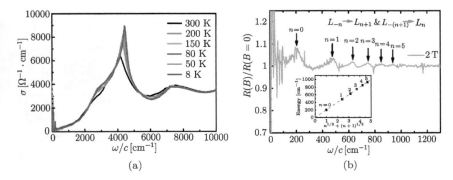

Fig. 7.12 (a) The optical conductivity in the Dirac semimetal ZrTe5 as a function of frequency at a few values of temperature. In agreement with the theoretical results in Sec. 7.3.1, the optical conductivity shows a characteristic linear dependence on frequency. (Adapted from [846].) (b) The frequency dependence of the relative reflectivity $R(B)/R(0)$ at $B = 2$ T. Numbers $n = 0, \ldots, 5$ label the positions of peaks in the reflectivity. The inset shows the dependence of transition energies between the Landau levels on $\sqrt{n} + \sqrt{n+1}$, which agrees well with the theoretical predictions in Sec. 7.3.2. (Adapted from [847].)

analysis of the optical conductivity in Sec. 7.3.2.[dd] Indeed, the data confirm that peaks in the reflectivity occur at frequencies $\propto \sqrt{n+1} + \sqrt{n}$, which can be interpreted as a signature of a relativistic-like linear dispersion relation.

Let us now turn to the case of Weyl semimetals. Among the first experimental studies, reporting a linear dependence of optical conductivity on frequency were [853, 854]. They found that pyrochlores $Nd_2Ir_{2-2x}Rh_{2x}O_7$ and $Eu_2Ir_2O_7$, which are Weyl semimetal candidates, indeed show such a linear dependence at certain values of doping and temperature. Later an optical response linear in frequency was reported in other materials, including transition metal monopnictides. As a typical example of the experimental dependence, the optical conductivity for the Weyl semimetal NbP from [858] is presented in Fig. 7.13(a). While the low-frequency behavior is complicated due to several different absorption mechanisms, there is a clear linear dependence at intermediate frequencies. As in ZrTe5, the optical conductivity in NbP also has a sharp peak feature that is associated most likely with the Van Hove singularity in the DOS. The results for the

---

[dd]By definition, $R = [(1 - \mathrm{Re}\, n_{\mathrm{ref}})^2 + (\mathrm{Im}\, n_{\mathrm{ref}})^2]/[(1 + \mathrm{Re}\, n_{\mathrm{ref}})^2 + (\mathrm{Im}\, n_{\mathrm{ref}})^2]$ is the reflectivity given in terms of the complex refractive index n_{ref}.

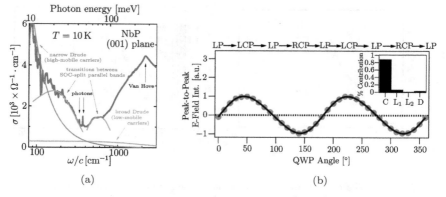

Fig. 7.13   (a) The optical conductivity in the Weyl semimetal NbP as a function of frequency. A few features related to different absorption mechanisms are identified. Among them is the Van Hove peak at large frequencies, linear optical conductivity, phonons, transitions between the bands split by a spin–orbit coupling (SOC), and the Drude modes. (Adapted from [858].) (b) Peak-to-peak electric field intensity plotted in TaAs as a function of the quarter wave plate (QWP) angles. Angles 0°, ±45°, and ±22.5° correspond to the linear, right/left circular, and right/left elliptic polarizations, respectively. Red dots correspond to the experimental data and the black line is a fit given in Eq. (7.141). The inset shows relative contributions of the terms in Eq. (7.141). (Adapted from [863])

optical conductivity in TaAs were reported in [856, 857] and appear to be qualitatively the same as in NbP. The conductivity has a characteristic linear dependence on frequency and a narrow Drude peak. The latter can be easily explained by a high mobility of charge carriers in Weyl semimetals. In addition, the Drude weight has a quadratic dependence on temperature, which agrees with the result in Eq. (7.123).

Let us also say a few words about the experimental evidence for the photogalvanic and second harmonic generation effects in Weyl semimetals [855, 859–861, 863, 864]. Overall, the experimental data agree qualitatively with the theoretical predictions considered in Sec. 7.3.4. Interestingly, Weyl semimetals could demonstrate a sufficiently strong nonlinear response. For example, in the configuration with the fields parallel to the polar axis, the SHG signal in TaAs [855, 862] was measured to be almost an order of magnitude larger than that in GaAs and ZnTe. It also exceeds the effect in $BaTiO_3$, $BiFeO_3$, and $LiNbO_3$ by two orders of magnitude. In TaAs, the linear PGE current (sometimes referred to as the bulk PGE current) is also large [861]. It exceeds the maximum PGE currents in $BaTiO_3$ by about an order of magnitude.

The optical response of Weyl semimetals to light waves of different polarizations was studied in several experiments [859, 860, 863, 864]. As was discussed in Sec. 7.3.4, the circular PGE currents change their sign when the helicity of the incoming light changes. The corresponding results for the $[1, -1, 0]$ crystallographic axis of TaAs obtained in [863] are shown in Fig. 7.13(b) and are fitted with the following expression for the photocurrent:

$$J(\theta) = C \sin(2\theta) + L_1 \sin(4\theta) + L_2 \cos(4\theta) + D. \tag{7.141}$$

Here θ is the quarter wave plate angle, which defines the initial pulse polarization. The helicity-dependent term is quantified by the coefficient C. The terms L_1 and L_2 describe small deviations from the ideal behavior. The coefficient D corresponds to a polarization-independent contribution. As is clear from the results in Fig. 7.13(b), the circular PGE dominates for a given crystallographic axis in TaAs. On the other hand, emission along $[1, 1, -1]$ crystallographic axis (noncentrosymmetric) does not depend on the polarization [863] and, most probably, is not related to the Weyl nature of quasiparticles in TaAs. Indeed, the corresponding effect could be explained by the fact that TaAs-like family of materials belongs to the group of polar metals.

The direct conversion of light into electric current is very important for novel sensors, including those able to detect infrared radiation. In addition, a sensitivity to polarization might open new opportunities for processing and transmitting information. Indeed, the photogalvanic effect in Weyl semimetals might allow for an efficient control of THz polarization-dependent photocurrents induced by infrared femtosecond laser pulses [863, 864]. In addition, Weyl semimetals could be utilized as a reliable platform for emitters of broadband elliptically polarized THz radiation [864].

Before finalizing this chapter, we present a few aspects of a novel transport regime where electrons are governed by hydrodynamics equations. While this field is new and rapidly developing, we find it useful to discuss its current status and future perspectives.

7.4 Hydrodynamic transport regime

7.4.1 *Historical remarks*

Hydrodynamics (or fluid dynamics) is a broad branch of physics that describes the motion of fluids. Its history dates back to ancient Greece. The mathematical foundations of classical hydrodynamics were laid down in the 17th–19th centuries. Generically, the hydrodynamic description applies to liquid and gaseous states of matter on sufficiently long space and time scales. It assumes that a local thermodynamic equilibrium is achieved everywhere in the system. This implies, in particular, that the range of validity of hydrodynamics is limited by the particle density and the rate of inter-particle collisions, among other things. In essence, the hydrodynamic equations describe the space–time evolution of conserved quantities, such as momentum, energy, particle number, electric charge, etc.

Usually, the Drude-like (diffusive) or ballistic regime of electronic transport is realized in metals and semimetals. However, as was first understood by Radii Gurzhi [240, 241], an unusual hydrodynamic regime of electron transport also could be potentially realized in clean crystals. This would require that the electron–electron scattering rate dominates over the electron–impurity and electron–phonon processes.[ee] In such a regime, electrons do not move independently but participate in a collective flow as a liquid. To describe such a flow, a new variable, i.e., the hydrodynamic flow velocity, is introduced. Its dynamics is governed by the Navier–Stokes equation, which is the core of any hydrodynamic approach.

The electron hydrodynamic regime could give rise to many interesting effects that differ qualitatively from the standard diffusive transport in the Drude regime. One of them is a nonmonotonic dependence of the resistivity on temperature, which is also known as the *Gurzhi effect* [240]. The other effects include a local electric current flow against the electric field, producing a "negative" resistance and vortices, and a nontrivial dependence of the resistivity on the system size, realizing the Poiseuille flow of the electron fluid.

Historically, a hydrodynamic electron flow was first observed experimentally in a 2D electron gas of high-mobility (Al, Ga)As heterostructures [764, 765]. In particular, it was confirmed that the resistivity decreased with temperature in agreement with Gurzhi's prediction [240]. Clear signatures of electron hydrodynamics were also observed in the ultra-pure 2D metal palladium cobaltate ($PdCoO_2$) [894], where the resistivity had a characteristic Poiseuille-like dependence on the channel size, as expected in the hydrodynamic regime [241]. Recently, the electron hydrodynamic regime

---

[ee]Since the electron–impurity interactions dominate at low temperature and the electron–phonon interactions become very important at high temperature, the realization of the electron hydrodynamic regime was believed to be very unlikely in usual metals.

was confirmed experimentally in graphene [261–267]. In this 2D Dirac semimetal, hydrodynamic effects cause a violation of the Wiedemann–Franz law and the Mott relation, give extra viscous contributions to the resistivity, as well as allow for higher-than-ballistic conduction in narrow constrictions. A more detailed discussion of hydrodynamic effects in graphene is given in Sec. 2.5.2.

Since the electron quasiparticles in Dirac and Weyl semimetals have relativistic-like dispersion relations, such materials could provide a unique possibility for testing relativistic-like hydrodynamic effects in condensed matter physics using tabletop experiments. Historically, relativistic hydrodynamics was formulated more than half a century ago (see, for example, [895]) and is widely used for the description of various phenomena in nuclear physics, astrophysics, and cosmology. Recently, the hydrodynamic framework was also generalized to the case of plasmas made of chiral fermions [896–899]. In these systems, the chiral charge is included as an additional degree of freedom, whose conservation is violated only by the chiral anomaly. This extension is very important for the hydrodynamic transport in Dirac and Weyl semimetals, where chirality plays a crucial role. Chiral hydrodynamics can be used to describe the negative magnetoresistance [812, 900], the anomalous thermoelectric transport [810], a wide range of collective modes [901, 902], and unusual profiles of electron hydrodynamic flow [903, 904] in Weyl semimetals.

Recently, the first experimental evidence of 3D relativistic-like hydrodynamic electron transport was reported in the Weyl semimetal tungsten diphosphide (WP_2) [905].[ff] The hydrodynamic nature of electron transport was supported by a characteristic dependence of the electrical resistivity on the sample width and by a strong violation of the Wiedemann–Franz law with the lowest value of the Lorenz number ever reported.

7.4.2 *Electron hydrodynamics in Weyl semimetals*

In the case of Weyl semimetals, the hydrodynamic equations describe the local conservation of momentum, energy, and electric charge. In addition, there is an extra equation that governs the evolution of the chiral charge. It is appropriate to also mention that the status of chiral charge is somewhat different from the other quantities since its conservation is violated by the

---

[ff] Note that, according to ARPES studies [593], WP_2 is a type-II Weyl semimetal, i.e., a Weyl semimetal with overtilted Weyl cones.

quantum chiral anomaly. By taking into account that the chirality flipping time τ_5 is much larger than the thermal equilibration time or, equivalently, the intravalley scattering time (see Appendix D for estimates of τ_5), one can treat the chiral charge as approximately conserved. As we will see, an added bonus of chiral hydrodynamics is its ability to reproduce the effects of the chiral anomaly.

According to the standard approach [739, 740], the equations of chiral hydrodynamics can be derived by calculating the corresponding moments of the Boltzmann equation (6.151) in the chiral kinetic theory, which was discussed in Sec. 6.6.2. For the sake of clarity, in this subsection only the key features needed to derive the hydrodynamic equations will be outlined. In particular, the distribution function is taken in the local equilibrium form

$$f_\lambda = \frac{1}{1 + e^{[\epsilon_{\mathbf{p}} - (\mathbf{u} \cdot \mathbf{p}) - \lambda \hbar (\mathbf{p} \cdot \boldsymbol{\omega})/(2p) - \mu_\lambda]/T}}, \qquad (7.142)$$

where $\mathbf{u}$ is the electron fluid velocity, T is temperature, and $\mu_\lambda = \mu + \lambda \mu_5$ is the effective chemical potential for quasiparticles of a given chirality λ. The latter is expressed in terms of the electric μ and chiral μ_5 chemical potentials. The term $\lambda \hbar (\mathbf{p} \cdot \boldsymbol{\omega}) / (2p)$ describes an energy correction due to nonzero vorticity $\boldsymbol{\omega} = [\boldsymbol{\nabla} \times \mathbf{u}] / 2$ of the flow [152, 737, 899, 906]. The Berry curvature $\boldsymbol{\Omega}_\lambda$, the quasiparticle energy $\epsilon_{\mathbf{p}}$, and the quasiparticle velocity $\mathbf{v}_{\mathbf{p}}$ are given by Eqs. (6.150), (6.153), and (6.154) in Sec. 6.6.2, respectively. For small magnetic field and fluid velocity, it is convenient to expand the distribution function in $\mathbf{u}$, $\mathbf{B}$, and $\boldsymbol{\omega}$ to linear order

$$f_\lambda \approx f_\lambda^{\text{eq}} - (\mathbf{p} \cdot \mathbf{u}) \frac{\partial f_\lambda^{\text{eq}}}{\partial \epsilon_{\mathbf{p}}} + \frac{e}{c} v_F p (\mathbf{B} \cdot \boldsymbol{\Omega}_\lambda) \frac{\partial f_\lambda^{\text{eq}}}{\partial \epsilon_{\mathbf{p}}} - \frac{\lambda \hbar (\mathbf{p} \cdot \boldsymbol{\omega})}{2p} \frac{\partial f_\lambda^{\text{eq}}}{\partial \epsilon_{\mathbf{p}}}, \quad (7.143)$$

where f_λ^{eq} is the standard Fermi–Dirac distribution function given in Eq. (6.172). For qualitative estimates, one can assume that the collision integral on the right-hand side of the Boltzmann equation (6.151) is approximated as $I_{\text{coll}} = - (f_\lambda - f_\lambda^{\text{eq}}) / \tau$, where τ is the relaxation time due to the scattering on defects and phonons. Note that the electron–electron scatterings do not contribute to the relaxation of momentum in the hydrodynamic regime.

Before presenting the hydrodynamic equations, it should be noted that electrons will be treated as a single component fluid. In other words, by assumption, the left- and right-handed quasiparticles form a single fluid with a common flow velocity $\mathbf{u}$. The corresponding local equilibrium state is enforced, for example, by the elastic Coulomb scattering without chirality flipping.[gg]

In Weyl semimetals, quasiparticles of opposite chiralities could also form separate fluids. Generically, a two-fluid regime can be realized when the relaxation rate of relative fluid motion is comparable to or smaller than the rate of processes separating the fluid components.[hh] As in graphene [245, 255, 256], one such regime with electron and hole fluids can be realized near the charge-neutrality point. The realization of a two-fluid regime could also be achieved in the presence of strain-induced pseudoelectromagnetic fields. Indeed, by their nature, such fields act differently on the left- and right-handed fluids (see also the discussion in Sec. 3.7). Nevertheless, even when pseudoelectromagnetic fields are applied, the use of a two-fluid regime is limited only to a finite window of frequencies. On the one hand, at small frequencies equilibration processes can be sufficiently fast to wash out any distinction between separate fluid components. On the other hand, at high enough frequencies a hydrodynamic description becomes unreliable. Yet, if realized, a two-fluid regime in topological semimetals could reveal many interesting properties. Of course, its theoretical description is considerably more complicated and its experimental detection is likely to be even more challenging. In what follows, we will consider only the single-fluid regime in Weyl semimetals without strain-induced fields.

As mentioned already, one of the simplest approaches to derive the hydrodynamics equations is to calculate the appropriate moments of the Boltzmann equation in the chiral kinetic theory. While tedious, such a calculation is straightforward. To the linear order in electromagnetic fields and fluid velocity, one obtains the following Euler equation [901]:

$$
\frac{1}{v_F}\partial_t\left(\frac{w\mathbf{u}}{v_F} + \sigma^{(\epsilon,\mathrm{B})}\mathbf{B} + \frac{\hbar\boldsymbol{\omega}n_5}{2}\right) + \boldsymbol{\nabla}P + \frac{c\left[\boldsymbol{\nabla}\times\mathbf{E}\right]\sigma^{(\epsilon,\mathrm{B})}}{3v_F}
$$

$$
+ \frac{4}{15v_F}\left[\sum_{j=x,y,z}B_j\boldsymbol{\nabla}u_j + (\mathbf{B}\cdot\boldsymbol{\nabla})\mathbf{u} + \mathbf{B}(\boldsymbol{\nabla}\cdot\mathbf{u})\right]\sigma^{(\epsilon,\mathrm{B})} + \frac{5\sigma^{(\epsilon,\mathrm{u})}\boldsymbol{\nabla}B^2}{2}
$$

$$
+ \frac{2}{3v_F}\sum_{j=x,y,z}u_j\boldsymbol{\nabla}B_j\sigma^{(\epsilon,\mathrm{B})} - \frac{4\sigma^{(\epsilon,\mathrm{B})}}{15v_F}\left[\sum_{j=x,y,z}u_j\boldsymbol{\nabla}B_j + (\mathbf{u}\cdot\boldsymbol{\nabla})\mathbf{B}\right]
$$

[gg]While thermal equilibrium is reached, the local chemical equilibrium with $\mu_+ = \mu_-$ is not necessarily established if the rate of chirality flipping processes is low.
[hh]It is assumed, of course, that both rates are sufficiently small to ensure the validity of a hydrodynamic description.

$$+ \left[(\mathbf{B} \cdot \boldsymbol{\nabla})\boldsymbol{\omega} + \sum_{j=x,y,z} B_j \boldsymbol{\nabla}\omega_j \right] \frac{\sigma^{(\epsilon,\mathrm{V})}}{5c} - \sum_{j=x,y,z} \frac{\sigma^{(\epsilon,\mathrm{V})}\omega_j \boldsymbol{\nabla}B_j}{2c}$$

$$- \frac{\sigma^{(\epsilon,\mathrm{V})}}{5c} \left[\sum_{j=x,y,z} \omega_j \boldsymbol{\nabla}B_j + (\boldsymbol{\omega} \cdot \boldsymbol{\nabla})\mathbf{B} \right] = -\frac{w\mathbf{u}}{v_F^2 \tau} - \frac{\hbar\boldsymbol{\omega} n_5}{2v_F T} - en\mathbf{E}$$

$$+ \frac{1}{c} \left[\mathbf{B} \times \left(en\mathbf{u} - \frac{\sigma^{(\mathrm{V})}\boldsymbol{\omega}}{3} \right) \right] + \frac{\sigma^{(\mathrm{B})}\mathbf{u}(\mathbf{E} \cdot \mathbf{B})}{3v_F^2} + \frac{5c\sigma^{(\epsilon,\mathrm{u})}(\mathbf{E} \cdot \mathbf{B})\boldsymbol{\omega}}{v_F},$$

$$(7.144)$$

where $w = \epsilon + P$ is the enthalpy density, ϵ is the energy density, and P is the pressure. The convention in Eq. (7.144) is that derivatives act on all terms to their right. By definition, the quasiparticle contributions to the electric and chiral charge densities are given by $-en$ and $-en_5$, respectively, and the explicit expressions for n and n_5 in terms of the electric and chiral chemical potentials are presented in Eqs. (6.177) and (6.178) in Sec. 6.6.4. The energy density ϵ is defined in Eq. (3.128) for a single Weyl node. In the case of two nodes, it reads

$$\epsilon = \sum_{\lambda=\pm} \frac{1}{8\pi^2 \hbar^3 v_F^3} \left(\mu_\lambda^4 + 2\pi^2 \mu_\lambda^2 T^2 + \frac{7\pi^4 T^4}{15} \right)$$

$$= \frac{\mu^4 + 6\mu^2 \mu_5^2 + \mu_5^4}{4\pi^2 \hbar^3 v_F^3} + \frac{T^2(\mu^2 + \mu_5^2)}{2\hbar^3 v_F^3} + \frac{7\pi^2 T^4}{60\hbar^3 v_F^3}. \qquad (7.145)$$

The corresponding pressure is given by the standard expression for a relativistic-like plasma, $P \simeq \epsilon/3$.

Note that the last two terms $\propto (\mathbf{E} \cdot \mathbf{B})$ are related to the chiral anomaly. Among the terms proportional to vorticity $\boldsymbol{\omega}$, there are several that come from an interplay of vorticity and magnetic field. In particular, there is a contribution due to the *chiral vortical effect* (CVE) $\propto \sigma^{(\mathrm{V})}\boldsymbol{\omega}$ [149–152, 737] to the electric current that modifies the Lorentz force. As expected, the conventional hydrodynamic equations for a relativistic system [895] can be easily recovered by setting $\mathbf{B} \to 0$ and $\boldsymbol{\omega} \to 0$.

In Eq. (7.144), we used the following shorthand notation for the anomalous transport coefficients:

$$\sigma^{(\mathrm{B})} = \frac{e^2 \mu_5}{2\pi^2 \hbar^2 c}, \qquad \sigma_5^{(\mathrm{B})} = \frac{e^2 \mu}{2\pi^2 \hbar^2 c}, \qquad (7.146)$$

$$\sigma^{(\epsilon,u)} = \frac{e^2 v_F}{120\pi^2 \hbar c^2}, \qquad \sigma^{(\epsilon,\mathrm{B})} = -\frac{e\mu\mu_5}{2\pi^2 \hbar^2 v_F c}, \qquad (7.147)$$

$$\sigma^{(V)} = -\frac{e\mu\mu_5}{\pi^2 v_F^2 \hbar^2}, \qquad \sigma_5^{(V)} = -\frac{e}{2\pi^2 \hbar^2 v_F^2}\left(\mu^2 + \mu_5^2 + \frac{\pi^2 T^2}{3}\right), \qquad (7.148)$$

$$\sigma^{(\epsilon,V)} = -\frac{e\mu}{6\pi^2 \hbar v_F}, \qquad \sigma_5^{(\epsilon,V)} = -\frac{e\mu_5}{6\pi^2 \hbar v_F}. \qquad (7.149)$$

These agree with the expressions obtained in the "no-drag" reference frame [907–909].

In order to describe viscous effects, one needs to include the following additional terms on the left-hand side of Eq. (7.144):

$$-\eta\Delta\mathbf{u} - \left(\zeta + \frac{\eta}{3}\right)\mathbf{\nabla}\left(\mathbf{\nabla}\cdot\mathbf{u}\right), \qquad (7.150)$$

where η and ζ are the shear and bulk viscosities. In relativistic-like systems, $\zeta = 0$ and $\eta = \eta_{\mathrm{kin}} w/v_F^2$, where $\eta_{\mathrm{kin}} \sim v_F^2 \tau_{\mathrm{ee}}$ is the kinematic shear viscosity and τ_{ee} is the electron–electron scattering time [739]. Formally speaking, the inclusion of the viscosity term (7.150) turns the Euler equation (7.144) into the Navier–Stokes one. Note also that the viscosity terms given in Eq. (7.150) were added phenomenologically rather than obtained from a microscopic collision integral.

In addition to the Navier–Stokes (or Euler) equation, which governs the fluid velocity, the following energy conservation relation should be satisfied:

$$\partial_t \epsilon + (\mathbf{\nabla}\cdot\mathbf{u})w + \sigma^{(\epsilon,\mathbf{u})}\left[\sum_{j=x,y,z} B_j\left(\mathbf{B}\cdot\mathbf{\nabla}\right)u_j - 2B^2(\mathbf{\nabla}\cdot\mathbf{u})\right]$$

$$-\frac{2c\left(\mathbf{E}\cdot\left[\mathbf{\nabla}\times\mathbf{u}\right]\right)\sigma^{(\epsilon,\mathbf{B})}}{3v_F} - 5c\sigma^{(\epsilon,\mathbf{u})}\left(\mathbf{E}\cdot\left[\mathbf{\nabla}\times\mathbf{B}\right]\right) + v_F\left(\mathbf{B}\cdot\mathbf{\nabla}\right)\sigma^{(\epsilon,\mathbf{B})}$$

$$-2\sigma^{(\epsilon,\mathbf{u})}\left[(\mathbf{u}\cdot\mathbf{\nabla})B^2 - 3\sum_{j=x,y,z}u_j\left(\mathbf{B}\cdot\mathbf{\nabla}\right)B_j\right] + \frac{\hbar v_F(\mathbf{\nabla}\cdot\boldsymbol{\omega})n_5}{2}$$

$$-\frac{\left(\mathbf{E}\cdot\left[\mathbf{\nabla}\times\boldsymbol{\omega}\right]\right)\sigma^{(\epsilon,V)}}{2} = -\mathbf{E}\cdot\left(en\mathbf{u} - \sigma^{(\mathbf{B})}\mathbf{B} - \frac{\sigma^{(V)}\boldsymbol{\omega}}{3}\right). \qquad (7.151)$$

As in the Navier–Stokes equation, to linear order in gradients, the viscosity and thermal conductivity effects will be described phenomenologically by including the following extra terms on the left-hand side [895]:

$$-\kappa\mathbf{\nabla}\cdot\left(\mathbf{\nabla}T - \frac{T}{w}\mathbf{\nabla}P\right), \qquad (7.152)$$

where κ is the thermal conductivity.

As one can see, the inclusion of numerous anomalous terms and vorticity effects makes the structure of the Euler (or Navier–Stokes) equation and the energy continuity relation complicated. It should be mentioned, however, that these equations are not affected directly by the topology of the band structure. This is clear, for example, from the fact that there are no terms with the chiral shift in Eqs. (7.144) and (7.151). In order to understand this, one should remember that the above hydrodynamic equations capture only the "matter" contributions from low-energy quasiparticles in a close vicinity of the Fermi surface. This is also consistent with the assumptions used in the derivation of the chiral kinetic theory governed by Eq. (6.151). Indeed, by utilizing a linear dispersion of quasiparticles, the chiral kinetic theory and, therefore, the resulting hydrodynamic equations carry no information about the separation between the Weyl nodes. As we discussed in Sec. 6.6.3, however, some of the topological effects originate from the filled electron states deeply below the Fermi surface. They are captured by the Chern–Simons terms in the currents and charge densities, i.e., in the constitutive relations.

It is interesting that the Euler equation (7.144) contains a dissipative term proportional to $\mathbf{u}/\tau$ on the right-hand side. Such a term originates from electron scattering on phonons and impurities and breaks explicitly the Galilean symmetry. The breaking of the Galilean symmetry should not be surprising since there is a preferred coordinate system in solids where the lattice ions are at rest.

As usual for systems with conserved charges, the hydrodynamic equations should be also supplemented by the corresponding continuity relations. In Dirac and Weyl semimetals, the latter must include the relations for the electric and chiral charges, i.e.,

$$\partial_t \rho + (\boldsymbol{\nabla} \cdot \mathbf{J}) = 0, \tag{7.153}$$

$$\partial_t \rho_5 + (\boldsymbol{\nabla} \cdot \mathbf{J}_5) = -\frac{e^3 (\mathbf{E} \cdot \mathbf{B})}{2\pi^2 \hbar^2 c}, \tag{7.154}$$

where ρ (ρ_5) and $\mathbf{J}$ ($\mathbf{J}_5$) denote the total electric (chiral) charge and current densities, respectively. By using Eqs. (6.155) and (6.156) the explicit expressions for the charge and current densities can be derived. In particular, the charge densities read

$$\rho = -en + \frac{\sigma^{(B)} (\mathbf{B} \cdot \mathbf{u})}{3v_F^2} + \frac{5c\sigma^{(\epsilon, u)} (\mathbf{B} \cdot \boldsymbol{\omega})}{v_F^2} + \rho_{\text{cs}}, \tag{7.155}$$

$$\rho_5 = -en_5 + \frac{\sigma_5^{(B)}(\mathbf{B}\cdot\mathbf{u})}{3v_F^2}. \tag{7.156}$$

Note that there are anomalous terms related to the hydrodynamic flow in the presence of a nonzero magnetic field.

The explicit expressions for the electric and chiral current densities are given by

$$\mathbf{J} = -en\mathbf{u} + \sigma\mathbf{E} + \kappa_e\boldsymbol{\nabla}T + \frac{\sigma_5}{e}\boldsymbol{\nabla}\mu_5 + \sigma^{(V)}\boldsymbol{\omega} + \sigma^{(B)}\mathbf{B}$$

$$+ \frac{c\sigma^{(B)}[\mathbf{E}\times\mathbf{u}]}{3v_F^2} + \frac{5c^2\sigma^{(\epsilon,u)}[\mathbf{E}\times\boldsymbol{\omega}]}{v_F^2} - \frac{[\mathbf{u}\times\boldsymbol{\nabla}]\sigma^{(V)}}{3}$$

$$+ \frac{[\boldsymbol{\nabla}\times\boldsymbol{\omega}]\sigma^{(\epsilon,V)}}{2} + \mathbf{j}_{cs}, \tag{7.157}$$

$$\mathbf{J}_5 = -en_5\mathbf{u} + \sigma_5\mathbf{E} + \kappa_{e,5}\boldsymbol{\nabla}T + \frac{\sigma}{e}\boldsymbol{\nabla}\mu_5 + \sigma_5^{(V)}\boldsymbol{\omega} + \sigma_5^{(B)}\mathbf{B}$$

$$+ \frac{c\sigma_5^{(B)}[\mathbf{E}\times\mathbf{u}]}{3v_F^2} - \frac{[\mathbf{u}\times\boldsymbol{\nabla}]\sigma_5^{(V)}}{3} + \frac{[\boldsymbol{\nabla}\times\boldsymbol{\omega}]\sigma_5^{(\epsilon,V)}}{2}. \tag{7.158}$$

As is easy to see, the chiral magnetic, the chiral separation, and the chiral vortical effects are reproduced by these currents. They also contain terms related to an interplay of fluid velocity and electric field. While these terms resemble the contributions obtained by the Lorentz transformation from the rest-frame currents in a magnetic field, the coefficients are different. This is not contradictory since there is no true Lorentz invariance in Weyl and Dirac semimetals.

The above currents include the effects of the intrinsic (also known as quantum critical or incoherent) electrical σ and thermoelectric κ_e conductivities, as well as their chiral counterparts σ_5 and $\kappa_{e,5}$. The intrinsic electrical conductivity σ was studied, in particular, in the holographic approach [274, 900, 910–914]. This is a contribution to conductivity due to incoherent and diffusive currents.

It is worth noting that intrinsic conductivities play an important role in the Dirac fluid regime, which occurs near the charge neutrality point $\mu = 0$ and was actively investigated in graphene (see Sec. 2.5.2). The realization of the Dirac fluid regime is a consequence of relativistic-like hydrodynamics that is absent in conventional solids. Physically, the Dirac fluid consists of counter-propagating populations of thermally excited electrons and holes with antiparallel momenta. The momentum of such a fluid is relaxed not only by the collisions with imperfections and phonons, but also by the electron–hole collisions. Such a mechanism plays little role in conventional Fermi fluids, however.

Unlike the Euler equation and the energy conservation relation, the charge and current densities include the topological Chern–Simons terms originating from the filled electron states deeply below the Fermi surface (see also the discussion in Secs. 6.3 and 6.6.2)

$$\rho_{\text{CS}} = \frac{e^2 (\mathbf{b} \cdot \mathbf{B})}{2\pi^2 \hbar c}, \tag{7.159}$$

$$\mathbf{j}_{\text{CS}} = \frac{e^2 b_0 \mathbf{B}}{2\pi^2 \hbar^2 c} - \frac{e^2 [\mathbf{b} \times \mathbf{E}]}{2\pi^2 \hbar}. \tag{7.160}$$

As in the consistent chiral kinetic theory, the Chern–Simons terms in hydrodynamics are important for the description of the anomalous Hall effect, the CME, as well as other topological effects. It might be justified to call the corresponding formulation with the Chern–Simons terms in constitutive relations as the *consistent hydrodynamics* [901]. In global equilibrium, the requirement of the vanishing CME current implies that $\mu_5 = -b_0$ (see the discussion in Secs. 6.4 and 6.6.2). In other words, the value of the chiral chemical potential is determined by the energy separation between the Weyl nodes.

Since an electron fluid is electrically charged, the hydrodynamic equations should be also amended with Maxwell's equations

$$\varepsilon_e \mathbf{\nabla} \cdot \mathbf{E} = 4\pi \left(\rho + \rho_{\text{b}} \right), \qquad \mathbf{\nabla} \times \mathbf{E} = -\frac{1}{c} \frac{\partial \mathbf{B}}{\partial t}, \tag{7.161}$$

$$\mathbf{\nabla} \times \mathbf{B} = \mu_{\text{m}} \frac{4\pi}{c} \mathbf{J} + \varepsilon_e \frac{1}{c} \frac{\partial \mathbf{E}}{\partial t}, \qquad \mathbf{\nabla} \cdot \mathbf{B} = 0. \tag{7.162}$$

Here ε_e and μ_{m} denote the background electric permittivity and the magnetic permeability, respectively. It is worth noting that Gauss's law includes not only the electric charge density ρ due to electron quasiparticles but also the background charge density $\rho_{\text{b}} = e n_{\text{b}}$ due to electrons in the inner shells and ions of the lattice. The total electric charge density equals

$$\rho + \rho_{\text{b}} = -e \left[n(\mathbf{r}) - n_{\text{b}} \right] + \frac{e^2 (\mathbf{b} \cdot \mathbf{B})}{2\pi^2 \hbar c}, \tag{7.163}$$

where $n(\mathbf{r})$ is the local quasiparticle fermion number density. Interestingly, the latter may deviate from the background charge density n_{b} even in a state of local equilibrium. This is the consequence of having an extra topological contribution $\propto (\mathbf{b} \cdot \mathbf{B})$ in the charge density. Indeed, enforcing the condition of local equilibrium $\rho + \rho_{\mathrm{b}} = 0$ in the presence of an external magnetic field induces a B-dependent correction to $n(\mathbf{r})$. This implies in turn that the equilibrium electric chemical potential becomes a function of a magnetic field.

In the following two subsections, we apply the hydrodynamic framework defined above to two cases: (i) electron flow in a slab and (ii) nonlocal flows in a vicinity setup.

7.4.3 *Hydrodynamic flow in slab geometry*

In this subsection, the anomalous features of the electron fluid flow are discussed by using the framework of the consistent hydrodynamics. For this purpose, it is convenient to use a simple setup in the form of a slab of finite thickness along the y-axis, $0 \leq y \leq L_y$, which is subject to a constant electric field along the x-axis. For simplicity, the slab is assumed infinite in the x- and z-directions. Then, in view of the translational symmetry, the flow profile has a nontrivial dependence only on the y coordinate. The corresponding schematic setup is presented in Fig. 7.14(a).

Before considering the hydrodynamic flow, it is important to specify the boundary conditions at the surfaces of the slab. Since electron quasiparticles are confined to the material, the components of both electric current and fluid velocity normal to the surface should vanish at the surfaces of the

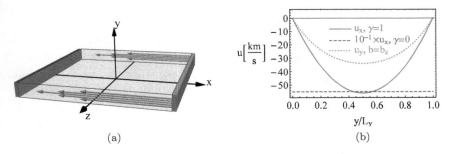

(a) (b)

Fig. 7.14 (a) Schematic setup for studying the hydrodynamic flow. The slab is finite in the y-direction, $0 \leq y \leq L_y$, and infinite in the x and z ones. Arrows depict the electron fluid flow in the slab. (b) The dependence of the fluid velocity components $u_x(y)$ and $u_y(y)$ on the spatial coordinate y, assuming either no-slip boundary conditions ($\gamma = 1$) or free-surface ones ($\gamma = 0$). It is assumed that $\mathbf{b} \parallel \hat{\mathbf{z}}$ and the characteristic values of the parameters in Eq. (7.179) are used together with $\mu = 10$ meV, $T = 10$ K, $E_x = 10$ V/m, and $L_y = 10$ μm.

slab, i.e.,

$$J_y(y)|_{y=0,L_y} = 0, \tag{7.164}$$

$$u_y(y)|_{y=0,L_y} = 0. \tag{7.165}$$

Note that, because of the Fermi arc states, the normal component of the electric current could in principle remain finite on the surface [915, 916]. For simplicity, these effects of the surface Fermi arcs on the flow will be neglected here. Because of electron fluid viscosity, one should also specify the boundary conditions for the components of the electron fluid velocity parallel to the surfaces. There are two main types of such boundary conditions [895]: (i) the no-slip boundary conditions and (ii) the free-surface (or no-stress) ones. They are defined by the following relations:

$$u_x(y)|_{y=0,L_y} = u_z(y)|_{y=0,L_y} = 0, \tag{7.166}$$

$$\partial_y u_x(y)|_{y=0,L_y} = \partial_y u_z(y)|_{y=0,L_y} = 0, \tag{7.167}$$

respectively. The free-surface boundary conditions imply that the tangential forces vanish at the surfaces. However, according to the experimental data for WP$_2$ [905], the most relevant boundary conditions for the electron transport are the no-slip ones given in Eq. (7.166) where the electron fluid sticks to the surfaces of slab. It is worth noting that in many materials the realistic boundary conditions are mixed ones, e.g.,

$$u_x(y)|_{y=0,L_y} = l_{\text{slip}} \partial_y u_x(y)|_{y=0,L_y}, \tag{7.168}$$

where l_{slip} is the slip length. The latter is determined by the atomically rough edges of material and surface disorder that allow for electron momentum dissipation. The detailed study of the boundary conditions for viscous flows in 2D systems was performed in [917, 918], where it was shown that the no-slip boundary conditions are appropriate only at intermediate temperatures and in large samples. The regime of ultralow temperatures $T \to 0$ should correspond to the free-surface boundary conditions [917]. In addition, the boundary conditions could be artificially engineered by changing the geometry of the boundary [918] providing a mesoscopic contribution to the slip length in addition to the microscopic one originating from rough edges.

By linearizing the hydrodynamics equations with respect to $\mathbf{E}(y)$, $\mathbf{B}(y)$, $\mathbf{u}(y)$, $\mu(y) - \mu$, and $T(y) - T$, as well as assuming the case of a $\mathcal{P}$ symmetric Weyl semimetal ($\mu_5 = -b_0 = 0$), one obtains the following three components of the steady-state Navier–Stokes equation:

$$\eta \partial_y^2 u_x(y) - enE_x - \frac{w}{v_F^2 \tau} u_x(y) = 0, \tag{7.169}$$

$$\eta_y \partial_y^2 u_y(y) - enE_y(y) - \frac{w}{v_F^2 \tau} u_y(y) = 0, \tag{7.170}$$

$$\eta \partial_y^2 u_z(y) - \frac{w}{v_F^2 \tau} u_z(y) = 0, \tag{7.171}$$

where, by definition, $\eta_y = 4\eta/3$. Note that the effects due to the pressure gradient ∇P were ignored, as they should be small compared to the driving force of the electric field.

In general, the Navier–Stokes equation should be also supplemented by the energy conservation relation (7.151). However, the latter can be neglected when the fluid velocity is much smaller than the speed of sound (which is $v_F/\sqrt{3}$ for electron quasiparticles in relativistic-like systems [895]). By assuming such a regime, one finds self-consistently that the spatial variations of temperature should be small. In other words, one can simply set $T(y) = T$ and ignore Eq. (7.151).

While the E_x component in Eq. (7.169) is fixed externally as a part of the electric field applied to the slab, the $E_y(y)$ component in Eq. (7.170) is dynamical and should be determined by the hydrodynamic flow. In order to find it, one can integrate Eq. (7.157) with the boundary conditions (7.164) and (7.165). Because of the intrinsic conductivity term $\sim \sigma$, the resulting field distribution takes the form

$$E_y(y) = E_{y,\text{hydro}}(y) + E_{y,\text{cs}}(y), \tag{7.172}$$

which contains both hydrodynamic and Chern–Simons contributions

$$E_{y,\text{hydro}}(y) = \frac{en}{\sigma} u_y(y), \tag{7.173}$$

$$E_{y,\text{cs}}(y) = \frac{e^2 E_x b_z}{2\pi^2 \hbar \sigma}, \tag{7.174}$$

respectively. Having found the electric field $E_y(y)$, one can solve the whole system of equations (7.169)–(7.171). As is easy to verify, Eq. (7.171) has only the trivial solution $u_z(y) = 0$ for both types of boundary conditions. The solution for the x component of the flow velocity can be easily found from Eq. (7.169), which decouples from the other equations. Its explicit solution reads

$$u_x(y) = -\frac{v_F^2 \tau e n E_x}{w} \left[1 - \gamma \frac{\cosh(\lambda_x y - \lambda_x L_y/2)}{\cosh(\lambda_x L_y/2)} \right], \qquad (7.175)$$

where

$$\lambda_x = \sqrt{\frac{w}{\eta v_F^2 \tau}} \qquad (7.176)$$

is an inverse length scale that determines the velocity spatial profile. Note that, formally, $\gamma = 1$ and $\gamma = 0$ correspond to the no-slip and free-surface boundary conditions given in Eqs. (7.166) and (7.167), respectively. In agreement with the classical fluid mechanics, the fluid velocity for $\gamma \neq 0$ shows a characteristic parabolic-like profile with the maximum in the middle of the slab. The latter is known as the Poiseuille flow in hydrodynamics. If $\lambda_x L_y \ll 1$, the viscous drag effects permeate the whole slab and the fluid velocity $u_x(y)$ shows a well-pronounced nonuniform profile. In the case of the free-surface boundary conditions, the velocity is uniform inside the slab. By noting that the hydrodynamical part of the current is proportional to $en\mathbf{u}$, one concludes that the electric current is affected in the same way as the fluid velocity ultimately leading to the characteristic $\int_0^{L_y} dy\,\sigma \propto L_y^3$ dependence of the conductance on the channel width in the case of the no-slip boundary conditions and small L_y.

Because of the presence of the chiral shift in Eq. (7.172), the solution to Eq. (7.170) is nontrivial, i.e.,

$$u_y(y) = -\frac{env_F^2 \tau}{e^2 n^2 v_F^2 \tau + \sigma w} \frac{e^2 b_z E_x}{2\pi^2 \hbar} \left[1 - \frac{\cosh(\lambda_y y - \lambda_y L_y/2)}{\cosh(\lambda_y L_y/2)} \right], \qquad (7.177)$$

where

$$\lambda_y = v_F \sqrt{\frac{\tau \sigma}{e^2 n^2 v_F^2 \tau + \sigma w}} \qquad (7.178)$$

is the corresponding inverse length scale. In order to clarify the origin of this unusual velocity profile, let us note that, according to the boundary condition (7.164), the vanishing electric current is achieved by compensating the constant Chern–Simons term with the electric field component $E_y(y)$ normal to the surface. However, this field also enters the Navier–Stokes equation (7.170), which becomes inhomogeneous and allows for a nontrivial normal component of the electron flow velocity $u_y(y)$. Therefore, the normal flow appears due to a self-consistent treatment of the electromagnetic (current conservation relation) and hydrodynamic (Navier–Stokes equation) sectors of the consistent hydrodynamics in Weyl semimetals.

The velocity profiles for u_x at $\gamma = 0$ and $\gamma = 1$ as well as u_y are shown in Fig. 7.14(b). In order to obtain numerical estimates, the following parameters are used[ii]

$$v_F \approx 1.4 \times 10^8 \text{ cm/s}, \quad b \approx 3 \text{ nm}^{-1}. \tag{7.179}$$

In general, the relaxation time τ is a function of the chemical potential as well as temperature. According to [905], it ranges from about $\tau \approx 0.6$ ns at $T = 2$ K to $\tau \approx 4$ ps at $T = 30$ K. In particular, we used $\tau = 0.4$ ns at $T = 5$ K, $\tau = 0.3$ ns at $T = 10$ K, and $\tau = 0.17$ ns at $T = 15$ K. In addition, it was assumed that the chemical potential depends weakly on temperature. For the intrinsic conductivity σ, one can use the result obtained in a holographic approach [900, 910–914], i.e.,

$$\sigma = \frac{3\pi^2 \hbar v_F^3}{2\pi} \left(\frac{\partial n}{\partial \mu} + \frac{\partial n_5}{\partial \mu_5} \right) \tau_{ee}, \tag{7.180}$$

where τ_{ee} is the electron–electron scattering time. The use of this expression is partially justified by the experimental findings in [905], where the momentum-conserving relaxation time is approximated well by the expression $\tau_{ee} = \hbar/T$. Such an estimate for τ_{ee} suggests that the electron fluid in WP$_2$ is strongly interacting and cannot be described by a conventional Fermi liquid, where $\tau_{ee} \sim \varepsilon_e v_F \hbar \mu / (e^2 T)^2$ is parametrically much larger than $\hbar/T$.

One should note that, in the setup at hand, the z component of electric current is determined by the chiral shift (cf. Eq. (7.160)). In essence, it is the anomalous Hall effect current. When $b_y \neq 0$, the corresponding current is induced by the applied electric field, i.e., $J_{\text{cs},z} \propto b_y E_x$. However, there is another interesting source of the anomalous Hall current in the z-direction that is triggered by the anomalous hydrodynamic flow, i.e., $J_{\text{cs},z} \propto b_x E_y(y)$, when both $b_x \neq 0$ and $b_z \neq 0$. Here a few words are in order regarding the origin of $E_y(y)$. According to Eqs. (7.172), (7.173), and (7.174), the normal component of the electric field $E_y(y)$ has two contributions. One of them, i.e., $E_{y,\text{hydro}}(y)$, is induced by the electron flow velocity $u_y(y)$, while the other, i.e., $E_{y,\text{cs}}(y) \propto b_z E_x$, comes from the anomalous Hall effect and is purely topological. Concerning $E_{y,\text{hydro}}(y) \propto u_y(y)$, one should note that the normal component of flow velocity $u_y(y)$ is driven by the anomalous Hall effect too. This is clear from Eq. (7.177), in which $u_y(y)$ is proportional

---

[ii]In order to better illustrate hydrodynamic features, v_F is taken an order of magnitude larger than its value in WP$_2$ measured in [919].

to the chiral shift component b_z. Yet, $u_y(y)$ has a nonuniform profile across the slab with the maximum value attained at $y = L_y/2$. Such a dependence is characteristic for the hydrodynamic regime. Therefore, the resulting normal component of the electric field also inherits a nontrivial hydrodynamic profile as a function of y. This is the *hydrodynamic anomalous Hall effect* (hAHE) in Weyl semimetals. In essence, it is the anomalous Hall effect current modified by hydrodynamic flow [903].

In principle, a nontrivial profile of the anomalous Hall current $J_z(y)$ can be extracted by measuring stray magnetic fields or nonlocal transport. However, the corresponding experimental studies might require a special design of devices and are difficult. A simpler indirect verification of the hydrodynamic anomalous Hall effect could be obtained from the scaling of transport properties with the slab thickness L_y. For example, the electric current density in the middle of the slab, $J_z(L_y/2)$, as a function of the slab thickness L_y is presented in Fig. 7.15(a). Note that the absolute value of $J_z(L_y/2)$ decreases with L_y. While such a dependence might seem surprising in the hydrodynamic regime, it has a simple explanation. The corresponding current $J_z \propto b_x E_y(y)$ is dominated by the boundary layers where $E_y(y)$ increases. Indeed, as is easy to check, the Chern–Simons term (7.174) dominates over the hydrodynamic one (7.173) in the expression for the total E_y. Moreover, the two terms have opposite signs (note that

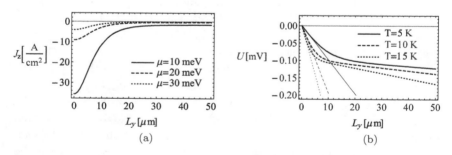

(a)   (b)

Fig. 7.15   (a) The electric current density in the middle of the slab, $J_z(L_y/2)$, as a function of slab thickness L_y for three different values of the chemical potential: $\mu = 10$ meV (solid lines), $\mu = 20$ meV (dashed lines), and $\mu = 30$ meV (dotted lines). The model parameters in Eq. (7.179) as well as $T = 10$ K, $E_x = 10$ V/m, and $\mathbf{b} = (1, 0, 1)\,|\mathbf{b}|/\sqrt{2}$ were used. (b) The hydrodynamic anomalous Hall effect voltage U as a function of L_y for three different temperatures: $T = 5$ K (solid lines), $T = 10$ K (dashed lines), and $T = 15$ K (dotted lines). In panel (b), the value of chemical potential is $\mu = 10$ meV and $\mathbf{b} \parallel \hat{\mathbf{z}}$ is assumed. Thin lines correspond to the purely topological contribution U_{CS} given in Eq. (7.183).

$u_y(y) < 0$). Therefore, the decrease of $u_y(y)$ near the boundary increases the total electric field there.

Another potentially interesting observable is the hAHE voltage between the opposite surfaces of the slab. It is defined as follows:

$$U = -\int_0^{L_y} E_y(y)\,dy = U_{\text{hydro}} + U_{\text{cs}}, \tag{7.181}$$

where, for convenience, the normal flow contribution was separated from the purely topological Chern–Simons one, i.e.,

$$U_{\text{hydro}} = \frac{e^2 b_z E_x}{2\pi^2 \hbar} \frac{\tau v_F^2 e^2 n^2}{\sigma \left(\tau v_F^2 e^2 n^2 + \sigma w\right)} \left[L_y - \frac{2}{\lambda_y} \tanh\left(\frac{\lambda_y L_y}{2}\right)\right], \tag{7.182}$$

$$U_{\text{cs}} = -\frac{L_y}{\sigma} \frac{e^2 b_z E_x}{2\pi^2 \hbar}. \tag{7.183}$$

The hAHE voltage U as a function of L_y is presented in Fig. 7.15(b), where the thin lines denote the Chern–Simons contribution (7.183). The dependence of U on the slab thickness is nonmonotonic and shows a knee-like behavior. The latter originates from the inhomogeneous profile of the electric field $E_y(y)$ that is enhanced near the surfaces of the slab. The non-monotonic dependence stems from the interplay of hydrodynamic and Chern–Simons contributions. As expected, electric potential U_{cs} linearly increases with the slab thickness. However, the hydrodynamic part U_{hydro} tends to compensate such a linear increase ultimately leading to a knee-like behavior at large L_y. The location of the knee-like feature is determined primarily by the second term in the square brackets in Eq. (7.182), which quickly reaches the constant value $\propto 2/\lambda_y$. Then the terms in U_{hydro} and U_{cs} linear in T_y cancel out approximately when $\sigma w \ll \tau v_F^2 e^2 n^2$. The knee-like behavior of the hAHE voltage could provide another way to investigate the viscosity of the electron fluid in Weyl semimetals.

Let us briefly summarize the key features related to electron hydrodynamic flow in a Weyl semimetal slab. Among them is the Poiseuille-like longitudinal flow, where the electron fluid velocity shows a parabolic-like profile in the slab. In addition, there are effects related to the interplay of the nontrivial topology and hydrodynamics, i.e., the hydrodynamic anomalous Hall effect. Its hydrodynamic nature is revealed in a saturating behavior of the corresponding voltage on the slab width. It is worth noting, however, that the hydrodynamic regime in Weyl and Dirac semimetals still needs to be carefully verified and studied experimentally.

7.4.4   *Nonlocal transport*

Another interesting method for studying the electron hydrodynamic regime is provided by a nonlocal transport. Unlike conventional local probes, measuring an electric potential drop at points where the currents are injected and collected, nonlocal ones refer to an electric potential at other points in a sample and, therefore, characterize currents going sideways. Note that, usually, the nonlocal transport cannot be understood in terms of motion of individual charge carriers. By analogy with the vicinity setup in graphene [249, 251], a nonlocal response in Weyl and Dirac semimetals can be studied in a slab with electric contacts attached on the same side. Such a setup is shown schematically in Fig. 7.16.

For simplicity, let us assume that a slab, which is made of a Weyl semimetal with broken $\mathcal{T}$ symmetry, is semi-infinite in the y-direction (i.e., $y \geq 0$), but infinite in the x- and z-directions. The source and drain of electric current are located at different positions in the x-direction, but on the same $y = 0$ surface. The contact bars will be assumed to be infinite in the z-direction and have a vanishing width in the x-direction (i.e., modeled by the δ-functions). As is clear, because of the translational symmetry in the z-direction, the hydrodynamic flow and electromagnetic fields will be independent of the z coordinate.

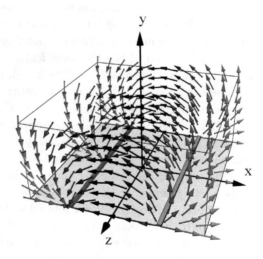

Fig. 7.16   Schematic illustration of the nonlocal setup for measuring hydrodynamic transport in a Weyl semimetal. The electric current source and drains are taken in the form of long thin bars. The electron flow velocity field is visualized by arrows.

Since the electric current is injected and collected at the surface of the slab, its normal component satisfies the following boundary condition:

$$J_y(x,y)\big|_{y=0} = I\delta(x+x_0) - I\delta(x-x_0), \qquad (7.184)$$

where I is the linear density of the external electric current. The source and drain of electrons are located at $x = x_0$ and $x = -x_0$, respectively.

The general expression for the hydrodynamic electric current density in Eq. (7.157) should be matched to the injected current density at $y = 0$ given by Eq. (7.184). Before that can be done, however, one should specify the boundary conditions for the fluid velocity and electromagnetic fields.

In general, the electric current density and the fluid velocity are not proportional to each other. Nevertheless, it is natural to assume that the y component of flow velocity across the $y = 0$ surface vanishes everywhere except at the locations of the source and drain, i.e.,

$$u_y(x,y)\big|_{y=0} = u_{y,1}\delta(x+x_0) + u_{y,2}\delta(x-x_0). \qquad (7.185)$$

The most general form of the electric field at $y = 0$ is given by the ansatz

$$E_y(x,y)\big|_{y=0} = E_{y,0}(x) + E_{y,1}\delta(x+x_0) + E_{y,2}\delta(x-x_0)$$
$$+ E_{y,3}\delta''(x+x_0) + E_{y,4}\delta''(x-x_0), \qquad (7.186)$$

where $E_{y,0}(x)$ is a nonsingular contribution. In a self-consistent treatment, the terms with the first derivatives are needed when a nonzero surface charge density is induced (they are ignored here). As for the terms with the second derivatives, they can be induced by the vorticity effects.

As is easy to verify, when the intrinsic conductivity σ is nonzero, the boundary condition (7.184) alone is not sufficient to determine both normal components of the fluid velocity and electric field at the surface. This is due to the fact that the current consists of two parts, i.e., a pure hydrodynamic flow and a diffusive current. One could use a phenomenological argument to resolve the problem. By assuming that the net current going across the surface into the bulk should match the measurable input current in the contacts, i.e., $\sigma_{\text{eff}}\mathbf{E}$, one derives

$$-enu_{y,i} + \sigma E_{y,i} = \sigma_{\text{eff}} E_{y,i}, \qquad (7.187)$$

where only the terms with δ-functions are taken into account. Here the two contacts are labeled by index $i = 1, 2$.

By performing the Fourier transform with respect to the x coordinate, one obtains the following expressions from Eqs. (7.184)–(7.187):

$$u_y(k_x, y)\big|_{y=0} = 2i \sin(k_x x_0) I \frac{\sigma - \sigma_{\text{eff}}}{e n \sigma_{\text{eff}}}, \tag{7.188}$$

$$E_y(k_x, y)\big|_{y=0} = 2i \sin(k_x x_0) I \left[1 + \frac{k_x^2 \sigma^{(\epsilon, V)}(\sigma_{\text{eff}} - \sigma)}{4 e n \sigma \sigma_{\text{eff}}} \right]$$

$$+ i k_x \frac{\sigma^{(V)}}{2\sigma} u_z(k_x, y)\big|_{y=0} - i k_x \frac{\sigma^{(\epsilon, V)}}{4\sigma} \partial_y u_x(k_x, y)\big|_{y=0}$$

$$+ \frac{e^2}{2\pi^2 \hbar \sigma} \left[b_z E_x(k_x, y) - b_x E_z(k_x, y) \right]\big|_{y=0}. \tag{7.189}$$

Note that temperature gradients were ignored here. As was discussed in Sec. 7.4.3, this is justified when the fluid velocity is much smaller than the speed of sound. An interesting feature of Eq. (7.189) is its nontrivial dependence on the chiral shift. This implies that the fluid velocity and the electric field are affected by the chiral shift not only directly by Chern–Simons terms in the bulk of the material, but also via the boundary conditions.

The above two relations should be supplemented also by the boundary conditions for the tangential components of the electron flow velocity. Similarly to Sec. 7.4.3, the standard no-slip boundary conditions

$$u_x(x, y)\big|_{y=0} = u_z(x, y)\big|_{y=0} = 0 \tag{7.190}$$

or the free-surface (no-stress) boundary conditions

$$[\partial_y u_x(x, y) + \partial_x u_y(x, y)]\big|_{y=0} = [\partial_y u_z(x, y) + \partial_z u_y(x, y)]\big|_{y=0} = 0, \tag{7.191}$$

can be considered.

In principle, one should also specify the boundary conditions at $y \to \infty$. It is natural to assume that there is no electron flow far away from the contacts, i.e.,

$$\lim_{y \to \infty} \mathbf{u}(x, y) = \mathbf{0}. \tag{7.192}$$

The same should be true also for the electric current and the electromagnetic fields.

In view of the boundary conditions (7.188) and (7.189), it is convenient to perform the Fourier transform in the linearized Navier–Stokes equation, see Eqs. (7.144) and (7.150). The result reads

$$\eta\left(-k_x^2 + \partial_y^2\right) u_x(k_x, y) - enE_x(k_x, y) - \frac{w}{v_F^2\tau} u_x(k_x, y)$$

$$+ \frac{\eta}{3}ik_x\left[ik_xu_x(k_x, y) + \partial_y u_y(k_x, y)\right] = 0, \tag{7.193}$$

$$\left(-\eta k_x^2 + \eta_y\partial_y^2\right) u_y(k_x, y) - enE_y(k_x, y) - \frac{w}{v_F^2\tau} u_y(k_x, y)$$

$$+ \frac{\eta}{3}ik_x\partial_y u_x(k_x, y) = 0, \tag{7.194}$$

$$\eta\left(-k_x^2 + \partial_y^2\right) u_z(k_x, y) - \frac{w}{v_F^2\tau} u_z(k_x, y) = 0, \tag{7.195}$$

where $\eta_y = 4\eta/3$. Because of the symmetry in the problem, the z component of the electric field vanishes.

The continuity relation $\nabla \cdot \mathbf{J} = 0$ takes the following explicit form:

$$-en\left[ik_xu_x(k_x, y) + \partial_y u_y(k_x, y)\right] + \sigma\left[ik_x E_x(k_x, y) + \partial_y E_y(k_x, y)\right] = 0. \tag{7.196}$$

In addition, Faraday's law implies that $ik_x E_y(k_x, y) = \partial_y E_x(k_x, y)$.

While the solution to Eq. (7.195) is trivial, $u_z(k_x, y) = 0$, the other equations are coupled and should be solved self-consistently. It is convenient to rewrite the remaining Navier–Stokes and continuity equations as the following first-order matrix equation

$$\partial_y\mathbf{w}(k_x, y) = \hat{M}\mathbf{w}(k_x, y), \tag{7.197}$$

where

$$\mathbf{w}(k_x, y) = \begin{pmatrix} u_x(k_x, y) \\ u_y(k_x, y) \\ \partial_y u_x(k_x, y) \\ E_x(k_x, y) \\ \partial_y E_x(k_x, y) \\ \partial_y^2 E_x(k_x, y) \end{pmatrix} \tag{7.198}$$

and

$$\hat{M} = \begin{pmatrix} 0 & 0 & 1 & 0 & 0 & 0 \\ -ik_x & 0 & 0 & ik_x\xi & 0 & \frac{\xi}{ik_x} \\ k_x^2 + P_\eta & 0 & 0 & \frac{en+k_x^2(\eta_y-\eta)\xi}{\eta} & 0 & -\frac{\xi(\eta_y-\eta)}{\eta} \\ 0 & 0 & 0 & 0 & 1 & 0 \\ 0 & 0 & 0 & 0 & 0 & 1 \\ 0 & \frac{ik_x\eta}{\xi\eta_y}\left(k_x^2 + P_\eta\right) & -\frac{k_x^2\eta}{\xi\eta_y} & 0 & k_x^2 + \frac{en}{\eta_y\xi} & 0 \end{pmatrix}.$$

$$(7.199)$$

Here we used the shorthand notation $\xi = \sigma/(en)$ and $P_\eta = w/(v_F^2\eta\tau)$. Formally, matrix (7.199) should have six sets of eigenvectors and eigenvalues. As is easy to check, however, only three of them correspond to physical solutions satisfying the boundary conditions (7.192) at $y \to \infty$. Therefore, the general solution for $\mathbf{w}(k_x, y)$ reads

$$\mathbf{w}(k_x, y) = C_1 V_1 e^{-|k_x|y} + C_2 V_2 e^{-\sqrt{k_x^2 + P_\eta}y} + C_3 V_3 e^{-\sqrt{k_x^2 + \frac{\eta}{\eta_y}P_\eta + \frac{en}{\eta_y\xi}}y},$$

$$(7.200)$$

where the three eigenvectors are given by

$$V_1 = \begin{pmatrix} -\frac{en}{k_x^2\eta P_\eta} \\ -i\frac{en}{k_x|k_x|\eta P_\eta} \\ \frac{en}{|k_x|\eta P_\eta} \\ \frac{1}{k_x^2} \\ -\frac{1}{|k_x|} \\ 1 \end{pmatrix}, \quad V_2 = \begin{pmatrix} -\frac{1}{\sqrt{k_x^2+P_\eta}} \\ -\frac{ik_x}{k_x^2+P_\eta} \\ 1 \\ 0 \\ 0 \\ 0 \end{pmatrix}, \quad V_3 = \begin{pmatrix} \xi g(k_x) \\ \frac{i\xi\sqrt{g(k_x)}}{k_x} \\ -\xi\sqrt{g(k_x)} \\ g(k_x) \\ -\sqrt{g(k_x)} \\ 1 \end{pmatrix},$$

$$(7.201)$$

and $g(k_x) = \eta_y\xi/[en + \xi(\eta_y k_x^2 + \eta P_\eta)]$.

Coefficient functions C_1, C_2, and C_3 can be determined from the boundary conditions in Eqs. (7.188) and (7.189), supplemented by either no-slip (7.190) or free-surface (7.191) boundary conditions, respectively.

In order to determine a spatial distribution of the flow velocity and the electric field, the inverse Fourier transform should be performed. The corresponding results for the fluid velocity distribution at $b = 0$ and $\mathbf{b} \parallel \hat{z}$ are presented in Figs. 7.17(a) and 7.17(b), respectively. While the results

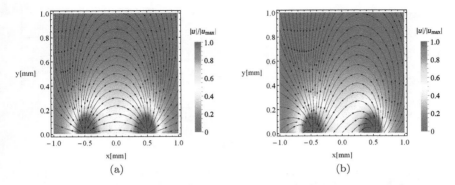

Fig. 7.17 The flow velocity lines in the $x - y$ plane for (a) $b = 0$ and (b) $\mathbf{b} \parallel \hat{\mathbf{z}}$. Different colors represent the absolute value of the electron flow velocity normalized by its maximum value. The model parameters are defined in Eqs. (7.179) and (7.203), as well as the no-slip boundary conditions are assumed. Additionally, $\mu = 10$ meV and $T = 20$ K.

at $b = 0$ are perfectly symmetric with respect to the y-axis, the case with nonzero $\mathbf{b}$ reveals a well-pronounced asymmetry. This should not be too surprising since the chiral shift breaks the corresponding mirror symmetry. Interestingly, while the flow pattern is strongly modified, the effect of the chiral shift on the absolute value of velocity is rather weak.

In order to test the unusual hydrodynamic transport in Weyl semimetals, one needs a convenient observable that can be easily extracted in experiment. As in studies of the hydrodynamic transport in graphene (see Sec. 2.5.2), one can use

$$R(x) = \frac{\phi(0,0) - \phi(x,0)}{I}, \qquad (7.202)$$

which measures the nonlocal surface resistance per unit length in the z-direction. Here $\phi(x,0)$ and $\phi(0,0)$ are the values of the electric potential measured on the surface of semimetal ($y = 0$) at an arbitrary x and at the midpoint between the contacts, $x = 0$, respectively.

Unlike the usual local resistance, the nonlocal resistance defined by Eq. (7.202) can be either positive or negative. The numerical results for $R(x)$ at $b = 0$ and $\mathbf{b} \parallel \hat{\mathbf{z}}$ are presented in Figs. 7.18(a) and 7.18(b), respectively. Similarly to the fluid velocity profile in Fig. 7.17, the absolute value of $R(x)$ becomes asymmetric when the chiral shift is nonzero. In fact, one

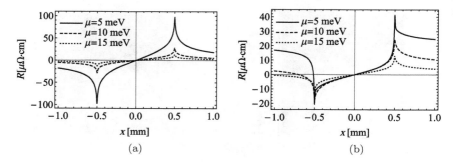

Fig. 7.18   The nonlocal surface resistance $R(x)$ per unit length in the z-direction for (a) $b = 0$ and (b) $\mathbf{b} \parallel \hat{\mathbf{z}}$ as a function of x. Here $\mu = 5$ meV (solid line), $\mu = 10$ meV (dashed line), and $\mu = 15$ meV (dotted line). The model parameters are defined in Eqs. (7.179) and (7.203), as well as $T = 20$ K. The results are calculated for the no-slip boundary conditions and $y = 5$ μm, but they will remain almost the same for the free-surface boundary conditions and depend only weakly on y.

finds that the latter could lead to a positive nonlocal resistance in the region where it should be negative (e.g., at $x < -x_0$). Unlike in graphene, here the sign change of resistance is primarily driven by the effects associated with the chiral shift rather than the electron viscosity. One might expect, however, that Weyl or Dirac semimetals with a different choice of model parameters could also demonstrate a graphene-like behavior.

In order to present the numerical results in Figs. 7.17 and 7.18, we used the same model parameters that were given in Eq. (7.179). The experimental value of effective electrical conductivity σ_{eff} ranges from about $\sigma_{\text{eff}} \approx 10^{10}$ S/m at $T = 2$ K to $\sigma_{\text{eff}} \approx 2 \times 10^8$ S/m at $T = 30$ K [905]. In particular, we used $\sigma_{\text{eff}} \approx 6.25 \times 10^8$ S/m at $T = 20$ K. We also assumed the following values for the linear electric current density I and the position of the source x_0:

$$I = 2 \times 10^{-7} \text{ A/cm}, \quad x_0 = 0.5 \text{ mm}, \tag{7.203}$$

respectively.

In summary, the nonlocal resistance in Weyl semimetals could be used to test the interplay of hydrodynamic flow and topological effects induced by the separation of Weyl nodes. One should remark, however, that the realization of a hydrodynamic regime in solids strongly depends on material parameters. Therefore, depending on the chemical composition, preparation conditions, and other experimental factors, the qualitative effects discussed above could be either suppressed or enhanced. The determination of the hydrodynamic window in Dirac and Weyl semimetals, i.e., the range of parameters where the hydrodynamic description could be applied, is still an open problem at the cutting edge of research.

Chapter 8

Collective Modes

A wave is never found alone, but is mingled with the other waves.

Leonardo da Vinci

Collective excitations are simple but very informative probes in all types of physical systems. They are particularly useful for the characterization of plasmas on large distance and long timescales. The studies of collective modes in plasmas are invaluable in fundamental research but also very useful for practical applications. Examples of various forms of plasmas include ionized gases, interstellar and intergalactic media, electron plasmas in solids, relativistic plasmas in heavy-ion collisions and the early Universe, as well as degenerate states of dense matter in compact stars. While chiral electron plasmas in Dirac and Weyl semimetals share many common properties with other plasmas, they do have a number of unusual features. Since electrons are electrically charged, collective modes of the corresponding plasma are accompanied by oscillating electromagnetic fields. Various types of such modes in chiral electron plasmas of Dirac and Weyl semimetals will be considered in this chapter. For simplicity, however, only waves with small amplitudes described by linearized equations will be presented. In the opposite regime of large amplitude waves, a diverse range of nonlinear phenomena can also be realized in a chiral matter. While this is also a fascinating topic, it has not been studied much in the literature yet and will not be covered here.

Generically, charge carriers in conductors form a very interesting type of plasma [920], where the electric charge of mobile electrons is neutralized by the positive charge of background ions. Since the average interparticle distance is usually much smaller than de Broglie's wavelength, the conduction electrons in metals form a quantum plasma even at room temperature.

This implies that quantum phenomena play a fundamental role in electron dynamics in solids. While the ion lattice determines the band structure of a material, the dynamics of the electron plasma can be considered as mostly decoupled from the lattice.[a] In other words, the behavior of an electron plasma is affected by the background field of ions, but does not modify much the lattice structure in return.

As we stated before, relativistic-like electron quasiparticles form a chiral plasma in Dirac and Weyl semimetals. Having such a plasma in solids opens the possibility of testing and utilizing many nontrivial phenomena of relativistic physics in tabletop experiments. Clearly, solid-state systems are much easier to control than the actual relativistic plasmas in cosmology, astrophysics, or heavy-ion collisions. It should be noted, however, that electron chiral plasmas in Dirac and Weyl semimetals have a number of unique features of their own. The latter stem primarily from a nontrivial topology of their electron states in a Brillouin zone. Additional interesting opportunities arise from the possibility of having strain-induced pseudo-electromagnetic fields and the surface Fermi arc states.

Unlike neutral gases, plasmas are extremely sensitive to magnetic fields. It is not surprising, therefore, that collective excitations in magnetized plasmas are often used as probes of the fundamental properties of such plasmas. In application to Dirac and Weyl semimetals, it is remarkable that their collective excitations (similarly to transport properties) in a magnetic field could reveal macroscopic implications of the quantum chiral anomaly. Indeed, when the Dirac mass of electron quasiparticles vanishes, the anomalous chiral charge appears naturally as an additional degree of freedom. It is responsible for the chiral magnetic effect (CME) [144–146, 156], the chiral separation effect (CSE) [144, 147, 148], and the chiral vortical effect (CVE) [149–152, 737]. Moreover, an interplay of these effects, together with oscillations of charge densities and electromagnetic fields, may lead to novel collective excitations. One of such possibilities is the chiral magnetic wave (CMW), which was argued to originate from an interplay of chiral and electric charge density oscillations induced by the CME and the CSE. If realized, it would be a new type of collective excitations that could exist

---

[a]Note that the statement about the decoupling of lattice and electron dynamics is correct for most but not all collective excitations. Electron–phonon interactions can be important, in particular, for some modes and effects [920]. Among them are the damping or amplification of sound waves, a modification of sound velocity, the appearance of local anomalies in dispersion relations, etc.

only in chiral plasmas. Another exotic collective mode is a hypothetical chiral vortical wave (CVW) that could be triggered by the CVE.

There are two common quasiclassical approaches for studying collective modes in plasmas. One of them is the kinetic theory,[b] while the other is hydrodynamics. Within the kinetic theory approach, one operates with a single-particle distribution function. The latter is hard to measure directly but provides a detailed description of a system. Although more complicated, such an approach allows one to describe a wide range of phenomena on long as well as short distance (time) scales. In particular, the kinetic approach is used routinely for the description of the electric charge and heat transport in solids. The hydrodynamic approach, on the other hand, is a macroscopic method that operates with local values of conserved quantities (e.g., the energy and local charge densities) and averaged characteristics such as the mean flow velocity, temperature, etc. A coarse-grained description of hydrodynamics is often much simpler and allows for a clear physical picture of many phenomena. Unlike the kinetic theory approach, however, it is less detailed and can easily miss some important effects (e.g., the Landau damping and certain plasma instabilities).

In this chapter, collective modes in a relativistic-like electron plasma of Dirac and Weyl semimetals will be discussed. Both kinetic theory and hydrodynamic approaches will be used to address the collective behavior of electron plasma. As will become clear, many properties of the corresponding collective modes are similar to those in conventional plasmas. Because of the chiral nature of an electron plasma in topological semimetals, however, some significant differences will also appear. Among the most interesting plasma modes in the long-wavelength limit are plasmons and helicons. The properties of these modes will be discussed in detail. Special attention will be paid to the effects of background pseudoelectromagnetic fields that can be induced by strains. While the emphasis will be made on applications in Dirac and Weyl semimetals, some qualitative findings will remain valid also for relativistic high-energy systems.

8.1 Kinetic approach

One of the most powerful and convenient approaches to study collective modes in a plasma is the kinetic theory. Since the electron quasiparticles

---

[b]Some collective modes can also be studied microscopically by using quantum field theoretical methods. The relevant information is usually extracted from various correlation functions in the long-wavelength and low-frequency limits. However, this is not the approach pursued in this chapter.

in Weyl and Dirac semimetals have a relativistic-like energy dispersion and nontrivial topological properties, one should use the consistent chiral kinetic theory (CKT) discussed in Sec. 6.6.2. The latter not only accounts for the Berry curvature but also captures the effects associated with a nonzero separation between the Weyl nodes and strain-induced pseudoelectromagnetic fields. On the downside, the calculations in the CKT are often tedious and difficult. The range of its validity is also restricted only to weak magnetic and pseudomagnetic fields. Finally, while offering a more detailed description, the CKT lacks the simplicity and intuitive clarity of a more coarse hydrodynamic approach.

Below we will start by reviewing the key notions of the consistent CKT. Then, by using this formalism, the properties of collective modes in the chiral electron plasma of Dirac and Weyl semimetals in background magnetic and pseudomagnetic fields will be discussed. The results will be compared with their counterparts in high energy physics.

8.1.1 *General formalism*

Let us start by recalling the main details of the consistent CKT that was introduced in Sec. 6.6.2. The central element of the formalism is the single-particle distribution function, which represents the average number of quasiparticles per unit volume of the phase space. In general, the distribution function depends on time t and space coordinate $\mathbf{r}$, as well as quasiparticle momentum $\mathbf{p}$.[c] The additional chirality degree of freedom is accounted for by using two independent distribution functions for the left-handed and right-handed quasiparticles, i.e., $f_\lambda = f_\lambda(t, \mathbf{r}, \mathbf{p})$ with $\lambda = \pm$.

One of the most distinctive features of the CKT is connected with a nontrivial Berry curvature. For chiral quasiparticles with a relativistic-like dispersion, the latter is given by the following monopole expression:

$$\mathbf{\Omega}_\lambda = \lambda \hbar \frac{\hat{\mathbf{P}}}{2p^2}. \tag{8.1}$$

The CKT analog of the Boltzmann equation reads

$$\partial_t f_\lambda + \frac{1}{1 - \frac{e}{c}(\mathbf{B}_\lambda \cdot \mathbf{\Omega}_\lambda)} \left\{ \left(-e\tilde{\mathbf{E}}_\lambda - \frac{e}{c}[\mathbf{v_p} \times \mathbf{B}_\lambda] + \frac{e^2}{c}(\tilde{\mathbf{E}}_\lambda \cdot \mathbf{B}_\lambda)\mathbf{\Omega}_\lambda \right) \cdot \partial_\mathbf{p} f_\lambda \right.$$

$$\left. + \left(\mathbf{v_p} - e[\tilde{\mathbf{E}}_\lambda \times \mathbf{\Omega}_\lambda] - \frac{e}{c}(\mathbf{v_p} \cdot \mathbf{\Omega}_\lambda)\mathbf{B}_\lambda \right) \cdot \boldsymbol{\nabla} f_\lambda \right\} = I_{\text{coll}}(f_\lambda), \tag{8.2}$$

---

[c]Here it is convenient to employ a relativistic-like notation for momentum $\mathbf{p} = \hbar\mathbf{k}$, where $\mathbf{k}$ is the wave vector. In this case, $v_F p$ has the units of energy.

where $\tilde{\mathbf{E}}_\lambda = \mathbf{E}_\lambda + (1/e)\boldsymbol{\nabla}\epsilon_\mathbf{p}$ is an effective electric field. The quasiparticles energy dispersion is given by

$$\epsilon_\mathbf{p} = v_F p \left[1 + \frac{e}{c}(\mathbf{B}_\lambda \cdot \boldsymbol{\Omega}_\lambda)\right], \tag{8.3}$$

where v_F is the Fermi velocity, c is the speed of light, and $-e$ is the electron charge. Note that the quasiparticle energy receives a nontrivial correction due to the Berry curvature, which could be interpreted as a potential energy of the quasiparticle magnetic moment in a background (pseudo)magnetic field. The quasiparticle velocity is defined by

$$\mathbf{v}_\mathbf{p} = \partial_\mathbf{p}\epsilon_\mathbf{p} = v_F\hat{\mathbf{p}}\left[1 - 2\frac{e}{c}(\mathbf{B}_\lambda \cdot \boldsymbol{\Omega}_\lambda)\right] + \frac{e v_F}{c}\mathbf{B}_\lambda(\hat{\mathbf{p}} \cdot \boldsymbol{\Omega}_\lambda). \tag{8.4}$$

Note that both electromagnetic $\mathbf{E}$ and $\mathbf{B}$ as well as strain-induced pseudo-electromagnetic $\mathbf{E}_5$ and $\mathbf{B}_5$ fields are easily included in the CKT by using the chiral electric and magnetic fields $\mathbf{E}_\lambda = \mathbf{E} + \lambda\mathbf{E}_5$ and $\mathbf{B}_\lambda = \mathbf{B} + \lambda\mathbf{B}_5$ (for a discussion of strain-induced gauge fields $\mathbf{E}_5$ and $\mathbf{B}_5$, see Sec. 3.7).

One of the simplest ways to account for the collision integral to linear order in deviations from the global equilibrium $I_{\text{coll}}(f_\lambda)$ is to employ the relaxation-time approximation, i.e.,

$$I_{\text{coll}} = -\frac{f_\lambda - f_\lambda^{\text{eq}}}{\tau}, \tag{8.5}$$

where τ is the intranode relaxation time[d] and

$$f_\lambda^{\text{eq}} = \frac{1}{e^{(\epsilon_\mathbf{p}-\mu_\lambda)/T} + 1} \tag{8.6}$$

is the equilibrium Fermi–Dirac distribution function, where T is temperature in the energy units and μ_λ is the effective chemical potential for quasiparticles of chirality λ. Note that this potential can also be expressed in terms of the electric chemical potential μ and the chiral chemical potential μ_5, i.e., $\mu_\lambda = \mu + \lambda\mu_5$.

In general, relaxation time τ could depend on the quasiparticle energy, material parameters of semimetals, the type and density of impurities, as well as the strength of background fields (see Sec. 7.1). As is clear, taking all such effects into account could complicate significantly the study of collective modes. It is of great interest, therefore, to consider collective modes in the so-called collisionless regime when the effects of collisions can be neglected. Formally, this is done by considering the limit $\tau \to \infty$.

---

[d]For a discussion about the internode scattering, see Sec. 6.6.2.

The resulting equation is known as the Vlasov equation [741, 742]. It can be justified when the frequencies of collective modes are much larger than $1/\tau$. In this connection, it is worth noting that one of the main effects of collisions is the damping of collective modes. Thus, the collisionless limit describes the situation when the damping of modes is negligible on the timescales of the order of the period of collective modes. Some dissipative effects beyond the collisionless regime will be briefly mentioned at the end of Sec. 8.1.3.3. As we will see in Sec. 8.2, the corresponding effects can be easily incorporated in the hydrodynamic description.

When studying collective excitations in a charged electron plasma, it is very important to take into account the evolution of electromagnetic fields $\mathbf{E}$ and $\mathbf{B}$, which is determined by Maxwell's equations

$$\mathbf{\nabla} \cdot \mathbf{E} = \frac{4\pi}{\varepsilon_e}(\rho + \rho_b), \quad \mathbf{\nabla} \times \mathbf{E} = -\frac{1}{c}\partial_t \mathbf{B}, \tag{8.7}$$

$$\mathbf{\nabla} \cdot \mathbf{B} = 0, \quad \frac{1}{\mu_m}\mathbf{\nabla} \times \mathbf{B} = \frac{4\pi}{c}\mathbf{J} + \frac{\varepsilon_e}{c}\partial_t \mathbf{E}. \tag{8.8}$$

Here ε_e and μ_m denote the electric permittivity and the magnetic permeability, respectively, due to the noniniterant electrons in the system. It should be emphasized that the electric charge density in Gauss's law includes the background contribution ρ_b due to the electrons bound in inner shells and lattice ions. Such a background contribution ensures that Weyl and Dirac semimetals are electrically neutral in equilibrium, i.e., $\rho + \rho_b = 0$. When collective modes are excited, the local charge density of the chiral electron plasma ρ may oscillate, but the background charge density ρ_b cannot.

Generically, the propagation of collective modes in an electron plasma is accompanied by local oscillations of electric charges and currents. Then, as is clear from Maxwell's equations, dynamical electromagnetic fields are necessarily induced as well. Since the corresponding fields usually play an important role, it is natural to ask whether there exists a similar backreaction also for the pseudoelectromagnetic fields $\mathbf{E}_5$ and $\mathbf{B}_5$. Indeed, collective modes of the electron plasma in Dirac and Weyl semimetals could also weakly perturb the lattice of ions, which would produce a nonzero strain. As it turns out, however, the corresponding effects are extremely small and the oscillating dynamical pseudoelectromagnetic fields are negligible in most cases. Intuitively, this can be explained by noting that the mass of the lattice ions is much larger than the mass of itinerant electrons. It is well justified, therefore, to set $\delta\mathbf{E}_5 = \delta\mathbf{B}_5 = \mathbf{0}$.

In the consistent CKT, the electric charge density $\rho = \rho_{\mathrm{cs}} + \sum_\lambda \rho_\lambda$ and the current density $\mathbf{J} = \mathbf{j}_{\mathrm{cs}} + \sum_\lambda \mathbf{j}_\lambda$ include the quasiparticle contributions of both left-handed ($\lambda = -$) and right-handed ($\lambda = +$) chiralities,

$$\rho_\lambda = -\sum_{\mathrm{e,h}} e \int \frac{d^3p}{(2\pi\hbar)^3} \left[1 - \frac{e}{c}(\mathbf{B}_\lambda \cdot \mathbf{\Omega}_\lambda)\right] f_\lambda, \tag{8.9}$$

$$\mathbf{j}_\lambda = -\sum_{\mathrm{e,h}} e \int \frac{d^3p}{(2\pi\hbar)^3} \left\{\mathbf{v_p} - \frac{e}{c}(\mathbf{v_p} \cdot \mathbf{\Omega}_\lambda)\mathbf{B}_\lambda - e[\tilde{\mathbf{E}}_\lambda \times \mathbf{\Omega}_\lambda]\right\} f_\lambda$$

$$- \sum_{\mathrm{e,h}} e \boldsymbol{\nabla} \times \int \frac{d^3p}{(2\pi\hbar)^3} f_\lambda v_F p \mathbf{\Omega}_\lambda, \tag{8.10}$$

as well as the following topological Chern–Simons terms:

$$\rho_{\mathrm{cs}} = \frac{e^2}{2\pi^2\hbar c}(\mathbf{b} \cdot \mathbf{B}), \tag{8.11}$$

$$\mathbf{j}_{\mathrm{cs}} = \frac{e^2}{2\pi^2\hbar^2 c}b_0\mathbf{B} - \frac{e^2}{2\pi^2\hbar}[\mathbf{b} \times \mathbf{E}]. \tag{8.12}$$

Note that b_0 and $\mathbf{b}$ describe the energy and momentum space separations between the Weyl nodes, respectively. These topological terms are very important for the consistency of the theory. Indeed, as discussed in Sec. 6.6.4, the first term in Eq. (8.12) is critical for the absence of the CME current in global equilibrium. Recall that the latter is reached when $\mu_5 = -b_0$. Additionally, as explained in Secs. 6.3 and 6.6.3, the second term in Eq. (8.12) describes the anomalous Hall effect (AHE) in Weyl semimetals with broken $\mathcal{T}$ symmetry.

Concerning Eqs. (8.9) and (8.10), it should be clear that $\sum_{\mathrm{e,h}}$ represents the summation over electron and hole states. In particular, one should replace $\mathbf{\Omega}_\lambda \to -\mathbf{\Omega}_\lambda$, $\mu_\lambda \to -\mu_\lambda$, and $e \to -e$ for the holes.

The chiral charge and current densities have the same structure, i.e., $\rho_5 = \rho_{\mathrm{cs},5} + \sum_{\lambda=\pm} \lambda\rho_\lambda$ and $\mathbf{J}_5 = \mathbf{j}_{\mathrm{cs},5} + \sum_{\lambda=\pm} \lambda\mathbf{j}_\lambda$, respectively. According to Eq. (6.162), the chiral counterparts of Chern–Simons terms are given by

$$\rho_{\mathrm{cs},5} = \frac{e^2}{6\pi^2\hbar c}(\mathbf{b} \cdot \mathbf{B}_5), \tag{8.13}$$

$$\mathbf{j}_{\mathrm{cs},5} = \frac{e^2}{6\pi^2\hbar^2 c}b_0\mathbf{B}_5 - \frac{e^2}{6\pi^2\hbar}[\mathbf{b} \times \mathbf{E}_5]. \tag{8.14}$$

As is easy to verify, the consistent currents satisfy the following relations:

$$\frac{\partial \rho}{\partial t} + (\mathbf{\nabla} \cdot \mathbf{J}) = 0, \tag{8.15}$$

$$\frac{\partial \rho_5}{\partial t} + (\mathbf{\nabla} \cdot \mathbf{J}_5) = -\frac{e^3}{2\pi^2\hbar^2 c}\left[(\mathbf{E} \cdot \mathbf{B}) + \frac{1}{3}(\mathbf{E}_5 \cdot \mathbf{B}_5)\right], \tag{8.16}$$

which mean that the electric charge is conserved locally, but the chiral charge conservation is violated because of the chiral anomaly.

In the rest of this subsection, we will use the consistent CKT to study plasmons and helicons in constant magnetic $\mathbf{B}_0$ and pseudomagnetic $\mathbf{B}_{0,5}$ fields. Note that, unlike electric and pseudoelectric fields, magnetic and pseudomagnetic fields can be nonvanishing in global equilibrium. For simplicity, we will assume that both magnetic and pseudomagnetic fields are parallel to the z-axis, i.e., $\mathbf{B}_0 \parallel \hat{\mathbf{z}}$ and $\mathbf{B}_{0,5} \parallel \hat{\mathbf{z}}$.

The global equilibrium state should be uniform in constant magnetic and pseudomagnetic fields. It is convenient to characterize such a state by fixed values of the electric chemical potential μ, the chiral chemical potential $\mu_5 = -b_0$, and temperature T. Then the energy density as well as the electric and chiral charge densities are given by the following standard expressions:

$$\epsilon = \sum_{e,h}\sum_{\lambda=\pm}\int\frac{d^3p}{(2\pi\hbar)^3}\epsilon_{\mathbf{p}}f_\lambda^{\text{eq}}$$

$$\simeq \frac{\mu^4 + 6\mu^2\mu_5^2 + \mu_5^4}{4\pi^2\hbar^3 v_F^3} + \frac{T^2(\mu^2 + \mu_5^2)}{2\hbar^3 v_F^3} + \frac{7\pi^2 T^4}{60\hbar^3 v_F^3}, \tag{8.17}$$

$$\rho = -\sum_{e,h} e\sum_{\lambda=\pm}\int\frac{d^3p}{(2\pi\hbar)^3}f_\lambda^{\text{eq}} \simeq -e\frac{\mu\left(\mu^2 + 3\mu_5^2 + \pi^2 T^2\right)}{3\pi^2\hbar^3 v_F^3}, \tag{8.18}$$

$$\rho_5 = -\sum_{e,h} e\sum_{\lambda=\pm}\lambda\int\frac{d^3p}{(2\pi\hbar)^3}f_\lambda^{\text{eq}} \simeq -e\frac{\mu_5\left(\mu_5^2 + 3\mu^2 + \pi^2 T^2\right)}{3\pi^2\hbar^3 v_F^3}. \tag{8.19}$$

The pressure and the enthalpy density are $P = \epsilon/3$ and $w = \epsilon + P = 4\epsilon/3$, respectively. Note that the quadratic corrections $O(B_0^2)$ were neglected in the above expressions.

It is interesting to point out that the electric charge density is modified by the Chern–Simons term in Eq. (8.11), which is linear in $\mathbf{B}_0$. As is clear, however, the external magnetic field by itself cannot violate the net electrical neutrality of the sample. A seeming contradiction is resolved if the equilibrium electric chemical potential changes in the presence of the magnetic field, $\mu \to \mu_B$. By using Eqs. (8.11) and (8.18), one obtains the following equation for μ_B valid up to linear order in $\mathbf{B}_0$:

$$\frac{\mu_B \left(\mu_B^2 + 3\mu_5^2 + \pi^2 T^2\right)}{3\pi^2 \hbar^3 v_F^3} = \frac{e(\mathbf{b} \cdot \mathbf{B}_0)}{2\pi^2 \hbar c} + \frac{\mu \left(\mu^2 + 3\mu_5^2 + \pi^2 T^2\right)}{3\pi^2 \hbar^3 v_F^3}. \qquad (8.20)$$

When $\mathbf{B}_0 \neq \mathbf{0}$, therefore, one should use chemical potential μ_B rather than μ to characterize the system. For simplicity of presentation, however, in the study of collective modes within the kinetic theory we will assume that the effects of chiral shift and magnetic field are already included in the electric chemical potential μ. This should be sufficient for the qualitative analysis below. When using the hydrodynamic approach in Sec. 8.2, however, we will restore the notation μ_B in order to emphasize its explicit magnetic field dependence.

8.1.2 *Chiral magnetic wave: Heuristic consideration*

As we discussed earlier, one of the most interesting characteristics of electron quasiparticles in Dirac and Weyl semimetals compared to that in usual metals is their chirality. This additional degree of freedom allows one to consider the corresponding chiral solid-state plasma as composed of two plasmas with right- and left-handed chirality. This fact immediately draws an analogy with superfluid helium II which can be viewed as a two-component fluid composed of normal and superfluid components. The corresponding model was suggested by László Tisza [921] and later refined by Lev Landau [895, 922, 923]. As is well known, one of the principal consequences of the two-component nature of superfluid helium II is a new collective excitation in this material. This is the famous second sound when the superfluid and normal components oscillate in antiphase keeping the local pressure in helium II unperturbed. Since only the normal component is related to the flow of entropy, the second flow can be described as a heat transfer. The heat transfer by means of a propagating wave is much more effective than the usual mechanism of diffusion. In fact, helium II has the highest thermal conductivity of any known material exceeding few hundred times that in cooper. Therefore, vaporization takes place only at the surface and no boiling occurs. This analogy with two-component helium II leads to a natural question whether there is a collective excitation in Dirac and Weyl semimetals connected with the two-component nature of their electron plasma?

In this subsection, we will demonstrate that such excitations, where chirality plays the key role, indeed exist. A paradigmatic example in high-energy physics is the *chiral magnetic wave* (CMW). To start with, we employ a heuristic approach and neglect the effects of dynamical electromagnetism. However, as will be explicitly shown in the following subsection, the corresponding effects are relevant and qualitatively change the properties of the CMW in solids.

To explain how the CMW arises, it is useful to recall its historical development that starts with the discovery of the chiral magnetic effect in chirally imbalanced plasma in a magnetic field [144–146, 156]. The corresponding analysis was performed in the context of quark–gluon plasma generated in heavy-ion collisions, where a very strong magnetic field $\mathbf{B}$ is produced by the motion of ultrarelativistic nuclei and a chiral imbalance was supposed to be generated by instantons. The latter are nonperturbative topologically nontrivial tunneling processes relating QCD states with different winding numbers. The corresponding chiral imbalance can be described phenomenologically by means of the chiral chemical potential μ_5. Then, as shown in Sec. 6.4.3, the CME leads to the electric current $\mathbf{j} = e^2 \mu_5 \mathbf{B}/(2\pi^2 \hbar^2 c)$ flowing in the direction of magnetic field. However, the realization of the chiral imbalance in heavy-ion collisions is still debated (see also the discussion in Sec. 6.4).

By studying the normal ground state of a dense relativistic plasma in a magnetic field, it was suggested that the CME could be realized even in the absence of an initial chiral imbalance [551]. The idea is simple. Since a dense magnetized relativistic plasma has a nonzero chemical potential μ, a chiral current flows due to the chiral separation effect. This produces an excess of opposite chiral charges around the polar regions of the fireball, see Fig. 8.1. Then, each of these chiral charges triggers a CME current whose direction is opposite at the opposite poles. The inward flows of these electric currents will diffuse inside the fireball, while the outward flows will lead to a distinct observational signal: an excess of same charges going back-to-back. For obvious reasons, the corresponding signal is known in the literature as the *quadrupole chiral magnetic effect*.

The key step in deriving the CMW was done by Dmitri Kharzeev and Ho-Ung Yee [924], who observed that an interplay of the chiral magnetic and separation effects producing the quadrupole effect could result in self-sustained cycle of oscillations of chiral and electric charges propagating in an infinite medium. The starting point in the corresponding analysis is the expression for the CME and CSE currents connected with fluctuations of

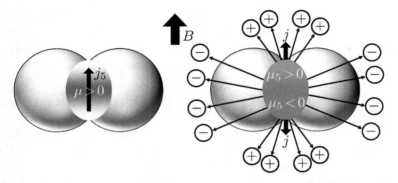

Fig. 8.1 Schematic illustration of the CSE and the CME in heavy-ion collisions. An axial current is initially driven by a nonzero baryon chemical potential, which leads to a chiral charge imbalance in polar regions. In turn, this imbalance results in two back-to-back electric currents produced by the CME.

electric and chiral chemical potentials

$$\delta\mathbf{j} = \frac{e^2\mathbf{B}}{2\pi^2\hbar^2 c}\delta\mu_5, \tag{8.21}$$

$$\delta\mathbf{j}_5 = \frac{e^2\mathbf{B}}{2\pi^2\hbar^2 c}\delta\mu. \tag{8.22}$$

Further, by using Eqs. (8.18) and (8.19) for the electric and chiral charge densities, one can easily express fluctuations $\delta\mu$ and $\delta\mu_5$ in terms of fluctuations of the corresponding densities δn and δn_5. In the limit of small chemical potentials, $|\mu| \ll T$ and $|\mu_5| \ll T$, one has

$$\delta\mu = \left(\frac{\partial n}{\partial\mu}\right)^{-1}\delta n = \frac{3v_F^3\hbar^3}{T^2}\delta n, \tag{8.23}$$

$$\delta\mu_5 = \left(\frac{\partial n_5}{\partial\mu_5}\right)^{-1}\delta n = \frac{3v_F^3\hbar^3}{T^2}\delta n_5. \tag{8.24}$$

Note that in a truly relativistic matter, the Fermi velocity v_F should be replaced with the speed of light c. Then by using the fact that both electric and axial currents are conserved in the absence of an electric field, one easily derives

$$-e\frac{T^2}{3v_F^3\hbar^3}\partial_t\delta\mu + \frac{e^2}{2\pi^2\hbar^2 c}(\mathbf{B}\cdot\boldsymbol{\nabla})\delta\mu_5 = 0, \tag{8.25}$$

$$-e\frac{T^2}{3v_F^3\hbar^3}\partial_t\delta\mu_5 + \frac{e^2}{2\pi^2\hbar^2 c}(\mathbf{B}\cdot\boldsymbol{\nabla})\delta\mu = 0. \tag{8.26}$$

Combing these equations and using the plane wave ansatz, $\delta\mu \propto e^{-i\omega t + \mathbf{k}\cdot\mathbf{r}}$ and $\delta\mu_5 \propto e^{-i\omega t + \mathbf{k}\cdot\mathbf{r}}$, it is straightforward to find that Eqs. (8.25) and (8.26) describe a propagating chiral magnetic wave with the following dispersion relation:

$$\omega_{CMW} = -\frac{3ev_F^3\hbar}{2\pi^2 cT^2}(\mathbf{B}\cdot\mathbf{k}), \qquad (8.27)$$

where $\mathbf{k}$ is a wave vector. An inclusion of a diffusion term in the relation for electric and current densities will lead to a dissipation of the CMW. In such a case, an additional term $\propto -ik^2$ appears in the dispersion relation.

Thus, we conclude that the chiral nature of (quasi)particles in relativistic and relativistic-like plasmas results in a collective excitation of a new type with oscillating electric and chiral charge densities. However, there is one important distinction compared to the case of the second sound in superfluid helium II connected with the fact that the electric charge oscillations $-e\delta n$ in a chiral plasma necessarily produce an oscillating electric field. Consequently, Maxwell's equations should be taken into account in the analysis too. As we will show below, similarly to the case of usual plasmas, the effects of dynamical electromagnetism generate a nonzero gap for the CMW transforming it into a chiral plasmon. The chiral plasmons in Dirac and Weyl semimetals will be analyzed within a kinetic approach in the following subsection.

Before concluding this subsection, it is worth recalling that, as was discussed in Chapter 4, many experimentally discovered Weyl and Dirac semimetals have multiple Weyl nodes and Dirac points. This fact enriches underlying physics and opens new possibilities. For example, by using a simple model of a Weyl semimetal with broken parity-inversion symmetry and four Weyl nodes, it was found that a new collective mode called the *chiral zero sound* can propagate when a background magnetic field is applied [925]. In essence, this mode is related to oscillations of different Fermi surfaces of the *same* chirality in antiphase, which is in contrast to the inphase oscillations in the CMW. Therefore, while the CME currents occur in the CMW, they cancel out in the chiral zero sound mode. Because these oscillations do not lead to any net electric currents, the dynamical electromagnetism does not open a gap by mixing this mode with plasmons.

8.1.3 *Chiral magnetic plasmons*

Plasmons are, perhaps, the best known collective excitations in plasma. They are gapped collective modes originating from oscillations of a local electric charge density and electromagnetic fields. The value of their energy gap is determined by the Langmuir frequency Ω_e,[e] which is also often called the plasma frequency. As will be shown below, the properties of plasmons in Dirac and Weyl semimetals have a number of distinctive features that stem from a relativistic-like energy dispersion relation of quasiparticles. In particular, the corresponding plasmons are driven by the oscillations of electric and chiral charge densities. Additionally, they are affected by the background magnetic and strain-induced pseudomagnetic fields already at linear order in the fields. Therefore, it is justified to call them *the chiral magnetic plasmons* [420, 421, 423].

8.1.3.1 *General approach*

In order to study the spectrum of plasmon excitations in Weyl and Dirac materials, let us use the kinetic framework outlined above. As usual, one seeks plasmon solutions in the form of plane waves. Then the oscillating parts of electric and magnetic fields can be written as

$$\delta \mathbf{E}(t, \mathbf{r}) = \delta \mathbf{E} e^{-i\omega t + i\mathbf{k}\cdot\mathbf{r}}, \quad \delta \mathbf{B}(t, \mathbf{r}) = \delta \mathbf{B} e^{-i\omega t + i\mathbf{k}\cdot\mathbf{r}}, \tag{8.28}$$

where ω is the frequency and $\mathbf{k}$ is the wave vector. By making use of Maxwell's equations (8.7) and (8.8), one can express $\delta \mathbf{B}$ in terms of the oscillating electric field, i.e., $\delta \mathbf{B} = c\left[\mathbf{k} \times \delta \mathbf{E}\right]/\omega$. As for the electric field, it satisfies the following equation:

$$\mathbf{k}\left(\mathbf{k} \cdot \delta \mathbf{E}\right) - k^2 \delta \mathbf{E} = -\frac{\omega^2}{c^2}\left(n_{\text{ref}}^2 \delta \mathbf{E} + 4\pi \mu_{\text{m}} \delta \mathbf{P}\right), \tag{8.29}$$

where $\delta \mathbf{P}(t, \mathbf{r}) = \delta \mathbf{P} e^{-i\omega t + i\mathbf{k}\cdot\mathbf{r}}$ denotes the polarization vector and $n_{\text{ref}} = \sqrt{\varepsilon_e \mu_{\text{m}}}$ is the background refractive index.[f] Since μ_{m} can differ substantially from unity only in ferromagnetic materials, it is reasonable to set $\mu_{\text{m}} = 1$. By using the electric susceptibility tensor χ_{mn} with $m, n = x, y, z$,[g] it is convenient to express the polarization vector in terms of the oscillating electric field as follows:

$$\delta P_m = i\frac{\delta J_m}{\omega} = \sum_n \chi_{mn} \delta E_n, \tag{8.30}$$

---

[e]For a nonrelativistic electron gas, the plasma frequency is $\Omega_e = \sqrt{4\pi e^2 n/m_e}$, where $n = -\rho/e$ is the particle number density and m_e is the electron mass.
[f]Note that the background refractive index is also a function of frequency. However, for the sake of simplicity, we assume that the corresponding dependence is weak.
[g]The electric susceptibility tensor χ_{mn} and the dielectric tensor ε_{mn} are related as follows: $\varepsilon_{mn} = n_{\text{ref}}^2 \delta_{mn} + 4\pi \chi_{mn}$, where δ_{mn} is the Kronecker delta.

where δJ_m is an oscillating part of the consistent electric current density. Then Eq. (8.29) takes the form

$$\left(n_{\text{ref}}^2\omega^2 - c^2k^2\right)\delta_{mn}\delta E_n = -c^2k_mk_n\delta E_n - 4\pi\omega^2\chi_{mn}\delta E_n, \qquad (8.31)$$

where δ_{mn} is the Kronecker delta. This system of linear equations admits nontrivial solutions only when the corresponding determinant vanishes

$$\det\left[\left(n_{\text{ref}}^2\omega^2 - c^2k^2\right)\delta_{mn} + c^2k_mk_n + 4\pi\omega^2\chi_{mn}\right] = 0. \qquad (8.32)$$

This characteristic equation defines the spectrum of collective modes.

The collective dynamics of Dirac/Weyl quasiparticles is encoded in the susceptibility tensor χ_{mn}. The latter can be determined by solving the kinetic equation (8.2) and calculating the corresponding current. Since the sought collective modes are connected with small perturbations on top of the global equilibrium state, the distribution function can be calculated perturbatively. In particular, one can assume that $f_\lambda(t, \mathbf{r}) = f_\lambda^{\text{eq}} + \delta f_\lambda(t, \mathbf{r})$, where f_λ^{eq} is the equilibrium distribution given by Eq. (8.6) and $\delta f_\lambda(t, \mathbf{r})$ is a small deviation. Similarly to the electromagnetic fields in Eq. (8.28), function $\delta f_\lambda(t, \mathbf{r})$ should take a plane-wave form, i.e.,

$$\delta f_\lambda(t, \mathbf{r}) = f_\lambda^{(1)}e^{-i\omega t + i\mathbf{k}\cdot\mathbf{r}}. \qquad (8.33)$$

At the *zeroth* order in perturbation theory, the kinetic equation (8.2) reduces to

$$[\hat{\mathbf{p}} \times \mathbf{B}_{0,\lambda}] \cdot \partial_{\mathbf{p}}f_\lambda^{\text{eq}} = 0. \qquad (8.34)$$

This equation is satisfied automatically by any distribution function that depends only on p and $(\mathbf{p} \cdot \mathbf{B}_{0,\lambda})$. This is indeed the case for the equilibrium distribution function (8.6) with the quasiparticle dispersion law given in Eq. (8.3).

At the *first* order in perturbation theory, Eq. (8.2) gives

$$i\left[(1 - \kappa_\lambda)\omega - (\mathbf{v_p} \cdot \mathbf{k}) + \frac{e}{c}(\mathbf{v_p} \cdot \mathbf{\Omega}_\lambda)(\mathbf{B}_{0,\lambda} \cdot \mathbf{k}) + \frac{e}{c}[\mathbf{v_p} \times \mathbf{B}_{0,\lambda}] \cdot \partial_{\mathbf{p}}\right]f_\lambda^{(1)}$$

$$= -e\left[(\delta\tilde{\mathbf{E}} \cdot \mathbf{v_p}) - \frac{e}{c}(\mathbf{v_p} \cdot \mathbf{\Omega}_\lambda)(\delta\tilde{\mathbf{E}} \cdot \mathbf{B}_{0,\lambda})\right]\frac{\partial f_\lambda^{\text{eq}}}{\partial\epsilon_{\mathbf{p}}}, \qquad (8.35)$$

where

$$\kappa_\lambda \equiv \frac{e}{c}(\mathbf{\Omega}_\lambda \cdot \mathbf{B}_{0,\lambda}) = \lambda\hbar\frac{e(\hat{\mathbf{p}} \cdot \mathbf{B}_{0,\lambda})}{2cp^2}, \qquad (8.36)$$

$$\delta\tilde{\mathbf{E}} = \delta\mathbf{E} + i\frac{\lambda v_F\hbar}{2\omega p}\mathbf{k}(\hat{\mathbf{p}} \cdot [\mathbf{k} \times \delta\mathbf{E}]). \qquad (8.37)$$

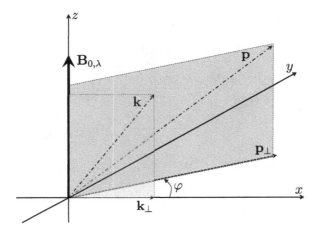

Fig. 8.2 The relative orientation of the (pseudo)magnetic field $\mathbf{B}_{0,\lambda}$, the wave vector $\mathbf{k}$, and the momentum $\mathbf{p}$ used in the analysis of collective modes. The azimuthal angle φ is measured from $\mathbf{k}_\perp$.

Note that the last term in $\delta\tilde{\mathbf{E}}$ originates from the magnetic moment term in the quasiparticle energy (8.3). The relative orientations of the (pseudo)magnetic field $\mathbf{B}_{0,\lambda}$, the wave vector $\mathbf{k}$, and the momentum $\mathbf{p}$ are shown in Fig. 8.2. By making use of the cylindrical coordinates, where φ is the azimuthal angle of momentum $\mathbf{p}$, the kinetic equation (8.35) can be rendered in the following form:

$$\frac{\partial f_\lambda^{(1)}}{\partial\varphi} - i\left[\frac{cp\omega/v_F - c(\mathbf{p}\cdot\mathbf{k})}{eB_{0,\lambda}} + \frac{\lambda\hbar\omega(\hat{\mathbf{p}}\cdot\mathbf{B}_{0,\lambda})}{2v_F p B_{0,\lambda}}\right] f_\lambda^{(1)} = \frac{c(\mathbf{p}\cdot\delta\tilde{\mathbf{E}})}{B_{0,\lambda}}\frac{\partial f_\lambda^{eq}}{\partial\epsilon_\mathbf{p}}.$$

$$(8.38)$$

This equation has the same form as its counterpart in a nonrelativistic plasma [739], i.e.,

$$\frac{\partial f_\lambda^{(1)}}{\partial\varphi} + i\left(a_1 + a_2\cos\varphi\right)f_\lambda^{(1)} = Q(\varphi).$$

$$(8.39)$$

In the problem at hand,

$$Q(\varphi) = a_3\cos\left(\varphi - \varphi_E\right) + a_4 + a_5\frac{p_\parallel k_\parallel + p_\perp k_\perp\cos\varphi}{p^2}$$

$$\times\left[E_\perp p_\perp k_\parallel\sin\left(\varphi - \varphi_E\right) + E_\perp p_\parallel k_\perp\sin\varphi_E - E_\parallel k_\perp p_\perp\sin\varphi\right],$$

$$(8.40)$$

where φ_E denotes the azimuthal angle of $\mathbf{E}$ measured from $\mathbf{k}_\perp$ in the plane perpendicular to the (pseudo)magnetic field (see also Fig. 8.2) and the coefficients a_i $(i = 1, 2, \ldots, 5)$ are given by

$$a_1 = -\frac{cp\omega/v_F - cp_\| k_\|}{eB_{0,\lambda}} - \frac{\lambda\hbar\omega p_\|}{2v_F p^2}, \quad a_2 = \frac{cp_\perp k_\perp}{eB_{0,\lambda}}, \tag{8.41}$$

$$a_3 = \frac{cp_\perp E_\perp}{B_{0,\lambda}}\frac{\partial f_\lambda^{eq}}{\partial \epsilon_{\mathbf{p}}}, \quad a_4 = \frac{cp_\| E_\|}{B_{0,\lambda}}\frac{\partial f_\lambda^{eq}}{\partial \epsilon_{\mathbf{p}}}, \quad a_5 = \frac{i\lambda\hbar v_F}{2\omega B_{0,\lambda}}\frac{\partial f_\lambda^{eq}}{\partial \epsilon_{\mathbf{p}}}. \tag{8.42}$$

The general solution to Eq. (8.39) reads

$$f_\lambda^{(1)}(\varphi) = e^{-ia_2 \sin\varphi} \int_0^{\varphi - C_0} e^{i[a_2 \sin(\varphi - \tau) - a_1 \tau]} Q(\varphi - \tau) d\tau$$

$$+ C_1 e^{-i(a_1\varphi + a_2 \sin\varphi)}. \tag{8.43}$$

Since the solution for $f_\lambda^{(1)}(\varphi)$ should be a periodic function, one must set $C_1 = 0$ and $C_0 = -s_B\infty$, where $s_B = \mathrm{sgn}(eB_{0,\lambda})$. Note that the correct sign of C_0 ensures that the integral is finite as $\omega \to \omega + i0$. From a physics viewpoint, the use of a frequency with a vanishingly small imaginary part, $\omega + i0$, represents a setup with oscillating fields gradually turning on. As is clear from Eq. (8.43), $f_\lambda^{(1)}(\varphi)$ is determined by the inhomogeneous term $Q(\varphi)$ defined in Eq. (8.39). Since $Q(\varphi)$ describes the driving forces associated with oscillating electromagnetic fields, it is crucial to take the effects of dynamical electromagnetism into account.

In the case of the chiral electron plasma in Dirac and Weyl semimetals, Eq. (8.43) takes the following explicit form:

$$f_\lambda^{(1)}(\varphi) = \int_0^{s_B\infty} d\tau \left\{ \frac{c(\mathbf{p} \cdot \delta\mathbf{E}_\tau)}{B_{0,\lambda}} + i\frac{\lambda\hbar v_F c}{2\omega B_{0,\lambda}}(\hat{\mathbf{p}} \cdot [\mathbf{K}_\tau^{(1)} \times \delta\mathbf{E}_\tau])(\hat{\mathbf{p}} \cdot \mathbf{K}_\tau^{(1)}) \right\}$$

$$\times \frac{\partial f_\lambda^{eq}}{\partial \epsilon_{\mathbf{p}}} e^{i[cp\tau\omega/v_F - c(\mathbf{p} \cdot \mathbf{K}_\tau)]/(eB_{0,\lambda})}, \tag{8.44}$$

where

$$\delta\mathbf{E}_\tau = \left(\delta\mathbf{E}_{\perp\tau}, \delta E_\|\right) = \delta\mathbf{E}\cos\tau + [\hat{\mathbf{z}} \times \delta\mathbf{E}]\sin\tau + \hat{\mathbf{z}}(\hat{\mathbf{z}} \cdot \delta\mathbf{E})(1 - \cos\tau), \tag{8.45}$$

$$\mathbf{K}_\tau = \mathbf{k}\sin\tau + [\hat{\mathbf{z}} \times \mathbf{k}](1 - \cos\tau) + \hat{\mathbf{z}}(\mathbf{k} \cdot \hat{\mathbf{z}})(\tau - \sin\tau) - \frac{\lambda\hbar\omega\tau}{2cv_F p^2}eB_{0,\lambda}. \tag{8.46}$$

and

$$\mathbf{K}_\tau^{(1)} = \mathbf{k}\cos\tau + [\hat{\mathbf{z}}\times\mathbf{k}]\sin\tau + \hat{\mathbf{z}}\,(\mathbf{k}\cdot\hat{\mathbf{z}})\,(1-\cos\tau). \tag{8.47}$$

Having determined the first-order correction to the distribution function, one can proceed with the calculation of the polarization vector $\mathbf{P}$ defined in Eq. (8.30). To linear order in oscillating fields, the latter is given by

$$\mathbf{P} = \mathbf{P}_0 + \mathbf{P}_\mathrm{M} + \mathbf{P}_\mathrm{CS}, \tag{8.48}$$

where

$$\mathbf{P}_0 = \sum_{e,h}\sum_{\lambda=\pm}\frac{ie^2}{\omega}\int\frac{d^3p}{(2\pi\hbar)^3}\left\{[\delta\tilde{\mathbf{E}}\times\boldsymbol{\Omega}_\lambda] + \frac{1}{\omega}\,(\mathbf{v_p}\cdot\boldsymbol{\Omega}_\lambda)\,[\mathbf{k}\times\delta\mathbf{E}]\right.$$
$$\left. + \frac{1}{c}\,(\delta\mathbf{v_p}\cdot\boldsymbol{\Omega}_\lambda)\,\mathbf{B}_{0,\lambda}\right\}f_\lambda^\mathrm{eq}$$
$$- \sum_{e,h}\sum_{\lambda=\pm}\frac{ie}{\omega}\int\frac{d^3p}{(2\pi\hbar)^3}v_F\hat{\mathbf{p}}\left[1 - 2\frac{e}{c}(\boldsymbol{\Omega}_\lambda\cdot\mathbf{B}_{0,\lambda})\right]f_\lambda^{(1)} \tag{8.49}$$

is the quasiparticle contribution,

$$\mathbf{P}_\mathrm{M} = \sum_{e,h}\sum_{\lambda=\pm}\frac{e}{\omega}\int\frac{d^3p}{(2\pi\hbar)^3}\epsilon_\mathbf{p}f_\lambda^{(1)}\,[\mathbf{k}\times\boldsymbol{\Omega}_\lambda] \tag{8.50}$$

is the magnetization term, and

$$\mathbf{P}_\mathrm{CS} = -i\frac{e^2}{2\pi^2\omega\hbar}\,[\mathbf{b}\times\delta\mathbf{E}] + i\frac{e^2b_0}{2\pi^2\omega^2\hbar}\,[\mathbf{k}\times\delta\mathbf{E}] \tag{8.51}$$

is the part that comes from the topological Chern–Simons contributions in Eq. (8.12).

The oscillating correction to velocity $\delta\mathbf{v_p}$ is connected with the dynamical (pseudo)magnetization given by the last term in the square brackets in Eq. (8.3). Its explicit form reads

$$\delta\mathbf{v_p} = -\frac{\lambda\hbar ev_F}{2\omega p^2}\left\{2\hat{\mathbf{p}}\,(\hat{\mathbf{p}}\cdot[\mathbf{k}\times\delta\mathbf{E}]) - [\mathbf{k}\times\delta\mathbf{E}]\right\}. \tag{8.52}$$

To linear order in $B_{0,\lambda}$, the explicit expression for the polarization tensor can be obtained by integrating over quasiparticle momentum $\mathbf{p}$ using the table integrals in Appendix B. It should be noted that, at small $B_{0,\lambda}$, the main contribution to the quasiparticle distribution function (8.44) comes from the region of small τ. Thus, it is justified to expand $\sin\tau$ and $\cos\tau$ in Eqs. (8.45)–(8.47) in powers of small τ and truncate the series at the leading order. Since intermediate calculations and the final expression for the electric susceptibility tensor are bulky, they will not be presented here. The interested reader can find the corresponding details in the original papers [420, 421, 423]. Below we will discuss several special cases and review the key qualitative features of plasmons in the electron plasmas of Dirac and Weyl semimetals.

8.1.3.2 *Gaps of chiral magnetic plasmons*

Let us start from discussing the physical properties of plasmons in the limit of the vanishing wave vector, $k \to 0$. This limit greatly simplifies the calculation of subleading correction $f_\lambda^{(1)}$ to the distribution function. The result reads [420]

$$f_\lambda^{(1)} = i\frac{ev_F}{p\omega}\frac{\partial f_\lambda^{\text{eq}}}{\partial \epsilon_{\mathbf{p}}}\left\{(\mathbf{p}\cdot\delta\mathbf{E})\left[1 - \frac{\lambda\hbar e\,(\mathbf{B}_{0,\lambda}\cdot\mathbf{p})}{2cp^3}\right] + i\,(\mathbf{p}\cdot[\mathbf{B}_{0,\lambda}\times\delta\mathbf{E}])\,\frac{ev_F}{cp\omega}\right\}.$$

$$(8.53)$$

The corresponding polarization vector is given as a sum of three terms

$$\delta\mathbf{P} = \frac{A_0}{4\pi}\delta\mathbf{E} + \frac{A_1}{4\pi}\left[\mathbf{b}\times\delta\mathbf{E}\right] + \frac{A_2}{4\pi}\left[\delta\mathbf{E}\times\hat{\mathbf{z}}\right], \qquad (8.54)$$

where

$$A_0 = -\frac{n_{\text{ref}}^2\Omega_e^2}{\omega^2}, \qquad (8.55)$$

$$A_1 = -i\frac{2n_{\text{ref}}^2\alpha v_F}{\pi\omega}, \qquad (8.56)$$

$$A_2 = i\frac{2en_{\text{ref}}^2\alpha v_F^2}{3\pi\omega c}\sum_{\lambda=\pm}\left[\frac{B_{0,\lambda}\mu_\lambda}{\hbar^2\omega^2} - \frac{B_{0,\lambda}}{4T}F\left(\frac{\mu_\lambda}{T}\right)\right]. \qquad (8.57)$$

Here we used the shorthand notation

$$\alpha = \frac{e^2}{\hbar v_F\varepsilon_e} \qquad (8.58)$$

for the effective coupling constant and

$$\Omega_e = \sqrt{\frac{4\alpha}{3\pi\hbar^2}\left(\mu^2 + \mu_5^2 + \frac{\pi^2 T^2}{3}\right)} \qquad (8.59)$$

for the plasma frequency. As in nonrelativistic plasmas, Ω_e^2 is determined by the electron density. In Eq. (8.57), the following function was introduced:

$$F\left(\frac{\mu_\lambda}{T}\right) = \frac{1}{4}\int \frac{dp}{p}\left[\frac{1}{\cosh\left(\frac{v_F p - \mu_\lambda}{2T}\right)} - \frac{1}{\cosh\left(\frac{v_F p + \mu_\lambda}{2T}\right)}\right]. \qquad (8.60)$$

In addition to the usual dependence on the electron density, the polarization vector in Eq. (8.54) is modified by topological contributions. Indeed, as is easy to verify, the second term in Eq. (8.54) originates from the anomalous Hall effect. The last term in Eq. (8.54) describes the Faraday rotation as well as its pseudomagnetic counterpart.

By comparing the explicit expression in Eq. (8.54) with the definition in Eq. (8.30), it is straightforward to extract the electric susceptibility tensor χ_{mn}. Then by substituting the result into the characteristic equation (8.32), one obtains

$$(n_{\text{ref}}^2 + A_0)[(n_{\text{ref}}^2 + A_0)^2 + A_1^2 b_\perp^2 + (A_2 - A_1 b_\parallel)^2] = 0, \qquad (8.61)$$

where $b_\perp = \sqrt{b_x^2 + b_y^2}$ and $b_\parallel = b_z$ are transverse and longitudinal (with respect to the field) components of the chiral shift. Note that the left-hand side of the spectral equation (8.61) factorizes into two factors. One of them describes longitudinal modes, while the other transverse. The corresponding approximate solutions read

$$\omega_\ell \approx \Omega_e, \qquad \omega_{\text{tr}}^\pm \approx \Omega_e\sqrt{1 \pm \frac{\Omega^{(1)}}{\Omega_e}}, \qquad (8.62)$$

where

$$\Omega^{(1)} = \frac{2\alpha v_F}{3\pi c}\left\{9c^2 b_\perp^2 + \left[\frac{2ev_F}{\hbar^2\Omega_e^2}(B_0\mu + B_{0,5}\mu_5)\right.\right.$$

$$\left.\left. + 3c b_\parallel - \frac{ev_F}{4T}\sum_{\lambda=\pm} B_{0,\lambda} F\left(\frac{\mu_\lambda}{T}\right)\right]^2\right\}^{1/2}. \qquad (8.63)$$

In the absence of the chiral shift, the collective modes with frequencies ω_ℓ and $\omega_{\text{tr}}^\pm$ in Eq. (8.62) correspond to the longitudinal (with $\delta\mathbf{E} \parallel \hat{\mathbf{z}}$)

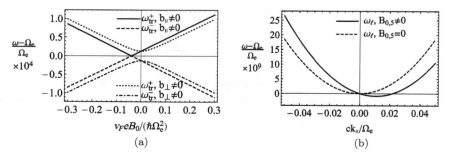

Fig. 8.3   (a) The gaps of the chiral magnetic plasmons at $b_{\parallel} = 0.2\,\Omega_e/c$ (solid and dashed lines) and $b_{\perp} = 0.2\,\Omega_e/c$ (dotted and dot-dashed lines). (b) The frequency of the longitudinal collective mode ω_ℓ as a function of the wave vector at $B_{0,5} = 0.5\,\hbar\Omega_e^2/(v_F e)$ (solid line) and $B_{0,5} = 0$ (dashed line). In both panels, we used $\mu = \hbar\Omega_e\sqrt{3\pi/(4\alpha)}$, $\mu_5 = 0$, $B_{0,5} = 0$, and $T \to 0$.

and transverse (with $\delta \mathbf{E} \perp \hat{\mathbf{z}}$) waves. Their nature is more complicated, however, when $\mathbf{b} \neq \mathbf{0}$.

Below we will discuss in more detail the case where $B_0 \neq 0$ and $\mu \neq 0$ but $\mathbf{B}_{0,5}$ and μ_5 vanish. It is worth noting, however, that a combined effect of the pseudomagnetic field $\mathbf{B}_{0,5}$ and the chiral chemical potential μ_5 on chiral magnetic plasmons is similar to that of the magnetic field $\mathbf{B}_0$ and the electric chemical potential μ.

The gaps of the chiral magnetic plasmons are presented in Fig. 8.3(a) for two special cases $b_{\perp} \neq 0$ and $b_{\parallel} \neq 0$. Note that here and below, we use the model parameters resembling those of the Dirac semimetal Cd_3As_2 (see also Appendix D). In particular, the background refractive index is $n_{\text{ref}} = \sqrt{\varepsilon_e \mu_{\text{m}}} \approx 6$ [926] and the Fermi velocity is $v_F \approx 1.5 \times 10^8$ cm/s [107].

From Fig. 8.3(a), one observes that there are two different transverse modes with nonequal frequencies $\omega_{\text{tr}}^{\pm}$. In the case $b_{\perp} \neq 0$, the frequency difference (or gap) grows with a magnetic field. The minimum value of the gap is given by $\omega_{\text{tr}}^+ - \omega_{\text{tr}}^- \approx \Omega_e^{(1)} = 2\alpha v_F b_{\perp}/\pi$ when background fields vanish $B_0 = B_{0,5} = 0$. The situation is slightly different in the case $b_{\parallel} \neq 0$. While the frequencies of transverse modes are still generically different, the degeneracy can be achieved at a certain magnetic field [420].

It should be noted that plasmons in Dirac and Weyl semimetals have several distinctive features absent in nonrelativistic plasmas. While conventional plasmons are related to oscillations of the electric charge density, chiral magnetic plasmons involve also oscillations of the chiral charge density [420, 421, 423]. This can be easily verified by calculating the local chiral charge density (8.9) and the current density (8.10) by using the distribution function (8.53). An interplay of oscillating electric and chiral charges resembles those of the CMW (see Sec. 8.1.2) obtained in the background field approximation (i.e., without taking dynamical electromagnetism into account). In contrast, however, the chiral magnetic plasmons are gapped modes obtained in a chiral plasma where dynamical electromagnetic fields are treated self-consistently.

8.1.3.3 *Chiral magnetic plasmons with nonzero wave vector*

Since the case of the collective modes with a nonzero wave vector is quite cumbersome, only several representative results will be outlined here. A more detailed analysis can be found in [421].

In order to illustrate the dependence on the wave vector $\mathbf{k}$, let us consider the case when the longitudinal $\delta\mathbf{E} \parallel \mathbf{k}$ and transverse $\delta\mathbf{E} \perp \mathbf{k}$ modes do not mix. This is possible for the longitudinal propagation $\mathbf{k} \parallel \mathbf{B}_0$ provided $\mathbf{b}_\perp = \mathbf{0}$. In such a case, the dispersion relation for the longitudinal mode is defined by the following spectral equation:

$$1 + \frac{3\Omega_e^2}{v_F^2 k^2}\left[1 - \frac{\omega}{2v_F k}\ln\left(\frac{\omega + v_F k}{\omega - v_F k}\right) + \frac{2\alpha e v_F^5 k^2 (\mathbf{B}_{0,5} \cdot \mathbf{k})}{3\pi\hbar c\omega\Omega_e^2(\omega^2 - v_F^2 k^2)}\right] = 0. \quad (8.64)$$

Note that the second term in the square brackets gives rise to an imaginary part when $\omega < v_F k$. Such an imaginary part describes the celebrated Landau damping [739, 927]. In the case of high-frequency modes such as plasmons, however, the Landau damping plays no role.

At $B_{0,5} = 0$, the spectral equation (8.64) has the same form as in a conventional nonrelativistic plasma [739]. The explicit expressions for the frequency of longitudinal modes can be obtained in the limit of long wavelength $ck \ll \Omega_e$, as well as small $B_{0,5}$ and $b_\parallel$, i.e.,

$$\omega_\ell \simeq \Omega_e\sqrt{1 - \frac{2\alpha e(\mathbf{B}_{0,5} \cdot \mathbf{k})v_F^3}{\pi c\hbar\Omega_e^3} + \frac{3}{5}\left(\frac{v_F k}{\Omega_e}\right)^2 + \cdots}. \quad (8.65)$$

The spectral equation for the transverse modes is more complicated, but it can also be solved analytically in the long-wavelength limit. To linear order in $B_{0,\lambda}$ and $b_\parallel$, the result reads

$$\omega_{\text{tr}}^\pm \simeq \sqrt{\left(\Omega_{e,B}^\pm\right)^2 + A_1^\mp v_F k + A_2^\pm (v_F k)^2 + \cdots}, \quad (8.66)$$

where

$$(\Omega_{e,B}^{\pm})^2 \simeq \Omega_e^2 \pm \frac{2\alpha\Omega_e}{\pi} \left\{ \sum_{\lambda=\pm} \frac{v_F^2 eB_{0,\lambda}}{3c\hbar^2\Omega_e^2} \left[\mu_\lambda - \frac{\hbar^2\Omega_e^2}{4T} F\left(\frac{\mu_\lambda}{T}\right) \right] - v_F b_{\parallel} \right\},$$

$$(8.67)$$

$$A_1^{\pm} \simeq \pm \frac{2\alpha\mu_5}{3\hbar\pi} - \frac{2\alpha^2 v_F^2 \mu_5}{9c\pi^2} \sum_{\lambda=\pm} \frac{eB_{0,\lambda}}{\hbar^3\Omega_e^3} \left[\mu_\lambda + \frac{\hbar^2\Omega_e^2}{4T} F\left(\frac{\mu_\lambda}{T}\right) \right]$$

$$+ \frac{2\alpha^2 b_{\parallel} v_F \mu_5}{3\pi^2\hbar\Omega_e},$$

$$(8.68)$$

$$A_2^{\pm} \simeq \frac{c^2}{v_F^2 n_{\text{ref}}^2} + \frac{1}{5} \pm \frac{\alpha b_{\parallel} v_F}{\pi\Omega_e} \left(\frac{c^2}{v_F^2 n_{\text{ref}}^2} - \frac{1}{5} - \frac{\alpha^2\mu_5^2}{9\pi^2\hbar^2\Omega_e^2} \right)$$

$$\mp \frac{\alpha\hbar v_F^2}{3c\pi} \sum_{\lambda=\pm} \frac{eB_{0,\lambda}}{\hbar^3\Omega_e^3} \left[\mu_\lambda \left(\frac{c^2}{v_F^2 n_{\text{ref}}^2} - \frac{3}{5} - \frac{\alpha^2\mu_5^2}{3\pi^2\hbar^2\Omega_e^2} \right) \right.$$

$$+ \frac{\hbar^2\Omega_e^2}{4T} F\left(\frac{\mu_\lambda}{T}\right) \left(\frac{c^2}{v_F^2 n_{\text{ref}}^2} - \frac{6}{5} - \frac{\alpha^2\mu_5^2}{9\pi^2\hbar^2\Omega_e^2} \right) \right],$$

$$(8.69)$$

and terms of order $(B_0^2, B_{0,5}^2, b_{\parallel}^2)$ were neglected. The main feature of plasmons, which is also evident from Eqs. (8.65) and (8.66), is a gapped spectrum. The underlying dynamics of plasmon modes in Dirac and Weyl semimetals is largely the same as in other plasmas. Thus, the value of the gap is of the order of the Langmuir frequency Ω_e. It is instructive to recall that plasmons are electromagnetic waves in a conducting medium that are accompanied by local oscillations of the electron charge density. It is the self-consistent treatment of such oscillations, coupled to dynamical electromagnetic fields, that produces the collective modes with the Langmuir frequency Ω_e. The chiral effects (i.e., the CME and the CSE) in topological semimetals, however, trigger additional oscillations of chiral charge density and, therefore, modify the properties of plasmons. In this connection, it should be mentioned that the CME current and its pseudomagnetic counterpart do not have to vanish out of equilibrium. In fact, they play an important role in the propagation of chiral plasmon modes. One of the corresponding implications is the terms linear in k in the dispersion relations (8.65) and (8.66).

The dispersion relation of the longitudinal mode ω_ℓ is shown in Fig. 8.3(b) for $B_{0,5} = 0$ and $B_{0,5} = 0.5\,\hbar\Omega_e^2/(v_F e)$. As one sees, the frequency of chiral magnetic plasmons is quadratic in the wave vector. As for

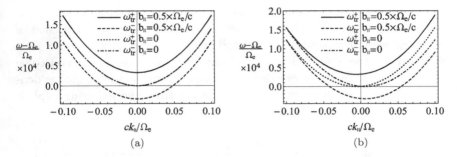

Fig. 8.4 (a) The frequencies of the transverse collective modes $\omega_{\mathrm{tr}}^{\pm}$ for a few values of $b_{\parallel}$ at $\mu = \sqrt{3\pi/(4\alpha)}\hbar\Omega_e$ and $\mu_5 = 0$. The solid and dotted lines correspond to ω_{tr}^{+}. The dashed and dot-dashed lines represent ω_{tr}^{-}. (b) The frequency of the transverse collective modes $\omega_{\mathrm{tr}}^{\pm}$ for a few values of $b_{\parallel}$ at $\mu_5 = \sqrt{3\pi/(4\alpha)}\hbar\Omega_e$ and $\mu = 0$. Here solid and dotted lines correspond to ω_{tr}^{+} as well as dashed and dot-dashed lines represent ω_{tr}^{-}. In both cases, we set $B_0 = B_{0,5} = 0$ and $T \to 0$.

the effect of a nonzero background pseudomagnetic field, it shifts the minimum to $k_{\parallel} \approx 5 v_F \alpha e B_{0,5}/(3\pi c\hbar\Omega_e)$.

The effects of the chiral shift **b** on the transverse modes can be seen from Figs. 8.4(a) and 8.4(b). The frequencies $\omega_{\mathrm{tr}}^{\pm}$ are generically split by **b**. This is consistent with the results at $k = 0$ discussed in Sec. 8.1.3.2. Another qualitative feature of the chiral magnetic plasmons is an asymmetry of the dispersion relation when the Weyl nodes are separated in energy $b_0 = -\mu_5$. The corresponding asymmetry can be seen in Fig. 8.4(b). One can verify that both features, i.e., the splitting of frequencies and the asymmetry, appear also in the case with **b** $\perp$ **B**$_0$ [421]. In the case of the transverse propagation of collective modes, i.e., **k** $\perp$ **B**$_0$, the effects of (pseudo)magnetic field and chiral shift are, in general, similar to those at **k** $\parallel$ **B**$_0$. For example, frequencies of the transverse modes are also split [421].

As stated earlier, the unique nature of the chiral magnetic plasmon is the oscillations of the local chiral charge density $\delta\rho_5(t, \mathbf{r})$. The corresponding explicit expression reads

$$\delta\rho_5(t, \mathbf{r}) = -\sin[\omega t - (\mathbf{k} \cdot \mathbf{r})]n_{\mathrm{ref}}^2 \left[\frac{2\alpha\mu\mu_5 \, (\delta\mathbf{E} \cdot \mathbf{k})}{\pi^2 \hbar^2 v_F^2 k^2} \left(1 - \frac{\omega}{2 v_F k} \ln\left| \frac{\omega + v_F k}{\omega - v_F k} \right| \right) \right.$$

$$\left. + \frac{\alpha v_F \omega e \, (\delta\mathbf{E} \cdot \mathbf{B}_0)}{2\pi^2 \hbar c(\omega^2 - v_F^2 k^2)} \right].$$

$$(8.70)$$

Certainly, such oscillations are absent for ordinary electromagnetic plasmons. In topological semimetals, they come from a dynamical version of the chiral electric separation effect [928], which is given by the term $\propto \mu\mu_5$. Similarly to other chiral effects, the chiral magnetic plasmon is an outcome of the chiral anomaly. Interestingly, the oscillations of electric charge density are also unusual. Indeed, as one can see from the following explicit expression:

$$\delta\rho(t,\mathbf{r}) = -\sin\left[\omega t - (\mathbf{k}\cdot\mathbf{r})\right]n_{\text{ref}}^2\left[\frac{3\Omega_e^2(\delta\mathbf{E}\cdot\mathbf{k})}{4\pi v_F^2 k^2}\left(1 - \frac{\omega}{2v_F k}\ln\left|\frac{\omega + v_F k}{\omega - v_F k}\right|\right)\right.$$
$$\left. + \frac{\alpha v_F \omega e\left(\delta\mathbf{E}\cdot\mathbf{B}_{0,5}\right)}{2\pi^2\hbar c(\omega^2 - v_F^2 k^2)}\right] - \cos\left[\omega t - (\mathbf{k}\cdot\mathbf{r})\right]n_{\text{ref}}^2\frac{\alpha v_F\left(\mathbf{b}\cdot[\mathbf{k}\times\delta\mathbf{E}]\right)}{2\pi^2\omega},$$

$$(8.71)$$

oscillations of the electric charge density depend on both background pseudomagnetic field and chiral shift.

The oscillating charge densities $\delta\rho_5(t,\mathbf{r})$ and $\delta\rho(t,\mathbf{r})$ associated with the propagation of plasmons suggest that their underlying dynamics is similar to the CMW (see Sec. 8.1.2). Recall that the latter was argued to be a gapless collective mode that is supported by oscillations of electric and chiral charge densities, which feed each other via the chiral separation and chiral magnetic effects. This appears to be a viable possibility in the background field approximation where the effects of dynamical electromagnetism are neglected. The corresponding approximation cannot be justified, however, in the complete dynamical theory. Indeed, local oscillations of electric charge density induce dynamical electric fields that have a profound effect on the collective mode. The effect is particularly strong in the long-wavelength limit where it is responsible for generating a nonzero gap in the spectrum. In essence, this implies that there is a chiral magnetic plasmon but not a gapless CMW.

It is important to remember that the above analysis was performed in the collisionless approximation. In such an approximation, all dissipative effects associated with nonzero electrical conductivity and charge diffusion were neglected. This is not a strong limitation for high-frequency and short-wavelength modes, where the effects of dissipation are small. However, both electrical conductivity and charge diffusion are detrimental for long-wavelength modes. In fact, as a hydrodynamic analysis reveals, the CMW or plasmon type modes become strongly overdamped [929].

Before concluding this subsection, let us note that the chiral kinetic theory cannot be used in the limit of a strong (pseudo)magnetic field. This is because the Landau-level quantization becomes important and a quasiclassical description in terms of quasiparticle momenta becomes meaningless. Instead, in the strong field limit, collective modes could be studied by using the methods of quantum field theory. In essence, the problem reduces to calculating the electric susceptibility tensor from the same polarization tensor that was used in the discussion of optical conductivity in Sec. 7.3.2. Another powerful approach is the Wigner function formalism [719, 723–728, 730, 732]. The plasmon dispersion relation in the long-wavelength limit is determined primarily by the LLL and, therefore, is rather unusual [721, 770, 930]. Indeed, the plasmon spectrum at $\mu = 0$ and $T \to 0$ is entirely determined by the chiral anomaly and its form is very different from the conventional plasmon spectrum. In particular, the plasmon gap is determined by a magnetic field $\Omega_e \propto |eB|$. Also, because of the chiral nature of quasiparticles in the LLL, this plasmon is not damped.

8.1.4 *Helicons*

For a long time, it was believed that low-frequency electromagnetic waves cannot propagate inside metals and conducting media. In the early 1960s, however, it was understood that gapless low-frequency electromagnetic modes can propagate inside uncompensated metals if a strong constant magnetic field is applied [931, 932].[h] The corresponding waves are called *helicons* [932]. They are sustained by the Lorentz force and have a specific circular polarization. Soon after the theoretical prediction, helicons were verified experimentally in metallic sodium [933].

Later it was understood that helicons are collective excitations of the same type as the atmospheric whistlers, i.e., the radio waves generated by lightnings that propagate along the magnetic field lines in the ionospheres of planets [934, 935]. The interest to helicons is driven by possible applications in research and technology [936–938]. As an example of industrial application, let us mention that helicons can be utilized for efficient heating of plasmas. Indeed, unlike other electromagnetic waves that penetrate and heat only a thin layer of conducting medium with the characteristic penetration depth $\sim c/\Omega_e$, helicons can propagate deep into the bulk and,

---

[h]Recall that the electron and hole densities are nonequal in uncompensated metals.

therefore, heat plasma more uniformly. In fact, by using the helicon heating, one can achieve an efficiency that is a few orders of magnitude higher than in other methods.

As in the case of plasmons, helicons in Dirac and Weyl semimetals have interesting unique physical properties, connected with a relativistic-like energy dispersion relation of quasiparticles as well as a nonzero chiral shift parameter **b** in Weyl semimetals [422, 939]. Moreover, it was found that a novel type of helicons can be realized in strain-induced pseudomagnetic fields **B**$_5$ [422].

To derive the dispersion relation of helicons in Dirac and Weyl semimetals, one can use the formalism of the consistent CKT. The characteristic equation as well as the general solution to the kinetic equation can be derived similarly to the case of the chiral magnetic plasmons in Sec. 8.1.3. In particular, the first-order correction to the distribution function follows from the same general equation (8.39) when a background (pseudo)magnetic field is nonzero. Indeed, the presence of such a field is crucial for the existence of helicons. As will be explicitly shown below, such collective modes are closely related to the cyclotron resonance [937]. By following the same approach as in nonrelativistic plasmas [739], it is convenient to use the following ansatz for the distribution function:

$$f_\lambda^{(1)}(\varphi) = g(\varphi)e^{-ia_2 \sin \varphi}. \qquad (8.72)$$

As is easy to check, the function $g(\varphi)$ satisfies the following equation:

$$\frac{\partial g}{\partial \varphi} + ia_1 g = e^{ia_2 \sin \varphi} Q(\varphi). \qquad (8.73)$$

Since $g(\varphi)$ is periodic in azimuthal angle φ, one can seek for the solution to Eq. (8.73) in the form of a Fourier series, i.e.,

$$g(\varphi) = \sum_{n=-\infty}^{\infty} g_n e^{in\varphi}, \qquad (8.74)$$

where Fourier coefficients g_n are determined by

$$g_n = -\frac{i}{2\pi(a_1 + n)} \int_0^{2\pi} e^{ia_2 \sin \tau - in\tau} Q(\tau)d\tau. \qquad (8.75)$$

For simplicity, one can use the local approximation [937], where the dependence of $f_\lambda^{(1)}$ and $g(\varphi)$ on the wave vector **k** is neglected. This approximation is valid in the long-wavelength limit $k \ll \Omega_c/v_F|_{p=p^*}$, where $p^* \sim \sqrt{\mu_5^2 + \mu^2 + \pi^2 T^2}/v_F$ is the characteristic momentum in a chiral plasma, and

$$\Omega_c \approx \frac{v_F|eB_{0,\lambda}|}{cp} \tag{8.76}$$

is the cyclotron frequency for massless fermions.

By using the definition of $Q(\varphi)$ in Eq. (8.40) as well as coefficients a_i in Eqs. (8.41)–(8.42), one derives the following expressions for the Fourier coefficients:

$$g_n = -\frac{i}{a_1 + n}\left[a_4 J_n(a_2) + \frac{a_3}{2}e^{-i\varphi_E}J_{n-1}(a_2) + \frac{a_3}{2}e^{i\varphi_E}J_{n+1}(a_2)\right], \tag{8.77}$$

where φ_E denotes the azimuthal angle of $\mathbf{E}$, which, similarly to φ, is measured from the direction of $\mathbf{k}_\perp$ in the plane perpendicular to the (pseudo)magnetic field, see Fig. 8.2. Note that, in order to perform the integration over τ in Eq. (8.75), we used the integral representation of the Bessel function of the first kind [281]

$$J_n(x) = \frac{1}{2\pi}\int_0^{2\pi} e^{i(n\theta - x\sin\theta)}d\theta. \tag{8.78}$$

To get an approximate solution, one can retain only the lowest few Fourier harmonics (i.e., $n = 0, \pm 1$) of the infinite Fourier series (8.74). Then the leading-order correction to the distribution function can be written as $f_\lambda^{(1)} = f_{\lambda,0}^{(1)} + f_{\lambda,-}^{(1)} + f_{\lambda,+}^{(1)}$, where [422, 939]

$$f_{\lambda,0}^{(1)} = i\frac{ev_Fp_\parallel(1-\kappa_\lambda)}{p\omega}(\delta\mathbf{E}\cdot\hat{\mathbf{z}})\frac{\partial f_\lambda^{\mathrm{eq}}}{\partial\epsilon_{\mathbf{p}}}, \tag{8.79}$$

$$f_{\lambda,\pm}^{(1)} = i\frac{ev_Fp_\perp(1-\kappa_\lambda)}{2p}\frac{\delta E_x \mp i\delta E_y}{\omega\pm\Omega_c}\frac{\partial f_\lambda^{\mathrm{eq}}}{\partial\epsilon_{\mathbf{p}}}e^{\pm i\varphi}. \tag{8.80}$$

The corresponding polarization vector can be calculated from the definition in Eqs. (8.48)–(8.51). Since helicons are gapless modes, it is sufficient to consider only the limit of small frequencies $\omega \ll \Omega_c|_{p=p^*}$. In this case, $\omega\pm\Omega_c$ in the denominator of Eq. (8.80) can be replaced with $\pm\Omega_c$.

Then, one arrives at the following explicit expression for the polarization vector [422]:

$$\delta\mathbf{P} = \frac{A_1}{4\pi}\left[\delta\mathbf{E} \times \hat{\mathbf{z}}\right] + \frac{A_2}{4\pi}\left[\mathbf{b} \times \delta\mathbf{E}\right] + \frac{A_3}{4\pi}\left[\mathbf{E} - \hat{\mathbf{z}}\left(\delta\mathbf{E} \cdot \hat{\mathbf{z}}\right)\right] + \frac{A_4}{4\pi}\hat{\mathbf{z}}\left(\delta\mathbf{E} \cdot \hat{\mathbf{z}}\right),$$

(8.81)

where

$$A_1 = i\frac{\tilde{A}_1}{\omega} = -\sum_{\lambda=\pm} i\frac{2ec\mu_\lambda}{3B_{0,\lambda}v_F^3\hbar^3\pi\omega}\left(\mu_\lambda^2 + \pi^2 T^2\right),$$

(8.82)

$$A_2 = i\frac{\tilde{A}_2}{\omega} = -i\frac{2e^2}{\pi\omega\hbar},$$

(8.83)

$$A_3 = \sum_{\lambda=\pm} \frac{2c^2}{3\pi\hbar^3 v_F^5 B_{0,\lambda}^2}\left(\mu_\lambda^4 + 2\pi^2\mu_\lambda^2 T^2 + \frac{7\pi^4 T^4}{15}\right),$$

(8.84)

$$A_4 = \frac{\tilde{A}_4}{\omega^2} = -\sum_{\lambda=\pm} \frac{2e^2}{3\pi\hbar^3 v_F\omega^2}\left(\mu_\lambda^2 + \frac{\pi^2 T^2}{3}\right).$$

(8.85)

Note that the global equilibrium condition $b_0 = -\mu_5$, which ensures that the CME current is absent, is assumed here.

Since helicons are absent at $\mathbf{k} \perp \mathbf{B}_0$ [739, 936, 937, 940], let us consider only the case $\mathbf{k} \parallel \mathbf{B}_0$. Then the characteristic equation (8.32) reads

$$\left(n_{\text{ref}}^2 + A_4\right)\left\{\left[\left(n_{\text{ref}}^2 + A_3\right)\omega^2 - c^2 k^2\right]^2 + \omega^4\left(A_1 - b_\parallel A_2\right)^2\right\}$$
$$- A_2^2\omega^4 b_\perp^2\left[c^2 k^2 - \left(n_{\text{ref}}^2 + A_3\right)\omega^2\right] = 0.$$

(8.86)

Before solving this equation, let us briefly discuss the role of coefficients A_i where $i = 1, \ldots, 4$. The existence of helicons, which are transverse collective modes, stems from nonzero off-diagonal components of the electric susceptibility tensor. Those components are determined by coefficients A_1 and A_2. The former captures the Hall conductivity generalized to the case of a pseudomagnetic field and a nonzero chiral chemical potential. The latter describes the anomalous Hall effect. The other two coefficients, i.e., A_3 and A_4, are not critical for the existence of helicons because they affect only the diagonal components of the electric susceptibility tensor.

The analysis of Eq. (8.86) can be further simplified by considering the case $\mathbf{b} \parallel \mathbf{B}_0$.[i] Then the characteristic equation for low-frequency collective

---

[i]Because of the large prefactor $n_{\text{ref}}^2 + A_4$, the quantitative effects of the chiral shift component $b_\perp$ are small in general.

modes becomes

$$\left(n_{\text{ref}}^2\omega^2 - c^2k^2 + \omega^2A_3\right)^2 + \omega^4\left(A_1 - A_2b_{\parallel}\right)^2 = 0. \tag{8.87}$$

Its gapless solution reads

$$\omega_h = \frac{-|\tilde{A}_1 - \tilde{A}_2b_{\parallel}| + \sqrt{4c^2k^2\left(n_{\text{ref}}^2 + A_3\right) + (\tilde{A}_2b_{\parallel} - \tilde{A}_1)^2}}{2\left(n_{\text{ref}}^2 + A_3\right)}$$

$$= \frac{c^2k^2}{|\tilde{A}_1 - \tilde{A}_2b_{\parallel}|} + O(k^3). \tag{8.88}$$

In the limit of vanishing temperature, in particular, one obtains

$$\omega_h = \frac{\pi\hbar k^2 c^2\left|B_0^2 - B_{0,5}^2\right|}{2\left|2\pi^2\hbar cB_0\rho_0 - 2\pi^2\hbar cB_{0,5}\rho_{0,5}b_{\parallel}(B_0^2 - B_{0,5}^2)\right|} + O(k^3), \tag{8.89}$$

where ρ_0 and $\rho_{0,5}$ are the electric and chiral charge densities defined in Eqs. (8.18) and (8.19), respectively. It is instructive to consider two special cases, i.e.,

$$\omega_h\big|_{B_{0,5}\to 0,\mu_5\to 0} = \frac{|eB_0|c^2\hbar\pi v_F^2k^2}{\left|\pi\hbar c\Omega_e^2\mu - 2e^3B_0v_F^2b_{\parallel}\right|} + O(k^3), \tag{8.90}$$

$$\omega_h\big|_{B_0\to 0,\mu\to 0} = \frac{|eB_{0,5}|c^2\hbar\pi v_F^2k^2}{\left|\pi\hbar c\Omega_e^2\mu_5 - 2e^3B_{0,5}v_F^2b_{\parallel}\right|} + O(k^3). \tag{8.91}$$

These dispersion relations correspond to a usual helicon [939] and a pseudomagnetic helicon [422] in Weyl semimetals, respectively. Note that the latter describes a gapless electromagnetic wave in the electron plasma of a strained semimetal in the absence of a background magnetic field.

To visualize the dispersion relation of a helicon in topological semimetals, one can use a set of model parameters for the Dirac semimetal Cd_3As_2 (see also Appendix D). In particular, the background refractive index is $n_{\text{ref}} = \sqrt{\varepsilon_e\mu_m} \approx 6$ [926], the Fermi velocity is $v_F \approx 1.5 \times 10^8$ cm/s [107], and the lattice spacing along the crystallographic axis [001] is $a_z \approx 25.480$ Å [107]. The value of the chiral shift is approximated by the measured distance between the Dirac nodes as $b^* = 0.3\pi/a_z$. The corresponding helicon frequencies for a few values of the chiral shift $b_{\parallel}$ are shown in Figs. 8.5(a) and 8.5(b) for $B_0 \neq 0$ and $B_{0,5} \neq 0$, respectively. Since the dispersion relations in Eqs. (8.90) and (8.91) remain unchanged when interchanging $B_0 \leftrightarrow B_{0,5}$ and $\mu \leftrightarrow \mu_5$ simultaneously, only the results for $\mu \neq 0$ are presented.

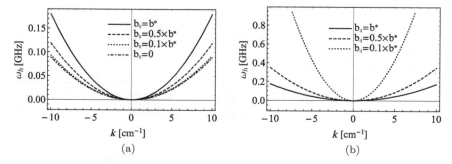

Fig. 8.5   The helicon dispersion relation ω_h given by Eq. (8.88) for (a) $B_0 = 10^{-2}$ T and $B_{0,5} = 0$ as well as (b) $B_{0,5} = 10^{-2}$ T and $B_0 = 0$. Chiral shift is $b_\parallel = b^*$ (solid line), $b_\parallel = 0.5\,b^*$ (dashed line), $b_\parallel = 0.1\,b^*$ (dotted line), and $b_\parallel = 0$ (dot-dashed line). In addition, $\mu_0 = 25$ meV, $\mu_5 = -b_0 = 0$, and $b^* \approx 0.3\pi/a_z$.

As is clear from Fig. 8.5(a), even the usual helicon is modified by the chiral shift. While it still has a quadratic dispersion, its frequency increases slightly when $b_\parallel > 0$ (for a given choice of signs of μ and eB_0). It is worth emphasizing, however, that the frequency could also decrease with increasing the magnitude of $b_\parallel$. This is evident also from Eq. (8.90) where the term with the chiral shift in the denominator can have either sign. The results for a pseudomagnetic helicon are shown in Fig. 8.5(b). As one can see, at nonzero $b_\parallel$, there exist helicons with the conventional quadratic dispersion. Their frequency is very sensitive to the value of the chiral shift.[j]

The simplified analysis presented above utilized a model of a Weyl semimetal with a single pair of Weyl nodes. In reality, however, Weyl materials usually have multiple pairs of Weyl nodes with opposite chiralities. A generalization to such a case can be done by introducing separate distribution functions for quasiparticles from different pairs of Weyl nodes, i.e., $f_\lambda \to f_\lambda^{(\xi)}$ where ξ is the pair index. Note that each pair of Weyl nodes can be characterized by different values of the chiral shift $\mathbf{b}^{(\xi)}$ and the energy separation $b_0^{(\xi)}$. Then the modified expression for the polarization vector can be derived by adding partial contributions from all nodes. The final result has the same form as Eq. (8.81) but contains an additional sum over the pair index ξ. For a purely magnetic field background (i.e., $B_0 \neq 0$ but $B_{0,5} = 0$), the helicon is similar to that discussed above. In the

---

[j]Formally, at $b_\parallel = 0$, there is also a solution with a linear dispersion relation. As is easy to check, the latter corresponds to a usual electromagnetic wave rather than a helicon. Also, neither conventional nor anomalous Hall effect plays any role in its propagation.

case of Weyl semimetals with strains, however, the modifications are more profound. Since pseudomagnetic fields are determined by the separation between the Weyl nodes, their direction and magnitude depend not only on the type of strain but also on the pair index ξ. This implies that quasi-particles from different Weyl nodes experience different pseudomagnetic fields. As a result, the study of helicons becomes more complicated [422].

Experimentally, the detection of helicons requires measuring the transmission amplitude of electromagnetic waves through a Weyl or Dirac crystal as a function of a (pseudo)magnetic field or frequency [937, 941]. Because of the interference of standing helicon waves inside the crystal, the resulting signal should be an oscillating function of a magnetic field or strain. Another conventional method is the so-called crossed-coil method, where the excitation and pickup coils are orthogonal. Without magnetic field, the mutual inductance between the coils is zero. However, the presence of a magnetic field allows for a helicon resonance, whose observation provides a convenient method to probe helicons.

8.2 Hydrodynamic approach

Another conventional approach to study long-wavelength collective excitations in plasmas is to use the language of hydrodynamics. The hydrodynamic formalism for the electron plasmas of Dirac and Weyl semimetals was reviewed in Sec. 7.4. While many features of such a formalism resemble relativistic chiral hydrodynamics derived in the context of high energy physics [812, 896–900], there are differences too. Most importantly, in the case of Weyl semimetals, one needs to utilize the consistent hydrodynamics that accounts properly for the topological effects associated with the nonvanishing separation between the Weyl nodes [901, 902].

Let us review briefly the key notions of electron hydrodynamics. The abbreviated versions of the Euler and energy conservation equations read

$$\frac{1}{v_F}\partial_t\left(\frac{w}{v_F}\mathbf{u}+\sigma^{(\epsilon,B)}\mathbf{B}\right)=-en\left(\mathbf{E}+\frac{1}{c}\left[\mathbf{u}\times\mathbf{B}\right]\right)-\frac{w}{\tau v_F^2}\mathbf{u}$$

$$+\frac{\sigma^{(B)}(\mathbf{E}\cdot\mathbf{B})}{3v_F^2}\mathbf{u}+O(\boldsymbol{\nabla_r}) \qquad (8.92)$$

and

$$\partial_t\epsilon=-\mathbf{E}\cdot\left(en\mathbf{u}-\sigma^{(B)}\mathbf{B}\right)+O(\nabla_{\mathbf{r}}), \qquad (8.93)$$

respectively. Here the terms with spatial derivatives are omitted for simplicity.

In the above hydrodynamic equations, $w = \epsilon + P$ is the enthalpy density, ϵ is the quasiparticle energy density, P is the pressure, n is the quasiparticle number density, $\mathbf{u}$ is the fluid velocity, τ is the intranode relaxation time, and v_F is the Fermi velocity. Note that the terms proportional to $\sigma^{(B)}$ and $\sigma^{(\epsilon,B)}$ capture the effects of the chiral anomaly. The explicit expressions for these and other transport coefficients are given in Eqs. (7.146)–(7.149).

As was already discussed in Sec. 7.4, Eqs. (8.92) and (8.93) should be supplemented by the electric and chiral charge continuity relations (8.15) and (8.16), as well as Maxwell's equations (8.7) and (8.8). The expressions for the charge and current densities are given in Eqs. (7.155) and (7.156) as well as Eqs. (7.157) and (7.158), respectively.

It is worth emphasizing that the Chern–Simons terms do not enter the Euler equation (or the Navier–Stokes equation in the viscous regime). Neither do they appear in the energy conservation equation (8.93). From a physics viewpoint, this is understandable since the Chern–Simons contributions come from quantum states deeply below the Fermi surface, rather than low-energy quasiparticles in the vicinity of the Fermi surface. Technically, the hydrodynamic equations could be obtained as moments of the Boltzmann equation in the chiral kinetic theory. Since the corresponding equations contain only derivatives of the Fermi–Dirac distribution functions, they are insensitive to details of the energy spectrum away from the Fermi surface. In the expressions for the charge and current densities, however, information about the separations between the Weyl nodes should be included. Thus, the Chern–Simons terms still affect hydrodynamics via a self-consistent treatment of dynamical electromagnetism.

An important feature of the electron hydrodynamics in semimetals is the dissipative term proportional to $\mathbf{u}/\tau$ on the right-hand side in Eq. (8.92). Such a term captures the dominant dissipative effects due to electron scattering on phonons and impurities [240].[k] It breaks explicitly the Galilean invariance. This is not surprising since there is a preferred frame in which the lattice of ions is at rest. When there are no external electromagnetic fields, the term $\mathbf{u}/\tau$ ensures that $\mathbf{u} = 0$ in the electron fluid in global equilibrium.

---

[k]For simplicity, the intrinsic electrical conductivity is neglected here.

It is worth emphasizing that the consistent hydrodynamics presented in Sec. 7.4 assumes the single-fluid approximation, where the left- and right-handed electron quasiparticles form a single fluid with a common flow velocity $\mathbf{u}$. While such an approach is simple and routinely used in conventional plasma studies [940], it could miss effects related to the intercomponent dynamics of the fluid. The single-fluid approximation breaks down, for example, when there are chirality-dependent forces. A strain-induced pseudomagnetic field provides a natural way to achieve the chirality-dependent Lorentz force, which can separate quasiparticles of opposite chiralities. In such a case, a two-fluid description with nonequal velocities $\mathbf{u}$ and $\mathbf{u}_5$ is needed. Below, for the sake of simplicity, the pseudomagnetic field will be set to zero and the validity of the single-fluid approximation will be assumed.

8.2.1 *Linearized consistent hydrodynamics*

For collective modes in a hydrodynamic approach, one can assume that deviations of the local thermodynamic parameters from their global equilibrium values are small. This justifies the use of linearized hydrodynamic equations and allows one to seek the solutions in the form of plane waves, i.e., $\delta X(x) = \delta X\, e^{-i\omega t + i\mathbf{k}\cdot\mathbf{r}}$, where ω is a frequency, $\mathbf{k}$ is a wave vector, and δX is a generic name for the oscillating variables such as the chemical potential ($\delta\mu$), temperature (δT), etc.

To linear order in perturbations, the electric and chiral charge continuity relations (8.15) and (8.16) are given by

$$\omega\delta\rho - (\mathbf{k}\cdot\delta\mathbf{J}) = 0, \tag{8.94}$$

$$\omega\delta\rho_5 - (\mathbf{k}\cdot\delta\mathbf{J}_5) = -i\frac{e^3}{2\pi^2\hbar^2 c}(\mathbf{B}_0\cdot\delta\mathbf{E}), \tag{8.95}$$

where

$$\delta\rho = -e\delta n + \frac{\sigma^{(B)}(\mathbf{B}_0\cdot\delta\mathbf{u})}{3v_F^2} + i\frac{5c^2\sigma^{(\epsilon,u)}(\mathbf{B}_0\cdot[\mathbf{k}\times\delta\mathbf{u}])}{2v_F^2} + \frac{e^2(\mathbf{b}\cdot\delta\mathbf{B})}{2\pi^2\hbar c}, \tag{8.96}$$

$$\delta\rho_5 = -e\delta n_5 + \frac{\sigma_5^{(B)}(\mathbf{B}_0\cdot\delta\mathbf{u})}{3v_F^2}, \tag{8.97}$$

and

$$\delta \mathbf{J} = -en\delta\mathbf{u} + \mathbf{B}_0\delta\sigma^{(B)} - \frac{e^2\,[\mathbf{b} \times \delta\mathbf{E}]}{2\pi^2\hbar} + \sigma^{(V)}\delta\boldsymbol{\omega} + i\frac{\sigma^{(\epsilon,V)}}{2}\,[\mathbf{k} \times \delta\boldsymbol{\omega}],$$

(8.98)

$$\delta \mathbf{J}_5 = -en_5\delta\mathbf{u} + \sigma_5^{(B)}\delta\mathbf{B} + \mathbf{B}_0\delta\sigma_5^{(B)} + \sigma_5^{(V)}\delta\boldsymbol{\omega} + i\frac{\sigma_5^{(\epsilon,V)}}{2}\,[\mathbf{k} \times \delta\boldsymbol{\omega}].$$

(8.99)

While no background vorticity is assumed, a dynamical vorticity $\delta\boldsymbol{\omega} = i\,[\mathbf{k} \times \delta\mathbf{u}]\,/2$ could be induced by collective modes.

The linearized forms of the Euler equation and the energy conservation relation are given by

$$\frac{\omega}{v_F}\left[\frac{w\delta\mathbf{u}}{v_F} + \mathbf{B}_0\delta\sigma^{(\epsilon,B)} + \sigma^{(\epsilon,B)}\delta\mathbf{B} + \frac{\hbar n_5\delta\boldsymbol{\omega}}{2}\right] - \mathbf{k}\delta P$$

$$-\frac{4\sigma^{(\epsilon,B)}}{15v_F}\,[\mathbf{k}\,(\mathbf{B}_0\cdot\delta\mathbf{u}) + \delta\mathbf{u}\,(\mathbf{B}_0\cdot\mathbf{k}) + \mathbf{B}_0\,(\mathbf{k}\cdot\delta\mathbf{u})] - \frac{c\sigma^{(\epsilon,B)}}{3v_F}\,[\mathbf{k} \times \delta\mathbf{E}]$$

$$-5\sigma^{(\epsilon,u)}\mathbf{k}\,(\mathbf{B}_0\cdot\delta\mathbf{B}) - \frac{\sigma^{(\epsilon,V)}}{2c}\,[(\mathbf{B}_0\cdot\mathbf{k})\delta\boldsymbol{\omega} + \mathbf{k}\,(\mathbf{B}_0\cdot\delta\boldsymbol{\omega})]$$

$$= -eni\delta\mathbf{E} + \frac{i}{c}\left[\mathbf{B}_0 \times \left(en\delta\mathbf{u} - \frac{\sigma^{(V)}\delta\boldsymbol{\omega}}{3}\right)\right] - \frac{iw\delta\mathbf{u}}{v_F^2\tau} - i\frac{\hbar n_5\delta\boldsymbol{\omega}}{2v_F\tau},$$

(8.100)

and

$$\omega\delta\epsilon - w(\mathbf{k}\cdot\delta\mathbf{u}) - \sigma^{(\epsilon,u)}\left[(\mathbf{B}_0\cdot\delta\mathbf{u})\,(\mathbf{B}_0\cdot\mathbf{k}) - 2B_0^2\,(\mathbf{k}\cdot\delta\mathbf{u})\right]$$

$$- v_F\,(\mathbf{B}_0\cdot\mathbf{k})\,\delta\sigma^{(\epsilon,B)} = i\sigma^{(B)}\,(\delta\mathbf{E}\cdot\mathbf{B}_0),$$

(8.101)

respectively. These should be supplemented by the equation for a fluctuating electric field, i.e.,

$$\left(n_{\text{ref}}^2\omega^2 - c^2k^2\right)\delta\mathbf{E} + c^2\mathbf{k}(\mathbf{k}\cdot\delta\mathbf{E}) + 4\pi i\omega\delta\mathbf{J} = 0,$$

(8.102)

which follows from Ampere's law after taking into account that $\delta\mathbf{B} = (c/\omega)\,[\mathbf{k} \times \delta\mathbf{E}]$ (Faraday's law). As before, here we assumed that $\mu_m = 1$.

Equations (8.94), (8.95), (8.100), (8.101), and (8.102) define a complete set of linearized equations for the chiral electron fluid in Dirac and Weyl semimetals. Depending on the values of model parameters, this system describes a wide range of different types of long-wavelength collective excitations. They include various electromagnetic waves, diffusive and sound modes, as well as other excitations [901, 902]. In what follows, only a few representative collective modes will be analyzed.

8.2.2 *Collective modes*

Let us start our discussion of collective modes with the case of vanishing external magnetic field, $\mathbf{B}_0 = \mathbf{0}$. When electric and chiral charge densities vanish (i.e., $\mu = \mu_5 = 0$), the linearized system of hydrodynamic equations can be easily solved. In the spectrum of collective modes, one finds a doubly degenerate diffusive mode with imaginary frequency, i.e.,

$$\omega_{\mathrm{d}} = -\frac{i}{\tau}, \tag{8.103}$$

a damped sound-like wave with the dispersion relation

$$\omega_{\mathrm{s},\pm} = \pm \frac{1}{\sqrt{3}} \sqrt{v_F^2 k^2 - \frac{3}{4\tau^2}} - \frac{i}{2\tau}, \tag{8.104}$$

as well as propagating modes with dispersion relations given by

$$\omega_{\mathrm{AHW},\pm} = \pm \frac{1}{n_{\mathrm{ref}}} \sqrt{c^2 k^2 + \frac{2e^4 b^2}{\pi^2 n_{\mathrm{ref}}^2 \hbar^2} - \frac{2e^2}{\pi n_{\mathrm{ref}} \hbar} \sqrt{\frac{e^4 b^4}{\pi^2 n_{\mathrm{ref}}^2 \hbar^2} + c^2 (\mathbf{k} \cdot \mathbf{b})^2}}, \tag{8.105}$$

$$\omega_{\mathrm{gAHW},\pm} = \pm \frac{1}{n_{\mathrm{ref}}} \sqrt{c^2 k^2 + \frac{2e^4 b^2}{\pi^2 n_{\mathrm{ref}}^2 \hbar^2} + \frac{2e^2}{\pi n_{\mathrm{ref}} \hbar} \sqrt{\frac{e^4 b^4}{\pi^2 n_{\mathrm{ref}}^2 \hbar^2} + c^2 (\mathbf{k} \cdot \mathbf{b})^2}}. \tag{8.106}$$

Recall that n_{ref} is the refractive index.

As one can verify, the damped sound mode with the frequency in Eq. (8.104) is sustained by oscillations of the fluid velocity but not accompanied by any local oscillations of electromagnetic fields. The damping rate of this sound mode becomes negligible in the limit $\tau \gg (v_F k)^{-1}$, i.e., when the scattering time is much longer than the inverse frequency. In this case, as expected, the standard dispersion relation for the sound waves in electron fluid is reproduced, $\omega_{\mathrm{s},\pm} \simeq \pm v_F k / \sqrt{3}$. One can also check that these modes transform into gapped plasmons when $\mu \neq 0$.

The propagating modes with the dispersion relations in Eqs. (8.105) and (8.106) are the in-medium electromagnetic waves that are modified by the anomalous Hall effect. As $b \to 0$, both modes reduce to the usual electromagnetic waves with the speed c/n_{ref}. When $\mathbf{b} \neq \mathbf{0}$, however, their properties in the long-wavelength limit are strongly affected by the anomalous Hall effect. It is natural, therefore, to call the modes with the frequencies $\omega_{\text{AHW},\pm}$ and $\omega_{\text{gAHW},\pm}$ the *anomalous Hall wave*(AHW) and the *gapped anomalous Hall wave* (gAHW), respectively.

Note that, depending on the direction of propagation, the anomalous Hall wave can transform into the usual in-medium electromagnetic wave at $\mathbf{k} \perp \mathbf{b}$ with $\omega_{\text{em},\pm} = \lim_{k_\parallel \to 0} \omega_{\text{AHW},\pm} = \pm ck/n_{\text{ref}}$ or the *anomalous helicon* (AH) at $\mathbf{k} \parallel \mathbf{b}$. The latter has the quadratic dispersion relation given by

$$\omega_{\text{AH},\pm} = \lim_{k_\perp \to 0} \omega_{\text{AHW},\pm} = \pm \frac{\pi c k_\parallel^2}{2eb} + O(k^3). \qquad (8.107)$$

This result is quite remarkable. Indeed, unlike the usual or pseudomagnetic helicons discussed in Sec. 8.1.4, this collective mode can exist without any background (pseudo)magnetic fields. Also, it is realized at the vanishing electric charge density. As for its origin, it is connected directly to a nonzero chiral shift $\mathbf{b}$, which is one of the signature characteristics of Weyl semimetals.

The spectrum of collective modes changes dramatically when the electric charge density ρ is nonzero (i.e., at $\mu \neq 0$). This can be seen already from the diffusive and sound modes. Indeed, instead of the dispersion relations in Eqs. (8.103) and (8.104), one finds

$$\omega_{\text{d},\pm} = \pm\Omega_{\text{hyd}} - \frac{i}{2\tau} + O(k), \qquad (8.108)$$

$$\omega_{\text{s},\pm} = \pm\Omega_{\text{hyd}}\sqrt{1 - \frac{1}{4\Omega_{\text{hyd}}^2\tau^2} + \frac{v_F^2 k^2}{3n_{\text{ref}}^2\Omega_{\text{hyd}}^2}} - \frac{i}{2\tau}. \qquad (8.109)$$

Both of these describe gapped modes. This is particularly evident in the limit $\tau \to \infty$ when the frequencies of these modes approach the hydrodynamic plasmon gap[1]

$$\Omega_{\text{hyd}} = \sqrt{\frac{4\pi e^2 v_F^2 n^2}{n_{\text{ref}}^2 w}}. \qquad (8.110)$$

---

[1]Note that, in general, the hydrodynamic plasma frequency Ω_{hyd} in Eq. (8.110) is different from the expression obtained in the kinetic approach, i.e., Ω_e in Eq. (8.59). From a technical viewpoint, the difference between the two approaches stem from the use of the distribution function that is either a solution of the Boltzmann equation in the kinetic theory or the ansatz in Eq. (7.142) in the hydrodynamic approach.

The anomalous Hall waves also change. Depending on the direction of propagation, they can become either completely diffusive modes ($\mathbf{k} \perp \mathbf{b}$) or transform into anomalous helicons ($\mathbf{k} \parallel \mathbf{b}$). The latter have the following dispersion relation:

$$\omega_{\text{AH},\pm} = \lim_{b_\perp \to 0} \omega_{\text{AHW},\pm} = \pm \frac{\pi c^2 \hbar w^2 b_\parallel k^2}{2e^2 \left(4\pi^4 v_F^4 \hbar^2 n^4 \tau^2 + w^2 b_\parallel^2 \right)}$$

$$- i\tau \frac{2\pi^3 c^2 v_F^2 \hbar^2 w n^2 k^2}{2e^2 \left(4\pi^4 v_F^4 \hbar^2 n^4 \tau^2 + w^2 b_\parallel^2 \right)} + O(k^3). \tag{8.111}$$

As expected, the anomalous helicons are dissipative when relaxation time τ and electric charge density n are nonzero. Nevertheless, there is still no gap in their energy spectrum.

8.2.3 *Magnetic field effects*

The presence of a background magnetic field significantly enriches the properties of collective modes in Dirac and Weyl semimetals. It also gives rise to new types of excitations. For simplicity, below we will discuss only two special cases of collective modes that propagate either along or perpendicularly to the magnetic field. The former (with $\mathbf{k} \parallel \mathbf{B}_0$) is called longitudinal and the latter (with $\mathbf{k} \perp \mathbf{B}_0$) transverse modes.

8.2.3.1 *Longitudinal collective modes*

At $\mathbf{B}_0 \neq \mathbf{0}$, the dispersion relation of the sound wave is given by

$$\omega_{\text{s},\pm} = \pm \frac{1}{\sqrt{3}} \sqrt{v_F^2 k^2 - \frac{v_F^2 \sigma^{(\epsilon,u)}}{w} [2k^2 B_0^2 - (\mathbf{k} \cdot \mathbf{B}_0)^2] - \frac{3}{4\tau^2}} - \frac{i}{2\tau}. \tag{8.112}$$

This expression remains valid for an arbitrary direction of propagation. As one can see, the sound velocity is modified by a magnetic field, which means that sound is transformed into a magnetoacoustic wave.

The spectrum of collective modes also contains a *gapped chiral magnetic wave* (gCMW). In the case of Dirac semimetals (i.e., $\mathbf{b} = \mathbf{0}$), its dispersion relation reads[m]

$$\omega_{\text{gCMW},\pm} = \pm \frac{e B_0 \sqrt{3 v_F^3 (4\pi e^2 T^2 + 3 n_{\text{ref}}^2 \hbar^3 v_F^3 k_\parallel^2)}}{2\pi^2 T^2 c n_{\text{ref}} \sqrt{\hbar}}. \tag{8.113}$$

---

[m]At $\mathbf{b} \neq \mathbf{0}$, there are additional terms that contribute to the gap.

This mode is closely related to the CMW. Recall that the latter is a gapless collective excitation propagating along the direction of a magnetic field. Its propagation is sustained by alternating oscillations of local electric and chiral charge densities that feed each other via the chiral anomaly (see Sec. 8.1.2). Mathematically, it results from the following linearized system of continuity relations:

$$\omega \delta \rho - k_\parallel \delta J_\parallel = 0, \tag{8.114}$$

$$\omega \delta \rho_5 - k_\parallel \delta J_{5,\parallel} = -i \frac{e^3 B_0}{2\pi^2 \hbar^2 c} \delta E_\parallel. \tag{8.115}$$

The chiral magnetic and the chiral separation effects imply that the oscillations of current densities are proportional to the charge densities, i.e., $\delta J_\parallel \propto |eB_0|\delta\rho_5$ and $\delta J_{5,\parallel} \propto |eB_0|\delta\rho$. Thus, if the effect of the chiral anomaly on the right-hand side of Eq. (8.115) were neglected, one would naively obtain a gapless dispersion relation of the CMW with $\omega_{\mathrm{CMW}} \propto |eB_0|k_\parallel$. Such an approximation is unjustified, however, because the anomalous term $\propto B_0 \delta E_\parallel$ gives a large contribution in the long-wavelength limit. Indeed, after taking into account Gauss's law, i.e., $\delta E_\parallel = -4\pi i \delta\rho/(\varepsilon_e k_\parallel)$, the corrected spectrum of the corresponding mode becomes gapped with the value of the gap proportional to $|eB_0|$ as in Eq. (8.113).

In high-energy physics applications, it was argued that the CMW should turn into a diffusive mode when its frequency becomes smaller than the chirality-flipping rate [743]. Most likely, this is not the biggest concern in the context of topological semimetals. A much more detrimental effect on the CMW comes from a high electrical conductivity, which causes a rapid screening of the electric charge oscillations $\delta\rho$ and, therefore, a strong damping of the mode [929]. Mathematically, the effects of electrical conductivity are captured by an extra term in the electric current density, i.e., $\delta J_\parallel = \sigma \delta E_\parallel$. This leads to a large imaginary part in the CMW dispersion relation in the long-wavelength limit.

It is should be emphasized that dynamical electromagnetic fields play a profound role in the underlying dynamics of the CMW. Indeed, by taking into account Maxwell's equations, one finds that the propagation of the CMW is necessarily accompanied by an oscillating electric field $\delta E_\parallel$. Moreover, such a dynamical field modifies dramatically the properties of the collective mode. It generates a nonzero gap in the spectrum and also causes strong dissipative effects. Taking this into account, one can argue that a high electrical conductivity makes the realization of the CMW very difficult in Dirac or Weyl semimetals. If a propagating CMW mode in solids is possible at all, the most promising regime is at low temperatures and in strong magnetic fields.

In addition to the CMW, there exist other interesting modes. In particular, in the presence of a nonzero chiral shift, assuming $\mathbf{b} \perp \mathbf{B}_0$, one finds that there is a *longitudinal anomalous Hall wave* (lAHW) with the following dispersion relation in the long-wavelength limit:

$$\omega_{\text{lAHW},\pm} = \pm \frac{\hbar B_0 k_\| \sqrt{3v_F^3 \left(\pi^3 c^2 T^2 + 3e^2 v_F^3 \hbar b_\perp^2\right)}}{T \sqrt{\pi^3 \left(3n_{\text{ref}}^2 v_F^3 \hbar^2 B_0^2 + 4\pi \hbar c^2 T^2 b_\perp^2\right)}} + O(k_\|^3). \qquad (8.116)$$

In order to clarify the origin of this mode, let us consider the equations that govern its dynamics. By setting $\mu_0 = \mu_{5,0} = 0$ and $\mathbf{b} \perp \mathbf{B}_0$, the corresponding linearized equations of consistent hydrodynamics read

$$\frac{T^2 \omega}{3v_F^3 \hbar} \delta\mu + \frac{eB_0 k_\|}{2\pi^2 c} \delta\mu_5 = 0, \qquad (8.117)$$

$$\frac{eB_0 k_\|}{2\pi^2 c} \delta\mu + \frac{T^2 \omega}{3v_F^3 \hbar} \delta\mu_5 - i\frac{e^2 B_0}{2\pi^2 c} \delta E_\| = 0, \qquad (8.118)$$

$$(n_{\text{ref}}^2 \omega^2 - c^2 k_\|^2)\delta\tilde{E}_\perp + i\frac{2e^2 \omega b_\perp}{\pi \hbar} \delta E_\| = 0, \qquad (8.119)$$

$$n_{\text{ref}}^2 \omega \delta E_\| + i\frac{2e^2}{\pi c \hbar^2}(B_0 \delta\mu_5 - c\hbar b_\perp \delta\tilde{E}_\perp) = 0, \qquad (8.120)$$

where $\delta\tilde{E}_\perp$ denotes the component of oscillating electric field parallel to $[\mathbf{B}_0 \times \mathbf{b}]$. While the first two equations are the continuity relations for the electric and chiral charges, respectively, the other two are Maxwell's equations. Here, it was taken into account that $\delta T = 0$ and $\delta\mathbf{u} = \mathbf{0}$ for this longitudinal anomalous Hall wave mode.

As is clear from the continuity equations (8.117) and (8.118), the oscillations of the electric and chiral chemical potentials are coupled to each other. Because of the chiral anomaly term in Eq. (8.118), they also induce oscillations of $\delta E_\|$. However, it is the oscillating anomalous Hall currents proportional to $b_\perp$ in Eqs. (8.119) and (8.120) that make a self-consistent solution for longitudinal anomalous Hall wave possible. Indeed, oscillations of $\delta E_\|$ drive the anomalous Hall effect current $\delta\mathbf{J}_{\text{AHE}} \propto [\mathbf{b} \times \delta\mathbf{E}]$, which then induces an oscillating electric field $\delta\tilde{E}_\perp$. Finally, $\delta\tilde{E}_\perp$ produces the component of $\delta\mathbf{J}_{\text{AHE}}$ parallel to $\mathbf{B}_0$, which together with $\delta E_\|$, leads to an oscillating CME current $\delta\mathbf{J}_{\text{CME}} \propto \mathbf{B}_0 \delta\mu_5$ and, therefore, closes the cycle.

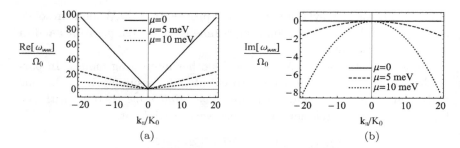

Fig. 8.6 The (a) real and (b) imaginary parts of the longitudinal anomalous Hall wave frequency. The results are shown for $\mu = 0$ (solid lines), $\mu = 5$ meV (dashed lines), and $\mu = 10$ meV (dotted lines). In addition, $b_0 = -\mu_5 = 0$, $T = 10$ K, $B_0 = 0.01$ T, and $b_\perp = 0.3\pi/a_z$. The values of wave vectors and frequencies are given in units of $K_0 = 10^{-6}\pi/a_z \approx 12.3$ cm^{-1} and $\Omega_0 = v_F K_0 \approx 1.84$ GHz.

While the longitudinal anomalous Hall wave is also driven by coupled oscillations of electric and chiral charge densities (or, alternatively, the corresponding chemical potentials), it is qualitatively different from the (gapped) CMW. In particular, unlike the CMW, the longitudinal anomalous Hall wave relies on the anomalous Hall effect and remains gapless even when dynamical electromagnetic fields are self-consistently taken into account. It should be said, however, that the effects of high electrical conductivity could lead to a strong damping of the longitudinal anomalous Hall wave too.

The real and imaginary parts of the longitudinal anomalous Hall wave frequency are shown schematically in Figs. 8.6(a) and 8.6(b), respectively. In addition to the $\mu = 0$ case, the dispersion relations for two nonzero values of μ are considered as well. To obtain the numerical results in Fig. 8.6, we used the model parameters reminiscent of those in the Dirac semimetal Cd$_3$As$_2$ (see also Appendix D). In particular, the background refractive index is $n_{\text{ref}} = \sqrt{\varepsilon_e \mu_m} \approx 6$ [926], the Fermi velocity is $v_F \approx 1.5 \times 10^8$ cm/s [107], the lattice spacing along the crystallographic axis [001] is $a_z \approx 25.480$ Å [107], and the characteristic value of the chiral shift equals $b = 0.3\pi/a_z$. By comparing the results in Fig. 8.6 for different values of μ, it is clear that the electric chemical potential tends to increase the damping of the mode.

For $\mathbf{b} \parallel \mathbf{B}_0$, one of the most interesting modes is an anomalous helicon (assuming $\mu = 0$) with the frequency given by

$$\omega_{\text{AH},\pm} = \pm\frac{\pi c^2 \hbar k_\parallel^2}{2e^2 b_\parallel} - i\tau \frac{v_F^2 B_0^2 k_\parallel^2}{4\pi w} + O(k_\parallel^3). \tag{8.121}$$

As one can see, the anomalous helicon becomes dissipative in the presence of an external magnetic field. At $\mu = 0$, this is connected with a nonvanishing electric chemical potential $\mu_B \propto (\mathbf{b} \cdot \mathbf{B}_0)$, which is induced by the Chern–Simons term in Eq. (8.20). A nonzero value of μ_B also causes diffusive and sound modes to acquire gaps proportional to $(\mathbf{b} \cdot \mathbf{B}_0)$. This underlying mechanism resembles the case with $\mu \neq 0$ that leads to the dispersion relations in Eqs. (8.108) and (8.109).

Note that, at nonzero μ, the hydrodynamic approach reproduces the same conventional helicon as in the kinetic theory in Sec. 8.1.4. At $\mathbf{b} = \mathbf{0}$, its frequency equals

$$\omega_{h,\pm} = \pm \frac{ck_\parallel^2 B_0}{4\pi e^2 n} - \frac{i}{\tau} \frac{c^2 k_\parallel^2}{n_{\text{ref}}^2 \Omega_{\text{hyd}}^2} + O(k_\parallel^3). \qquad (8.122)$$

As in the case of anomalous helicons in Eq. (8.121), a finite value of relaxation time τ leads to a damping of the mode.

8.2.3.2 *Transverse collective modes*

Now let us consider the case of transverse propagation of collective modes, i.e., $\mathbf{k} \perp \mathbf{B}_0$. For the sound wave, the general form of the dispersion relation in Eq. (8.112) remains valid. In fact, the sound waves in the transverse and longitudinal directions are similar, but the speeds of their propagation are slightly different. The properties of the diffusive mode and the gapped CMW remain nearly the same as well. In particular, at $\mathbf{b} = \mathbf{0}$, the dispersion relation has the same form, but $k_\parallel$ should be replaced with $k_\perp$. There are no helicons, however, when $\mathbf{k} \perp \mathbf{B}_0$.

At $\mathbf{b} \parallel \mathbf{k}$, there exists a *transverse anomalous Hall wave* (tAHW). It is a gapless mode with the following dispersion relation:

$$\omega_{\text{tAHW},\pm} = \pm \frac{ck_\perp \sqrt{3 v_F^3 \hbar B_0}}{\sqrt{4\pi c^2 T^2 b_\perp^2 + 3 n_{\text{ref}}^2 v_F^3 \hbar B_0^2}} + O(k_\perp^2). \qquad (8.123)$$

This mode looks similar to the longitudinal anomalous Hall wave discussed in Sec. 8.2.3.1. However, its physical origin and governing equations are slightly different. The corresponding linearized hydrodynamic equations read

$$\frac{T^2 \omega}{3 v_F^3 \hbar} \delta\mu_5 - \frac{ie^2 B_0}{2\pi^2 c} \delta E_\parallel = 0, \qquad (8.124)$$

$$\left(n_{\text{ref}}^2 \omega^2 - c^2 k_\perp^2 \right) \delta\tilde{E}_\perp + \frac{2ie^2 \omega b_\perp}{\pi \hbar} \delta E_\parallel = 0, \qquad (8.125)$$

$$\left(n_{\text{ref}}^2\omega^2 - c^2k_\perp^2\right)\delta E_\parallel + \frac{2ie^2\omega}{\pi c\hbar^2}(B_0\delta\mu_5 - \hbar cb_\perp\delta\tilde{E}_\perp) = 0, \qquad (8.126)$$

$$\frac{k_\perp}{T}\delta T - \frac{i+\omega\tau}{v_F^2\tau}\delta u_\perp + \frac{5ck_\perp^2 B_0\sigma^{(\epsilon,u)}}{\omega w}\delta\tilde{E}_\perp = 0, \qquad (8.127)$$

$$\frac{4\epsilon\omega}{T}\delta T - k_\perp(w - 2B_0^2\sigma^{(\epsilon,u)})\delta u_\perp = 0, \qquad (8.128)$$

where $\tilde{E}_\perp$ is the component of an oscillating electric field in the direction perpendicular to both $\mathbf{k}$ and $\mathbf{B}_0$. As is easy to check, the relevant equations include the chiral charge continuity relation, Maxwell's equations, as well as the Euler equation and the energy conservation relation.

As in the case of the longitudinal anomalous Hall wave, the propagation of the transverse anomalous Hall wave is sustained by a dynamical version of the anomalous Hall effect. There are, however, some important differences between the two. For example, the hydrodynamic fluid velocity plays a nontrivial role in the case of the transverse mode. In addition, there are no oscillations of the electric chemical potential, i.e., $\delta\mu = 0$.

As one can see from Eq. (8.124), the chiral chemical potential $\delta\mu_5$ drives oscillations of the electric field $\delta E_\parallel$ via the chiral anomaly term. Then the anomalous Hall effect in Eq. (8.125) induces $\delta\tilde{E}_\perp$, which in turn causes oscillations of δT and $\delta u_\perp$ via the remaining set of coupled equations. As is clear, the Euler equation (8.127) and the energy conservation relation (8.128) are also important for the propagation of this mode. Therefore, the transverse anomalous Hall wave is a hybrid collective mode that involves both dynamical electromagnetism and hydrodynamic motion. It is possible because of the chiral anomaly and the effects of the chiral shift.

The real and imaginary parts of the transverse anomalous Hall wave frequencies are shown in Figs. 8.7(a) and 8.7(b), respectively. Note that the results for three different values of the electric chemical potential, $\mu = 0$, $\mu = 5$ meV, and $\mu = 10$ meV, are presented. At $\mu = 0$, the transverse anomalous Hall wave has a linear dispersion relation. At $\mu \neq 0$, on the other hand, the corresponding mode becomes completely diffusive for sufficiently small wave vectors $|k_\perp| \lesssim K_0/2$.

In summary, the hydrodynamic approach provides a convenient way to study long-wavelength collective excitations in Dirac and Weyl semimetals. Among the most interesting modes obtained in the hydrodynamic framework are anomalous Hall waves. The latter include the longitudinal anomalous Hall wave, which is primarily electromagnetic in origin, and the

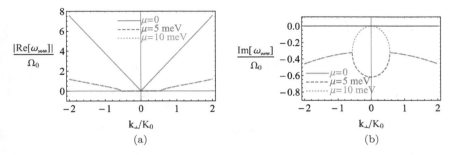

Fig. 8.7 The (a) real and (b) imaginary parts of the transverse anomalous Hall wave frequency. While solid lines correspond to the case $\mu = 0$, dashed and dotted lines represent positive and negative frequencies at $\mu = 10$ meV. In addition, $b_0 = -\mu_5 = 0$, $T_0 = 10$ K, $B_0 = 0.01$ T, and $b_\perp = 0.3\pi/a_z$. The values of the wave vectors and frequencies are given in units of $K_0 = 10^{-6}\pi/a_z \approx 12.3$ cm^{-1} and $\Omega_0 = v_F K_0 \approx$ 1.84 GHz.

transverse anomalous Hall wave, which is a hybridized electromagnetic and hydrodynamic excitation.

As we saw in Chapters 4 and 5, the electron dynamics at the surface of Dirac and Weyl semimetals is very interesting. In particular, there are topologically protected Fermi arc states, which affect the transport of charge and heat in these materials. Therefore, it is worth considering in detail the collective modes in Dirac and Weyl semimetals such as surface plasmons and polaritons. In the spirit of this chapter, these surface modes will be considered in both kinetic and hydrodynamic regimes.

8.3 Surface collective modes

As was first predicted by Rufus Ritchie [942] in the late 1950s, the surface plasmons are collective excitations associated with electromagnetic field and electric charge oscillations that propagate along surfaces (or interfaces) of metals, semimetals, and semiconductors [943–947]. Since these plasmons involve oscillations of charge density in a metal and electromagnetic waves in a surrounding dielectric, it is appropriate to call them the *surface plasmon-polaritons*(SSPs) [948–950]. Indeed, surface plasmon-polariton is a particular case of polaritons[n] that is confined to a metal-dielectric

---

[n]Polaritons are quasiparticles related to the coupling of electromagnetic waves with any resonance in a material. They were first theoretically predicted by Kirill Tolpygo in 1950 [951].

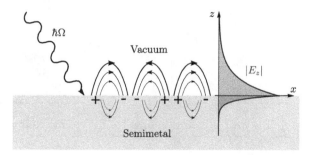

Fig. 8.8　Schematic illustration of the surface plasmon-polariton propagating along an interface of a semimetal and vacuum. An external electromagnetic field couples to the oscillations of the surface charge allowing for a mode confined to the interface.

interface. In what follows, however, we will use the terms surface plasmon-polariton and surface plasmon interchangeably. A schematic illustration of SPPs is given in Fig. 8.8. The surface plasmons can be experimentally probed by the scattering-type near-field optical spectroscopy [952, 953] or the momentum-resolved electron energy loss spectroscopy [954, 955].

The surface plasmons have already found numerous practical applications in chemical catalysis, optics, sensors, microscopy, solar cells, data transfer, and more. The basic research in surface plasmons gave rise to a new field of study known as plasmonics [956]. In essence, the latter merges the fields of photonics and electronics to create nanoscale surface-plasmon-based circuits. These circuits promise great advantages in information processing and create fertile soil for new emerging technologies. For example, a substantial shrinking of light wavelength enables subwavelength optics and lithography beyond the diffraction limit [946]. Further, field enhancement and strong dependence on material parameters make SPPs very sensitive to external fields and nonlinear effects providing excellent conditions for the creation of nanoscale devices connected with optical switching and biosensing.

The use of topological materials opens new opportunities in the field of plasmonics [957]. The list of relevant 3D materials includes topological insulators as well as Weyl and some Dirac semimetals. Because of the spin-momentum locking in topological insulators, for example, the corresponding surface plasmons are accompanied by transverse oscillations of spin density. This could be invaluable in emerging spintronic devices. One can also expect unusual surface plasmons in Weyl and Dirac semimetals related to their remarkable properties. They include a relativistic-like dispersion relation of quasiparticles, a nonzero separation between Weyl nodes in energy and/or momentum, and the existence of topologically protected surface Fermi arc states.

It was found that Weyl semimetals with broken $\mathcal{T}$ symmetry have surface plasmons similar to magnetoplasmon modes in usual metals [874, 958–960]. In particular, the chiral shift makes the modes nonreciprocal, i.e., the corresponding plasmons propagate only in one direction even in the absence of a magnetic field. It is worth noting, however, that the anomalous Hall effect (as well as other related effects) due to the topological Chern–Simons terms is not the only feature that distinguishes Weyl semimetals from their topologically trivial counterparts. The existence of unusual surface states, i.e., the Fermi arcs, could also play an important role. In particular, when the chiral shift has a nonzero component parallel to the surface, the properties of surface plasmons are determined by a subtle interplay of bulk and surface states. The properties of collective modes in such a regime were analyzed in [915, 916, 961, 962]. Below we will review some properties of surface collective modes in Weyl semimetals starting with the SPPs.

8.3.1 *Surface plasmon-polaritons*

In the context of the present book, it is a natural question what are the physical properties of SPPs in Dirac and Weyl semimetals. It was shown that the topological charge of Weyl nodes and their separation in energy and momentum space essentially affect the dispersion relation of SPPs [874, 958–960]. The chiral shift acts as an effective magnetic field leading to a dispersion relation of surface plasmons similar to magnetoplasmons in ordinary metals (see also Sec. 8.1.3). Remarkably, there exist modes that strongly depend on the direction of propagation, which are known in the literature as nonreciprocal surface modes.

As we will explicitly demonstrate in this section, the study of SPPs is rather direct and requires the knowledge of the dielectric function and the boundary conditions of a material. It is common in the literature to consider Maxwell's equations with a simple dielectric function accounting only for intraband electronic transitions. A more refined approach would be to obtain this function self-consistently and take into account the underlying dynamics of surface states. While this procedure will be considered in Secs. 8.3.2 and 8.3.2.2, here we employ a simpler approach. For our purposes, it suffices to consider the minimal model of a Weyl semimetal with a single pair of Weyl nodes separated by $2\mathbf{b}$ in momentum (for simplicity, we set $b_0 = 0$) given in Eq. (3.5). In addition to the usual Drude-like term proportional to the electric field, the displacement field $\mathbf{D}$ in this Weyl semimetal contains also a term connected with the anomalous Hall effect (for the discussion of the anomalous Hall effect, see Sec. 6.3), which is perpendicular to an electric field $\mathbf{E}$

$$\mathbf{D} = \left[\varepsilon(\omega) + \frac{4\pi i}{\omega}\sigma\right]\mathbf{E} - \frac{2ie^2}{\pi\hbar\omega}[\mathbf{b} \times \mathbf{E}]. \tag{8.129}$$

Here σ is the electrical conductivity and $\varepsilon(\omega)$ is the dielectric function of the semimetal, which also depends on the frequency ω. For simplicity, we assumed that the conductivity tensor is isotropic.

As we discussed at the beginning of this section, SPPs are solutions of Maxwell's equations localized at the interface of two media. Therefore, it is convenient to assume a semi-infinite model of a Weyl semimetal where the semimetal is situated at $z > 0$ and vacuum is at $z < 0$. In view of the translational invariance along the interface, the electric field is sought as a plane wave with the frequency ω and the wave vector $\mathbf{k} = (k_x, k_y, 0)$

$$\mathbf{E} = \mathbf{E}_0\, e^{-i\omega t + ik_x x + ik_y y}\, e^{-qz}, \tag{8.130}$$

which decays exponentially away from the boundary for $q > 0$. The electric field in vacuum is sought in the same form as above albeit with a different negative $q < 0$. The decay constant q is determined as a solution of the wave equation

$$\mathbf{\nabla} \times [\mathbf{\nabla} \times \mathbf{E}] = -\frac{1}{c^2}\frac{\partial^2}{\partial t^2}\mathbf{D}, \tag{8.131}$$

and the same equation with $\mathbf{D} \to \mathbf{E}$ in vacuum. Then, by substituting ansatz (8.129) into Eq. (8.131), one obtains a homogeneous system of linear algebraic equations. The zeros of the determinant of this system define q. Furthermore, demanding the continuity of the parallel components of the electric field and the normal component of the displacement field leads to an equation that determines the dispersion relation of SPPs. As to the dielectric function, it is assumed that $\varepsilon(\omega)$ does not depend on the wave vector $\mathbf{k}$. This approximation implies that the inverse wave vector of surface plasmon-polaritons is larger than the inverse Fermi wave vector. Then the dependence of $\varepsilon(\omega)$ on frequency has the standard form $\varepsilon(\omega) = \varepsilon_\infty(1 - \Omega_e^2/\omega^2)$, where ε_∞ is the high-frequency dielectric constant and $\Omega_e^2 = 4\alpha\mu^2/(3\pi\hbar^2)$ is the bulk plasmon frequency (see also Eq. (8.59) at $\mu_5 = 0$ and $T \to 0$).

Let us discuss first the case of a Dirac semimetal with $\mathbf{b} = \mathbf{0}$. It is easy to find that the dispersion of surface plasmon-polaritons coincides with that in standard metals [944, 945, 947, 950]

$$\varepsilon(\omega)q_0 + q_D = 0, \qquad (8.132)$$

where $q_0 = \sqrt{k^2 - \omega^2/c^2}$ and $q_D = \sqrt{k^2 - \varepsilon(\omega)\omega^2/c^2}$ determine the decay length of SPP in vacuum and Dirac semimetal, respectively. The major difference from usual metals is the dependence of the dielectric function $\varepsilon(\omega)$ on parameters such as temperature and electric chemical potential.

Let us present now the results for Weyl semimetals with broken $\mathcal{T}$ symmetry ($\mathbf{b} \neq \mathbf{0}$). As one can see from Eq. (8.129), the presence of a chiral shift allows for the off-diagonal components of the dielectric tensor. Therefore, qualitatively, the situation is the same as for the surface magneto-plasmon polaritons in conventional metals [950, 963–965]. The strength of the off-diagonal components of the dielectric tensor can be parameterized as $\varepsilon_b(\omega) = \varepsilon_\infty \omega_b/\omega$, where $\omega_b = 2e^2 b/(\pi\hbar\epsilon_\infty)$. Due to the presence of the boundary and a chiral shift, the symmetry of the system is quite low and the SPP dispersion is cumbersome in a general case. Therefore, we consider only three practically important configurations: (i) $\mathbf{b} \perp \mathbf{k}$ and $\mathbf{b} \parallel \hat{\mathbf{z}}$, (ii) $\mathbf{b} \parallel \mathbf{k}$ and $\mathbf{b} \perp \hat{\mathbf{z}}$ (Faraday configuration), and (iii) $\mathbf{b} \perp \mathbf{k}$ and $\mathbf{b} \perp \hat{\mathbf{z}}$ (Voigt configuration).

In the first case ($\mathbf{b} \perp \mathbf{k}$ and $\mathbf{b} \parallel \hat{\mathbf{z}}$), we have the following solutions for the decay constant:

$$\left(q_\pm^{(i)}\right)^2 = k^2 - \frac{\omega^2}{c^2}\varepsilon(\omega) \pm \frac{\omega\sqrt{\varepsilon(\omega)}|\varepsilon_b(\omega)|}{c}\sqrt{\frac{\omega^2\varepsilon(\omega)}{c^2} - k^2}, \qquad (8.133)$$

where the corresponding characteristic equation reads

$$\varepsilon(\omega)q_0 q_+^{(i)} q_-^{(i)}(q_+^{(i)} + q_-^{(i)}) + q^2\left\{ q_+^{(i)} q_-^{(i)} + q_0(q_+^{(i)} + q_-^{(i)}) \right.$$

$$\left. + [(q_+^{(i)})^2 + q_+^{(i)} q_-^{(i)} + (q_-^{(i)})^2]\varepsilon(\omega) \right\}$$

$$- \frac{\omega^2}{c^2}\varepsilon(\omega)\left\{ (q_+^{(i)} + q_-^{(i)})(q_0 + q_+^{(i)} + q_-^{(i)}) + k^2[1 - \varepsilon(\omega)] \right\} = 0.$$

$$(8.134)$$

Its solutions are shown in Fig. 8.9(a) for a few values of ω_b. As one can see, the frequency of the SSPs is reduced in the presence of the chiral shift.

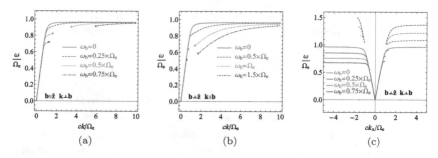

Fig. 8.9   The dispersion relation of the surface plasmon-polaritons in a $\mathcal{T}$ symmetry broken Weyl semimetal for a few different orientations of wave vector **k** and the chiral shift **b**. In particular, we consider the following configurations: (a) **b** ⊥ **k** and **b** ∥ **ẑ**, (b) the Faraday configuration **b** ∥ **k** and **b** ⊥ **ẑ**, and (c) the Voigt configuration **b** ⊥ **k** and **b** ⊥ **ẑ**. In panels (a) and (c), we set $\omega_b = 0.25\,\Omega_e$ (red solid lines), $\omega_b = 0.5\,\Omega_e$ (blue dashed lines), and $\omega_b = 0.75\,\Omega_e$ (green dot-dashed lines). In panel (b), $\omega_b = 0.5\,\Omega_e$ (red solid lines), $\omega_b = \Omega_e$ (blue dashed lines), and $\omega_b = 1.5\,\Omega_e$ (green dot-dashed lines). Dotted lines represent the frequencies corresponding to a hybridization of the surface plasmon-polaritons with bulk plasmons. In all panels, we used $\varepsilon_\infty = 13$ [854, 874].

Further, the surface excitations can hybridize with bulk plasmons, which is manifested in an imaginary part of the frequencies. This explains the discontinuities in the dispersion relations in Figs. 8.9(a) and 8.9(b).

In the case of the Faraday configuration (**b** ∥ **k** and **b** ⊥ **ẑ**), one obtains

$$(q_\pm^{(ii)})^2 = k^2 + \frac{\omega^2}{c^2}\left[\frac{\varepsilon_b(\omega)}{2\varepsilon(\omega)} - \varepsilon(\omega)\right] \pm \frac{|\omega\varepsilon_b(\omega)|}{2c\varepsilon(\omega)}\sqrt{4k^2\varepsilon(\omega) + \frac{\omega^2}{c^2}\left[\varepsilon_b(\omega)\right]^2}. \tag{8.135}$$

The corresponding characteristic equation reads

$$q_+^{(ii)}q_-^{(ii)}(q_0 + q_+^{(ii)} + q_-^{(ii)}) + q_0[(q_+^{(ii)})^2 + q_+^{(ii)}q_-^{(ii)} + (q_-^{(ii)})^2]\varepsilon(\omega)$$

$$+ k^2\{q_0\left[1 - \varepsilon(\omega)\right] + (q_+^{(ii)} + q_-^{(ii)})\varepsilon(\omega)\}$$

$$- \frac{\omega^2}{c^2}\varepsilon(\omega)\{q_+^{(ii)} + q_-^{(ii)} + q_0\left[1 - \varepsilon(\omega)\right]\} = 0. \tag{8.136}$$

The SPPs for the Faraday configuration are shown in Fig. 8.9(b) for a few values of ω_b. As in the case (i), the SPPs become dissipative for sufficiently large values of the chiral shift and their frequency is reduced at $\omega_b \neq 0$.

Finally, let us discuss the results for the Voigt configuration (**b** ⊥ **k** and **b** ⊥ **ẑ**). The decay constant reads

$$(q^{(iii)})^2 = k^2 + \frac{\omega^2}{c^2}\left[\frac{\left[\varepsilon_b(\omega)\right]^2}{\varepsilon(\omega)} - \varepsilon(\omega)\right]. \tag{8.137}$$

Compared to the two previous cases, the corresponding characteristic equation is more simple and reads

$$\varepsilon(\omega)q^{(iii)} + q_0\{[\varepsilon(\omega)]^2 - [\varepsilon_b(\omega)]^2\} + k\varepsilon_b(\omega)[\hat{\mathbf{k}} \times \hat{\mathbf{b}}]_z = 0. \qquad (8.138)$$

As one can already see from Eq. (8.138), the SPPs in Weyl semimetals for the Voigt configuration are *nonreciprocal*. The latter means that the velocity of the SPPs depends on the direction of the wave vector **k**. Indeed, the last term in the above equation changes sign at $\mathbf{k} \to -\mathbf{k}$ or $\mathbf{b} \to -\mathbf{b}$. The dispersion relation of the SPPs in the Voigt configuration is shown in Fig. 8.9(c). For definiteness, we assumed that $\mathbf{k} \parallel \hat{\mathbf{x}}$ and $\mathbf{b} \parallel \hat{\mathbf{y}}$. Let us first discuss the case $k_x > 0$. It is interesting that the dispersion relation consists of two branches separated by a gap. In particular, the dashed lines correspond to high-frequency branches and start at momenta defined by $\omega = \Omega_e$. The endpoints of the branches occur at the intersection with bulk plasmon solutions defined by $q^{(iii)} = 0$. The case $k_x < 0$ is clearly different from that at $k_x > 0$. While there is the full low-energy branch, the high-energy one quickly terminates at the bulk plasmon frequency. Thus, due to the effects of the chiral shift, the SPPs in the Voigt configuration are strongly nonreciprocal.

Finally, let us comment on the SPP properties in finite samples of Weyl and Dirac semimetals [874, 960]. Here we briefly discuss only the results in the case of the Voigt geometry. The corresponding dispersion relation reads

$$e^{-2q^{(iii)}d} = \frac{Q_{++}Q_{+-}}{Q_{-+}Q_{--}}, \qquad (8.139)$$

where d is the thickness of a film and

$$Q_{s_1s_2} = q_0\{[\varepsilon(\omega)]^2 - [\varepsilon_b(\omega)]^2\} + s_1\{\varepsilon(\omega)q^{(iii)} + s_2k\varepsilon_b(\omega)\left[\hat{\mathbf{k}} \times \hat{\mathbf{b}}\right]_z\}. \qquad (8.140)$$

It is clear that SPPs on the opposite surfaces for $d \to \infty$ can be considered as independent. In such a case, their dispersion relation is given in Fig. 8.9(c). Top and bottom surfaces, however, differ by replacing $\mathbf{k} \to -\mathbf{k}$. Therefore, in order to investigate the nonreciprocity induced by the chiral shift, one needs to apply nonlocal probes. As for the usual surface plasmons, the nonreciprocity can be enhanced by creating a high optical contrast with a substrate. In particular, this can be achieved by placing Weyl semimetals on a substrate with a high dielectric constant. Next, let us briefly discuss the case of a thin film. When the film thickness is sufficiently small, the

SPPs start to hybridize. For $q_0 d \ll 1$, the dependence on the off-diagonal components of the dielectric tensor $\varepsilon_b(\omega)$ becomes negligible and the SPPs become reciprocal. In applications, the nonreciprocal SSPs in $\mathcal{T}$ symmetry broken Weyl semimetals could be used to create tunable and compact nonreciprocal optical elements.

Having reviewed the surface plasmon-polaritons in a simple approach, let us turn our attention to the properties of surface modes when the dynamics of the surface states is explicitly taken into account. In particular, the following subsection will be devoted to the surface modes in the hydrodynamic regime.

8.3.2 *Hydrodynamic study*

As we already saw in Secs. 7.4 and 8.2, while providing a coarse description, the hydrodynamic approach allows one to study various properties of certain systems in a simple and clear way. Therefore, let us investigate surface collective modes by using the approximation known in the literature as hydrodynamic [944, 945, 947]. It neglects the retardation effects in Maxwell's equations.° A more rigorous albeit complicated analysis can be done by utilizing a quantum-mechanical nonlocal description [961].

8.3.2.1 *General framework*

In the hydrodynamic approximation, the effects of oscillating magnetic fields are neglected and oscillating electric potential $\phi(t, \mathbf{r})$ is governed by the following Poisson's equation:

$$\Delta\phi(t, \mathbf{r}) = \frac{4\pi e}{\varepsilon_e(y)} \delta n(t, \mathbf{r}), \qquad (8.141)$$

where $\delta n(t, \mathbf{r})$ describes deviations of the particle number density from its equilibrium value and $\varepsilon_e(y)$ is the coordinate-dependent electric permittivity of the system. Let us assume that a Weyl semimetal with broken $\mathcal{T}$ symmetry is located in the upper half-space ($y > 0$) and vacuum is in the lower half-space ($y < 0$). Then $\varepsilon_e(y) = \theta(y)\varepsilon_e + \theta(-y)$, where $\theta(y)$ is the unit step function. In addition, we assume that the chiral shift is parallel to the surface and points in the z-direction, $\mathbf{b} = b\hat{\mathbf{z}}$.

Since here we are interested only in surface-localized modes, solutions for oscillating variables will be sought in the form of plane waves propagating

---

°Formally, this can be achieved by taking the limit $c \to \infty$.

in the x- and z-directions, i.e., $\delta\mu(t,\mathbf{r}) = \delta\mu(y)\,e^{-i\omega t + i\mathbf{k}\cdot\mathbf{r}}$, where ω is the frequency and $\mathbf{k} = (k_x, 0, k_z)$ is the wave vector along the surface.

Additional simplifications in the hydrodynamic approximation come from the fact that the energy conservation relation plays a little role and temperature oscillations are negligible [944, 945, 947]. Taking this into account, one can express all oscillating thermodynamical variables in terms of the electric charge density

$$\delta P = \tilde{P}\delta n, \tag{8.142}$$

$$\delta w = \tilde{w}\delta n, \tag{8.143}$$

where

$$\tilde{P} = \frac{\partial P}{\partial \mu}\left(\frac{\partial n}{\partial \mu}\right)^{-1} = \mu\frac{\mu^2 + \pi^2 T^2}{3\mu^2 + \pi^2 T^2} \overset{T\to 0}{=} \frac{\mu}{3}, \tag{8.144}$$

$$\tilde{w} = \frac{\partial w}{\partial \mu}\left(\frac{\partial n}{\partial \mu}\right)^{-1} = 4\mu\frac{\mu^2 + \pi^2 T^2}{3\mu^2 + \pi^2 T^2} \overset{T\to 0}{=} \frac{4\mu}{3}. \tag{8.145}$$

Furthermore, one can neglect the effects of vorticity $\boldsymbol{\omega} = [\boldsymbol{\nabla}\times\mathbf{u}]/2$, which are usually very small. Then oscillations of the corresponding gradient flow velocity can be expressed in terms of velocity potential $\psi(t,\mathbf{r})$ as follows:

$$\delta\mathbf{u}(t,\mathbf{r}) = -\boldsymbol{\nabla}\psi(t,\mathbf{r}). \tag{8.146}$$

For simplicity, let us also assume that there is no external magnetic field and that the dissipative effects due to viscosity and intrinsic conductivity are negligible. Then, by using the Euler equation (8.100), we derive the following equation for the oscillating variables:

$$i\omega\frac{w}{v_F^2}\boldsymbol{\nabla}\psi + \tilde{P}\boldsymbol{\nabla}\delta n - en\boldsymbol{\nabla}\phi = 0. \tag{8.147}$$

By employing the electric charge continuity relation (8.94), one can relate the oscillating fermion number density to the fluid velocity potential as follows:

$$\delta n = \frac{in}{\omega}\Delta\psi, \tag{8.148}$$

where the electric charge and current densities were taken in the form

$$\delta\rho = -e\delta n, \tag{8.149}$$

$$\delta\mathbf{J} = -en\delta\mathbf{u} - \frac{e^2\,[\mathbf{b}\times\delta\mathbf{E}]}{2\pi^2\hbar}. \tag{8.150}$$

With the help of Eq. (8.148), the Poisson equation (8.141) and the Euler equation (8.147) can be rewritten as

$$\Delta\phi = i\frac{4\pi en}{\varepsilon_e\omega}\Delta\psi, \tag{8.151}$$

$$\Delta\left[\omega^2\frac{w}{v_F^2} + \tilde{P}n\Delta - \frac{4\pi e^2 n^2}{\varepsilon_e}\right]\psi = 0. \tag{8.152}$$

Note that the last relation is obtained by applying the ∇ operator to the Euler equation (8.147). A surface-localized solution for $\psi(y)$ that decreases into the bulk of the semimetal should have the following form:

$$\psi = C_0^\psi e^{-ky} + \sum_{j=\pm} C_j^\psi e^{-\lambda_j y}, \tag{8.153}$$

where the explicit expressions for $\lambda_\pm$ are given by

$$\lambda_\pm = \pm\frac{\sqrt{\Omega_{\text{hyd}}^2 + K^2 k^2 - \omega^2}}{K}. \tag{8.154}$$

Here Ω_{hyd} is the plasma frequency in the hydrodynamic approximation given in Eq. (8.110) and the shorthand notation $K^2 = v_F^2 n\tilde{P}/w$ was used.[P] The imaginary coefficients $\lambda_\pm$ correspond to hybridized surface–bulk excitations. On the other hand, when coefficients $\lambda_\pm$ are real, the wave function with λ_+ describes a mode localized at the surface, while λ_- produces an unphysical solution that diverges as $y \to \infty$. Therefore, one should set $C_-^\psi = 0$.

By substituting Eq. (8.153) into Eq. (8.151), one obtains the following expressions for the electric potentials at $y > 0$ (semimetal) and $y < 0$ (vacuum):

$$\phi^{y>0} = C^\phi e^{-ky} + i\frac{4\pi en}{\varepsilon_e\omega}C_+^\psi e^{-\lambda_+ y}, \tag{8.155}$$

$$\phi^{y<0} = \tilde{C}^\phi e^{ky}. \tag{8.156}$$

The corresponding oscillating fermion number density δn can be obtained from Eq. (8.148). The explicit result reads

$$\delta n = \frac{in}{\omega}\left(\lambda_+^2 - k^2\right)C_+^\psi e^{-\lambda_+ y}. \tag{8.157}$$

[P]Coefficient $K = v_F\sqrt{n\tilde{P}/w}$ approaches the relativistic-like sound velocity $v_F/\sqrt{3}$ in the limit of vanishing temperature, $T \to 0$.

In order to determine the fluid velocity and fix the integration constants in the solutions for electric potential (8.155) and (8.156), one needs to first specify the boundary conditions on the surface. Since the electron fluid is present only inside the semimetal, it is natural to set

$$\delta u_y(0) = -\partial_y \psi(0) = 0. \tag{8.158}$$

The boundary conditions for the electric potential have the standard form, i.e.,

$$\phi^{y>0}(0) = \phi^{y<0}(0), \tag{8.159}$$

$$\varepsilon_e \partial_y \phi^{y>0}(0) - \partial_y \phi^{y<0}(0) = 4\pi e \delta n^{(\mathrm{FA})}, \tag{8.160}$$

where $\delta n^{(\mathrm{FA})}$ is the surface electric charge density due to the Fermi arc states. As usual, a nonzero surface charge density causes a discontinuity in the normal component of the electric field. The value of $\delta n^{(\mathrm{FA})}$ can be obtained from the total electric charge conservation of surface and bulk charge carriers. Since the surface and bulk charges are not necessarily conserved separately, the source and drain terms should be added on the right-hand sides of the surface and bulk conservation relations

$$\partial_t \rho^{(\mathrm{FA})} + \boldsymbol{\nabla}_\perp \cdot \mathbf{J}^{(\mathrm{FA})} = Q^{(\mathrm{FA})}, \tag{8.161}$$

$$\partial_t \rho + \boldsymbol{\nabla} \cdot \mathbf{J} = -Q^{(\mathrm{FA})} \delta(y), \tag{8.162}$$

where $\rho^{(\mathrm{FA})} \approx -e n^{(\mathrm{FA})}$ is the surface charge density and $\mathbf{J}^{(\mathrm{FA})} = v_F \hat{\mathbf{x}} \rho^{(\mathrm{FA})}$ is the surface current density.[q] Then for oscillating variables on the surface, one derives

$$ie(\omega - v_F k_x)\, \delta n^{(\mathrm{FA})} = -\delta J_y(0) = -\frac{i k_x b e^2 \phi(0)}{2\pi^2 \hbar}, \tag{8.163}$$

which gives

$$\delta n^{(\mathrm{FA})} = -\frac{k_x \mu b e \phi(0)}{2\pi^2 \hbar (\omega - v_F k_x)}. \tag{8.164}$$

Another boundary condition comes from the y component of Eq. (8.147) after re-expressing $\Delta \psi$ in terms of δn. Its explicit form reads

$$\tilde{P} \partial_y \delta n(0) = e n \partial_y \phi^{y>0}(0). \tag{8.165}$$

---

[q]The expression for the surface current density follows from the quasiclassical relation $\mathbf{J}^{(\mathrm{FA})} = \mathbf{v}^{(\mathrm{FA})} \rho^{(\mathrm{FA})}$, where $\mathbf{v}^{(\mathrm{FA})} = \partial \epsilon_s / (\hbar \partial \mathbf{k})$ and ϵ_s is the energy of surface Fermi arc state given by Eq. (5.17).

In the end, the dispersion relations for the surface modes can be obtained by satisfying the complete set of boundary conditions in Eqs. (8.158), (8.159), (8.160), (8.164), and (8.165). In the case of hybridized surface–bulk modes, the dispersion relations follow from Eq. (8.154), where λ is a continuous variable [945].

8.3.2.2   *Dispersion relation of surface plasmons*

By using the general framework presented in Sec. 8.3.2.1, here we discuss the properties of the surface collective modes. Let us start from the simplest case without the Fermi arcs on the surface. Physically, this is realized when the chiral shift is absent (e.g., in topologically trivial Dirac semimetals) or its direction is perpendicular to the surface. After satisfying all boundary conditions, Eq. (8.165) gives the following spectral equation for modes localized at the surface:

$$K^2\lambda_+ \left(\lambda_+ + k\right) = \frac{\Omega_{\text{hyd}}^2}{1 + \varepsilon_e}. \tag{8.166}$$

Its positive solution is given by

$$
\begin{aligned}
\omega &= \frac{1}{\sqrt{2}}\sqrt{\frac{2\varepsilon_e\Omega_{\text{hyd}}^2}{1 + \varepsilon_e} + K^2k^2 + Kk\sqrt{\frac{4\Omega_{\text{hyd}}^2}{1 + \varepsilon_e} + K^2k^2}} \\
&= \frac{\Omega_{\text{hyd}}\sqrt{\varepsilon_e}}{\sqrt{1 + \varepsilon_e}} + \frac{Kk}{2\sqrt{\varepsilon_e}} + \frac{K^2k^2\sqrt{1 + \varepsilon_e}(2\varepsilon_e - 1)}{8\Omega_{\text{hyd}}\varepsilon_e^{3/2}} + O(k^3). \quad (8.167)
\end{aligned}
$$

Note that the long-wavelength approximation is justified for this surface mode only when $\omega < ck$, which is consistent with the nonretarded regime assumed in the analysis. Taking into account that the speed of light c is very large, however, the range of validity extends down to rather small values of wave vector of the order of $k \simeq \Omega_{\text{hyd}}/c$.

As one can see, the spectrum of surface plasmons is isotropic and gapped. Also, the value of the gap is of the order of Ω_{hyd}. The surface plasmon frequency ω in Eq. (8.167) qualitatively agrees with the results for usual metals [944, 945, 947]. There are only a few differences in the expressions for the gap and the velocity, which have a relativistic-like dependence on the chemical potential and temperature.

The presence of the Fermi arcs changes qualitatively the properties of surface collective modes. Indeed, because of a nonzero surface charge density given in Eq. (8.160), the spectral equation becomes more complicated

$$K^2\lambda_+(k+\lambda_+) = \frac{\Omega_{\text{hyd}}^2\left[\omega_b - (\omega - v_F k_x)\right]}{\omega_b - (1+\varepsilon_e)(\omega - v_F k_x)}, \tag{8.168}$$

where $\omega_b = 2e^2bk_x/(\pi\hbar k)$. To leading order in the wave vector, the solutions to Eq. (8.168) are given by

$$\omega_\pm = \frac{1}{2(1+\varepsilon_e)}\left(\omega_b \pm \sqrt{\omega_b^2 + 4\varepsilon_e(1+\varepsilon_e)\Omega_{\text{hyd}}^2}\right) + O(k), \tag{8.169}$$

$$\omega^{(\text{FA})} = v_F k_x + \frac{kK\omega_b}{\varepsilon_e\Omega_{\text{hyd}}} + O(k^2). \tag{8.170}$$

The gapped modes with $\omega_\pm$ can be identified with surface plasmons [915, 916, 961]. It is interesting to note that the dispersion relations in Eq. (8.169) have discontinuities $\propto \text{sign}(k_x)$ at $k_x = k_z = 0$, namely, $(\lim_{k_x \to +0} - \lim_{k_x \to -0})\lim_{k_z \to 0}\omega_\pm = |\omega_b|/(1+\varepsilon_e)$. The discontinuities disappear, however, at $k_z \neq 0$. Depending on the value of the chiral shift, ω_+ (ω_-) at $k_x > 0$ $(k_x < 0)$ could become larger than $\Omega_{\text{hyd}}^2 + K^2k^2$. In such a case, the characteristic root defined in Eq. (8.154) becomes purely imaginary. The corresponding excitations are, therefore, hybridized surface–bulk modes [945, 947].

In contrast to the surface plasmons in Eq. (8.169), which exist even in usual nontopological materials, the mode with the dispersion relation in Eq. (8.170) originates exclusively from the Fermi arcs and has an unconventional directional dependence [915, 916]. In particular, such a mode is nonreciprocal meaning that it propagates only in a certain direction defined by the Fermi arc dispersion relation. In passing, it might be appropriate to mention here that the Landau damping, which is not captured in the hydrodynamic approach, could become relevant for this gapless Fermi arc mode. In particular, the damping is expected to be substantial when the phase velocity is smaller than the quasiparticle velocity v_F. As is easy to check, the mode with the dispersion relation in Eq. (8.170) is not overdamped at least for sufficiently small values of k_x because $\omega^{(\text{FA})}/k_x > v_F$. In general, however, it is important to keep in mind that the Landau damping could cause additional dissipation, which is missed in the hydrodynamic description.

The frequencies of the surface plasmons and the Fermi arc mode are presented in Figs. 8.10(a) and 8.10(b), respectively. As one can see from Fig. 8.10(a), the contour lines of the gapped surface plasmon are closed

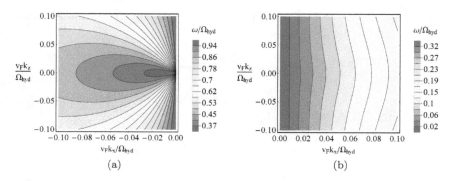

Fig. 8.10   Contour plots for the frequencies ($\omega > 0$) of (a) the surface plasmon and (b) the Fermi arc mode. The group velocity is normal to contour lines. Model parameters similar to those in Cd_3As_2, as well as $\mu = 25$ meV and $T = 10$ K were used.

ellipses elongated in the direction of the Fermi arcs, i.e., k_x. The contours of the Fermi arc mode in Fig. 8.10(b) are bell-shaped. Taking into account that the group velocity of the surface collective mode is given by the derivative of its frequency with respect to momentum, it can be represented by a vector normal to the contour lines. Then, as is clear from Figs. 8.10(a) and 8.10(b), the surface plasmon waves tend to spread and propagate primarily parallel to k_x. The gapless Fermi arc modes, on the other hand, propagate in one direction, although some spreading is noticeable at large values of k_x.

One can identify two qualitative features that are related to the effects of the Fermi arcs on the surface collective modes in Weyl semimetals. Firstly, the Fermi arc surface plasmons are described by a strongly anisotropic dispersion relation, which is different from the conventional surface plasmons described by Eq. (8.167). Secondly, there is a new gapless collective mode that appear only when the surface Fermi arc states are taken into account. If experimentally observed, this excitation could provide information about the chiral shift and the dispersion relation of the Fermi arcs. Moreover, such a unidirectional mode can be used for creating nonreciprocal devices (see, also, the discussion in Sec. 8.3.1).

It should be mentioned that an analysis by a different method [915, 961] revealed that the constant frequency contour lines for the surface plasmons have a hyperbolic shape. It is plausible that this feature is not reproduced in the hydrodynamic approximation because a hybridization of surface and bulk states is neglected at large frequencies. A more careful treatment requires to evaluate the overlap of surface and bulk wave functions [961]. The same approach can also be used to show that the surface plasmons

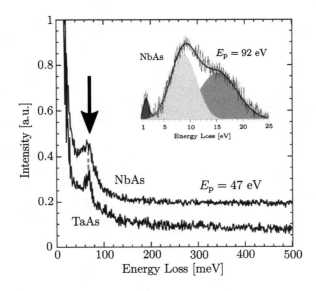

Fig. 8.11 High-resolution energy loss spectrum of the (001)-oriented Weyl semimetals TaAs and NbAs at room temperature. The position of the plasmon resonance is indicated by the arrow. The inset shows the excitation spectrum in an extended energy range probed for NbAs. Here E_p denotes the impinging energy of the beam. (Adapted from [967].)

are highly directional and have a small intrinsic damping only along the direction of the Fermi arcs.

In conclusion, let us briefly mention that the surface plasmons can be observed in experiment via the energy loss spectrum. The electron energy loss spectra in Weyl semimetals TaAs and NbAs were recently studied in [966, 967]. The representative results are shown in Fig. 8.11 for the (001)-oriented Weyl semimetals TaAs and NbAs at room temperature. In the same studies, it was also shown that the surface plasmon frequencies exhibit a great tunability due to the presence of chemical adsorbates. Such an effect can be used to construct Weyl plasmonic sensors operating in the mid-infrared frequency interval.

Energy loss (eV)

Chapter 9

Topological Superconducting Phases

> *In these days the angel of topology and the devil of abstract algebra fight for the soul of each individual mathematical domain.*
>
> Hermann Weyl

The phenomenon of superconductivity was observed experimentally in mercury by Heike Kamerlingh Onnes more than a century ago in 1911. In essence, superconductivity implies that an electric current flows without any resistance. The appearance of superconductivity is indicated by a phase transition at sufficiently low temperatures. This transition is marked by a jump of specific heat, which is related to a gap in the spectrum of low-energy electron excitations in a superconductor. In addition, magnetic fields are expelled from the bulk of superconducting material. This phenomenon is known as the Meissner effect [968].

The first phenomenologically successful attempt to describe superconductivity was proposed by brothers Fritz and Heinz London in 1935 [969]. While their eponymous equations were able to formally explain the Meissner effect, they shed little light on the origin of superconductivity. The first truly successful phenomenological theory was proposed by Vitaly Ginzburg and Lev Landau in 1950 [970]. The cornerstone of their eponymous theory is an effective action for a superconducting-order parameter.

The microscopic theory of superconductivity was developed by John Bardeen, Leon Cooper, and John Schrieffer in 1957 [140]. They argued that superconductivity comes from a condensation of Cooper pairs. The latter are weakly coupled bound states of electron quasiparticles with opposite spins and momenta. In the Bardeen–Cooper–Schrieffer (BCS) theory, the underlying mechanism of attraction between electrons is provided by a phonon-exchange interaction. The condensation of bosonic Cooper pairs leads to a nonperturbative ground state that is characterized by a nonzero energy gap at the Fermi level. The expectation value of pairing gap function in the BCS theory can be identified with the superconducting order parameter in the Ginzburg–Landau description. It decreases with temperature and vanishes at some critical value T_c, where a phase transition to a normal metal state occurs. In 1958, Nikolay Bogolyubov proposed another useful method for studying superconductivity that relies on a special canonical transformation that diagonalizes the Hamiltonian [971]. The method is often referred to as the Bogolyubov transformation.

Superconductivity is indeed one of the most amazing phenomena in condensed matter physics that continues to attract a significant interest of scientific community. It is also a paradigmatic example of a quantum phenomenon realized on a macroscopic scale. Quite importantly, superconductivity is not only interesting from the viewpoint of fundamental physics but finds numerous practical applications. Many of them rely on the nondissipative nature of superconducting currents. Furthermore, over the long period of its history, superconductivity grew into a rather broad field of research that covers many diverse phenomena. Their partial list includes the Meissner effect [968], the Josephson effect [972], the Andreev reflection [973], the formation of vortex lattices [974], as well as various manifestations of the proximity effect [975].

With the discovery of Dirac and Weyl semimetals, it was natural to ask whether superconductivity is possible in these novel materials and whether their nontrivial topology plays any important role in the superconducting properties [617–622]. Notably, the general idea about a possible interplay of topology and superconductivity was raised even before. Historically, this line of research was also predated by studies of superfluidity with a nontrivial topology in the superfluid B-phase of ^{3}He [73].[a] The milestone in the study of topological superconductivity was made by Nicholas Read and Dmitry Green [976] as well as Alexei Kitaev [977] who conceived a possibility of realizing Majorana zero modes. Later Liang Fu and Charles Kane suggested that Majorana fermions can also be created on a surface of topological insulators via the proximity effect of a conventional s-wave superconductor with a ferromagnetic material [978, 979].

---

[a]Note that superconductivity is similar to superfluidity but involves a condensation of charged Cooper pairs instead of neutral bosons. In the case of ^{3}He, the similarity is particularly close since its superfluid state is induced by a condensation of composite bosons made of pairs of fermionic atoms.

Because of a nontrivial bulk topology, the resulting Majorana modes are topologically protected [617–622]. Such robustness is invaluable for practical applications. For example, it could be used for preventing decoherence and errors in quantum computations [980].

Unconventional superconducting states of matter are usually realized in the presence of strong spin–orbit coupling. Since electron spins of chiral quasiparticles in the vicinity of Weyl nodes are aligned either parallel or antiparallel to their momenta, the spin–orbit locking in Dirac and Weyl semimetals could be qualified as the strongest possible. This suggests that a topologically nontrivial superconductivity with exotic properties is likely to be realized in such materials. Since Weyl nodes in topological semimetals occur in pairs of opposite chirality, the multiband superconductivity [981] could be also realized. The corresponding multiband structure is encoded in the Cooper pair wave function. Additional unusual properties of superconducting states in Weyl and certain Dirac semimetals might be induced by the surface Fermi arc states. Everything considered, therefore, it is clear that Weyl and Dirac semimetals make a promising platform for investigating an interplay of topology and superconductivity. It should be noted, however, that the study of superconductivity in Dirac and Weyl semimetals is a relatively new and developing field of research. While substantial progress has already been made, many unanswered questions remain. In this chapter, we review only some of the main ideas.

Before proceeding further, it is worth noting that superconductivity in Weyl semimetals should not be confused with Weyl superconductors [982]. In the latter, Weyl nodes appear in the spectrum only when the material (not necessarily a Weyl or Dirac semimetal) becomes superconducting [620, 622]. Although Weyl superconductors have interesting properties too, they are not discussed here.

9.1 Superconductivity in Weyl semimetals: intranode and internode pairing

Generically, there exist two distinctive types of superconducting pairing of Weyl fermions that could be realized in Weyl semimetals [983–991]. The first type is the *internode* pairing of quasiparticles from the Weyl nodes of *opposite* chirality. It resembles the pairing of electrons with opposite spins in the BCS theory. Therefore, the resulting superconducting state is sometimes referred to as a BCS-type ground state. The other type of pairing in Weyl materials is the *intranode* one. It involves quasiparticles from the *same* Weyl node of a given chirality. Such intranode pairing leads to spin-singlet Cooper pairs with nonzero momenta and, consequently, produces a ground state of the Larkin–Ovchinnikov–Fulde–Ferrell (LOFF) type, which was first considered in conventional materials in [992, 993].

While Cooper pairs have the vanishing center-of-mass momentum in the BCS superconducting state, this is not the case in the LOFF phase. Interestingly, the LOFF state was proposed theoretically in a system with an imbalance between spin-up and spin-down Fermi momenta more than 50 years ago [992, 993]. However, it gained the experimental support only recently in heavy fermion superconductors and organic superconductors in a magnetic field [994–996]. An analogue of a LOFF state was also observed in ^{3}He under confinement in a microfluidic cavity, where the superfluid order parameter forms a spatially modulated ("polka dot") pattern [997]. It is possible that the LOFF superconducting states can be realized also in relativistic forms of matter in high energy physics. In particular, the LOFF types of superconducting phases of dense quark matter may exist inside neutron stars [998].

As is clear, the question of the pairing type is very important as it can affect the physical properties of a Weyl semimetal in a superconducting ground state. However, the competition between the two pairing types is subtle. It is not easy to decide unambiguously which one is realized. This depends not only on the specifics of the effective electron–electron interaction but often also on other properties of a topological semimetal (e.g., details of the band structure, interaction effects, disorder, etc.).

Within a class of models with the phonon-mediated interaction, one can argue that the internode pairing is often more favorable energetically than the intranode one [986], even though the former has point nodes in the gap function [982, 983]. This conclusion could change, however, when the inversion symmetry is broken or a different interaction mechanism is considered. In the case of a simplified relativistic-like model with a point-like interaction, for example, the internode pairing with the vanishing momentum of Cooper pairs is disfavored [984]. In what follows, we discuss the structure of the superconducting gaps by using the *Bogolyubov–de Gennes* (BdG) formalism [971, 999].

9.1.1   *Bogolyubov–de Gennes Hamiltonian*

In order to describe qualitatively possible superconducting states, let us start with a minimal relativistic-like model of a Weyl semimetal with broken

$\mathcal{T}$ and a single pair of Weyl nodes separated by $2\mathbf{b}$ in momentum space. The explicit form of the low-energy Hamiltonian reads

$$H = \int d^3r \Psi^\dagger(\mathbf{r})\hat{H}\Psi(\mathbf{r}), \tag{9.1}$$

where

$$\hat{H} = \begin{pmatrix} H_+ & 0 \\ 0 & H_- \end{pmatrix} \tag{9.2}$$

and

$$H_\lambda = -\mu + \lambda\hbar v_F \boldsymbol{\sigma} \cdot (-i\boldsymbol{\nabla} - \lambda\mathbf{b}). \tag{9.3}$$

Here $\lambda = \pm$ denotes the chirality of Weyl nodes, μ is the electric chemical potential, v_F is the Fermi velocity, and $\boldsymbol{\sigma}$ is the vector made of the three Pauli matrices that act in spin (or, in general, pseudospin) space. Note that $\mathcal{T}$ symmetry is broken by a nonzero chiral shift $\mathbf{b}$.

Let us discuss what superconducting states are possible in this simplified model of a Weyl semimetal. For this purpose, we use the method of the Bogolyubov–de Gennes Hamiltonian [971, 999, 1000]. In essence, the latter is an effective Hamiltonian that acts in the Nambu space, spanned by electron states and their charge-conjugate (hole) states. The general form of the BdG Hamiltonian in momentum space is given by

$$H_{\text{BdG}}(\mathbf{k}) = \begin{pmatrix} \hat{H}(\mathbf{k}) & \hat{\Delta} \\ \hat{\Delta}^\dagger & -\hat{\Theta}\hat{H}(\mathbf{k})\hat{\Theta}^{-1} \end{pmatrix}, \tag{9.4}$$

where $\mathbf{k}$ is the momentum and $\hat{\Delta}$ is the gap matrix. Note that we use the following time-reversal operator:

$$\hat{\Theta} = i\mathbb{1}_2 \otimes \sigma_y \hat{K}\Pi_{\mathbf{k}\to-\mathbf{k}}. \tag{9.5}$$

By definition, $\mathbb{1}_2$ is a 2×2 matrix acting in the chirality space, $i\sigma_y$ is the Pauli matrix that describes the spin flip, $\hat{K}$ is the complex conjugation operator, and $\Pi_{\mathbf{k}\to-\mathbf{k}}$ is the inversion operator that changes the sign of $\mathbf{k}$. It is easy to show then that $\hat{\Theta}\hat{H}(\mathbf{k})\hat{\Theta}^{-1} = -\mu + \lambda\hbar v_F\boldsymbol{\sigma} \cdot (\mathbf{k} + \lambda\mathbf{b})$. The 8×8 BdG Hamiltonian (9.4) acts in the space of the Nambu–Gor'kov spinors [1000]

$$\Psi_{\text{BdG}} = (\Psi, \Psi_{\Theta})^T, \tag{9.6}$$

where

$$\Psi = [\psi_{\uparrow}^{+}(\mathbf{k}), \psi_{\downarrow}^{+}(\mathbf{k}), \psi_{\uparrow}^{-}(\mathbf{k}), \psi_{\downarrow}^{-}(\mathbf{k})]^T. \tag{9.7}$$

Here the subscripts of the wave function components denote the (pseudo)spin and the superscripts correspond to chirality. The time-reversal conjugate spinor is given by

$$\Psi_{\Theta} = [\psi_{\downarrow}^{+}(-\mathbf{k}), -\psi_{\uparrow}^{+}(-\mathbf{k}), \psi_{\downarrow}^{-}(-\mathbf{k}), -\psi_{\uparrow}^{-}(-\mathbf{k})]^{\dagger}. \tag{9.8}$$

The structure of the gap matrix $\hat{\Delta}$ depends on a Cooper pairing channel. Note that a nontrivial spin texture around Weyl nodes plays an important role in determining the gap. A schematic representation of the spin texture for Hamiltonian (9.2) is shown in Fig. 9.1(a). Similarly to the Berry curvature in Fig. 3.5(a), it shows a characteristic monopole–antimonopole form.

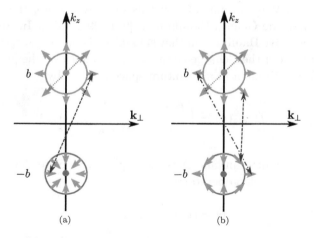

Fig. 9.1   Schematic representation of the spin texture around Weyl nodes in momentum space for the model Hamiltonians in (a) Eq. (9.2) and (b) Eq. (9.22). The upper (lower) Weyl nodes are right-handed (left-handed) and large green arrows denote spin directions. While partner states of intranode pairing in the upper Weyl nodes are shown by dotted magenta double-headed arrows, those corresponding to the internode pairing are denoted by brown dashed and dot-dashed double-headed arrows. The latter distinguish the spin-triplet (dashed) and spin-singlet (dot-dashed) pairings.

In the case of the *internode* pairing, which involves quasiparticles from Weyl nodes of opposite chiralities, the corresponding gap matrix is given by

$$\hat{\Delta}_{\text{inter}} = \begin{pmatrix} 0 & \hat{\Delta}_+ \\ \hat{\Delta}_- & 0 \end{pmatrix}, \qquad (9.9)$$

where $\Delta_\pm = \pm (\boldsymbol{\Delta} \cdot \boldsymbol{\sigma})$ and the gap function may depend on momentum **k**, i.e., $\boldsymbol{\Delta} = \boldsymbol{\Delta}(\mathbf{k})$. As is easy to check, the order parameter $\boldsymbol{\Delta}$ describes a spin-triplet gap that corresponds to a pairing of electron quasiparticles with opposite chiralities. Indeed, as one can infer from Fig. 9.1(a), in order to form Cooper pairs with the total zero momentum, electrons with parallel spins should couple. As a result, only a spin-triplet gap $\boldsymbol{\Delta}$ is present in Eq. (9.9), but a spin-singlet gap Δ_0 is absent.[b] This is also consistent with the canonical anticommutation relations of fermionic fields, $\{\psi_\alpha^{\lambda_1}, \psi_\beta^{\lambda_2}\} = 0$, which can be viewed as a formal representation of the Pauli principle. Note that the same fermion anticommutation relation also implies that the sign of gap $\hat{\Delta}_\lambda$ should flip when the sign of λ changes. This is consistent with the BdG Hamiltonian (9.4), where $\Psi_{\text{BdG}}^\dagger H_{\text{BdG}} \Psi_{\text{BdG}}$ becomes trivial when $\hat{\Delta}_\lambda$ is an even function of λ.

Next, let us discuss the gap matrix for the *intranode* pairing. Such pairing involves quasiparticles from Weyl nodes of the same chirality,

$$\hat{\Delta}_{\text{intra}} = \begin{pmatrix} \Delta_0 & 0 \\ 0 & -\Delta_0 \end{pmatrix}. \qquad (9.10)$$

In this case, the Pauli principle implies that spin-triplet terms are absent. This can be verified either by using the BdG Hamiltonian (9.4) or the fermion anticommutation relations. In addition, the spin-singlet nature of the gap is evident from the schematic representation of pairing in Fig. 9.1(a). Indeed, because of the nontrivial spin texture of the Fermi surface, it is clear that only quasiparticles with opposite spins can pair.

9.1.2 Structure of gaps

By considering the simplest case of a Weyl semimetal with a moderately large Fermi energy, it is natural to assume that only states near the Fermi

---

[b]Formally, the spin-singlet Δ_0 would appear in the gap matrix via the combination $(\boldsymbol{\Delta} \cdot \boldsymbol{\sigma}) + \Delta_0 \mathbb{1}_2$.

surface participate in Cooper pairing. In this case, the analysis of the BdG Hamiltonian (9.4) and its low-energy spectrum greatly simplifies.

9.1.2.1 *Internode pairing*

In the case of a superconducting state with internode pairing, it is sufficient to consider the following 4×4 effective BdG Hamiltonian [982, 983]:

$$
\hat{H}_{\text{BdG}}(\mathbf{k}) = \begin{pmatrix} \hat{H}_\lambda(\mathbf{k}) & \hat{\Delta}_\lambda \\ \hat{\Delta}_\lambda^\dagger & i\sigma_y \hat{H}_{-\lambda}^*(-\mathbf{k})i\sigma_y \end{pmatrix}
$$

$$
= \begin{pmatrix} -\mu + \lambda\hbar v_F (\boldsymbol{\sigma} \cdot \mathbf{k}_\lambda) & \lambda(\boldsymbol{\Delta} \cdot \boldsymbol{\sigma}) \\ \lambda(\boldsymbol{\Delta} \cdot \boldsymbol{\sigma})^\dagger & \mu + \lambda\hbar v_F (\boldsymbol{\sigma} \cdot \mathbf{k}_\lambda) \end{pmatrix}, \quad (9.11)
$$

where $\mathbf{k}_\lambda = \mathbf{k} - \lambda\mathbf{b}$. To separate the low- and high-energy modes, it is convenient to diagonalize the kinetic part of the effective BdG Hamiltonian (9.11) by using a suitable unitary transformation, i.e.,

$$
\begin{pmatrix} U_\lambda^\dagger & 0 \\ 0 & U_\lambda^\dagger \end{pmatrix} \hat{H}_{\text{BdG}}(\mathbf{k}) \begin{pmatrix} U_\lambda & 0 \\ 0 & U_\lambda \end{pmatrix}
$$

$$
= \left(\begin{array}{cc|cc} -\mu + \hbar v_F k_\lambda & 0 & & \\ 0 & -\mu - \hbar v_F k_\lambda & \multicolumn{2}{c}{U_\lambda^\dagger \hat{\Delta}_\lambda U_\lambda} \\ \hline \multicolumn{2}{c|}{U_\lambda^\dagger \hat{\Delta}_\lambda^\dagger U_\lambda} & \mu + \hbar v_F k_\lambda & 0 \\ & & 0 & \mu - \hbar v_F k_\lambda \end{array} \right). \quad (9.12)
$$

Here $k_\lambda = |\mathbf{k}_\lambda|$ and U_λ is composed of the eigenvectors corresponding to the kinetic part of Hamiltonian, i.e.,

$$
U_\lambda = \frac{1}{\sqrt{2}} \begin{pmatrix} \lambda e^{-i\varphi_\lambda}\sqrt{1 + \lambda\cos\theta_\lambda} & -\lambda e^{-i\varphi_\lambda}\sqrt{1 - \lambda\cos\theta_\lambda} \\ \sqrt{1 - \lambda\cos\theta_\lambda} & \sqrt{1 + \lambda\cos\theta_\lambda} \end{pmatrix}, \quad (9.13)
$$

where the spherical coordinates in momentum space were used, i.e., $\mathbf{k}_\lambda = k_\lambda (\cos\varphi_\lambda \sin\theta_\lambda, \sin\varphi_\lambda \sin\theta_\lambda, \cos\theta_\lambda)$. The explicit form of the transformed gap reads

$$
U_\lambda^\dagger \hat{\Delta}_\lambda U_\lambda = \lambda(\Delta_1 \cos\varphi_\lambda \cos\theta_\lambda + \Delta_2 \sin\varphi_\lambda \cos\theta_\lambda - \Delta_3 \sin\theta_\lambda)\sigma_x
$$

$$
- (\Delta_1 \sin\varphi_\lambda - \Delta_2 \cos\varphi_\lambda)\sigma_y
$$

$$
+ (\Delta_1 \cos\varphi_\lambda \sin\theta_\lambda + \Delta_2 \sin\varphi_\lambda \sin\theta_\lambda + \Delta_3 \cos\theta_\lambda)\sigma_z. \quad (9.14)
$$

Without loss of generality, let us assume that $\mu > 0$. Then, by keeping only the two dominant low-energy modes near the Fermi level in Eq. (9.12), i.e., $k_\lambda \simeq k_F = \mu/(\hbar v_F)$, one derives the following reduced BdG Hamiltonian for the internode pairing:

$$\hat{H}_{\text{BdG}}^{(\text{inter})}(\mathbf{k}_\lambda) = \begin{pmatrix} -\mu + \hbar v_F k_\lambda & \Delta_\lambda \\ \Delta_\lambda^\dagger & \mu - \hbar v_F k_\lambda \end{pmatrix}, \qquad (9.15)$$

where the effective gap Δ_λ is given by

$$\Delta_\lambda = \Delta_1 \left(\lambda \cos\varphi_\lambda \cos\theta_\lambda + i\sin\varphi_\lambda\right)$$
$$+\Delta_2 \left(\lambda\sin\varphi_\lambda\cos\theta_\lambda - i\cos\varphi_\lambda\right) - \Delta_3\lambda\sin\theta_\lambda. \qquad (9.16)$$

It is worth noting that the effective gap function in the reduced BdG Hamiltonian (9.15) has a nontrivial dependence on the angular coordinates of momentum $\mathbf{k}_\lambda$. This is true even if the initial ansatz for the gap in Eq. (9.9) is momentum-independent. As is easy to check, such a dependence is introduced by the unitary transformation U_λ, that was used for diagonalizing the kinetic part of the Hamiltonian.

The internode spin-triplet pairing considered above results in a gapless energy spectrum. For example, in the special case $\Delta_1 = \Delta_2 = 0$, the energy spectrum of Hamiltonian (9.15) is given by

$$\epsilon_{\mathbf{k}_\lambda} = \pm\sqrt{(\mu - \hbar v_F k_\lambda)^2 + |\Delta_3|^2 \sin^2\theta_\lambda}, \qquad (9.17)$$

which indeed has gapless nodes at $\theta_\lambda = 0$ and $\theta_\lambda = \pi$.

The appearance of gapless nodes in the spectrum of superconducting Weyl semimetals could also be inferred from general arguments based on symmetry and topology [622, 991]. Indeed, the pairing between two separate Fermi surfaces with opposite Chern numbers produces Cooper pairs with a nontrivial Berry phase structure. One can also argue that the phase of Cooper pairing cannot be well defined over the entire Fermi surface. Then, similarly to the arguments used in the derivation of the Berry phase and the Berry curvature in Sec. 3.3, this implies that the gap function must necessarily have a topologically protected nodal structure. The gapless energy spectrum of the reduced Hamiltonian (9.15) for the internode pairing is presented schematically in Fig. 9.2(a) for $\Delta_1 \neq 0$ and $\Delta_3 \neq 0$.

9.1.2.2 *Intranode pairing*

Let us also discuss briefly the intranode pairing. By diagonalizing the kinetic part of the Hamiltonian, removing the chiral shift dependence via

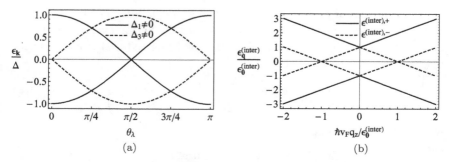

Fig. 9.2 (a) The energy spectrum of the reduced Hamiltonian (9.15) for $\Delta = \Delta_1$ (solid lines) and $\Delta = \Delta_3$ (dashed lines). For simplicity, $\hbar v_F k_\lambda = \mu$ and $\varphi_\lambda = 0$. The case $\Delta = \Delta_2$ is similar to $\Delta = \Delta_1$ but the nodes appear at $\varphi_\lambda = \pi/2$. (b) The energy spectrum (9.43) of the BdG Hamiltonian at $k_x = k_y = 0$ for the internode pairing. While the energy branch $\epsilon_{\mathbf{k}}^{(\text{inter}),-}$ has two Weyl nodes, the other branch, $\epsilon_{\mathbf{k}}^{(\text{inter}),+}$, is gapped.

the chiral transformation $\psi_\lambda \to e^{i\lambda\mathbf{b}\cdot\mathbf{r}}\psi_\lambda$ and $\psi_\lambda^\dagger \to e^{-i\lambda\mathbf{b}\cdot\mathbf{r}}\psi_\lambda^\dagger$, as well as projecting out high-energy modes, one can obtain the following reduced Hamiltonian:

$$\hat{H}_{\text{BdG}}^{(\text{intra})}(\mathbf{k}) = \begin{pmatrix} -\mu + \hbar v_F k & \lambda\Delta_0 e^{2i\lambda\mathbf{b}\cdot\mathbf{r}} \\ \lambda\Delta_0^* e^{-2i\lambda\mathbf{b}\cdot\mathbf{r}} & \mu - \hbar v_F k \end{pmatrix}, \tag{9.18}$$

where, for simplicity, the chirality index λ is omitted in the notation of momentum k. Note that the additional phase factors in the gaps come from the chiral transformation. As is easy to check, the energy spectrum of the reduced Hamiltonian (9.18) is given by

$$\epsilon_{\mathbf{k}} = \pm\sqrt{(\mu - \hbar v_F k)^2 + |\Delta_0|^2}, \tag{9.19}$$

which describes a fully gapped spin-singlet state.

It is clear from comparing the energy spectra in Eqs. (9.17) and (9.19) that the spin-singlet intranode pairing produces a superconducting state that is qualitatively different from that with the internode spin-triplet pairing. Indeed, while the latter has nodes, the intranode state is gapped.

9.1.3 *Intranode and internode pairing in two-band model*

In Secs. 9.1.1 and 9.1.2, the general form of a gap function was discussed by using a relativistic-like low-energy model of a Weyl semimetal. In order to clarify the physical origin of the intranode and internode pairing states, however, it is instructive to also consider a more realistic two-band model with a finite band width. In particular, we use a minimal cubic-lattice two-band model of a Weyl semimetal with broken $\mathcal{T}$ symmetry [389, 983]. The corresponding kinetic part of the Hamiltonian in momentum space reads

$$H_0(\mathbf{k}) = t\left[\sigma_x \sin(ak_x) + \sigma_y \sin(ak_y)\right] + t_z \sigma_z \left[\cos(ak_z) - \cos(ab)\right]$$
$$+ m\sigma_z \left[2 - \cos(ak_x) - \cos(ak_y)\right] - \mu, \tag{9.20}$$

where a is the lattice constant. This model describes a Weyl semimetal with two Weyl nodes located at $\mathbf{q}_0^\pm = (0, 0, \mp b)$.[c] The Fermi velocities are defined in terms of hopping parameters t and t_z as follows: $v_x = v_y = ta/\hbar$ and $v_z = t_z a \sin(ab)/\hbar$. For simplicity, we assume an isotropic model with $t_z = t/\sin(ab)$, which gives $v_x = v_y = v_z = v_F$.

At sufficiently small electric chemical potential $|\mu/t| \ll ab$, there are two separate and well-resolved Fermi surface pockets, as shown in Fig. 9.1(b). As one can see, they exhibit a nontrivial spin-momentum locking. Indeed, the spin texture is different from that in the relativistic-like model discussed in Sec. 9.1.1 and presented in Fig. 9.1(a). In order to see the origin of such a difference explicitly, let us expand Hamiltonian (9.20) in the vicinity of Weyl nodes, $\mathbf{k} = \mathbf{q}_0^\pm + \mathbf{q}$, where $\mathbf{q}$ is small. Then, to linear order in $\mathbf{q}$, one derives the following effective low-energy Hamiltonian:

$$H_0 \approx \sum_{\lambda=\pm} \psi_\lambda^\dagger H_\lambda(\mathbf{q}) \psi_\lambda, \tag{9.21}$$

where Weyl fermions of chirality λ are described by

$$H_\lambda(\mathbf{q}) = -\mu + \hbar v_F \left(q_x \sigma_x + q_y \sigma_y + \lambda q_z \sigma_z\right). \tag{9.22}$$

By comparing Hamiltonians (9.3) and (9.22), we see that the difference in the spin–orbital texture stems from a different account of chirality, which is now present only in the last term $\propto q_z \sigma_z$.

In order to study superconducting pairing states, it is sufficient to consider the case of a simple short-ranged interaction Hamiltonian, defined by

$$H_{\text{int}} = \sum_{\mathbf{k}} V(\mathbf{k}) n_{\mathbf{k}} n_{-\mathbf{k}}, \tag{9.23}$$

---

[c] Note that the Weyl node of positive chirality is located at $k_z = -b$ in model (9.20).

where $n_{\mathbf{k}} = \sum_s c^\dagger_{\mathbf{k},s} c_{\mathbf{k},s}$ is the number operator for electrons with momentum $\mathbf{k}$, $c^\dagger_{\mathbf{k},s}$ and $c_{\mathbf{k},s}$ are the creation and annihilation operators, and $\sum_s$ denotes the summation over the spin degrees of freedom. The interaction potential in momentum space is given by

$$V(\mathbf{k}) = V_0 + V_1 \left[\cos(ak_x) + \cos(ak_y) + \cos(ak_z) \right], \qquad (9.24)$$

where $V_0 < 0$ represents an attractive on-site interaction and V_1 describes the nearest-neighbor interaction.

By using Eq. (9.24), let us derive an explicit form of the interaction potential for states in the vicinity of Weyl nodes. The low-energy interaction Hamiltonian (9.23) for chiral quasiparticles reads

$$H_{\text{int}} = \sum_{s_1,s_2} \sum_{\mathbf{k},\mathbf{p},\mathbf{q}} V(\mathbf{k}) c^\dagger_{s_1}(\mathbf{k}+\mathbf{p}) c^\dagger_{s_2}(\mathbf{q}-\mathbf{k}) c_{s_2}(\mathbf{q}) c_{s_1}(\mathbf{p}). \qquad (9.25)$$

In the case of the internode pairing with the zero total momentum, i.e., $\mathbf{q} = -\mathbf{p}$, the Hamiltonian can be rewritten as

$$H_{\text{int}} = \sum_{s_1,s_2} \sum_{\mathbf{l},\mathbf{p}} V(\mathbf{l}-\mathbf{p}) c^\dagger_{s_1}(\mathbf{l}) c^\dagger_{s_2}(-\mathbf{l}) c_{s_2}(-\mathbf{p}) c_{s_1}(\mathbf{p}). \qquad (9.26)$$

This form is particularly suitable in the vicinity of Weyl nodes. By making the substitutions $\mathbf{l} = \mathbf{k} + \mathbf{q}_0^{\lambda_1}$ and $\mathbf{p} = \mathbf{q} + \mathbf{q}_0^{\lambda_2}$, where $\mathbf{k}$ and $\mathbf{q}$ are small momentum deviations from Weyl nodes, one finds

$$H_{\text{int}}^{(\text{inter})} = \sum_{\lambda_1,\lambda_2} \sum_{s_1,s_2} \sum_{\mathbf{k},\mathbf{q}} V_{\text{int}}^{\lambda_1,-\lambda_1;-\lambda_2,\lambda_2}(\mathbf{k}-\mathbf{q}) \psi^\dagger_{\lambda_1,s_1}(\mathbf{k}) \psi^\dagger_{-\lambda_2,s_2}(-\mathbf{k})$$

$$\times \psi_{-\lambda_1,s_2}(-\mathbf{q}) \psi_{\lambda_2,s_1}(\mathbf{q}) + \sum_{\lambda} \sum_{s_1,s_2} \sum_{\mathbf{k},\mathbf{q}} V_{\text{int}}^{\lambda,\lambda;-\lambda,-\lambda}(\mathbf{k}-\mathbf{q})$$

$$\times \psi^\dagger_{\lambda,s_1}(\mathbf{k}) \psi^\dagger_{-\lambda,s_2}(-\mathbf{k}) \psi_{\lambda,s_2}(-\mathbf{q}) \psi_{-\lambda,s_1}(\mathbf{q}). \qquad (9.27)$$

From the general form of the result, it is possible to identify several different pairing channels characterized by specific combinations of spins and chiralities of wave functions. For example, one of the terms is given by

$$V(\mathbf{k} + \mathbf{q}_0^+ - \mathbf{q} - \mathbf{q}_0^+) \psi^\dagger_{+,s_1}(\mathbf{k}) \psi^\dagger_{-,s_2}(-\mathbf{k}) \psi_{-,s_2}(-\mathbf{q}) \psi_{+,s_1}(\mathbf{q}). \qquad (9.28)$$

This allows one to identify $V_{\text{int}}^{\lambda_1,-\lambda_1;-\lambda_2,\lambda_2}(\mathbf{k}-\mathbf{q}) = V(\mathbf{k}+\mathbf{q}_0^{\lambda_1} - \mathbf{q} - \mathbf{q}_0^{\lambda_2})$. Then, by using Eq. (9.24), the following expressions are obtained [983]:

$$V_{\text{int}}^{\lambda,-\lambda;-\lambda,\lambda}(\mathbf{k}-\mathbf{q}) = V_{\text{int}}^{\lambda,\lambda;-\lambda,-\lambda}(\mathbf{k}-\mathbf{q}) = V_0 + 3V_1 - V_1 \frac{a^2(\mathbf{k}-\mathbf{q})^2}{2},$$

$$\qquad (9.29)$$

$$V_{\text{int}}^{\lambda,-\lambda;\lambda,-\lambda}(\mathbf{k} - \mathbf{q}) = V_0 + 2V_1 + V_\perp + V_\parallel^\lambda, \tag{9.30}$$

where

$$V_\perp = -V_1 \frac{a^2(\mathbf{k}_\perp - \mathbf{q}_\perp)^2}{2}, \tag{9.31}$$

$$V_\parallel^\lambda = V_1 \left[1 - \frac{a^2(k_z - q_z)^2}{2} \right] \cos(2ab) + \lambda V_1 a(k_z - q_z) \sin(2ab). \tag{9.32}$$

Note that the transverse momentum $\mathbf{k}_\perp = (k_x, k_y, 0)$ is defined with respect to the direction of the chiral shift $\mathbf{b} = (0, 0, b)$.

In the case of the intranode pairing, the relevant part of the interaction Hamiltonian is given by

$$H_{\text{int}}^{(\text{intra})} = \sum_\lambda \sum_{s_1,s_2} \sum_{\mathbf{k},\mathbf{q}} V_{\text{int}}^{\lambda,\lambda;\lambda,\lambda}(\mathbf{k} - \mathbf{q}) \psi_{\lambda,s_1}^\dagger(\mathbf{k}) \psi_{\lambda,s_2}^\dagger(-\mathbf{k}) \psi_{\lambda,s_2}(-\mathbf{q}) \psi_{\lambda,s_1}(\mathbf{q}),$$

$$\tag{9.33}$$

where the interaction potential $V_{\text{int}}^{\lambda,\lambda;\lambda,\lambda}(\mathbf{k}, \mathbf{q})$ for states in the vicinity of Weyl nodes is given by the same expression as in Eq. (9.29).

The general form of the gap equation reads

$$\Delta_{s_1 s_2}^{\lambda_1 \lambda_2}(\mathbf{p}) = \sum_{\lambda_3, \lambda_4} \sum_{\mathbf{q}} V^{\lambda_1 \lambda_2; \lambda_3 \lambda_4}(\mathbf{p}, \mathbf{q}) \langle \psi_{\lambda_3, s_2}(-\mathbf{q}) \psi_{\lambda_4, s_1}(\mathbf{q}) \rangle. \tag{9.34}$$

It allows one to obtain any type of the pairing gap function $\Delta_{s_1 s_2}^{\lambda_1 \lambda_2}(\mathbf{p})$ in a self-consistent manner. Before proceeding to the analysis of this equation, let us further discuss the spin and chiral structure for both intranode and internode pairing channels.

The most general spin structure of the gap function includes a spin-singlet channel $\propto i\sigma_y$ and a spin-triplet one $\propto i\boldsymbol{\sigma}\sigma_y$. We assume that $V_0 < 0$ and $|V_0| \gg |V_1|$. In this case, the spin-singlet channel is preferred for both intranode and internode pairings since it is determined both by V_0 and V_1 rather than V_1 alone.[d] As to the chiral structure, we consider the internode pairing described by τ_x and the intranode pairing described by $\mathbb{1}_2$.[e]

By taking into account that the internode pairing is described by a spin-singlet gap matrix $\Delta_{s_1 s_2}^{\lambda_1 \lambda_2}(\mathbf{p}) = \delta_{\lambda_1, -\lambda_2}(-1)^{(1+s_1)/2} \delta_{s_1, -s_2} \Delta_{\text{inter}}(\mathbf{p})$, we can

---

[d]Note that a different spin texture of the Fermi surfaces allows for the internode pairing in the spin-singlet channel in model (9.22), see Fig. 9.1(b).
[e]The kinetic part of the Hamiltonian is diagonal in chiral space, i.e., $H_0(\mathbf{q}) = \text{diag}\,[H_+(\mathbf{q}), H_-(\mathbf{q})]$.

rewrite the gap equation (9.34) as

$$\Delta_{\text{inter}}(\mathbf{p}) = \frac{1}{4} \sum_{\lambda_1,\lambda_2,\lambda_3,\lambda_4} \sum_{s_1,s_2} \sum_{\mathbf{q}} \delta_{\lambda_1,-\lambda_2} V^{\lambda_1\lambda_2;\lambda_3\lambda_4}(\mathbf{p},\mathbf{q})$$

$$\times \langle \psi_{\lambda_3,s_2}(-\mathbf{q})(-1)^{(1+s_2)/2} \delta_{s_1,-s_2} \psi_{\lambda_4,s_1}(\mathbf{q}) \rangle. \qquad (9.35)$$

Furthermore, by using Eqs. (9.29) and (9.30) and replacing the sum over $\mathbf{q}$ with an integral, we arrive at the following explicit form of the gap equation for the internode pairing [983]:

$$\Delta_{\text{inter}} = -\frac{2V_0 + V_1[5 + \cos(2ab)]}{2} \sum_{\lambda,s=\pm} s \int \frac{d^3q}{(2\pi)^3} \langle \psi_{\lambda,s}(\mathbf{q})\psi_{-\lambda,-s}(-\mathbf{q}) \rangle.$$

$$(9.36)$$

For the intranode pairing, one can assume that the pairing dynamics in the vicinity of each Weyl node is decoupled from other nodes. This allows one to consider the gap equation for each node independently. By taking into account the spin-single structure of the gap matrix for the intranode pairing, i.e., $\Delta^{\lambda_1\lambda_2}_{s_1s_2}(\mathbf{p}) = \delta_{\lambda_1,\lambda_2}(-1)^{(1+s_1)/2}\delta_{s_1,-s_2}\Delta_{\text{intra}}(\mathbf{p})$, we can rewrite the gap equation (9.34) in the following form [983]:

$$\Delta_{\text{intra}} = -\frac{V_0 + 3V_1}{2} \sum_{\lambda=\pm} \sum_{s=\pm} s \int \frac{d^3q}{(2\pi)^3} \langle \psi_{\lambda,s}(\mathbf{q})\psi_{\lambda,-s}(-\mathbf{q}) \rangle. \qquad (9.37)$$

The intranode and internode pairing terms enter the BdG Hamiltonian as

$$H^{(\text{intra})}_{\text{pair}} = \Delta_{\text{intra}} \sum_{\lambda=\pm} \sum_{s=\pm} \sum_{\mathbf{p}} s\psi^\dagger_{\lambda,s}(\mathbf{p})\psi^\dagger_{\lambda,-s}(-\mathbf{p}) + h.c. \qquad (9.38)$$

and

$$H^{(\text{inter})}_{\text{pair}} = \Delta_{\text{inter}} \sum_{\lambda=\pm} \sum_{s=\pm} \sum_{\mathbf{p}} s\psi^\dagger_{\lambda,s}(\mathbf{p})\psi^\dagger_{-\lambda,-s}(-\mathbf{p}) + h.c., \qquad (9.39)$$

respectively. By making use of these pairing terms and following the same considerations as in Sec. 9.1.2, we can obtain the energy spectrum in the corresponding superconducting states with the intranode and internode pairings.

Let us start from the intranode pairing. As is easy to verify, in terms of the Nambu–Gor'kov spinors $\Psi_{\text{BdG}}(\mathbf{q}) = [\psi_{\lambda=+}(\mathbf{q}), i\sigma_y\psi^\dagger_{\lambda=+}(-\mathbf{q})]^T$, the corresponding BdG Hamiltonian is given by

$$H_{\text{BdG}}^{(\text{intra})}(\mathbf{q}) = \begin{pmatrix} H_+(\mathbf{q}) & \Delta_{\text{intra}}\mathbb{1}_2 \\ \Delta_{\text{intra}}\mathbb{1}_2 & -H_+(\mathbf{q}) \end{pmatrix}. \tag{9.40}$$

This leads to the following gapped dispersion relation for the quasiparticle excitations:

$$[\epsilon_{\mathbf{q}}^{(\text{intra}),\pm}]^2 = (\hbar v_F q \pm \mu)^2 + |\Delta_{\text{intra}}|^2 . \tag{9.41}$$

As is clear, for $\mu > 0$, low-energy excitations are described by $\epsilon_{\mathbf{q}}^{(\text{intra}),-}$. Note that this dispersion is the same as in a relativistic-like model considered in Sec. 9.1.2 and is given in Eq. (9.19).

Let us now turn to the case of the internode pairing. The corresponding BdG Hamiltonian reads

$$H_{\text{BdG}}^{(\text{inter})}(\mathbf{q}) = \begin{pmatrix} H_+(\mathbf{q}) & \Delta_{\text{inter}}\mathbb{1}_2 \\ \Delta_{\text{inter}}\mathbb{1}_2 & -H_-(\mathbf{q}) \end{pmatrix}. \tag{9.42}$$

It gives the following dispersion relation of excitations:

$$[\epsilon_{\mathbf{q}}^{(\text{inter}),\pm}]^2 = \hbar^2 v_F^2 q^2 + |\Delta_{\text{inter}}|^2 + \mu^2 \pm 2\hbar v_F \sqrt{q_z^2 |\Delta_{\text{inter}}|^2 + q^2\mu^2}. \tag{9.43}$$

As is easy to see, the low-energy excitation branch $\epsilon_{\mathbf{k}}^{(\text{inter}),-}$ has nodes at $q_x = q_y = 0$ and $q_z = \pm\sqrt{|\Delta_{\text{inter}}|^2 + \mu^2}/(\hbar v_F)$. In the vicinity of these nodes, the approximate dispersion relations are given by

$$[\epsilon_{\mathbf{q},\zeta}^{(\text{inter}),-}]^2 \approx (\hbar v_F q_z - \zeta\sqrt{|\Delta_{\text{inter}}|^2 + \mu^2})^2 + \hbar^2 v_F^2 q_\perp^2 \frac{|\Delta_{\text{inter}}|^2}{\mu^2 + |\Delta_{\text{inter}}|^2}, \tag{9.44}$$

where $\zeta = \pm$ labels two distinct zero-energy nodes in momentum space. This result agrees qualitatively with that in Sec. 9.1.2 as well as with the findings in [982] at $\mu = 0$. Curiously, the low-energy excitation spectrum in Eq. (9.44) resembles an electron quasiparticle spectrum in anisotropic Weyl semimetals. Its physical meaning as the energy of bosonic excitations in a superconducting state is quite different, however.

Similarly to Weyl nodes in the normal (nonsuperconducting) phase of Weyl semimetals, the zero-energy nodes in the low-energy branch of the dispersion relation in Eq. (9.44) are associated with a nontrivial topology of the internode pairing (see also the discussion at the end of Sec. 9.1.2.1). This implies, in particular, that the existence of such nodal points is topologically protected against small perturbations. In view of the bulk-boundary correspondence discussed in Sec. 5.2, this also suggests that there should exist nontrivial states on the surface of such superconductors. Indeed, the latter were found theoretically [982, 986, 991, 1001] and were identified with Majorana modes. We will discuss Majorana modes in detail in Sec. 9.2.4. Here let us mention, however, that surface states in the superconducting state with the internode pairing contain not only Majorana modes, but also the Fermi arcs inherited from the normal phase of Weyl semimetal [986, 1001] and protected by bulk topology.

Since there exist several possible superconducting phases with different Cooper pairing channels in Weyl semimetals, it is important to figure out which one of them is the most favorable energetically. Naively, the energy gain from a Cooper pair condensation should scale as a square of the gap function. This suggests, therefore, that a state with gapless nodes in the energy spectrum (9.43) of a superconductor with internode pairing is not the lowest energy (ground) state. Such a conclusion was indeed reconfirmed by direct calculations of the energy density in the model with Hamiltonian (9.20) [983].

This result appears to be also supported by a qualitative analysis of spin-momentum locking in pairing around the Fermi surface. Indeed, as one can see from Fig. 9.1(b), the intranode pairing connects electron states with momenta $\mathbf{k}$ and $-\mathbf{k}$, which always have opposite spins. The situation is very different for the internode pairing, however. Indeed, because of the specifics of the low-energy Hamiltonian in Eq. (9.22), Weyl nodes of opposite chiralities differ only by the sign of z component of their quasiparticle momenta. As a result, the internode pairing involves quasiparticles with opposite spins only for the states with $k_z = \pm b$. For other momenta, the quasiparticle spins are not exactly antiparallel and could even become parallel near the poles ($k_\perp = 0$). This means that such pairs of quasiparticles cannot fully contribute to the formation of a spin-singlet superconducting gap, reducing the free energy gain of the superconducting state with the internode pairing. It is not surprising, therefore, that the intranode superconducting state is found to be the ground state in the case of short-ranged interaction. However, this conclusion may change in the case of long-ranged interaction and/or in more sophisticated models of Weyl semimetals [986]. In fact, as advocated in [986], despite the presence of nodes in the spectrum of excitations, the superconducting phase with the internode pairing could have a lower free energy than the one with the intranode pairing.

9.1.4 Effects of (pseudo)magnetic fields

An interplay of magnetic fields and superconductivity is an important research topic that is motivated not only by interesting underlying physics, but also by practical applications. As is well known, magnetic fields generically inhibit superconductivity due to the Meissner effect [968]. It is possible, in principle, that the situation may change in the limit of very strong fields [1002, 1003]. This was argued based on the fact that quenching of the kinetic energy due to the formation of Landau levels could greatly enhance the electron pairing. Such a possibility has not been verified rigorously, however, since the analysis also requires a self-consistent treatment of the Meissner effect. Indeed, the superconducting currents cause a back-reaction and tend to increase the energy of the superconducting state. The final outcome of the two competing effects could be quite nontrivial.

As we discussed in Sec. 3.7, in addition to usual magnetic fields, there is an intriguing possibility to realize pseudomagnetic (or axial magnetic) fields in Weyl and Dirac semimetals. Such fields can be induced by mechanical deformations (e.g., by bending or twisting samples) and hold a great promise for applications. Recall that the key property of the pseudomagnetic field $\mathbf{B}_5$ is that its coupling to Weyl fermions is proportional to the chirality. This implies that Cooper pairs in the case of the internode pairing are neutral with respect to $\mathbf{B}_5$. It does not mean, however, that $\mathbf{B}_5$ has no effect on pairing. The situation with the intranode pairing is expected to be even more complicated because Cooper pairs from the Weyl nodes of the same chirality feel the same effective magnetic field. One might naively expect, therefore, that there is an analogue of the Meissner effect for the intranode pairing. However, the dynamics of pseudoelectromagnetic fields is governed by equations for mechanical deformations rather than Maxwell's equations. Therefore, the conventional textbook derivation of the Meissner effect no longer applies.

Several studies of superconducting phases in Weyl (semi-)metals in weak background pseudomagnetic fields were performed in [1004, 1005] by using the quasiclassical Eilenberger equation approach [1006, 1007]. The latter allows one to obtain the Ginzburg–Landau action for the superconducting order parameter. Its analysis shows that the pseudomagnetic field causes a spatial modulation of the gap function. Moreover, no pseudomagnetic Meissner effect occurs [1004] in such an approach. In the case of Weyl semimetals with broken $\mathcal{T}$ symmetry, it was found [1005] that the internode spin-triplet pairing with the spins of Cooper pairs parallel to the pseudomagnetic field is the most favorable channel among all spin-triplet options. Moreover, both electric and chiral supercurrents vanish in such a superconducting state. On the other hand, the pairing with the spins of Cooper pairs normal to $\mathbf{B}_5$ is inhibited by the field. Unfortunately, the use of the quasiclassical Eilenberger equation is restricted only to weak pseudo-magnetic fields and cannot capture the dynamics when the pseudo-Landau levels form.

9.1.5 *Experimental results for intrinsic superconductivity*

In this subsection, we discuss the key experimental signatures of intrinsic superconductivity in Dirac and Weyl semimetals. Let us start with the results for Dirac semimetals. Initially, two groups [1008, 1009] reported an evidence of unconventional superconductivity in the Dirac semimetal Cd_3As_2 by using the point contact technique. It was argued that the observation of the zero biased peak in the conductance as well as the symmetric structure of other peaks around the zero bias are the indicators of p-wave superconductivity (see also the experimental data in Fig. 9.5(a) as well as the discussion in Sec. 9.2.2). However, such an observation cannot be attributed unambiguously to the intrinsic superconductivity in Dirac semimetals. Indeed, it was suggested that either the point contact modifies a local carrier density in favor of superconductivity or it alters the local band structure by introducing mechanical deformation and lowering the crystal symmetry. The defining property of superconductivity, i.e., a vanishing direct current (DC) resistance, was observed in [1010] by applying high pressure to the Dirac semimetal Cd_3As_2. The corresponding experiential data is shown in Fig. 9.3(a). The critical temperature of the superconducting phase transition increases with pressure P and reaches approximately 2 K at $P \approx 8.5$ GPa. However, the X-ray diffraction measurements showed that, as in the case of the tip-induced superconductivity, the observed superconducting transition under pressure is accompanied by a structure transition [1010, 1011] approximately at $P \approx 4.7$ GPa. Moreover, pressure might even lead to a breakdown of Dirac semimetal state [1011]. Therefore, additional study is needed to clarify whether the induced superconductivity is indeed associated with the Cooper pairing in a Dirac semimetal or other phase.

A study similar to the tip-induced superconductivity in Cd_3As_2 was also performed for the Weyl semimetal TaAs [1013, 1014]. As in the case of the

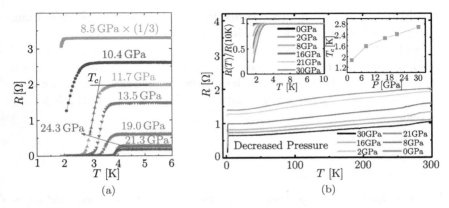

Fig. 9.3   The dependence of resistance on temperature in (a) the Dirac semimetal Cd_3As_2 and (b) the Weyl semimetal TaP at several fixed values of pressure. (Adapted from [1010, 1012], respectively.)

Dirac semimetal Cd_3As_2, the differential conductivity dI/dV curve exhibits a zero-biased peak and two additional peaks situated symmetrically with respect to zero (see also the experimental data in Fig. 9.5(b) as well as the discussion in Sec. 9.2.2). Since superconductivity was observed only for a "hard" tip, it is, most probably, related to a structural phase transition. An interplay of the structural transition and superconductivity in the Weyl semimetal TaP was also investigated in [1012]. Unlike the case of the Dirac semimetal Cd_3As_2, pressure has to increase significantly ($P \gtrsim 70$ GPa) before the structural phase transition is reached. It was argued, however, that while a structural phase transition indeed takes place, the new crystalline structure should also support Weyl semimetal spectrum. The experimental results for the temperature dependence of resistance are shown in Fig. 9.3(b). The appearance of superconductivity is evident from a sharp drop of resistance at about $P \approx 30$ GPa. As in Cd_3As_2, the corresponding critical temperature T_c is enhanced by pressure. Superconductivity was also observed in the type-II Weyl semimetal $MoTe_2$. The latter exhibits a superconducting phase transition even at ambient pressure [1015–1018]. The corresponding critical temperature, however, could be significantly enhanced by applying high pressure.

While there is a clear experimental evidence supporting the realization of superconductivity in Dirac and Weyl semimetals, a nontrivial interplay of pairing and structural phase transitions complicates the interpretation. Therefore, additional studies are highly desirable.

So far, we discussed only the intrinsic superconductivity in Dirac and Weyl semimetals. However, there are other effective means to realize a superconducting state. In particular, superconductivity can be induced by various proximity effects. The latter will be discussed in the following section.

9.2   Proximity effects

In this section, we discuss several effects associated with junctions and interfaces. They play a very important role in applications of superconductivity. Let us start by noting that the *proximity effect*, which is characterized by the permeation of Cooper pairs into a nonsuperconducting medium [975], can be used to induce superconductivity in many materials. This could be particularly important, for example, in the case of those materials where superconducting state is hard or even impossible to achieve intrinsically.

The electron transport through superconducting-normal (SN) heterostructures is another unusual implication of the proximity effect. A paradigmatic example is the *Josephson junction*, which is made of a pair of superconducting contacts coupled by a weak link, usually made of a nonsuperconducting metal or an insulator. The Josephson current in such a junction provides another manifestation of a macroscopic quantum phenomenon [972, 1019]. It finds useful applications in precise measurements and novel computing devices.

A closely related effect is the *Andreev reflection*. In essence, it is an unusual form of reflection where electrons in a normal material are reflected from the surface of a superconductor as holes [973]. Such a surprising outcome is possible when the electron quasiparticle energy (or the electric chemical potential) lies inside the energy gap of a superconductor. The Andreev reflection implies that incoming electrons on the normal side produce Cooper pairs that propagate into the superconducting region.

The physics of interfaces of topological superconductors has another hidden but very important implication. They could host the so-called *Majorana zero modes*, known also as Majorana bound states, which are highly sought-after but elusive quasiparticles with unusual physical properties. Arguably, their topological protection could be invaluable in applications to quantum computing. Since neutrinos could be also Majorana fermions [32], the experimental realization of Majorana modes in solid-state systems could provide yet another link between high energy and condensed matter physics.

9.2.1 Proximity induced superconductivity: Qualitative description

The proximity effect, also known as the Holm–Meissner effect [975], describes the superconducting phenomena that occur when a superconductor is placed in contact with a normal (semi)metal, insulator, or another superconductor. Generically, the superconducting order parameter does not change abruptly at the boundary. The Cooper pairs from the superconductor penetrate partially into the normal conducting material leading to an induced superconductivity there. There is also a reciprocal effect, meaning that the order parameter in the superconductor weakens near the contact. Qualitatively, the reduction of the gap is proportional to $\exp\left(-2|z - z_0|/\xi\right)$, where $z = z_0$ denotes the coordinate of the boundary. The corresponding length scale is the coherence length ξ. Its counterpart inside the normal conducting material is called the penetration depth ξ_n. It characterizes how rapidly the superconducting order parameter fades as a function of the distance from the boundary, i.e., $\Delta \propto \exp\left(-2|z - z_0|/\xi_n\right)$. The proximity effect and the coherence lengths are schematically shown in Fig. 9.4(a). This effect provides a convenient way to induce superconducting states in materials in which intrinsic superconductivity is hard to achieve.

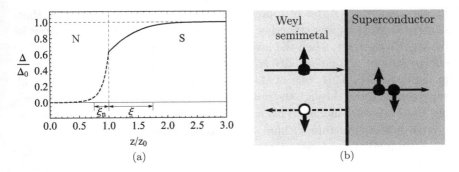

(a) (b)

Fig. 9.4 (a) Schematic illustration of the proximity effect with Cooper pairs exhibiting a gradual decoherence in the vicinity of a superconducting-normal conductor interface at $z = z_0$. The solid and dashed lines represent the superconducting gap Δ inside the superconducting (S) and normal conducting (N) material, respectively. Here ξ_n is the Cooper pair penetration depth in a normal conducting material, ξ is the coherence length in a superconductor, and Δ_0 is the superconducting gap deeply in the superconducting region $z \to \infty$. (b) Schematic representation of the Andreev reflection. An incident electron can penetrate into a superconductor as a Cooper pair only when a hole appears in a normal conducting material (e.g., a Weyl semimetal). Here filled and empty circles represent electrons and holes, respectively. Spins are denoted by arrows.

The easiest description of proximity effects is given by the phenomenological Ginzburg–Landau theory. The microscopic mechanism can be understood in terms of the Andreev reflection, which provides a single-particle picture. This mechanism will be discussed in Sec. 9.2.2. Proximity effects can also be studied by using the semiclassical approach. The corresponding analysis for 2D and 3D Dirac semimetals was done, for example, in [1020]. As a direct method, one can diagonalize the BdG Hamiltonian consisting of the three main parts: the conventional Weyl/Dirac BdG Hamiltonian $H_{W/D}$, the BdG Hamiltonian of a superconductor H_S, and the tunneling part H_{tun}, which mixes the states in normal conducting and superconducting materials. While the diagonalization method provides a microscopic treatment of the problem, very often the corresponding calculations can be done only numerically.

Proximity effects are very natural in various heterostructure models. In this connection, it is worth recalling that one of the original theoretical models of Weyl semimetals with broken $\mathcal{T}$ symmetry was formulated as a heterostructure made of alternating topological and normal insulator layers [391] (see Sec. 3.5.2 for the details of the model). Similarly, Tobias Meng and Leon Balents studied [982] the realization of a Weyl superconducting state in heterostructures comprised of layers of superconductors and topological insulators. In this case, the proximity effect induces superconductivity for the surface states of topological insulator layers. It is notable that nontrivial Weyl superconducting phase is realized only if $\mathcal{T}$ symmetry is broken. In contrast, when the inversion symmetry is broken, the proximity effect leads to a topologically trivial gapped phase. As will be discussed in detail in Sec. 9.2.4, one of the key advantages of inducing topological superconductivity through the proximity effect is the possibility of engineering the Majorana modes. This can be done by using commonly available s-wave superconductors without the need to obtain intrinsical topological superconductivity in topological materials with strong spin–orbit coupling. Proximity-induced superconductivity could also be utilized to affect the pairing of surface states in Weyl semimetals.

On the experimental side, an interesting approach to induce superconductivity in the Weyl semimetal NbAs (which belongs to the same family of monopnictides as TaAs, TaP, and NbP) was proposed [1021]. It was shown that the focused ion-beam technique can be used to deplete surface layers of NbAs crystals of arsenic and, therefore, create a niobium-rich coating. Niobium is a well-known superconductor with a sizable critical temperature ($T_c = 9.2$ K). In contrast, when a superconducting material is deposited on a crystal surface by using a conventional method, the corresponding heterostructure is often plagued by lattice mismatches and the formation of interface layers. The scaling of the critical current, i.e., the current at which the superconducting state is destroyed, supports the superconducting shell scenario and is also consistent with supercurrents in an extended proximity-induced superconducting layer in the bulk of a semimetal. In addition, proximity-induced superconductivity in type-II Dirac and Weyl semimetals was experimentally probed in $PdTe_2$ [1022, 1023] and WTe_2 [1024].[f] In the former, a two-step dependence of the alternating current (AC) magnetic susceptibility on temperature provides an evidence for nontrivial surface superconductivity. Indeed, the first step could be attributed to surface superconductivity and the second one, where a full diamagnetic screening is reached, could be attributed to bulk superconductivity.

9.2.2 *Andreev reflection*

From a microscopic viewpoint, proximity effects could be studied in terms of the electron scattering on SN interfaces. Naively, since electrons with subgap energies cannot propagate in a superconductor, they should be completely reflected at the interface. If this were the only possibility, there would be no electric current through the SN junction. However, as was first shown by Alexander Andreev [973] an electric current can, in fact, flow through the junction. However, electrons must form Copper pairs in order to enter into the superconductor. Because of the electric charge conservation, this also requires the creation of a hole propagating back into the normal conductor. This phenomenon is known as the *Andreev reflection* [973]. A schematic illustration of the Andreev reflection is presented in Fig. 9.4(b). It should be noted that the formation of a spin-singlet Cooper pair in the superconductor also implies that the incident electron and the corresponding reflected hole should have opposite spins.

From the kinematics of the Andreev reflection, it is clear that the process can be strongly affected by the spin-momentum locking of quasiparticles. This is indeed confirmed by a theoretical consideration of the Andreev reflection at the interface of a Weyl semimetal and an *s*-wave superconductor [1026–1028]. For example, in the special case of Weyl semimetals with broken $\mathcal{T}$ symmetry that is described by model (9.3), the chirality of the electron states can hinder completely the Andreev reflection [1028].

---

[f]Superconductivity could be also achieved by doping of WTe_2 [1025].

This phenomenon is called the *chirality blockade*. In addition, similarly to the topological Imbert–Fedorov shift[g] in normal conducting Weyl semimetals [1031], there is a transverse spatial shift in the Andreev reflection related to the spin–orbit coupling in the normal phase of a Weyl semimetal [1032]. In such a case, incident electrons and reflected holes are spatially separated.

In some cases, the chirality blockade is not necessarily perfect but may depend on the details of spin-momentum locking in Weyl semimetals. This can be easily seen by comparing the spin texture for models with and without the inversion symmetry in Figs. 9.1(a) and 9.1(b), respectively. Generically, the Andreev reflection of an electron with momentum $\mathbf{k}$ should produce a hole with the opposite momentum, $-\mathbf{k}$. In the inversion symmetric case in Fig. 9.1(a), the hole belongs to an opposite chirality node and has the same spin as the electron. This is incompatible with the spin-singlet pairing in a superconductor, however. Therefore, the chirality blockade takes place [1028]. The situation is somewhat different in the model with broken inversion symmetry represented by Fig. 9.1(b). In this case, the polar regions of the Fermi surface do not contribute to the Andreev reflection because the formation of only spin-triplet Cooper pairs is allowed there. However, there are regions in momentum space, namely near the equatorial part of the Fermi surface, where the spins of electrons and holes are compatible with the spin-singlet pairing inside the superconductor. This implies that the chirality blockade is generically lifted in models with broken inversion symmetry.

A natural way to describe the Andreev reflection is provided by the scattering matrix (S-matrix) theory. In order to illustrate it, let us consider a simple model for a junction of normal and superconducting Weyl semimetals. Without loss of generality, let us assume that the z-axis points in the direction perpendicular to the interface. The normal Weyl semimetal lies in the region $z < 0$ and the superconducting one is in the region $z > 0$. For illustrative purposes, only the intranode channel with $\lambda = +$ will be considered. Then the corresponding low-energy BdG Hamiltonian reads [cf. with Eq. (9.40)]

$$H_{\text{BdG}}^{(\text{intra})}(\mathbf{k}) = \begin{pmatrix} \hbar v_F (\boldsymbol{\sigma} \cdot \mathbf{k}) - \mu \mathbb{1}_2 & e^{i\phi} \Delta_{\text{intra}}(z) \mathbb{1}_2 \\ e^{-i\phi} \Delta_{\text{intra}}(z) \mathbb{1}_2 & -\hbar v_F (\boldsymbol{\sigma} \cdot \mathbf{k}) + \mu \mathbb{1}_2 \end{pmatrix}, \qquad (9.45)$$

---

[g]The Imbert–Fedorov shift is the optical phenomenon that describes the transverse lateral shift of the reflected beam with respect to the incident one [1029, 1030].

where ϕ is the phase of the superconducting gap and $\Delta_{\text{intra}}(z) = \Delta_{\text{intra}}^{\dagger}(z)$. In order to use the same model on both sides of the SN interface, it is assumed that $\Delta_{\text{intra}}(z) = \theta(z)\Delta$.

The Andreev reflection in model (9.45) can be analyzed by using the original method proposed by Andreev [973] and further developed in the Blonder–Tinkham–Klapwijk theory [1033]. Since the corresponding details are rather cumbersome for the 4×4 Hamiltonian at hand, here we outline only the key steps of the derivation.

Before considering the scattering problem, one needs to determine the set of basis wave functions describing the propagation of electrons and holes. In a normal Weyl semimetal, there are four states: $\psi_{e,+}^{(N)}$, $\psi_{e,-}^{(N)}$, $\psi_{h,+}^{(N)}$, and $\psi_{h,-}^{(N)}$, which describe electrons and holes propagating in the positive $(+)$ and negative $(-)$ directions of the z-axis. A similar set of wave functions, $\psi_{e,+}^{(S)}$, $\psi_{e,-}^{(S)}$, $\psi_{h,+}^{(S)}$, and $\psi_{h,-}^{(S)}$, can be also obtained in the superconducting phase. Then the incident and outgoing wave functions can be decomposed as follows:

$$\psi_{\text{in}} = \psi_{e,+}^{(N)} + C_1 \psi_{e,-}^{(N)} + C_2 \psi_{h,-}^{(N)}, \qquad (9.46)$$

$$\psi_{\text{out}} = C_3 \psi_{e,+}^{(S)} + C_4 \psi_{h,+}^{(S)}. \qquad (9.47)$$

Coefficient C_1 describes the amplitude of the normal electron reflection and C_2 is the amplitude of the Andreev reflection. The amplitudes of the electron and hole transmission are given by C_3 and C_4, respectively. All four coefficients can be calculated by matching the wave functions at the SN interface, i.e., at $z = 0$. As usual in the scattering theory, in order to obtain the probabilities of the normal reflection R_{ee}, the Andreev reflection R_{eh}, as well as the transmission probabilities T_{ee} and T_{eh}, one should calculate the corresponding current densities and divide them by the absolute value of the incoming electron current density. Note that the sum of all probabilities should be equal unity, i.e., $R_{\text{ee}} + R_{\text{eh}} + T_{\text{ee}} + T_{\text{eh}} = 1$.

Having determined the reflection coefficients, one can proceed with the calculation of the differential conductance G. In the limit of vanishing temperature, the differential conductance at the interface is given by [1033]

$$G = \frac{dI}{dV} = \frac{e^2}{2\pi\hbar} \int \frac{d^2 k_{\parallel}}{(2\pi)^2} \left[1 - R_{\text{ee}}\left(eV, \mathbf{k}_{\parallel}\right) + R_{\text{eh}}\left(eV, \mathbf{k}_{\parallel}\right)\right], \qquad (9.48)$$

where the reflection probabilities R_{ee} and R_{eh} depend on the momentum components parallel to the interface and the bias electric potential V that sets a characteristic energy scale. Equation (9.48) has a transparent physical meaning, namely the normal electron reflection hinders the conductivity and the Andreev reflection enhances it.

It is instructive to review some qualitative features of the Andreev reflection. To start with, we note that the transmission of individual electrons and holes into a superconductor is forbidden for the subgap excitation energy ϵ (or, equivalently, the bias electric potential V). This means that $T_{ee} = T_{eh} = 0$ for $|eV| < |\Delta|$. However, because of the Andreev reflection, the conductance of an SN contact is nonzero. Moreover, it could become almost twice as large as in a normal junction. Therefore, the observation of the corresponding conductance peaks provides a clear signature of the Andreev reflection and the realization of a superconducting state. Additionally, because of the spin-momentum locking in Weyl semimetals, the normal electron reflection R_{ee} must vanish exactly in the case of $|eV| > |\Delta|$ and the normal incidence $k_x = k_y = 0$. Finally, for large enough V, the usual normal state conductance of a contact applies.

In agreement with our discussion before, the differential conductance at a contact of a Weyl semimetal and a usual superconducting metal with a quadratic energy spectrum depends on the orientation of the chiral shift [1027]. In general, the conductance has a peak at $|eV| = |\Delta|$ and diminishes away from it. For $|eV| < |\Delta|$, a suppression of the conductance is especially strong when the chiral shift is normal to the interface, $\mathbf{b} \perp \hat{\mathbf{z}}$. Such an effect could be interpreted as the suppression of the Andreev reflection due to a nontrivial spin texture of electron states in Weyl semimetals [1026–1028, 1034–1036].

Experimentally, the Andreev reflection in Dirac and Weyl semimetals can be studied by using the point-contact spectroscopic methods [1037]. In particular, a characteristic enhancement of the conductance at low electric potential was observed in Cd_3As_2 [1008, 1009] and TaAs [1013, 1014]. In both topological materials, superconductivity was not intrinsic but induced by a contact with a nonsuperconducting tip. While the exact mechanism was not identified, a tip-induced superconductivity could be a result of local pressure, local doping or confinement effects at the contact (see also the discussion in Sec. 9.1.5). The experimental results for the normalized differential conductance dI/dV spectra in Cd_3As_2 and TaAs are shown in Figs. 9.5(a) and 9.5(b), respectively. In both figures, the Andreev peaks in the conductance spectra are observed at about $V \approx 1$ mV. It is worth noting that the Andreev reflection in TaAs is suppressed compared to Cd_3As_2. Indeed, by comparing the scale of relative conductances in Figs. 9.5(a)

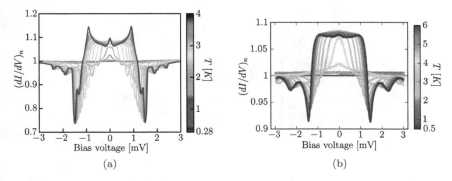

Fig. 9.5 The normalized differential conductance dI/dV spectra at different temperatures as a function of bias electric potential V in (a) the Dirac semimetal Cd_3As_2 and (b) the Weyl semimetal TaAs. (Adapted from [1009, 1013], respectively.)

and 9.5(b), one can easily see that the key features are much more pronounced in Cd_3As_2. As suggested in [1014], the suppression of the Andreev reflection in TaAs could be related to the spin-polarized texture of its quasiparticle states.

9.2.3 *Josephson effect*

Another hallmark phenomenon associated with SN junctions is the *Josephson effect*. It was predicted theoretically by Brian Josephson, a 22-year old graduate student at that time, in 1962 [972, 1019]. The first experimental evidence of the Josephson effect was found by Philip Anderson and John Rowell already in 1963 [1038]. In essence, the effect is a flow of superconducting current across a junction formed by two or more superconductors coupled by a weak link. The weak link can be made of a usual metal, an insulator, or a narrow constriction. The corresponding heterostructures are usually called the *Josephson junctions*. As we discuss below, the Josephson current is caused by a difference of phases of the complex order parameters in the two superconductors composing the junction. A schematic illustration of the Josephson junction is shown in Fig. 9.6(a), where the heterostructure is formed by a nonsuperconducting material placed between two superconducting contacts. The Josephson effect provides a clear signature of the superconducting pairing and plays an important role in applications. Among the most important of them are the superconducting quantum interference devices (SQUIDs), metrology, superconducting qubits, and others.

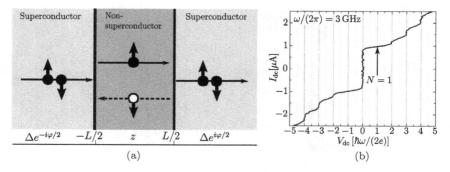

Fig. 9.6   (a) Schematic representation of the Josephson junction, where a nonsuperconducting material (middle region) is placed between two superconducting ones (outer regions). (b) The current-voltage characteristics measured in Al-Cd$_3$As$_2$-Al Josephson junction using the microwave irradiation with the frequency $\omega/(2\pi) = 3$ GHz. The first Shapiro step marked by the arrow is missing indicating the 4π-periodic Josephson effect. (Adapted from [1039].)

9.2.3.1   *DC Josephson effect*

In the simplest case, the tunneling current $J_J \propto \sin\phi$ is determined by a phase difference across the junction ϕ, i.e., the difference between the phases of superconducting order parameters on the two sides of the junction [288, 481, 999, 1040]. Clearly, it is an intrinsically quantum phenomenon. Within the Ginzburg–Landau theory, the Josephson current is defined as

$$J_J \propto \mathrm{Im}\left[t_0 \int d^2 r_\parallel \Delta_L^\dagger(\mathbf{r}_\parallel)\Delta_R(\mathbf{r}_\parallel)\right] = J_0 \sin\left(\phi_R - \phi_L\right), \qquad (9.49)$$

where t_0 is the phenomenological coupling strength due to Cooper pairs tunneling through the link, $\Delta_{R/L}(\mathbf{r}_\parallel)$ are the complex order parameters (or "gap functions"), and $\phi_{R/L}$ are the corresponding phases in the left and right superconducting contacts. The current is obtained by integrating over the coordinate $\mathbf{r}_\parallel$ of the junction surface. The resulting amplitude of the Josephson current, $J_0 \propto t_0|\Delta_L\Delta_R|$, is often called the "critical current" of the junction. It should be noted that the general expression for the Josephson current in Eq. (9.49) is modified in the case of the LOFF Cooper pairing [1041]. This is easy to understand since Cooper pairs in a LOFF superconductor carry a nonzero momentum. The latter affects the Josephson current in several ways. In particular, it causes a spatial modulation of the current density that, in turn, could result in vanishing or highly suppressed net current across the junction. Note that the similar effect could also arise in an s-wave superconductor when Cooper pairs have a nonzero center-of-mass momentum due to a supercurrent running through the junction.

One of the powerful theoretical methods for studying the Josephson effect is based on the scattering matrix formula for a supercurrent [1042–1044]. In such an approach, the excitation spectrum ϵ, the density of states $\nu(\epsilon)$, and the grand canonical potential Ω at the Josephson junction are expressed in terms of the scattering matrix in the normal region. In the simplest model setup, the junction is composed of two SN interfaces, which are similar to those discussed in Sec. 9.2.2 and shown in Fig. 9.6(a). While superconductors could be, in general, different in the left ($z < -L/2$) and right ($z > L/2$) contacts, it is sufficient to assume that they have the same absolute value of the gap. The phases of their order parameters are different, however. The corresponding intranode BdG Hamiltonian is given by Eq. (9.45), where ϕ should be replaced by $\mathrm{sgn}(z)\,\phi/2$ in the two super-conducting regions with $|z| > L/2$. For simplicity, one can assume that the weak link in the region $|z| < L/2$ is made of a nonsuperconducting Weyl semimetal. The latter is described by the same Hamiltonian (9.45), but with the vanishing gap function. Therefore, one can describe the whole junction by Hamiltonian (9.45) with $\Delta_{\mathrm{intra}}(z) = \Delta\theta(|z| - L/2)$. It can be also assumed that the spatial dependence of the order parameters is negligible near the boundaries of the junction.

A set of basis wave functions, needed for the description of the Josephson junction, is similar to that in the discussion of the Andreev reflection in Sec. 9.2.2. It should be expanded, however, in order to describe two separate superconducting contacts. Therefore, instead of $\psi_{e/h,\pm}^{(S)}$ one should use $\psi_{e/h,\pm}^{(L/R)}$, where the superscripts L and R label the left and right contacts, respectively. For small excitation energies, $\epsilon < \Delta$, one can define the usual scattering matrix $\hat{S}$ that describes the normal reflection of electrons and holes without any intermixing. Because of the Andreev reflection at the SN interfaces, there should also exist the transformation of electrons into holes and vice versa. The corresponding Andreev processes are described by an off-diagonal scattering matrix $\hat{A}$. For large energies, $\epsilon > \Delta$, one should also take into account a quasiparticle propagation in superconducting regions. In this case, it is convenient to calculate the scattering matrix of the whole SNS junction [1042]. While such a calculation is straightforward in a simple model, the resulting expressions are usually cumbersome. Therefore, here only the underlying physics and general details are outlined. (For detailed derivations of the scattering matrices for Weyl semimetal Josephson junctions, see, for example, [1028, 1045–1047].)

Before providing a qualitative description of the Josephson current, it is instructive to mention a few limitations of the standard scattering matrix approach.[h] In simplified model studies, one usually makes the following two assumptions: (i) the order parameter Δ is spatially uniform in superconducting regions and vanishes abruptly in a normal conductor and (ii) the pairing gap function is unaffected by a supercurrent. Both assumptions break down in a more realistic treatment. In particular, because of a supercurrent running through a superconductor, each Cooper pair carries a nonzero momentum, $2\mathbf{k}$. As a result, the pairing potential acquires a nontrivial phase factor, i.e., $\Delta \to \Delta e^{2i\mathbf{k}\cdot\mathbf{r}}$. This should be carefully taken into account when self-consistently solving the gap equation for the order parameter. In general, since the structure of the equation itself is affected by a change of the wave functions across the SN interfaces, the resulting Δ is expected to have a complicated nonuniform profile. While such calculations are difficult, they should be done in a rigorous analysis. In fact, without such a treatment, the electric charge conservation formally breaks down. This is seen from the modified form of the corresponding continuity equation, which acquires an additional unphysical source term [1049]. One must conclude, therefore, that the superconducting current cannot be predicted reliably in superconducting contacts of the junction if a simplified approximation is used. The problem can be avoided by evaluating the current in the normal region, which can be then used to infer its counterpart in superconducting regions.

The Josephson current can be obtained by calculating the variation of the grand canonical potential with respect to the phase difference ϕ of the superconducting order parameters in the two regions [1040, 1042–1044]. In order to see this, let us assume that the phase ϕ varies slowly in time. Then the time derivative of the grand canonical potential Ω can be expressed as follows:

$$\frac{d\Omega}{dt} = \frac{\partial\Omega(\phi)}{\partial\phi}\frac{d\phi}{dt}, \qquad (9.50)$$

---

[h]A clear qualitative discussion of the transport in superconducting junctions is given in [1048].

where, in view of the gauge symmetry, $d\phi/dt = -2eV/\hbar$ is determined by the electric potential V at the junction. By noting that the rate of change of the energy is nothing else but the dissipated power $V J_J(\phi)$, we derive the following expression for the Josephson current:

$$J_J(\phi) = \frac{2e}{\hbar} \frac{\partial \Omega(\phi)}{\partial \phi}. \tag{9.51}$$

Since the absolute value of the order parameter $|\Delta|$ is independent of ϕ, the phase-dependent part of the grand canonical potential can be written in terms of the density of states $\nu(\epsilon)$ as

$$\Omega_\phi = -2T \int_0^\infty d\epsilon \, \nu(\epsilon) \ln \left[2 \cosh \left(\frac{\epsilon}{2T} \right) \right], \tag{9.52}$$

where, by definition,

$$\nu(\epsilon) = -\frac{1}{\pi} \text{Im} \frac{d}{d\epsilon} \int \frac{d^2 k_\parallel}{(2\pi)^2} \left\{ \ln \det[1 - \hat{A}(\mathbf{k}_\parallel, \epsilon + i0)\hat{S}(\mathbf{k}_\parallel, \epsilon + i0)] \right.$$

$$\left. - \frac{1}{2} \ln \det[\hat{A}(\mathbf{k}_\parallel, \epsilon + i0)\hat{S}(\mathbf{k}_\parallel, \epsilon + i0)] \right\}. \tag{9.53}$$

Note that the usual and off-diagonal scattering matrices, $\hat{S}$ and $\hat{A}$, respectively, depend on the momentum components parallel to the SN interfaces.

It is worth noting that the subgap energy spectrum with $\epsilon < |\Delta|$ contains discrete levels ϵ_n that describe the *Andreev bound states*. The corresponding energies are defined by the solutions of the characteristic equation $\det(1 - \hat{A}\hat{S}) = 0$ [1042]. The latter is nothing else but the condition for the existence of bound states that satisfy $\psi_{\text{in}} = \hat{A}\hat{S}\psi_{\text{in}}$. Conceptually, one can understand the formation of the Andreev bound states as the result of multiple Andreev scattering of electrons and holes in the junction. The confined motion and, therefore, the formation of bound states occurs because neither electrons nor holes with subgap energies could escape into the superconducting contacts.

Generically, the density of states should be independent of the phase ϕ for $\epsilon \gg |\Delta|$. Therefore, only the states with sufficiently low energies (i.e., of order $|\Delta|$ or less) are expected to contribute to the Josephson current. Moreover, in some special cases (e.g., the short-junction limit), one can show that the continuous spectrum does not contribute [1042]. Then, by taking into account only the bound states and performing integration by parts with the contour closed in the upper half of the complex plane, the Josephson current density is given by

$$J_J(\phi) = -\frac{2eT}{\hbar} \sum_{m=0}^{\infty} \int \frac{d^2 k_\parallel}{(2\pi)^2} \frac{d}{d\phi} \ln \det[1 - \hat{A}(\mathbf{k}_\parallel, i\omega_m)\hat{S}(\mathbf{k}_\parallel, i\omega_m)], \quad (9.54)$$

where $\omega_m = \pi T(2m + 1)$ are the fermionic Matsubara frequencies. Note that the scattering matrices $\hat{A}$ and $\hat{S}$ depend on the momentum components parallel to the SN interfaces. At vanishing temperature, the Josephson current density takes a simpler form, i.e.,

$$J_J(\phi) = \frac{2e}{\hbar} \sum_n \int \frac{d^2 k_\parallel}{(2\pi)^2} \frac{d\epsilon_n}{d\phi}. \quad (9.55)$$

As is clear, in this case, the calculation of the Josephson current requires only knowledge of the eigenvalues but not the eigenfunctions.

9.2.3.2   *AC Josephson effect and Shapiro steps*

The Josephson current discussed at the beginning of Sec. 9.2.3.1 corresponds to the DC Josephson effect. When a constant electric potential V is applied to the junction, the phase ϕ changes linearly with time, $\phi = -2eVt/\hbar$, leading to the *AC Josephson effect*. In such a case, the current oscillates with the frequency $2|eV|/\hbar$. The AC Josephson current with the emission of photons was observed in [1050]. Such a result is rather striking since one would naively expect an enhancement of the DC current by a DC electric potential. As for practical applications, the AC Josephson effect allows one to create precise electric potential-frequency converters.

A specific type of the Josephson response appears when electromagnetic waves are applied to the junction. This was first done by Sidney Shapiro [1051, 1052], who found that the Josephson current forms characteristic steps in the current-voltage characteristics when the junction is subject to a microwave irradiation. The characteristic *Shapiro steps* appear when the electric potential is multiple of $\hbar\omega/(2e)$, where ω is the angular frequency of the radiation. Physically, this means that Cooper pairs exchange energy with the external electromagnetic field by absorbing and emitting quanta with the energies multiple of $2|eV|$. The corresponding experimental results obtained in [1039] for an Al-Cd_3As_2-Al heterostructure are shown in Fig. 9.6(b). Note that the first Shapiro step is missing in this figure. It is believed that the physical reason for its absence is related to Majorana zero energy modes, which will be discussed in Sec. 9.2.4.

In order to get a deeper insight into the AC Josephson effect let us consider the appearance of the Shapiro steps more rigorously. Let us assume that the electric potential V oscillates with the angular frequency ω, i.e.,

$$V(t) = V_0 + V_1 \cos(\omega t). \tag{9.56}$$

Then the time dependence of the superconducting phase is given by

$$\phi(t) = \phi_0 - \frac{2e}{\hbar}\left[V_0 t + \frac{V_1}{\omega}\sin(\omega t)\right]. \tag{9.57}$$

This gives the AC Josephson current

$$J_{\mathrm{J}}(t) = J_{\max}\sin\phi(t) = J_{\max}\sin\left(\phi_0 - \frac{2eV_0 t}{\hbar}\right)\cos\left[\frac{2eV_1}{\hbar\omega}\sin(\omega t)\right]$$

$$-J_{\max}\cos\left(\phi_0 - \frac{2eV_0 t}{\hbar}\right)\sin\left[\frac{2eV_1}{\hbar\omega}\sin(\omega t)\right]. \tag{9.58}$$

After simple algebraic manipulations, one can show that this current contains time-independent terms for $V_0 = n\hbar\omega/(2e)$, where n is an integer. Indeed, by using the generating functions for the Bessel functions [281],

$$e^{\pm ix\sin y} = J_0(x) + 2\sum_{k=1}^{\infty} J_{2k}(x)\cos(2ky) \pm 2i\sum_{k=1}^{\infty} J_{2k-1}(x)\sin[(2k-1)y], \tag{9.59}$$

one can express $\cos[2eV_1\sin(\omega t)/(\hbar\omega)]$ and $\sin[2eV_1\sin(\omega t)/(\hbar\omega)]$ in terms of the Bessel functions. Then the Josephson current can be rewritten as

$$J_{\mathrm{J}}(t) = J_{\max}\sin\left(\phi_0 - \frac{2eV_0 t}{\hbar}\right)\left[J_0\left(\frac{2eV_1}{\hbar\omega}\right)\right.$$

$$\left. + 2\sum_{k=1}^{\infty} J_{2k}\left(\frac{2eV_1}{\hbar\omega}\right)\cos(2k\omega t)\right]$$

$$- 2J_{\max}\cos\left(\phi_0 - \frac{2eV_0 t}{\hbar}\right)\sum_{k=1}^{\infty} J_{2k-1}\left(\frac{2eV_1}{\hbar\omega}\right)\sin[(2k-1)\omega t]. \tag{9.60}$$

By making use of identity $2\sin x\cos y = \sin(x-y)\sin(x+y)$, it is easy to check that one of the terms in the series becomes independent of time when $V_0 = n\hbar\omega/(2e)$ and n is a nonnegative integer. This leads to a nonvanishing expression for the time-averaged current

$$\langle J_{\mathrm{J}}\rangle = (-1)^n J_{\max}\sin\phi_0 J_n\left(\frac{2eV_1}{\hbar\omega}\right). \tag{9.61}$$

In terms of observables, this implies that a resonance peak is expected in the current-voltage characteristics at $V_0 = n\hbar\omega/(2e)$. In experiments with Josephson junctions, it is often an electric potential rather than a current that is usually measured. Then the current-voltage characteristics shows plateaus at $V_0 = n\hbar\omega/(2e)$. Each time the current exceeds the maximum value in Eq. (9.61), a new plateau in V_0 is reached leading to a step-like profile. It is common, however, to rotate the measured current-voltage characteristics and show the dependence of current on an electric potential (see Fig. 9.6(b)). The Shapiro steps are extremely useful in metrology providing an exact conversion between electric potential and the frequency units in terms of the ratio of the two fundamental constants, $\hbar/(2e)$.

The Josephson junctions with Dirac and Weyl semimetals were investigated theoretically in many studies [990, 1028, 1034, 1036, 1045–1047, 1053–1055]. Because of the specifics of Cooper pairing in topological semimetals, the underlying physics can be rather unusual. As an example, let us consider a Josephson junction with one superconducting contact made of a Weyl semimetal and the other of a usual s-wave superconductor [990]. Because of the LOFF-type intranode pairing in the Weyl semimetal, Sec. 9.1.2.2, the corresponding Cooper pairs carry a nonzero total momentum $2\mathbf{b}$, where $\mathbf{b}$ is the chiral shift. The spatial modulation of the order parameter, $\Delta_{\mathrm{LOFF}} \propto e^{\pm 2i\mathbf{b}\cdot\mathbf{r}}$, in this case implies that the amplitude of the Josephson current averaged out over the junction should be suppressed or even vanish. The suppression can be avoided by driving an electric current through the conventional superconductor. Such a current allows for the s-wave Cooper pairs to have a nonzero center-of-mass momentum $\mathbf{q}$ that can be tuned to match the momentum of the LOFF pairing state. When $\mathbf{q}$ and chiral shift $\mathbf{b}$ coincide, the suppression of the Josephson current is lifted.

In the case of the internode pairing in a Weyl semimetal contact, the Josephson current should be qualitatively similar to that in a usual junction made of a pair of conventional s-wave superconductors. Nevertheless, one can still probe the topologically protected nodal structure of the pairing gap function that results from the underlying Fermi surface topology [991].[i] Signatures of such nodal superconductivity could be observed by applying a current through a superconducting system [990]. Such a supercurrent can be used to drive a topological phase transition connected with the annihilation of Weyl–Majorana nodes. Thus, the observation of the corresponding

---

[i]Note that nodal points carry a nonzero vorticity and behave as Weyl–Majorana points of the Bogoliubov-de Gennes pairing Hamiltonian [991].

abrupt changes in the density of states or the differential conductance may serve as signatures of nodal superconductivity [990].

9.2.4 *Majorana zero modes*

One of the defining aspects of topological superconductors is their potential ability of hosting the *Majorana zero modes* [617–622]. By definition, Majorana particles are their own antiparticles [1056]. In topological superconductors, such modes can appear as zero-energy Bogolyubov quasiparticles made of an equal-weight superposition of electron and hole states [976].

The use of proximity effects is currently one of the main approaches for realizing Majorana zero modes in solid-state systems. The origin of these modes is related to the Andreev bound states discussed in Sec. 9.2.2. Indeed, when the lowest Andreev bound state at the junction crosses the zero energy it can be represented as two unpaired Majorana zero modes [621]. Since Majorana zero modes are protected by the bulk topology, their properties are not affected much by disorder. Such an inherent robustness against decoherence from environment makes Majorana zero modes very desirable as building blocks of topological quantum computers.

Given that Majorana zero modes in solids are being actively investigated, we discuss only the most general properties. In particular, we consider the appearance of Majorana zero modes in a simple model and their manifestation in Josephson junctions. Finally, a few experimental manifestations in Dirac and Weyl semimetals will be discussed.

9.2.4.1 *Kitaev model*

In order to explicitly show how Majorana modes can appear in solids, let us consider one of the simplest toy models, namely the Kitaev lattice model [977]. It describes spinless fermions on an N-site chain with the p-wave superconductivity

$$H = -\frac{t}{2} \sum_{n=1}^{N-1} c_n^\dagger c_{n+1} - \frac{\mu}{2} \sum_{n=1}^{N} c_n^\dagger c_n - \frac{\Delta e^{i\phi}}{2} \sum_{n=1}^{N-1} c_n c_{n+1} + h.c., \qquad (9.62)$$

where t is the hopping parameter, μ is the electric chemical potential, Δ is the magnitude of the superconducting order parameter, and ϕ is the corresponding superconducting phase.

It is convenient to rewrite the Hamiltonian in terms of the fermion operators in momentum space, i.e., $c_n = \sum_k c_k e^{ink}/\sqrt{N}$. If the periodic boundary conditions are assumed, the values of k are discretized in units of $2\pi/N$ and lie in the range between $-\pi$ and π. Then the Hamiltonian of Kitaev's model (with the periodic boundary conditions) takes the following explicit expression in momentum space:

$$H = \frac{1}{2}\sum_k C_k^\dagger \begin{pmatrix} \tilde{\epsilon}_k & \tilde{\Delta}_k \\ \tilde{\Delta}_k^* & -\tilde{\epsilon}_k \end{pmatrix} C_k = \frac{1}{2}\sum_k C_k^\dagger \left(\mathbf{h}_k \cdot \boldsymbol{\tau}\right) C_k, \qquad (9.63)$$

where $\mathbf{h}_k = (\mathrm{Re}[\tilde{\Delta}_k], -\mathrm{Im}[\tilde{\Delta}_k], \tilde{\epsilon}_k)$, $\tilde{\epsilon}_k = -t\cos k - \mu$, $\tilde{\Delta}_k = -i\Delta e^{i\phi}\sin k$, $C_k = (c_k, c_{-k}^\dagger)^T$ is the Nambu spinor, and $\boldsymbol{\tau} = (\tau_x, \tau_y, \tau_z)$ are the Pauli matrices in the Nambu space. From Eq. (9.63), one can easily find that the quasiparticle energies are given by

$$\epsilon_k = \pm\sqrt{\tilde{\epsilon}_k^2 + |\tilde{\Delta}_k|^2}. \qquad (9.64)$$

Generically, this equation describes a gapped spectrum of bulk excitations. The value of the gap vanishes, however, when $\mu = \pm t$.

While both ranges of the model parameters, $|\mu| < t$ and $|\mu| > t$, describe gapped phases, the topology of the corresponding Hamiltonians are different. This can be seen explicitly by considering the mapping of the momentum interval between $k = 0$ and $k = \pi$ onto the unit sphere represented by $\hat{\mathbf{h}}_k = \mathbf{h}_k/|\mathbf{h}_k|$. Since the sign of $\tilde{\epsilon}_k$ does not change when $|\mu| > t$, the image of the corresponding interval describes a closed curve on the unit sphere. Since such a curve can be shrunk to a single point by continuous deformations, the mapping is topologically trivial. The situation is dramatically different when $|\mu| < t$ because the sign of $\tilde{\epsilon}_k$ changes when the value of k sweeps the interval from 0 to π. In this case, the image on the unit sphere is given by an open curve that starts on the south pole ($\hat{\mathbf{h}}_k = -\hat{\mathbf{z}}$) and ends on the north pole ($\hat{\mathbf{h}}_k = \hat{\mathbf{z}}$). The latter cannot be shrunk to a single point without closing the gap in the spectrum. This is a clear indication of the topologically nontrivial $\mathbb{Z}_2$ phase.

In the case with $|\mu| < t$, the nontrivial topology of Hamiltonian (9.62) leads to the appearance of Majorana zero modes when a chain with the open boundary conditions is used. In order to see this explicitly, it is convenient to perform the following decomposition of the fermionic operators c_n into the pairs of Majorana fermions:

$$c_n = \frac{e^{-i\phi/2}}{2}\left(\gamma_{\mathrm{B},n} + i\gamma_{\mathrm{A},n}\right). \qquad (9.65)$$

Here $\gamma_{a,n}$ (with $a = A, B$) are the Majorana operators that obey the following relations:

$$\gamma_{a,n} = \gamma_{a,n}^{\dagger} \quad \text{and} \quad \{\gamma_{a,n}, \gamma_{a',m}\} = 2\delta_{aa'}\delta_{nm}. \tag{9.66}$$

In term of these Majorana operators, Hamiltonian (9.62) can be rewritten as follows:

$$H = -\frac{i}{4}\sum_{n=1}^{N-1}(\Delta + t)\gamma_{B,n}\gamma_{A,n+1} - \frac{i}{4}\sum_{n=1}^{N-1}(\Delta - t)\gamma_{A,n}\gamma_{B,n+1}$$

$$-\frac{\mu}{2}\sum_{n=1}^{N}(1 + i\gamma_{B,n}\gamma_{A,n}) + h.c. \tag{9.67}$$

In the special case with $\mu = 0$ and $\Delta = t$, the above Hamiltonian reduces down to

$$H = -i\frac{\Delta}{2}\sum_{n=1}^{N-1}\gamma_{B,n}\gamma_{A,n+1} + h.c., \tag{9.68}$$

where only Majorana modes on the adjacent lattice sites are coupled. Since the operator on the first site, $\gamma_1 = \gamma_{A,1}$, and the operator on the last site, $\gamma_2 = \gamma_{B,N}$, are absent from Hamiltonian (9.68), the creation of the corresponding modes costs zero energy. These are the Majorana zero modes. The operators γ_1 and γ_2 can be combined into a highly nonlocal fermion operator $c_{nonl.} = (\gamma_1 + i\gamma_2)/2$, which produces a doubly degenerate ground state spanned by $|0\rangle$ and $|1_{nonl.}\rangle = c_{nonl.}^{\dagger}|0\rangle$.

Similarly to the Fermi arc surface states, the presence of Majorana zero modes is protected by the nontrivial $\mathbb{Z}_2$ topology of the bulk Hamiltonian, which was discussed before. Indeed, since they belong to different topological classes, the trivial vacuum and nontrivial bulk states cannot be smoothly connected without closing the gap. The above arguments hold for all values of μ smaller than t in absolute value. Note also that, at a nonzero μ, the wave functions of Majorana modes decay exponentially into the bulk. As a result, the double degeneracy of the ground state is lifted. For a long chain, however, the energy splitting is negligible and can be often ignored.

It should be mentioned that the inclusion of the spin degree of freedom could have a detrimental role on the existence of Majorana modes. This is connected with the fact that the number of Majorana zero modes should formally double at the ends of the chain when fermion spins are taken into account. The corresponding pairs of modes can form new fermionic states with nonzero energies, which cannot be of the Majorana type. In turn, this means that the spin degree of freedom should be excluded in order to realize Majorana zero modes in electronic systems. Materials with strong spin–orbital coupling, such as topological insulators as well as Weyl and Dirac semimetals, are, therefore, among the most promising candidates for achieving this goal.

9.2.4.2 *Majorana modes in Josephson junctions*

It was proposed that Majorana zero modes, if realized, could lead to a fractionalization of Cooper pairs in Josephson junctions, where the Josephson current is carried by particles with the charge $-e$ rather than by Cooper pairs with the charge $-2e$. Therefore, the tunneling current becomes a 4π periodic function of the superconducting phase difference across the junction [977–979, 1057–1060]. In order to understand why the period should be doubled, let us consider a simple model of a junction made of two topological Kitaev chains (labeled by $j = \mathrm{R}, \mathrm{L}$) separated by a barrier

$$H = -\frac{\Delta}{2} \sum_{j=\mathrm{R},\mathrm{L}} \sum_{n=1}^{N-1} (c_{j,n}^\dagger c_{j,n+1} + e^{i\phi_j} c_{j,n} c_{j,n+1} + h.c.) + H_{\mathrm{tun}}, \quad (9.69)$$

where $\mu = 0$, $\Delta = t$, and

$$H_{\mathrm{tun}} = -T_{LR}(c_{\mathrm{L},N}^\dagger c_{\mathrm{R},1} + c_{\mathrm{R},1}^\dagger c_{\mathrm{L},N}) \quad (9.70)$$

describes the tunneling across the barrier with the coupling strength T_{LR}. Since the model (9.69) is effectively made of two copies of the model (9.62), it should support a total of four Majorana modes at the ends of the chains. At the junction, the electron operators are given by

$$c_{\mathrm{L},N} = \frac{e^{-i\phi_\mathrm{L}/2}}{2} (\gamma_{\mathrm{L},1} + i\gamma_{\mathrm{L},2}), \quad (9.71)$$

$$c_{\mathrm{R},1} = \frac{e^{-i\phi_\mathrm{R}/2}}{2} (\gamma_{\mathrm{R},1} + i\gamma_{\mathrm{R},2}), \quad (9.72)$$

where $\gamma_{\mathrm{L},2}$ and $\gamma_{\mathrm{R},1}$ hybridize with the operators in the left and right regions. Therefore, one can drop them from Eqs. (9.71) and (9.72). Then the tunneling Hamiltonian (9.70) reads

$$H_{\mathrm{tun}} = -\frac{T_{LR}}{2} \cos\left(\frac{\phi}{2}\right) i\gamma_{\mathrm{L},1}\gamma_{\mathrm{R},2}, \quad (9.73)$$

where $\phi = \phi_R - \phi_L$ is the phase difference. It is convenient to introduce the quasiparticle occupation number of a fused state made of Majorana zero modes at the junction

$$n = d^\dagger d = \frac{1 + i\gamma_{L,1}\gamma_{R,2}}{2}, \qquad (9.74)$$

where we defined an operator $d = (\gamma_{L,1} + i\gamma_{R,2})/2$. Clearly, Hamiltonian (9.73) commutes with operator (9.74). Therefore, one can define the contribution to the Josephson current operator related to the Majorana modes as

$$J_M = \frac{2e}{\hbar}\frac{\partial H_{\text{tun}}}{\partial \phi} = \frac{eT_{LR}}{\hbar}\sin\left(\frac{\phi}{2}\right)\left(n - \frac{1}{2}\right). \qquad (9.75)$$

Obviously, the corresponding current has the 4π periodicity. This is related to the fact that it is effectively carried by quasiparticles with the charge $-e$ instead of $-2e$.

The 4π periodicity of the Josephson current leads to the absence of odd-integer Shapiro steps (see the discussion in Sec. 9.2.3.2). This implies that the AC Josephson effect could provide a compelling signature of topological superconductivity [1057, 1060, 1061]. Indeed, as was discussed in Sec. 9.2.3, the constant electric potential of the Shapiro step in the current-voltage characteristics $I - V$ is given by $V = n\hbar\omega/(2e)$, where n is an integer and ω is the angular frequency of microwave radiation. When Majorana modes are present in the junction, the size of the step becomes exactly two times larger because the corresponding Josephson current is 4π-periodic, i.e., $J_M \propto \sin(\phi/2)$. It should be noted that an experimental verification of the unusual 4π-periodic Josephson current could be challenging. Indeed, due to a finite length of topological superconductors, additional Majorana modes appear at their ends and could hybridize with those in the junction. As argued in [1062], however, the 4π periodicity can be measured in a transient regime supported by a properly chosen bias electric potential that induces the Landau–Zener transitions. If realized, such a long-lived transient AC Josephson current can be an effective method for detecting the Majorana bound states.

Experimentally, the absence of odd-integer Shapiro steps was first observed in a hybrid semiconductor–superconductor InSb/Nb nanowire junction [1063]. Later the same effect was confirmed in HgTe-based topological Josephson junctions [1064]. In addition to the missing Shapiro steps, the observation of the zero-biased conductivity peak in superconductor heterostructures [1065–1068] and vortex cores [1069] provides another piece of evidence supporting the existence of Majorana modes.

9.2.4.3 *Majorana modes in Weyl and Dirac semimetals*

Now let us discuss the status of Majorana zero modes in Weyl and Dirac semimetals. To start with, let us consider a simple model of proximity induced superconductivity on the surface of a Weyl semimetal with broken $\mathcal{P}$ symmetry. The effective Hamiltonian describing a pair of Fermi arcs is given by

$$H_{\mathrm{arcs}}(\mathbf{k}) = \hbar v_F k_x \sigma_z, \tag{9.76}$$

where the Fermi arcs have finite length $|k_z| < b_z$ (see also Sec. 5.3). As in the case of topological insulators [978, 979], one can couple the surface of the Weyl semimetal to an s-wave superconductor. In this case, the Fermi arc states with opposite momenta can form s-wave Cooper pairs. The corresponding BdG Hamiltonian reads [1070]

$$H_{\mathrm{BdG}}(\mathbf{k}) = \hbar v_F k_x \tau_z \otimes \sigma_z - \mu \tau_z \otimes \mathbb{1}_2 + \Delta \tau_x \otimes \mathbb{1}_2, \tag{9.77}$$

where $\boldsymbol{\tau} = (\tau_x, \tau_y, \tau_z)$ is the vector of the Pauli matrices that acts in the Nambu space and the imaginary part of the gap was neglected.

A simple π-junction, i.e., a junction where the phases of the gaps differ by π, is sufficient to achieve a pair of Majorana zero modes on the surface of a Weyl semimetal [1070]. Indeed, by diagonalizing the kinetic part of Hamiltonian (9.77) and replacing $k_x \to -i\partial_x$, one obtains $H_{\mathrm{BdG}}(\mathbf{k}) = \mathrm{diag}\,(H_+, H_-)$, where

$$H_{\pm}(\mathbf{k}) = \begin{pmatrix} \mp i\hbar v_F \partial_x - \mu & \Delta(x) \\ \Delta(x) & \pm i\hbar v_F \partial_x + \mu \end{pmatrix}. \tag{9.78}$$

Here it is assumed that the gap $\Delta(x)$ has a kink-like profile, where $\lim_{x \to \pm\infty} \Delta(x) \to \pm\Delta_0$. In this case, Hamiltonian (9.78) supports the exact Jackiw–Rebbi zero modes [69]. The corresponding wave functions are given by

$$\psi_{\pm} \propto \begin{pmatrix} i \\ \pm 1 \end{pmatrix} \exp\left\{ -\frac{1}{\hbar v_F} \int_0^x dx'\, [\Delta(x') \mp i\mu] \right\}. \tag{9.79}$$

These modes are localized at the surface of the Weyl semimetal and have a flat dispersion relation. They can be identified with Majorana zero energy modes and their existence is robust against a nonmagnetic disorder. Similar solutions are also present in a periodic model of a Weyl semimetal [1070].

Finally, let us discuss experimental evidence related to Majorana zero modes in Weyl and Dirac semimetals. As was shown in Sec. 9.2.1, topological superconductivity could also be induced in the Dirac semimetal Cd_3As_2 by the proximity effect that couples this semimetal to an s-wave Josephson junction. Then Majorana states should appear at the surface of Dirac semimetals [989, 1054] and, as was discussed in Sec. 9.2.4.2, lead to a 4π-periodic Josephson current. The bulk states could also contribute to a tunneling current with, however, a 2π-periodic contribution. In addition, it was proposed [1054] that the current–phase relation should exhibit a discontinuous current jump at $\phi = \pi$ in Dirac semimetals. The latter appears exactly at the phase where Majorana bands emerge in the spectrum. Recently, superconducting currents with π and 4π periodicity were observed [1039] in Al-Cd_3As_2-Al Josephson junctions via the half-integer Shapiro steps and the missing first step, respectively. The corresponding experimental results are shown in Fig. 9.6(b). It was suggested that the observed 4π-periodic current is a manifestation of topological superconductivity and Majorana zero modes. On the other hand, the π-periodic part might stem from an interference of induced bulk and Fermi arc surface superconductivities.

Before finalizing this chapter, it is worth noting that the study of superconductivity in topological materials is still under active development. In particular, a definite example of intrinsic superconductivity in Weyl and Dirac semimetals without any structural phase transitions is lacking. In this situation it is clear that an additional input is needed to fill the gaps in knowledge and to form a more complete picture of the topological superconductivity in Dirac and Weyl semimetals.

Appendix A

Green's Functions

A.1 Definition

Green's functions are versatile and ubiquitous across many fields of modern theoretical physics, including condensed matter [136, 138, 139, 235, 236] and high energy physics [42–45, 357].

Our consideration in this appendix is restricted to two-point Green's functions of noninteracting fermions in the vicinity of Weyl nodes. The corresponding effective Hamiltonian takes the following canonical form in coordinate space:

$$\hat{H}_{\mathrm{W}} = -i\lambda\hbar v_F \left(\boldsymbol{\sigma} \cdot \boldsymbol{\nabla}\right), \tag{A.1}$$

where $\lambda = \pm$ labels the quasiparticle chirality, v_F is the Fermi velocity, $\boldsymbol{\sigma}$ is the vector of Pauli matrices, and $\boldsymbol{\nabla}$ is the spatial gradient.

There exist several types of Green's functions that differ by the time ordering of quantum fields in the definition of the two-point correlators. In this book, we use primarily the retarded and advanced Green's functions. They are defined by

$$G^{\mathrm{R}}\left(t, \mathbf{r}; t', \mathbf{r}'\right) = \theta(t - t') \left[\left\langle \psi\left(t, \mathbf{r}\right) \psi^{\dagger}\left(t', \mathbf{r}'\right)\right\rangle + \left\langle \psi^{\dagger}\left(t', \mathbf{r}'\right) \psi\left(t, \mathbf{r}\right)\right\rangle\right], \tag{A.2}$$

$$G^{\mathrm{A}}\left(t, \mathbf{r}; t', \mathbf{r}'\right) = -\theta(t' - t) \left[\left\langle \psi^{\dagger}\left(t', \mathbf{r}'\right) \psi\left(t, \mathbf{r}\right)\right\rangle + \left\langle \psi\left(t, \mathbf{r}\right) \psi^{\dagger}\left(t', \mathbf{r}'\right)\right\rangle\right]. \tag{A.3}$$

Here $\theta(x)$ is the unit step function, $\psi(t, \mathbf{r})$ is the fermion field, and $\langle\ldots\rangle$ denotes the ground-state expectation value. Both Green's functions satisfy the following inhomogeneous differential equation:

$$\left[i\hbar\partial_t + i\lambda v_F \hbar \left(\boldsymbol{\sigma} \cdot \boldsymbol{\nabla}\right) + \mu\right] (-i)G^{\mathrm{R,A}}\left(t, \mathbf{r}; t', \mathbf{r}'\right) = \delta(t - t')\delta^3(\mathbf{r} - \mathbf{r}'), \tag{A.4}$$

where μ is the electric chemical potential and $\delta(x)$ is the Dirac δ-function. It is often convenient to use the Fourier transform to obtain Green's functions in the reciprocal space, i.e.,

$$G^{\mathrm{R,A}}\left(\omega, \mathbf{k}\right) = \int d(t - t') \int d^3(\mathbf{r} - \mathbf{r}') e^{i\omega(t-t') - i\mathbf{k}\cdot(\mathbf{r}-\mathbf{r}')} G^{\mathrm{R,A}}\left(t - t', \mathbf{r} - \mathbf{r}'\right).$$
(A.5)

In an explicit form, the latter are given by

$$G^{\mathrm{R}}\left(\omega, \mathbf{k}\right) = \frac{i}{\hbar\omega + \mu - \lambda\hbar v_F\left(\mathbf{k}\cdot\boldsymbol{\sigma}\right) + i0},$$
(A.6)

$$G^{\mathrm{A}}\left(\omega, \mathbf{k}\right) = \frac{i}{\hbar\omega + \mu - \lambda\hbar v_F\left(\mathbf{k}\cdot\boldsymbol{\sigma}\right) - i0}.$$
(A.7)

Note that these differ only by the $i0$ prescription used in the denominator. This prescription determines how one should circumvent the poles of the propagator when integrating along the real axis of energy or frequency.

When using a diagrammatic technique or the Dyson perturbation theory, the quantum-mechanical evolution is described by the time-ordered exponent $U(t, t_0) = \mathcal{T}\exp[-i/\hbar \int_{t_0}^{t} H_{\mathrm{int}}(t')\, dt']$, where $\mathcal{T}$ is the time-ordering operator and $H_{\mathrm{int}}(t)$ is the interaction Hamiltonian. In this case, it is convenient to introduce the Feynman propagator defined by the following time-ordered product of fields:

$$\begin{aligned} G_{\mathrm{F}}\left(t, \mathbf{r}; t', \mathbf{r}'\right) &= \left\langle \mathcal{T}\left[\psi\left(t, \mathbf{r}\right)\psi^\dagger\left(t', \mathbf{r}'\right)\right]\right\rangle \\ &= \left\langle \theta(t - t')\psi\left(t, \mathbf{r}\right)\psi^\dagger\left(t', \mathbf{r}'\right) - \theta(t' - t)\psi^\dagger\left(t', \mathbf{r}'\right)\psi\left(t, \mathbf{r}\right)\right\rangle. \end{aligned}$$
(A.8)

Similarly to the retarded and advanced Green's functions, the Feynman propagator satisfies the inhomogeneous differential equation (A.4). In momentum space, it differs from the retarded and advanced Green's functions only by the prescription used in its denominator, i.e.,

$$G_{\mathrm{F}}(\omega, \mathbf{k}) = \frac{i}{\hbar\omega + \mu - \lambda\hbar v_F\left(\mathbf{k}\cdot\boldsymbol{\sigma}\right) + i0\,\mathrm{sgn}\left(\omega\right)}.$$
(A.9)

The special prescription in the Feynman propagator, which is also known as the casual one, ensures that electron and hole quasiparticles propagate forward and backward in time, respectively.

The spectral function $A(\omega, \mathbf{k})$ is defined by the difference of the retarded and advanced Green's functions at the vanishing chemical potential,

$$A(\omega, \mathbf{k}) = \frac{1}{2\pi} \left[G^{\mathrm{R}}(\omega, \mathbf{k}) - G^{\mathrm{A}}(\omega, \mathbf{k}) \right]_{\mu=0}. \tag{A.10}$$

The retarded and advanced Green's function is related to the spectral function as follows:

$$G^{\mathrm{R,A}}(\omega, \mathbf{k}) = i \int_{-\infty}^{\infty} \hbar d\omega' \frac{A(\omega', \mathbf{k})}{\hbar\omega + \mu - \hbar\omega' \pm i0}, \tag{A.11}$$

where the sign $\pm$ corresponds to the retarded $(+)$ and advanced $(-)$ Green's function, respectively. The spectral function satisfies the following sum rule [139, 234]:

$$\int_{-\infty}^{\infty} \hbar d\omega A(\omega, \mathbf{k}) = 1, \tag{A.12}$$

which can be useful in many applications. Note also that the trace of the spectral function determines the density of states, i.e., $\nu(\omega, \mathbf{k}) = \mathrm{tr}\left[A(\omega, \mathbf{k})\right]$.

A.2   Green's function in magnetic field

Here we derive the quasiparticle Green's function in an external magnetic field. The starting point is the following relativistic-like Hamiltonian of a Weyl semimetal:

$$H = \int d^3 r \bar{\psi}(\mathbf{r}) \hat{H} \psi(\mathbf{r}), \tag{A.13}$$

where $\bar{\psi}(\mathbf{r}) = \psi^\dagger(\mathbf{r})\gamma^0$ is the Dirac-conjugate bispinor and the one-particle Hamiltonian is given by

$$\hat{H} = v_F (\boldsymbol{\pi} \cdot \boldsymbol{\gamma}) - \hbar v_F (\mathbf{b} \cdot \boldsymbol{\gamma})\gamma_5. \tag{A.14}$$

This model describes a Weyl semimetal with broken $\mathcal{T}$ symmetry and two Weyl nodes separated by $2\mathbf{b}$ in momentum space, where $\mathbf{b}$ is the chiral shift. By definition, $\boldsymbol{\pi} = -i\hbar\boldsymbol{\nabla} + e\mathbf{A}/c$ is the canonical momentum. In addition, $\boldsymbol{\gamma}$ and γ_5 are the Dirac γ-matrices, whose explicit form is not important here.

The inverse Green's function is defined as

$$iG^{-1}(u, u') = \left[(i\hbar\partial_t + \mu)\gamma^0 - v_F(\boldsymbol{\pi} \cdot \boldsymbol{\gamma}) + \hbar v_F(\mathbf{b} \cdot \boldsymbol{\gamma})\gamma_5 \right] \delta^4(u - u'), \tag{A.15}$$

where $u = (t, \mathbf{r})$. For the sake of simplicity, let us assume that both magnetic field and chiral shift are parallel to the z-axis, $\mathbf{B} \parallel \mathbf{b} \parallel \hat{\mathbf{z}}$. For the vector potential describing the external magnetic field, it is convenient to use the Landau gauge $\mathbf{A} = (0, xB, 0)$, where B is the field strength.

By using the definition in Eq. (A.15) and noting that $\langle u|u' \rangle = \delta^4(u - u')$, we derive the following representation for Green's function:

$$
\begin{aligned}
G(u, u') &= i\langle u| \left[(i\hbar\partial_t + \mu)\,\gamma^0 - v_F(\boldsymbol{\pi} \cdot \boldsymbol{\gamma}) + i\hbar v_F b_z \gamma^0 \gamma_x \gamma_y \right]^{-1} |u'\rangle \\
&= i\langle u| \left[(i\hbar\partial_t + \mu)\,\gamma^0 - v_F(\boldsymbol{\pi} \cdot \boldsymbol{\gamma}) - i\hbar v_F b_z \gamma^0 \gamma_x \gamma_y \right] \\
&\quad \times \left\{ \left[(i\hbar\partial_t + \mu)\,\gamma^0 - v_F(\boldsymbol{\pi} \cdot \boldsymbol{\gamma}) + i\hbar v_F b_z \gamma^0 \gamma_x \gamma_y \right] \right. \\
&\quad \times \left. \left[(i\hbar\partial_t + \mu)\,\gamma^0 - v_F(\boldsymbol{\pi} \cdot \boldsymbol{\gamma}) - i\hbar v_F b_z \gamma^0 \gamma_x \gamma_y \right] \right\}^{-1} |u'\rangle \\
&= i\Big\langle u \Big| \left[(i\hbar\partial_t + \mu)\,\gamma^0 - v_F(\boldsymbol{\pi} \cdot \boldsymbol{\gamma}) - i\hbar v_F b_z \gamma^0 \gamma_x \gamma_y \right] \\
&\quad \times \Big[(i\hbar\partial_t + \mu)^2 - v_F^2 \boldsymbol{\pi}_\perp^2 - v_F^2(\pi_z)^2 - (\hbar v_F b_z)^2 \\
&\quad - i\frac{\hbar v_F^2 eB}{c}\gamma_x \gamma_y - 2\hbar v_F^2 b_z \pi_z \gamma_5 \Big]^{-1} \Big| u' \Big\rangle,
\end{aligned} \tag{A.16}
$$

where $\boldsymbol{\pi}_\perp = (\pi_x, \pi_y, 0)$ and we used the equality $i\gamma^0 \gamma_x \gamma_y = \gamma_z \gamma_5$.

Let us use Eq. (A.16) to derive the quasiparticle propagator in the Landau-level representation [322, 551]. Since the translation invariance in the x- and y-directions is broken by the external magnetic field, it is convenient to perform the Fourier transform of Green's function with respect to time t and spacial z coordinate only, i.e.,

$$
\begin{aligned}
G(\omega, k_z; \mathbf{r}_\perp, \mathbf{r}'_\perp) &= \int dt\, dz\, e^{i\omega(t - t') - ik_z(z - z')} G(u, u') \\
&= i[\hat{W} - v_F(\boldsymbol{\pi}_\perp \cdot \boldsymbol{\gamma})] \\
&\quad \Big\langle \mathbf{r}_\perp \Big| \left(\hat{M} - v_F^2 \boldsymbol{\pi}_\perp^2 - i\frac{\hbar v_F^2 eB}{c}\gamma_x \gamma_y \right)^{-1} \Big| \mathbf{r}'_\perp \Big\rangle,
\end{aligned} \tag{A.17}
$$

where $\mathbf{r}_\perp = (x, y, 0)$ and

$$
\hat{W} = (\hbar\omega + \mu)\,\gamma^0 - i\hbar v_F b_z \gamma^0 \gamma_x \gamma_y - \hbar v_F k_z \gamma_z, \tag{A.18}
$$

$$
\hat{M} = (\hbar\omega + \mu)^2 - (\hbar v_F b_z)^2 - (\hbar v_F k_z)^2 - 2\hbar^2 v_F^2 b_z k_z \gamma^5. \tag{A.19}
$$

Since all three operators $\hat{M}$, $v_F^2 \boldsymbol{\pi}_\perp^2$, and $i\hbar v_F^2 eB\gamma_x\gamma_y/c$ inside the matrix element on the right-hand side in Eq. (A.17) commute, it is possible to find a common basis of eigenfunctions. The eigenvalues of the operator $v_F^2 \boldsymbol{\pi}_\perp^2$ are given by [285] $(2n+1)\hbar v_F^2|eB|/c$ where $n = 0, 1, 2, \ldots$. The corresponding normalized wave functions read

$$\psi_{n,k_y}(\mathbf{r}_\perp) = \frac{1}{\sqrt{2\pi l_{\mathrm{B}}}} \frac{1}{\sqrt{2^n n! \sqrt{\pi}}} H_n(\xi) e^{-\xi^2/2} e^{is_\mathrm{B} k_y y}, \tag{A.20}$$

where $\xi = x/l_\mathrm{B} + k_y l_\mathrm{B}$, $H_n(\xi)$ are the Hermite polynomials [281], $l_\mathrm{B} = \sqrt{\hbar c/|eB|}$ is the magnetic length, and $s_\mathrm{B} = \mathrm{sgn}\,(eB)$. These wave functions satisfy the standard conditions of normalizability and completeness,

$$\int d^2 r_\perp \psi^*_{n,k_y}(\mathbf{r}_\perp)\psi_{n',k'_y}(\mathbf{r}_\perp) = \delta_{nn'}\delta(k_y - k'_y), \tag{A.21}$$

$$\sum_{n=0}^{\infty} \int_{-\infty}^{\infty} dk_y \psi_{n,k_y}(\mathbf{r}_\perp)\psi^*_{n,k_y}(\mathbf{r}'_\perp) = \delta(\mathbf{r}_\perp - \mathbf{r}'_\perp), \tag{A.22}$$

respectively.

By making use of the completeness condition (A.22), the matrix element on the right-hand side of Eq. (A.17) can be rewritten as follows:

$$\langle \mathbf{r}_\perp | \left(\hat{M} - v_F^2 \boldsymbol{\pi}_\perp^2 - is_\mathrm{B} \frac{\hbar^2 v_F^2}{l_\mathrm{B}^2}\gamma_x\gamma_y \right)^{-1} |\mathbf{r}'_\perp\rangle$$

$$= \sum_{n=0}^{\infty} \int_{-\infty}^{\infty} dk_y \psi_{n,k_y}(\mathbf{r}_\perp)$$

$$\times \left[\hat{M} - (2n+1)\frac{\hbar^2 v_F^2}{l_\mathrm{B}^2} - is_\mathrm{B}\frac{\hbar^2 v_F^2}{l_\mathrm{B}^2}\gamma_x\gamma_y \right]^{-1} \psi^*_{n,k_y}(\mathbf{r}'_\perp)$$

$$= \frac{e^{i\Phi(\mathbf{r}_\perp,\mathbf{r}'_\perp)}}{2\pi l_\mathrm{B}^2} e^{-\zeta/2} \sum_{n=0}^{\infty} \frac{L_n(\zeta)}{\hat{M} - (2n+1)\hbar^2 v_F^2/l_B^2 - is_B \hbar^2 v_F^2 \gamma_x\gamma_y/l_B^2}, \tag{A.23}$$

where $\zeta = (\mathbf{r}_\perp - \mathbf{r}'_\perp)^2/(2l_\mathrm{B}^2)$ and $\Phi(\mathbf{r}_\perp, \mathbf{r}'_\perp)$ is the Schwinger phase [311], which is defined by

$$\Phi(\mathbf{r}_\perp, \mathbf{r}'_\perp) = -\frac{e}{\hbar c} \int_{\mathbf{r}'_\perp}^{\mathbf{r}_\perp} d\mathbf{r}'' \cdot \mathbf{A}(\mathbf{r}'') = -s_\mathrm{B}\frac{(x+x')(y-y')}{2l_\mathrm{B}^2}. \tag{A.24}$$

It is worth noting that the Schwinger phase has a universal form for charged particles in a constant magnetic field. The integral over k_y in Eq. (A.23) was calculated by making use of the following formula (see expression 7.377 in [281]):

$$\int_{-\infty}^{\infty} e^{-x^2} H_m(x+y) H_n(x+z) dx = 2^n \sqrt{\pi} m! z^{n-m} L_m^{n-m}(-2yz),$$

(A.25)

where $m \leq n$. By definition, L_n^α are the generalized Laguerre polynomials, and $L_n \equiv L_n^0$ [281].

In order to remove the matrix structure $\gamma_x \gamma_y$ from the denominator, let us introduce the following (pseudo)spin projectors:

$$\mathcal{P}_\pm = \frac{1 \pm i s_B \gamma_x \gamma_y}{2}.$$

(A.26)

Then the nth term in the sum in Eq. (A.23) is decomposed as[a]

$$\frac{L_n(\xi)}{\hat{M} - (2n+1)\hbar^2 v_F^2 / l_B^2 - i s_B \hbar^2 v_F^2 \gamma_x \gamma_y / l_B^2} = \frac{\mathcal{P}_- L_n(\xi)}{\hat{M} - 2n\hbar^2 v_F^2 / l_B^2}$$

$$+ \frac{\mathcal{P}_+ L_n(\xi)}{\hat{M} - 2(n+1)\hbar^2 v_F^2 / l_B^2}.$$

(A.27)

By using Eqs. (A.23) and (A.27), and redefining the summation index $n \to n-1$ in the term with $\mathcal{P}_+$, the matrix element in Eq. (A.17) can be written as follows:

$$\langle \mathbf{r}_\perp | \left[\hat{M} - v_F^2 \boldsymbol{\pi}_\perp^2 - i s_B \frac{\hbar^2 v_F^2}{l_B^2} \gamma_x \gamma_y \right]^{-1} | \mathbf{r}_\perp' \rangle$$

$$= \frac{e^{i\Phi(\mathbf{r}_\perp, \mathbf{r}_\perp')}}{2\pi l_B^2} e^{-\zeta/2} \sum_{n=0}^{\infty} \frac{\mathcal{P}_- L_n(\zeta) + \mathcal{P}_+ L_{n-1}(\zeta)}{\hat{M} - 2n\hbar^2 v_F^2 / l_B^2},$$

(A.28)

where $L_{-1} \equiv 0$ by definition.

It is clear that the only term that breaks the translation invariance in Green's function given in Eq. (A.17) is the Schwinger phase $\Phi(\mathbf{r}_\perp, \mathbf{r}_\perp')$.

---

[a]Note that the ordering of $\mathcal{P}_\pm$ and $(\hat{M} - 2n\hbar^2 v_F^2 / l_B^2)^{-1}$ is not important because they commute.

Fortunately, since

$$\pi_x e^{i\Phi(\mathbf{r}_\perp, \mathbf{r}'_\perp)} = \hbar e^{i\Phi(\mathbf{r}_\perp, \mathbf{r}'_\perp)} \left(-i\partial_x - s_{\mathrm{B}} \frac{y - y'}{2l_{\mathrm{B}}^2} \right), \tag{A.29}$$

$$\pi_y e^{i\Phi(\mathbf{r}_\perp, \mathbf{r}'_\perp)} = \hbar e^{i\Phi(\mathbf{r}_\perp, \mathbf{r}'_\perp)} \left(-i\partial_y + s_{\mathrm{B}} \frac{x - x'}{2l_{\mathrm{B}}^2} \right), \tag{A.30}$$

the Schwinger phase can be factorized

$$G(\omega, k_z; \mathbf{r}_\perp, \mathbf{r}'_\perp) = e^{i\Phi(\mathbf{r}_\perp, \mathbf{r}'_\perp)} \bar{G}(\omega, k_z; \mathbf{r}_\perp - \mathbf{r}'_\perp). \tag{A.31}$$

The translation invariant part of Green's function is

$$\bar{G}(\omega, k_z; \mathbf{r}_\perp - \mathbf{r}'_\perp) = i \left[\hat{W} - \hbar v_F \gamma_x \left(-i\partial_x - s_{\mathrm{B}} \frac{y - y'}{2l_{\mathrm{B}}^2} \right) \right.$$

$$\left. - \hbar v_F \gamma_y \left(-i\partial_y + s_{\mathrm{B}} \frac{x - x'}{2l_{\mathrm{B}}^2} \right) \right] \frac{e^{-\zeta/2}}{2\pi l_{\mathrm{B}}^2} \sum_{n=0}^{\infty} \frac{L_n(\zeta)\mathcal{P}_- + L_{n-1}(\zeta)\mathcal{P}_+}{\hat{M} - 2n\hbar^2 v_F^2 / l_{\mathrm{B}}^2}. \tag{A.32}$$

Note that the ordering of the matrix factors in the above expression is important because the expression in the square brackets does not commute with matrix $\hat{M}$.

Finally, let us perform the Fourier transform of the translation invariant part of Green's function given in Eq. (A.32), i.e.,

$$\bar{G}(\omega, \mathbf{k}) = \int d^2 r_\perp\, e^{-i\mathbf{k}_\perp \cdot \mathbf{r}_\perp} \bar{G}(\omega, k_z; \mathbf{r}_\perp). \tag{A.33}$$

By using the following table integral (see formula 8.411.1 in [281]):

$$\int_0^{2\pi} d\varphi\, e^{-ik_\perp r_\perp \cos(\varphi - \varphi_k)} = 2\pi J_0(k_\perp r_\perp), \tag{A.34}$$

where $J_0(x)$ is the Bessel function, and the following formula (see expression 7.421.1 in [281]):

$$\int_0^{\infty} x e^{-\frac{1}{2}\alpha x^2} L_n\left(\frac{1}{2}\beta x^2 \right) J_0(xy)\,dx = \frac{(\alpha - \beta)^n}{\alpha^{n+1}} e^{-\frac{1}{2\alpha}y^2} L_n\left(\frac{\beta y^2}{2\alpha(\beta - \alpha)} \right), \tag{A.35}$$

which is valid for $y > 0$ and $\mathrm{Re}\,\alpha > 0$, the resulting expression reads

$$\bar{G}(\omega, \mathbf{k}) = i e^{-k_\perp^2 l_{\mathrm{B}}^2} \sum_{n=0}^{\infty} (-1)^n D_n(\omega, \mathbf{k}) \frac{1}{\hat{M} - 2n\hbar^2 v_F^2 / l_{\mathrm{B}}^2}. \tag{A.36}$$

Here the nth Landau-level contribution is determined by

$$D_n(\omega, \mathbf{k}) = 2\hat{W} \left[\mathcal{P}_- L_n \left(2k_\perp^2 l_{\mathrm{B}}^2 \right) - \mathcal{P}_+ L_{n-1} \left(2k_\perp^2 l_{\mathrm{B}}^2 \right) \right]$$
$$+ 4\hbar v_F (\mathbf{k}_\perp \cdot \boldsymbol{\gamma}) L_{n-1}^1 \left(2k_\perp^2 l_{\mathrm{B}}^2 \right). \tag{A.37}$$

The last matrix factor in Eq. (A.36) can be rewritten in a more convenient form, i.e.,

$$\frac{1}{\hat{M} - 2n\hbar^2 v_F^2 / l_B^2} = \frac{(\hbar\omega + \mu)^2 - (\hbar v_F b_z)^2 + 2\hbar^2 v_F^2 b_z k_z \gamma_5 - (\hbar v_F k_z)^2}{U_n}$$
$$- \frac{2n\hbar^2 v_F^2}{U_n l_B^2}, \tag{A.38}$$

where

$$U_n = \left[(\hbar\omega + \mu)^2 - (\hbar v_F b_z)^2 - (\hbar v_F k_z)^2 - 2n\frac{\hbar^2 v_F^2}{l_B^2} \right]^2 - 4(\hbar^2 v_F^2 k_z b_z)^2. \tag{A.39}$$

The fermion dispersion relation is determined by the poles of Green's function defined in Eq. (A.36), which are equivalent to the zeros of U_n.

Appendix B

Useful Formulas

In this appendix, we give several types of integrals and identities encountered in the chiral kinetic theory calculations. By making use of the shorthand notation $f_\lambda^{(0)} = 1/[e^{(v_F p - \mu_\lambda)/T} + 1]$, where v_F is the Fermi velocity, $p = |\mathbf{p}|$ is the absolute value of quasiparticle momentum, μ_λ is the effective chemical potential of the quasiparticles with the chirality λ, and T is temperature, it is straightforward to derive the following results:

$$\int \frac{d^3 p}{(2\pi\hbar)^3} p^{n-2} f_\lambda^{(0)} = -\frac{T^{n+1}\Gamma(n+1)}{2\pi^2 \hbar^3 v_F^{n+1}} \text{Li}_{n+1}(-e^{\mu_\lambda/T}), \qquad (B.1)$$

$$\int \frac{d^3 p}{(2\pi\hbar)^3} p^{n-2} \frac{\partial f_\lambda^{(0)}}{\partial p} = \frac{T^n \Gamma(n+1)}{2\pi^2 \hbar^3 v_F^n} \text{Li}_n(-e^{\mu_\lambda/T}). \qquad (B.2)$$

Here $\Gamma(n)$ is the gamma function, $T\partial f_\lambda^{(0)}/\partial p = -v_F T \partial f_\lambda^{(0)}/\partial \mu_\lambda = -v_F e^{(v_F p - \mu_\lambda)/T}/[e^{(v_F p - \mu_\lambda)/T} + 1]^2$, $n \geq 0$, and $\text{Li}_n(x)$ is the polylogarithm function [443]. The polylogarithm function at $n = 0, 1$ can be rewritten in terms of the elementary functions

$$\text{Li}_0(-e^x) = -\frac{1}{1 + e^{-x}}, \qquad (B.3)$$

$$\text{Li}_1(-e^x) = -\ln(1 + e^x). \qquad (B.4)$$

The following identities for the polylogarithm functions are useful when taking into account holes or, equivalently, the antiparticles contributions:

$$\text{Li}_1(-e^x) - \text{Li}_1(-e^{-x}) = -x, \qquad (B.5)$$

$$\text{Li}_2(-e^x) + \text{Li}_2(-e^{-x}) = -\frac{1}{2}\left(x^2 + \frac{\pi^2}{3}\right), \qquad (B.6)$$

$$\text{Li}_3(-e^x) - \text{Li}_3(-e^{-x}) = -\frac{x}{6}\left(x^2 + \pi^2\right), \tag{B.7}$$

$$\text{Li}_4(-e^x) + \text{Li}_4(-e^{-x}) = -\frac{1}{24}\left(x^4 + 2\pi^2 x^2 + \frac{7}{15}\pi^4\right). \tag{B.8}$$

By integrating over the angular coordinates, one can derive the following general relations:

$$\int \frac{d^3 p}{(2\pi\hbar)^3}\mathbf{p} f(p^2) = 0, \tag{B.9}$$

$$\int \frac{d^3 p}{(2\pi\hbar)^3}\mathbf{p}(\mathbf{p}\cdot\mathbf{a}) f(p^2) = \frac{\mathbf{a}}{3}\int \frac{d^3 p}{(2\pi\hbar)^3} p^2 f(p^2), \tag{B.10}$$

$$\int \frac{d^3 p}{(2\pi\hbar)^3}\mathbf{p}(\mathbf{p}\cdot\mathbf{a})(\mathbf{p}\cdot\mathbf{b}) f(p^2) = 0, \tag{B.11}$$

$$\int \frac{d^3 p}{(2\pi\hbar)^3}\mathbf{p}(\mathbf{p}\cdot\mathbf{a})(\mathbf{p}\cdot\mathbf{b})(\mathbf{p}\cdot\mathbf{c}) f(p^2) = \frac{\mathbf{a}(\mathbf{b}\cdot\mathbf{c}) + \mathbf{b}(\mathbf{a}\cdot\mathbf{c}) + \mathbf{c}(\mathbf{a}\cdot\mathbf{b})}{15}$$

$$\times \int \frac{d^3 p}{(2\pi\hbar)^3} p^2 f(p^2). \tag{B.12}$$

Further, the following formulas are useful:

$$\int \frac{d^3 p}{(2\pi\hbar)^3}\mathbf{p} G(p^2, \mathbf{p}\cdot\mathbf{B}) = \frac{\mathbf{B}}{B}\int \frac{d\xi d\varphi dp}{(2\pi\hbar)^3} p^3 \xi G(p^2, pB\xi), \tag{B.13}$$

$$\int \frac{d^3 p}{(2\pi\hbar)^3}\mathbf{p}(\mathbf{p}\cdot\mathbf{a}) G(p^2, \mathbf{p}\cdot\mathbf{B})$$

$$= \int \frac{d\xi d\varphi dp}{(2\pi\hbar)^3} p^4 \left[\mathbf{a}\frac{1 - \xi^2}{2} + \frac{\mathbf{B}(\mathbf{a}\cdot\mathbf{B})}{B^2}\frac{3\xi^2 - 1}{2}\right] G(p^2, pB\xi), \tag{B.14}$$

where φ is the azimuthal angle and $\xi = \cos\theta$ is the cosine of the angle between vectors $\mathbf{B}$ and $\mathbf{p}$.

Appendix C

Fundamental Constants and Units

Throughout the book, we use mostly the CGS system of units. By following common conventions, however, occasionally we also use other units, e.g., the electronvolt (eV) as a unit of energy and the angstrom (Å) as a unit of length.

The CGS values of the key fundamental constants are given below [1071]. The (reduced) Planck constant is

$$\hbar = 1.0546 \times 10^{-27} \, \text{erg} \cdot \text{s}, \tag{C.1}$$

the speed of light is

$$c = 2.9979 \times 10^{10} \, \text{cm} \cdot \text{s}^{-1}, \tag{C.2}$$

the absolute value of the electron charge is

$$e = \sqrt{\alpha_0 c \hbar} = 4.8032 \times 10^{-10} \, \text{erg}^{1/2} \cdot \text{cm}^{1/2}, \tag{C.3}$$

the Bohr magneton is

$$\mu_{\text{B}} = 9.2740 \times 10^{-21} \, \text{erg} \cdot \text{G}^{-1}, \tag{C.4}$$

and the fine structure constant is

$$\alpha_0 = \frac{e^2}{c\hbar} = 0.007297 \approx (137.0360)^{-1}. \tag{C.5}$$

It is also convenient to have the product of the speed of light and the Planck constant, which is given by

$$c\hbar = 3.1615 \times 10^{-17} \, \text{erg} \cdot \text{cm} = 1.9733 \times 10^3 \, \text{eV} \cdot \text{Å}. \tag{C.6}$$

Note that the fine structure constant is related to the effective coupling constant α in Dirac and Weyl semimetals as follows:

$$\alpha = \frac{c}{v_F \varepsilon_e} \alpha_0 = \frac{e^2}{\hbar v_F \varepsilon_e}, \tag{C.7}$$

where v_F is the quasiparticle Fermi velocity and ε_e is the dielectric permittivity of a material.

It is useful to keep in mind the following conversion of non-CGS units:

$$1 \text{ eV} = 1.6022 \times 10^{-12} \text{ erg}, \tag{C.8}$$

$$1 \text{ K} = 1.3806 \times 10^{-16} \text{ erg} = 8.6173 \times 10^{-5} \text{ eV}, \tag{C.9}$$

$$1 \text{ T} = 10^4 \text{ G}, \tag{C.10}$$

$$1 \text{ V} = 3.3356 \times 10^{-3} \text{ statV}, \tag{C.11}$$

$$1 \, \Omega \cdot \text{m} = 1.1126 \times 10^{-10} \text{ s}^{-1}. \tag{C.12}$$

The following standard notation for the key physics quantities are used throughout the monograph:

t is time,

$\mathbf{r}$ is the coordinate vector,

v_F is the Fermi velocity,

$\mathbf{u}$ is the electron fluid flow velocity,

k is the wave vector,

$\mathbf{b}$ is the chiral shift,

ω is the frequency,

p is the momentum,

$\epsilon_{\mathbf{p}}$ is the quasiparticle energy,

μ (μ_5) is the electric (chiral) chemical potential,

b_0 is the energy separation parameter,

T is temperature,

$\mathbf{B}$ ($\mathbf{B}_5$) is the magnetic (pseudomagnetic) field,

$\mathbf{E}$ ($\mathbf{E}_5$) is the electric (pseudoelectric) field,

l_B is the magnetic length,

$\mathbf{\Omega}$ is the Berry curvature,

$\mathbf{A}$ is the vector potential,

n (n_5) is the (chiral) fermion number density,

J (**J**$_5$) is the electric (chiral) current density,

σ is the conductivity,

τ (τ_5) is the (chiral) relaxation time,

ϵ is the energy density,

P is the pressure,

ω is the vorticity,

η_{kin} is the kinematic shear viscosity,

η is the dynamic shear viscosity.

In the CGS system, these quantities have the following units:

$[c] = [v_F] = [u] = \text{cm} \cdot \text{s}^{-1},$

$[k] = [b] = \text{cm}^{-1},$

$[\omega] = [ck] = \text{s}^{-1},$

$[p] = [\hbar k] = \text{erg} \cdot \text{s} \cdot \text{cm}^{-1},$

$[\epsilon_{\mathbf{p}}] = [\hbar\omega] = [b_0] = [\mu] = \text{erg},$

$[T] = [cp] = \text{erg},$

$[B] = [E] = \text{erg}^{1/2} \cdot \text{cm}^{-3/2},$

$[eB] = [eE] = \text{erg} \cdot \text{cm}^{-1},$

$l_B = \left[\sqrt{\dfrac{c\hbar}{eB}}\right] = \text{cm},$

$[\hbar ceB] = \text{erg}^2,$

$[\Omega] = \left[\dfrac{\hbar}{p^2}\right] = \text{erg}^{-1} \cdot \text{s}^{-1} \cdot \text{cm}^2,$

$[A] = \left[\dfrac{B}{k}\right] = \text{erg}^{1/2} \cdot \text{cm}^{-1/2},$

$[n] = [n_5] = \text{cm}^{-3},$

$[\rho] = [\rho_5] = [en]$
$= \text{erg}^{1/2} \cdot \text{cm}^{-5/2},$

$[J] = [J_5] = [\rho u] = \text{erg}^{1/2} \cdot \text{s}^{-1} \cdot \text{cm}^{-3/2},$

$[\sigma] = \left[\dfrac{1}{\tau}\right] = \text{s}^{-1},$

$[\epsilon] = [P] = \text{erg} \cdot \text{cm}^{-3},$

$[\omega] = [\mathbf{\nabla} \times \mathbf{u}] = \text{s}^{-1},$

$[\eta_{\text{kin}}] = [v_F^2 \tau] = \text{cm}^2 \cdot \text{s}^{-1},$

$[\eta] = \left[\dfrac{\eta_{\text{kin}}\epsilon}{v_F^2}\right] = \text{erg} \cdot \text{cm}^{-3} \cdot \text{s}.$

Appendix D

Numerical Values of Material Parameters

Representative values of parameters in Dirac and Weyl semimetals are given in Table D.1.

For numerical values of parameters obtained from transport experiments and quantum oscillations in several other materials, see [16]. The data for the Dirac semimetal Cd_3As_2 is compiled in [23].

In Table D.1, v_F is the Fermi velocity, b is the absolute value of the chiral shift (for the Dirac semimetals Cd_3As_2 and Na_3Bi, it is a half of the distance between the Dirac points), μ is the electric chemical potential, μ_e is the electron mobility, τ and τ_5 are the intra- and internode transport relaxation times, respectively, a, b, and c are the lattice constants along the [100], [010], and [001] directions, respectively, m^* is the cyclotron (effective) mass, and $\varepsilon_e(\omega = 0)$ is the static electric permittivity. Note that in the transition metal monopnictides family of Weyl semimetals (TaAs, TaP, NbAs, and NbP) there are two types of Weyl nodes: weakly separated W1 and strongly separated W2.

Table D.1 Representative values of parameters in Dirac and Weyl semimetals.

Parameter	Material	Typical values	References
v_F	Cd_3As_2	8×10^7–1.5×10^8 cm·s^{-1}	[23, 107, 445, 447, 1072]
	Na_3Bi	9.55×10^6–8×10^7 cm·s^{-1}	[104, 1073, 1074]
	TaAs	W1: 4.15×10^7–4.3×10^7 cm·s^{-1}	[62, 118]
	TaAs	W2: 3.51×10^6–3.63×10^7 cm·s^{-1}	[62]
b	Cd_3As_2[a]	4×10^5–1.6×10^7 cm^{-1}	[23, 106, 573, 893, 1072, 1075]
	Na_3Bi	9.5×10^6–1×10^7 cm^{-1}	[104, 1073]
	TaAs[b]	W1: 1.3×10^6 cm^{-1}	[112, 113]
	TaAs	W2: 4×10^6–5.5×10^6 cm^{-1}	[64, 122]
	TaP[b]	W1: 9.5×10^5 cm^{-1}	[124]
	TaP	W2: 2.8×10^6 cm^{-1}	[120, 124]
μ	Cd_3As_2	100–286 meV	[23, 445, 447, 448, 1072, 1076]
	Na_3Bi	16–420 meV	[57, 850, 1074]
	TaAs	W1: 0.5–30 meV	[64, 118, 456]
	TaAs	W2: (−12.5)–17 meV	[64, 118, 456]
	TaP	W1: 13–24 meV	[120, 454]
	TaP	W2: (−41)–(−40) meV	[120, 454]
μ_e	Cd_3As_2	10^3–10^7 cm^2·V^{-1}·s^{-1}	[107, 820, 1072, 1076]
	Na_3Bi	2.6×10^3–7.9×10^4 cm^2·V^{-1}·s^{-1}	[1074]
	TaAs	2.5×10^3–4.8×10^5 cm^2·V^{-1}·s^{-1}	[62, 118, 456]
	TaP	3×10^4–10^7 cm^2·V^{-1}·s^{-1}	[65, 451, 452, 454]
τ	Cd_3As_2[c]	0.6×10^{-13}–2.5×10^{-10} s	[107, 447, 1072, 1076]
	Na_3Bi[c]	1.5×10^{-12}–6.7×10^{-12} s	[1074]
	TaAs	W1: 3.8×10^{-13}–2.6×10^{-11} s	[118, 456]
τ_5	Cd_3As_2	$\gtrsim 10 \times \tau$	[60]
	Na_3Bi	$40 \times \tau$–$60 \times \tau$	[57]
	TaAs	5.96×10^{-11} s	[64]
a_x, a_y, a_z	Cd_3As_2[d]	$a_x = a_y = 12.670$ Å, $a_z = 25.480$ Å	[107, 926]
	Na_3Bi[e]	$a_x = a_y = 5.448$ Å, $a_z = 9.655$ Å	[926]
	TaAs[f]	$a_x = a_y = 3.435$ Å, $a_z = 11.641$ Å	[1077]
	TaP[f]	$a_x = a_y = 3.318$ Å, $a_z = 11.347$ Å	[1077]
m^*/m_e	Cd_3As_2	0.029–0.05	[16, 23, 445, 447, 448, 1076]
	Na_3Bi	0.11	[16, 1074]
	TaAs	W1: 0.057–0.16	[118, 456]
	TaP	0.04–0.4	[452, 454]
$\varepsilon_e(\omega = 0)$	Cd_3As_2	36–42	[926]
	Na_3Bi	120	[850]
	TaAs[b]	93–592	[1078, 1079]
	TaP[b]	140	[1079]

[a]The lower bound for b might correspond to a different regime where Kane fermions are realized instead of Dirac ones [573, 893, 1075].
[b]Only numerical results are available.
[c]Relaxation time is estimated based on the electron mobility, which is usually measured at low temperature $T \approx 1.5 - 5$ K.
[d]$I4_1cd$ body-centered tetragonal structure.
[e]$P6_3/mmc$ hexagonal structure.
[f]$I4_1md$ body-centered tetragonal structure.

Bibliography

[1] A. M. Turner and A. Vishwanath, Beyond band insulators: topology of semimetals and interacting phases, in M. Franz and L. Molenkamp (eds.), *Topological Insulators*, Chap. 11. Elsevier Science, 293 (2013).

[2] O. Vafek and A. Vishwanath, *Annu. Rev. Condens. Matter Phys.* **5**, 83 (2014), arXiv:1306.2272.

[3] T. Wehling, A. Black-Schaffer and A. Balatsky, *Adv. Phys.* **63**, 1 (2014), arXiv:1405.5774.

[4] B. Duplantier, V. Rivasseau and J.-N. Fuchs (eds.), *Dirac Matter*. Birkhäuser Basel (2017).

[5] N. P. Armitage, E. J. Mele and A. Vishwanath, *Rev. Mod. Phys.* **90**, 015001 (2018), arXiv:1705.01111.

[6] P. Hosur and X. Qi, *C. R. Phys.* **14**, 857 (2013), arXiv:1309.4464.

[7] A. Burkov, *J. Phys.: Condens. Matter* **27**, 113201 (2015).

[8] H.-Z. Lu and S.-Q. Shen, *Front. Phys.* **12**, 127201 (2017), arXiv:1609.01029.

[9] S. V. Syzranov and L. Radzihovsky, *Annu. Rev. Condens. Matter Phys.* **9**, 35 (2018), arXiv:1609.05694.

[10] H.-D. Song *et al.*, *Chin. Phys. B* **26**, 037301 (2017), arXiv:1701.04983.

[11] A. Burkov, *Annu. Rev. Condens. Matter Phys.* **9**, 359 (2018), arXiv:1704.06660.

[12] E. V. Gorbar, V. A. Miransky, I. A. Shovkovy and P. O. Sukhachov, *Low Temp. Phys.* **44**, 487 (2018), arXiv:1712.08947.

[13] S. Wang *et al.*, *Adv. Phys.: X* **2**, 518 (2017).

[14] H. Wang and J. Wang, *Chin. Phys. B* **27**, 107402 (2018), arXiv:1809.03282.

[15] H.-P. Sun and H.-Z. Lu, *Front. Phys.* **14**, 33405 (2019), arXiv:1812.10120.

[16] J. Hu, S.-Y. Xu, N. Ni and Z. Mao, *Annu. Rev. Mater. Sci.* **49**, 207 (2019), arXiv:1904.04454.

[17] H. Weng, X. Dai and Z. Fang, *J. Phys.: Condens. Matter* **28**, 303001 (2016), arXiv:1603.04744.

[18] M. Z. Hasan, S.-Y. Xu and G. Bian, *Phys. Scr.* **2015**, 014001 (2015).

[19] M. Z. Hasan, S.-Y. Xu, I. Belopolski and S.-M. Huang, *Annu. Rev. Condens. Matter Phys.* **8**, 289 (2017), arXiv:1702.07310.

[20] B. Yan and C. Felser, *Annu. Rev. Condens. Matter Phys.* **8**, 337 (2017), arXiv:1611.04182.

[21] H. Zheng and M. Z. Hasan, *Adv. Phys.: X* **3**, 1466661 (2018), arXiv:1805.10590.

[22] P. K. Das *et al.*, *Electron. Struct.* **1**, 014003 (2019), arXiv:1812.07215.

[23] I. Crassee *et al.*, *Phys. Rev. Materials* **2**, 120302 (2018), arXiv:1810.03726.

[24] E. Witten, *Riv. del Nuovo Cim.* **39**, 313 (2016).

[25] C.-K. Chiu, J. C. Y. Teo, A. P. Schnyder and S. Ryu, *Rev. Mod. Phys.* **88**, 035005 (2016), arXiv:1505.03535.

[26] V. A. Miransky and I. A. Shovkovy, *Phys. Rep.* **576**, 1 (2015), arXiv:1503.00732.

[27] W. Witczak-Krempa, G. Chen, Y. B. Kim and L. Balents, *Annu. Rev. Condens. Matter Phys.* **5**, 57 (2014), arXiv:1305.2193.

[28] L. Šmejkal, T. Jungwirth and J. Sinova, *Phys. Status Solidi Rapid Res. Lett.* **11**, 1770317 (2017), arXiv:1702.07788.

[29] G. P. Mikitik and Y. V. Sharlai, *J. Low Temp. Phys.* **197**, 272 (2019), arXiv:1903.12208.

[30] P. A. Dirac, *Proc. Royal Soc. Lond.* **A117**, 610 (1928).

[31] H. Weyl, *Z. Phys.* **56**, 330 (1929).

[32] K. Zuber, *Neutrino Physics*. CRC Press (2011).

[33] C. Herring, *Phys. Rev.* **52**, 365 (1937).

[34] A. A. Abrikosov and S. Beneslavskii, *Sov. Phys. JETP* **32**, 699 (1971).

[35] S. Groves and W. Paul, *Phys. Rev. Lett.* **11**, 194 (1963).

[36] V. Ivanov-Omskii, B. Kolomietz, L. Kleshchinskii and K. P. Smekalova, *Soviet Phys. Solid State* **10**, 2447 (1969).

[37] B. A. Bernevig, T. L. Hughes and S.-C. Zhang, *Science* **314**, 1757 (2006), arXiv:cond-mat/0611399.

[38] M. König *et al.*, *J. Phys. Soc. Jpn.* **77**, 031007 (2008), arXiv:0801.0901.

[39] J. S. Bell and R. Jackiw, *Nuovo Cim. A* **60**, 47 (1969).

[40] S. L. Adler, *Phys. Rev.* **177**, 2426 (1969).

[41] R. Bertlmann, *Anomalies in Quantum Field Theory*. Clarendon Press (2000).

[42] M. Peskin and D. V. Schroered, *An Introduction To Quantum Field Theory*. CRC Press (2019).

[43] S. Weinberg, *The Quantum Theory of Fields: Foundations*. Cambridge University Press (2005).

[44] S. Weinberg, *The Quantum Theory of Fields: Modern Applications*. Cambridge University Press (2005).

[45] C. Itzykson and J. B. Zuber, *Quantum Field Theory*. Dover Publications (2005).

[46] K. G. Wilson, *Phys. Rev. D* **10**, 2445 (1974).

[47] M. Creutz, *Quarks, Gluons and Lattices*. Cambridge University Press (1985).

[48] T. DeGrand and C. DeTar, *Lattice Methods for Quantum Chromodynamics*. World Scientific (2006).

[49] C. Gattringer and C. Lang, *Quantum Chromodynamics on the Lattice: An Introductory Presentation*. Springer-Verlag (2010).

[50] L. H. Karsten and S. Jan, *Nucl. Phys. B* **183**, 103 (1981).

[51] H. Nielsen and M. Ninomiya, *Nucl. Phys. B* **185**, 20 (1981).

[52] H. Nielsen and M. Ninomiya, *Nucl. Phys. B* **193**, 173 (1981).

[53] H. Nielsen and M. Ninomiya, *Phys. Lett. B* **105**, 219 (1981).

[54] J. Ambøjrn, J. Greensite and C. Peterson, *Nucl. Phys. B* **221**, 381 (1983).

[55] H. Nielsen and M. Ninomiya, *Phys. Lett. B* **130**, 389 (1983).

[56] H.-J. Kim *et al.*, *Phys. Rev. Lett.* **111**, 246603 (2013), arXiv:1307.6990.

[57] J. Xiong *et al.*, *Science* **350**, 413 (2015), arXiv:1503.08179.

[58] J. Feng *et al.*, *Phys. Rev. B* **92**, 081306 (2015), arXiv:1405.6611.

[59] C.-Z. Li *et al.*, *Nat. Commun.* **6**, 10137 (2015), arXiv:1504.07398.

[60] H. Li *et al.*, *Nat. Commun.* **7**, 10301 (2016a), arXiv:1507.06470.

[61] Q. Li *et al.*, *Nat. Mater.* **12**, 550 (2016b), arXiv:1412.6543.

[62] X. Huang *et al.*, *Phys. Rev. X* **5**, 031023 (2015), arXiv:1503.01304.

[63] X. Yang *et al.*, arXiv:1506.03190 (2015).

[64] C.-L. Zhang *et al.*, *Nat. Commun.* **7**, 10735 (2016), arXiv:1601.04208.

[65] J. Du *et al.*, *Sci. China Phys. Mech. Astron.* **59**, 657406 (2016), arXiv:1507.05246.

[66] Z. Wang *et al.*, *Phys. Rev. B* **93**, 121112 (2016), arXiv:1506.00924.

[67] Y. Li *et al.*, *Front. Phys.* **12**, 127205 (2017), arXiv:1612.04031.

[68] M. Berry, *Proc. R. Soc. Lond. A* **392**, 45 (1984).

[69] R. Jackiw and C. Rebbi, *Phys. Rev. D* **13**, 3398 (1976).

[70] Y. Hatsugai, *Phys. Rev. Lett.* **71**, 3697 (1993).

[71] Y. Hatsugai, *Phys. Rev. B* **48**, 11851 (1993).

[72] G. E. Volovik, *JETP Lett.* **46**, 98 (1987).

[73] G. Volovik, *The Universe in a Helium Droplet*. Oxford University Press (2009).

[74] T. D. C. Bevan *et al.*, *J. Low Temp. Phys.* **109**, 423 (1997).

[75] K. S. Novoselov *et al.*, *Science* **306**, 666 (2004), arXiv:cond-mat/0410550.

[76] G. W. Semenoff, *Phys. Rev. Lett.* **53**, 2449 (1984).

[77] D. P. DiVincenzo and E. J. Mele, *Phys. Rev. B* **29**, 1685 (1984).

[78] R. E. Peierls, *Ann. Inst. Henri Poincare* **5**, 177 (1935).

[79] L. D. Landau, *Phys. Z. Sowjetunion* **11**, 26 (1937).

[80] L. Landau and E. Lifshitz, *Statistical Physics, Part 1*. Butterworth-Heinemann (2013).

[81] M. I. Katsnelson, *Graphene: Carbon in Two Dimensions*. Cambridge University Press (2012).

[82] M. I. Katsnelson, K. S. Novoselov and A. K. Geim, *Nat. Phys.* **2**, 620 (2006), arXiv:cond-mat/0604323.

[83] M. I. Katsnelson, *Eur. Phys. J. B* **51**, 157 (2006), arXiv:cond-mat/0512337.

[84] A. V. Shytov, M. I. Katsnelson and L. S. Levitov, *Phys. Rev. Lett.* **99**, 236801 (2007), arXiv:0705.4663.

[85]   Y. Wang *et al.*, *Science* **340**, 734 (2013), arXiv:1510.02890.
[86]   N. Vandecasteele *et al.*, *Phys. Rev. B* **82**, 045416 (2010), arXiv:1003.2072.
[87]   X.-L. Qi, Y.-S. Wu and S.-C. Zhang, *Phys. Rev. B* **74**, 045125 (2006), arXiv:cond-mat/0604071.
[88]   L. Fu, C. L. Kane and E. J. Mele, *Phys. Rev. Lett.* **98**, 106803 (2007), arXiv:cond-mat/0607699.
[89]   L. Fu and C. L. Kane, *Phys. Rev. B* **76**, 045302 (2007), arXiv:cond-mat/0611341.
[90]   J. E. Moore and L. Balents, *Phys. Rev. B* **75**, 121306 (2007), arXiv: cond-mat/0607314.
[91]   R. Roy, *Phys. Rev. B* **79**, 195322 (2009), arXiv:cond-mat/0607531.
[92]   H. Zhang *et al.*, *Nat. Phys.* **5**, 438 (2009).
[93]   D. Hsieh *et al.*, *Nature* **452**, 970 (2008), arXiv:0910.2420.
[94]   Y. L. Chen *et al.*, *Science* **325**, 178 (2009).
[95]   Y. Xia *et al.*, *Nat. Phys.* **5**, 398 (2009), arXiv:0908.3513.
[96]   B. A. Volkov and O. A. Pankratov, *JETP Lett.* **42**, 4 (1985).
[97]   M. Z. Hasan and C. L. Kane, *Rev. Mod. Phys.* **82**, 3045 (2010), arXiv:1002.3895.
[98]   X.-L. Qi and S.-C. Zhang, *Rev. Mod. Phys.* **83**, 1057 (2011), arXiv: 1008.2026.
[99]   M. Z. Hasan and J. E. Moore, *Annu. Rev. Condens. Matter Phys.* **2**, 55 (2011), arXiv:1011.5462.
[100] B. A. Bernevig and T. L. Hughes, *Topological Insulators and Topological Superconductors.* Princeton University Press (2013).
[101] S.-Q. Shen, *Topological Insulators: Dirac Equation in Condensed Matters.* Springer-Verlag (2012).
[102] X. Wan, A. M. Turner, A. Vishwanath and S. Y. Savrasov, *Phys. Rev. B* **83**, 205101 (2011), arXiv:1007.0016.
[103] F. D. M. Haldane, arXiv:1401.0529 (2014).
[104] Z. K. Liu *et al.*, *Science* **343**, 864 (2014), arXiv:1310.0391.
[105] S.-Y. Xu *et al.*, *Science* **347**, 294 (2015), arXiv:1501.01249.
[106] Z. K. Liu *et al.*, *Nat. Mater.* **13**, 677 (2014).
[107] M. Neupane *et al.*, *Nat. Commun.* **5**, 3786 (2014), arXiv:1309.7892.
[108] S. Borisenko *et al.*, *Phys. Rev. Lett.* **113**, 027603 (2014), arXiv:1309.7978.
[109] H. Yi *et al.*, *Sci. Rep.* **4**, 6106 (2014), arXiv:1405.5702.
[110] Z. Wang *et al.*, *Phys. Rev. B* **85**, 195320 (2012), arXiv:1202.5636.
[111] Z. Wang *et al.*, *Phys. Rev. B* **88**, 125427 (2013), arXiv:1305.6780.
[112] H. Weng *et al.*, *Phys. Rev. X* **5**, 011029 (2015), arXiv:1501.00060.
[113] S.-M. Huang *et al.*, *Nat. Commun.* **6**, 7373 (2015).
[114] C.-C. Lee *et al.*, *Phys. Rev. B* **92**, 235104 (2015), arXiv:1508.05999.
[115] Y. Sun, S.-C. Wu and B. Yan, *Phys. Rev. B* **92**, 115428 (2015), arXiv:1508.06649.
[116] Z. K. Liu *et al.*, *Nat. Mater.* **15**, 27 (2015).
[117] I. Belopolski *et al.*, *Phys. Rev. Lett.* **116**, 066802 (2016), arXiv:1601.04327.
[118] F. Arnold *et al.*, *Phys. Rev. Lett.* **117**, 146401 (2016), arXiv:1603.08846.

[119] S.-Y. Xu *et al.*, *Science* **349**, 613 (2015a), arXiv:1502.03807.

[120] S.-Y. Xu *et al.*, *Sci. Adv.* **1** (2015b), arXiv:1508.03102.

[121] B. Q. Lv *et al.*, *Phys. Rev. X* **5**, 031013 (2015a), arXiv:1502.04684.

[122] B. Q. Lv *et al.*, *Nat. Phys.* **11**, 724 (2015b), arXiv:1503.09188.

[123] L. X. Yang *et al.*, *Nat. Phys.* **11**, 728 (2015).

[124] N. Xu *et al.*, *Nat. Commun.* **7**, 11006 (2016), arXiv:1507.03983.

[125] S.-Y. Xu *et al.*, *Nat. Phys.* **11**, 748 (2015).

[126] S.-Y. Xu *et al.*, *Phys. Rev. Lett.* **116**, 096801 (2016), arXiv:1510.08430.

[127] A. A. Soluyanov *et al.*, *Nature* **527**, 495 (2015), arXiv:1507.01603.

[128] Y. Xu, F. Zhang and C. Zhang, *Phys. Rev. Lett.* **115**, 265304 (2015), arXiv:1411.7316.

[129] C. Fang, M. J. Gilbert, X. Dai and B. A. Bernevig, *Phys. Rev. Lett.* **108**, 266802 (2012), arXiv:1111.7309.

[130] B. Bradlyn *et al.*, *Science* **353** (2016), arXiv:1603.03093.

[131] B. J. Wieder, Y. Kim, A. M. Rappe and C. L. Kane, *Phys. Rev. Lett.* **116**, 186402 (2016), arXiv:1512.00074.

[132] G. Chang *et al.*, *Nature* **567**, 500 (2019), arXiv:1812.04466.

[133] D. Takane *et al.*, *Phys. Rev. Lett.* **122**, 076402 (2019), arXiv:1809.01312.

[134] N. B. M. Schröter *et al.*, *Nat. Phys.* **15**, 759 (2019), arXiv:1812.03310.

[135] Z. Rao *et al.*, *Nature* **567**, 496 (2019), arXiv:1901.03358.

[136] A. A. Abrikosov, L. P. Gorkov and I. E. Dzyaloshinski, *Methods of Quantum Field Theory in Statistical Physics*. Dover Publications (1963).

[137] E. Lifshitz and L. Pitaevskii, *Statistical Physics: Theory of the Condensed State*. Butterworth-Heinemann (2013).

[138] L. S. Levitov and A. V. Shytov, *Green's Functions. Theory and Practice*. FizMatLit-Nauka (2003) (in Russian).

[139] H. Bruus and K. Flensberg, *Many-Body Quantum Theory in Condensed Matter Physics: An Introduction*. Oxford University Press (2004).

[140] J. Bardeen, L. N. Cooper and J. R. Schrieffer, *Phys. Rev.* **108**, 1175 (1957).

[141] K. G. Wilson, *Phys. Rev. B* **4**, 3174 (1971).

[142] K. G. Wilson, *Phys. Rev. B* **4**, 3184 (1971).

[143] L. P. Kadanoff, *Phys. Phys. Fiz.* **2**, 263 (1966).

[144] A. Vilenkin, *Phys. Rev. D* **22**, 3080 (1980).

[145] K. Fukushima, D. E. Kharzeev and H. J. Warringa, *Phys. Rev. D* **78**, 074033 (2008), arXiv:0808.3382.

[146] K. Fukushima, Views of the chiral magnetic effect, in *Strongly Interacting Matter in Magnetic Fields*. Springer (2013), p. 241.

[147] M. A. Metlitski and A. R. Zhitnitsky, *Phys. Rev. D* **72**, 045011 (2005), arXiv:hep-ph/0505072.

[148] G. M. Newman and D. T. Son, *Phys. Rev. D* **73**, 045006 (2006), arXiv:hep-ph/0510049.

[149] A. Vilenkin, *Phys. Rev. D* **20**, 1807 (1979).

[150] N. Erdmenger *et al.*, *J. High Energy Phys.* **01**, 055 (2009), arXiv:0809.2488.

[151] N. Erdmenger, M. Haack, M. Kaminski and A. Yarom, *J. High Energy Phys.* **01**, 094 (2011), arXiv:0809.2596.

[152] M. A. Stephanov and Y. Yin, *Phys. Rev. Lett.* **109**, 162001 (2012), arXiv:1207.0747.

[153] J. Liao, *Pramana* **84**, 901 (2015), arXiv:1401.2500.

[154] D. E. Kharzeev, *Annu. Rev. Nucl. Part. Sci.* **65**, 193 (2015), arXiv:1501.01336.

[155] X.-G. Huang, *Rep. Prog. Phys.* **79**, 076302 (2016), arXiv:1509.04073.

[156] D. Kharzeev, J. Liao, S. A. Voloshin and G. Wang, *Prog. Part. Nucl. Phys.* **88**, 1 (2016), arXiv:1511.04050.

[157] M. Joyce and M. Shaposhnikov, *Phys. Rev. Lett.* **79**, 1193 (1997), arXiv:astro-ph/9703005.

[158] A. Boyarsky, J. Fröhlich and O. Ruchayskiy, *Phys. Rev. Lett.* **108**, 031301 (2012), arXiv:1109.3350.

[159] K. I. Bolotin *et al.*, *Solid State Commun.* **146**, 351 (2008), arXiv:0802.2389.

[160] L. Šmejkal, J. Železný, J. Sinova and T. Jungwirth, *Phys. Rev. Lett.* **118**, 106402 (2017), arXiv:1610.08107.

[161] D. E. Kharzeev and H.-U. Yee, *Phys. Rev. B* **88**, 115119 (2013), arXiv:1207.0477.

[162] Q. Wang *et al.*, *Nano Lett.* **17**, 834 (2017), arXiv:1609.00768.

[163] C. Zhu *et al.*, *Nat. Commun.* **8**, 14111 (2017), arXiv:1602.08951.

[164] N. Yavarishad *et al.*, *Appl. Phys. Express* **10**, 052201 (2017).

[165] L. M. Schoop, F. Pielnhofer and B. V. Lotsch, *Chem. Mater.* **30**, 3155 (2018), arXiv:1804.10649.

[166] K. S. Novoselov *et al.*, *Proc. Natl. Acad. Sci. U.S.A.* **102**, 10451 (2005), arXiv:cond-mat/0503533.

[167] K. S. Novoselov *et al.*, *Nature* **438**, 197 (2005b), arXiv:cond-mat/0509330.

[168] Y. Zhang, Y.-W. Tan, H. L. Stormer and P. Kim, *Nature* **438**, 201 (2005), arXiv:cond-mat/0509355.

[169] S.-J. Han *et al.*, *Nat. Commun.* **5**, 3086 (2014).

[170] M. Engel *et al.*, *Nat. Commun.* **9**, 4095 (2018), arXiv:1802.07599.

[171] M. Wang and E.-H. Yang, *Nano-Structures & Nano-Objects* **15**, 107 (2018).

[172] D. A. Bandurin *et al.*, *Nat. Commun.* **9**, 5392 (2018), arXiv:1807.04703.

[173] R. R. Nair *et al.*, *Science* **320**, 1308 (2008), arXiv:0803.3718.

[174] I. Crassee *et al.*, *Nat. Phys.* **7**, 48 (2011), arXiv:1007.5286.

[175] I. W. Frank, D. M. Tanenbaum, A. M. van der Zande and P. L. McEuen, *J. Vac. Sci. Technol. B* **25**, 2558 (2007).

[176] C. Lee, X. Wei, J. W. Kysar and J. Hone, *Science* **321**, 385 (2008).

[177] K. S. Kim *et al.*, *Nature* **457**, 706 (2009).

[178] C. A. Marianetti and H. G. Yevick, *Phys. Rev. Lett.* **105**, 245502 (2010), arXiv:1004.1849.

[179] R. R. Nair *et al.*, *Science* **335**, 442 (2012), arXiv:1112.3488.

[180] D. Cohen-Tanugi and J. C. Grossman, *Nano Lett.* **12**, 3602 (2012).

[181] S. P. Surwade *et al.*, *Nat. Nanotechnol.* **10**, 459 (2015).

[182] J. R. Werber, C. O. Osuji and M. Elimelech, *Nat. Rev. Mater.* **1**, 16018 (2016).

[183] F. Schedin *et al.*, *Nat. Mater.* **6**, 652 (2007), arXiv:cond-mat/0610809.

[184] M. I. Katsnelson, *Mater. Today* **10**, 20 (2007), arXiv:cond-mat/0612534.

[185] A. K. Geim and K. S. Novoselov, *Nat. Mater.* **6**, 183 (2007), arXiv:cond-mat/0702595.

[186] C. W. J. Beenakker, *Rev. Mod. Phys.* **80**, 1337 (2008), arXiv:0710.3848.

[187] A. H. Castro Neto *et al.*, *Rev. Mod. Phys.* **81**, 109 (2009), arXiv:0709.1163.

[188] D. Abergel *et al.*, *Adv. Phys.* **59**, 261 (2010), arXiv:1003.0391.

[189] N. M. R. Peres, *Rev. Mod. Phys.* **82**, 2673 (2010), arXiv:1007.2849.

[190] S. Das Sarma, S. Adam, E. H. Hwang and E. Rossi, *Rev. Mod. Phys.* **83**, 407 (2011), arXiv:1003.4731.

[191] M. O. Goerbig, *Rev. Mod. Phys.* **83**, 1193 (2011), arXiv:1004.3396.

[192] V. N. Kotov *et al.*, *Rev. Mod. Phys.* **84**, 1067 (2012), arXiv:1012.3484.

[193] J. H. Warner, F. Schaffel, M. Rummeli and A. Bachmatiuk, *Graphene: Fundamentals and Emergent Applications*. Elsevier (2012).

[194] H. Aoki and M. Dresselhaus, *Physics of Graphene*. Springer International Publishing (2014).

[195] P. R. Wallace, *Phys. Rev.* **71**, 622 (1947).

[196] S. Reich, J. Maultzsch, C. Thomsen and P. Ordejón, *Phys. Rev. B* **66**, 035412 (2002).

[197] T. Ando, T. Nakanishi and R. Saito, *J. Phys. Soc. Jpn.* **67**, 2857 (1998).

[198] E. McCann and V. I. Fal'ko, *Phys. Rev. Lett.* **96**, 086805 (2006), arXiv:cond-mat/0510237.

[199] S. Zhou, G.-H. Gweon and A. Lanzara, *Ann. Phys. (N. Y.)* **321**, 1730 (2006), arXiv:cond-mat/0609028.

[200] T. Ohta *et al.*, *Phys. Rev. Lett.* **98**, 206802 (2007), arXiv:cond-mat/0612173.

[201] S. Zhou *et al.*, *Nat. Mater.* **7**, 259 (2008), arXiv:0804.1818.

[202] J. Cserti and G. Dávid, *Phys. Rev. B* **74**, 172305 (2006), arXiv:cond-mat/0604526.

[203] D. Allor, T. D. Cohen and D. A. McGady, *Phys. Rev. D* **78**, 096009 (2008), arXiv:0708.1471.

[204] B. Rosenstein, M. Lewkowicz, H. C. Kao and Y. Korniyenko, *Phys. Rev. B* **81**, 041416 (2010), arXiv:0909.2663.

[205] E. Schrödinger, *Sitzungsber. Preuss. Akad. Wiss. Phys. Math. Kl.* **24**, 418 (1930).

[206] O. Klein, *Z. Phys.* **53**, 157 (1929).

[207] A. V. Shytov, M. S. Rudner and L. S. Levitov, *Phys. Rev. Lett.* **101**, 156804 (2008), arXiv:0808.0488.

[208] P. E. Allain and J. N. Fuchs, *Eur. Phys. J. B* **83**, 301 (2011), arXiv:1104.5632.

[209] N. Stander, B. Huard and D. Goldhaber-Gordon, *Phys. Rev. Lett.* **102**, 026807 (2009), arXiv:0806.2319.

[210] A. F. Young and P. Kim, *Nat. Phys.* **5**, 222 (2009), arXiv:0808.0855.

[211] E. V. Gorbar, V. P. Gusynin, V. A. Miransky and I. A. Shovkovy, *Phys. Rev. B* **66**, 045108 (2002), arXiv:cond-mat/0202422.

[212] V. P. Gusynin and S. G. Sharapov, *Phys. Rev. Lett.* **95**, 146801 (2005), arXiv:cond-mat/0506575.

[213] Y. V. Skrypnyk and V. M. Loktev, *Low Temp. Phys.* **45**, 1310 (2019).

[214] V. G. Veselago, *Sov. Phys. Usp.* **10**, 509 (1968).

[215] J. B. Pendry, *Phys. Rev. Lett.* **85**, 3966 (2000).

[216] R. A. Shelby, D. R. Smith and S. Schultz, *Science* **292**, 77 (2001).

[217] C. G. Parazzoli *et al.*, *Phys. Rev. Lett.* **90**, 107401 (2003).

[218] A. A. Houck, J. B. Brock and I. L. Chuang, *Phys. Rev. Lett.* **90**, 137401 (2003).

[219] V. V. Cheianov, V. Falko and B. L. Altshuler, *Science* **315**, 1252 (2007), arXiv:cond-mat/0703410.

[220] F. Libisch *et al.*, *J. Phys. Condens. Matter* **29**, 114002 (2017).

[221] G.-H. Lee, G.-H. Park and H.-J. Lee, *Nat. Phys.* **11**, 925 (2015), arXiv:1506.06281.

[222] P. L. McEuen *et al.*, *Phys. Rev. Lett.* **83**, 5098 (1999), arXiv:cond-mat/9906055.

[223] H. Suzuura and T. Ando, *Phys. Rev. Lett.* **89**, 266603 (2002).

[224] I. L. Aleiner, D. E. Kharzeev and A. M. Tsvelik, *Phys. Rev. B* **76**, 195415 (2007), arXiv:0708.0394.

[225] M. Kharitonov, *Phys. Rev. B* **85**, 155439 (2012), arXiv:1103.6285.

[226] J.-N. Fuchs and P. Lederer, *Phys. Rev. Lett.* **98**, 016803 (2007), arXiv:cond-mat/0607480.

[227] V. Miransky, *Dynamical Symmetry Breaking In Quantum Field Theories.* World Scientific (1994).

[228] V. P. Gusynin, S. G. Sharapov and J. P. Carbotte, *Int. J. Mod. Phys.* **B21**, 4611 (2007), arXiv:0706.3016.

[229] R. Jackiw and S. Templeton, *Phys. Rev. D* **23**, 2291 (1981).

[230] F. D. M. Haldane, *Phys. Rev. Lett.* **61**, 2015 (1988).

[231] E. V. Gorbar, V. A. Miransky and I. A. Shovkovy, *Phys. Rev. C* **80**, 032801 (2009), arXiv:0904.2164.

[232] A. Shapere and F. Wilczek, *Geometric Phases in Physics.* World Scientific, Singapore (1989).

[233] J. N. Fuchs, F. Piéchon, M. O. Goerbig and G. Montambaux, *Eur. Phys. J. B* **77**, 351 (2010), arXiv:1006.5632.

[234] A. Altland and B. D. Simons, *Condensed Matter Field Theory*, 2nd edn. Cambridge University Press (2010).

[235] G. Mahan, *Many-Particle Physics.* Springer US (2000).

[236] P. Coleman, *Introduction to Many-Body Physics.* Cambridge University Press (2015).

[237] V. P. Gusynin and S. G. Sharapov, *Phys. Rev. B* **73**, 245411 (2006), arXiv:cond-mat/0512157.

[238] V. P. Gusynin, S. G. Sharapov and J. P. Carbotte, *Phys. Rev. Lett.* **96**, 256802 (2006), arXiv:cond-mat/0603267.

[239] Z. Q. Li *et al.*, *Nat. Phys.* **4**, 532 (2008), arXiv:0807.3780.

[240] R. Gurzhi, *Sov. Phys. JETP* **17**, 521 (1963).

[241] R. Gurzhi, *Sov. Phys. Usp.* **11**, 255 (1968).

[242] E. H. Hwang and S. Das Sarma, *Phys. Rev. B* **77**, 115449 (2008), arXiv:0711.0754.

[243] T. Sohier *et al.*, *Phys. Rev. B* **90**, 125414 (2014), arXiv:1407.0830.

[244] D. Y. H. Ho, I. Yudhistira, N. Chakraborty and S. Adam, *Phys. Rev. B* **97**, 121404 (2018), arXiv:1710.10272.

[245] M. S. Foster and I. L. Aleiner, *Phys. Rev. B* **79**, 085415 (2009), arXiv:0810.4342.

[246] M. Schütt, P. M. Ostrovsky, I. V. Gornyi and A. D. Mirlin, *Phys. Rev. B* **83**, 155441 (2011), arXiv:1011.5217.

[247] D. Svintsov *et al.*, *J. Appl. Phys.* **111**, 083715 (2012), arXiv:1201.0592.

[248] B. N. Narozhny *et al.*, *Phys. Rev. B* **91**, 035414 (2015), arXiv:1411.0819.

[249] I. Torre, A. Tomadin, A. K. Geim and M. Polini, *Phys. Rev. B* **92**, 165433 (2015), arXiv:1508.00363.

[250] U. Briskot *et al.*, *Phys. Rev. B* **92**, 115426 (2015), arXiv:1507.08946.

[251] F. M. D. Pellegrino, I. Torre, A. K. Geim and M. Polini, *Phys. Rev. B* **94**, 155414 (2016), arXiv:1607.03726.

[252] L. Levitov and G. Falkovich, *Nat. Phys.* **12**, 672 (2016), arXiv:1508.00836.

[253] G. Falkovich and L. Levitov, *Phys. Rev. Lett.* **119**, 066601 (2017), arXiv:1607.00986.

[254] H. Guo, E. Ilseven, G. Falkovich and L. S. Levitov, *Proc. Natl. Acad. Sci. USA* **114**, 3068 (2017), arXiv:1607.07269.

[255] P. S. Alekseev *et al.*, *Phys. Rev. B* **95**, 165410 (2017), arXiv:1612.02439.

[256] D. Svintsov, *Phys. Rev. B* **97**, 121405 (2018), arXiv:1710.05054.

[257] P. S. Alekseev *et al.*, *Phys. Rev. B* **97**, 085109 (2018a), arXiv:1711.03523.

[258] P. S. Alekseev *et al.*, *Phys. Rev. B* **98**, 125111 (2018b), arXiv:1805.10321.

[259] B. N. Narozhny and M. Schütt, *Phys. Rev. B* **100**, 035125 (2019), arXiv:1905.11424.

[260] S. Danz and B. N. Narozhny, arXiv:1910.14473 (2019).

[261] J. Crossno *et al.*, *Science* **351**, 1058 (2016), arXiv:1509.04713.

[262] F. Ghahari *et al.*, *Phys. Rev. Lett.* **116**, 136802 (2016), arXiv:1601.05859.

[263] R. Krishna Kumar *et al.*, *Nat. Phys.* **13**, 1182 (2017), arXiv:1703.06672.

[264] A. I. Berdyugin *et al.*, *Science* **364**, 162 (2019), arXiv:1806.01606.

[265] D. A. Bandurin *et al.*, *Nat. Commun.* **9**, 4533 (2018), arXiv:1806.03231.

[266] M. J. H. Ku *et al.*, arXiv:1905.10791 (2019).

[267] A. Rozen *et al.*, *Nature* **576**, 75 (2019), arXiv:1905.11662.

[268] M. Dyakonov and M. Shur, *Phys. Rev. Lett.* **71**, 2465 (1993).

[269] A. Tomadin, G. Vignale and M. Polini, *Phys. Rev. Lett.* **113**, 235901 (2014), arXiv:1401.0938.

[270] P. S. Alekseev, *Phys. Rev. Lett.* **117**, 166601 (2016), arXiv:1603.04587.

[271] C. Hodges, H. Smith and J. W. Wilkins, *Phys. Rev. B* **4**, 302 (1971).

[272] G. F. Giuliani and J. J. Quinn, *Phys. Rev. B* **26**, 4421 (1982).

[273] T. Jungwirth and A. H. MacDonald, *Phys. Rev. B* **53**, 7403 (1996), arXiv:cond-mat/9603001.

[274] A. Lucas and K. C. Fong, *J. Phys. Condens. Matter* **30**, 053001 (2018), arXiv:1710.08425.

[275] L. Fritz, J. Schmalian, M. Müller and S. Sachdev, *Phys. Rev. B* **78**, 085416 (2008), arXiv:0802.4289.

[276] M. Müller, L. Fritz and S. Sachdev, *Phys. Rev. B* **78**, 115406 (2008), arXiv:0805.1413.

[277] M. Müller, J. Schmalian and L. Fritz, *Phys. Rev. Lett.* **103**, 025301 (2009), arXiv:0903.4178.

[278] P. K. Kovtun, D. T. Son and A. O. Starinets, *Phys. Rev. Lett.* **94**, 111601 (2005), arXiv:hep-th/0405231.

[279] M. Mendoza, H. J. Herrmann and S. Succi, *Phys. Rev. Lett.* **106**, 156601 (2011), arXiv:1201.6590.

[280] A. Gabbana *et al.*, *Phys. Rev. Lett.* **121**, 236602 (2018), arXiv:1807.07117.

[281] I. Gradshteyn and I. Ryzhik, *Table of Integrals, Series, and Products.* Academic Press (2014).

[282] G. P. Mikitik and Y. V. Sharlai, *Phys. Rev. Lett.* **82**, 2147 (1999).

[283] M. F. Atiyah and I. M. Singer, *Bull. Amer. Math. Soc.* **69**, 422 (1963).

[284] Y. Aharonov and A. Casher, *Phys. Rev. A* **19**, 2461 (1979).

[285] L. Landau and E. Lifshitz, *Quantum Mechanics: Non-relativistic Theory.* Pergamon (1977).

[286] R. Prange and S. Girvin, *The Quantum Hall Effect.* Springer-Verlag (1990).

[287] D. Shoenberg, *Magnetic Oscillations in Metals.* Cambridge University Press (1984).

[288] A. Abrikosov, *Fundamentals of the Theory of Metals.* Courier Dover Publications (2017).

[289] I. Lifshits, M. Azbel and M. Kaganov, *Electron Theory of Metals.* Springer US (1973).

[290] L. A. Ponomarenko *et al.*, *Phys. Rev. Lett.* **105**, 136801 (2010), arXiv:1005.4793.

[291] S. G. Sharapov, V. P. Gusynin and H. Beck, *Phys. Rev. B* **69**, 075104 (2004), arXiv:cond-mat/0308216.

[292] N. M. R. Peres, F. Guinea and A. H. Castro Neto, *Phys. Rev. B* **73**, 125411 (2006), arXiv:cond-mat/0512091.

[293] K. von Klitzing, G. Dorda and M. Pepper, *Phys. Rev. Lett.* **45**, 494 (1980).

[294] T. J. B. M. Janssen *et al.*, *Metrologia* **49**, 294 (2012), arXiv:1202.2985.

[295] D. J. Thouless, M. Kohmoto, M. P. Nightingale and M. den Nijs, *Phys. Rev. Lett.* **49**, 405 (1982).

[296] Y. Zheng and T. Ando, *Phys. Rev. B* **65**, 245420 (2002).

[297] B. Simon, *Phys. Rev. Lett.* **51**, 2167 (1983).

[298] Y. Zhang *et al.*, *Phys. Rev. Lett.* **96**, 136806 (2006), arXiv:cond-mat/0602649.

[299]   D. A. Abanin *et al.*, *Phys. Rev. Lett.* **98**, 196806 (2007), arXiv:cond-mat/0702125.

[300]   Z. Jiang, Y. Zhang, H. L. Stormer and P. Kim, *Phys. Rev. Lett.* **99**, 106802 (2007), arXiv:0705.1102.

[301]   J. G. Checkelsky, L. Li and N. P. Ong, *Phys. Rev. Lett.* **100**, 206801 (2008), arXiv:0708.1959.

[302]   A. J. M. Giesbers *et al.*, *Phys. Rev. B* **80**, 201403 (2009), arXiv:0904.0948.

[303]   J. G. Checkelsky, L. Li and N. P. Ong, *Phys. Rev. B* **79**, 115434 (2009), arXiv:0808.0906.

[304]   X. Du, I. Skachko *et al.*, *Nature* **462**, 192 (2009), arXiv:0910.2532.

[305]   K. I. Bolotin *et al.*, *Nature* **462**, 196 (2009), arXiv:0910.2763.

[306]   L. Zhang *et al.*, *Phys. Rev. Lett.* **105**, 046804 (2010), arXiv:1003.2738.

[307]   Y. Zhao, P. Cadden-Zimansky, F. Ghahari and P. Kim, *Phys. Rev. Lett.* **108**, 106804 (2012), arXiv:1201.4434.

[308]   A. F. Young *et al.*, *Nat. Phys.* **8**, 550 (2012), arXiv:1201.4167.

[309]   A. F. Young *et al.*, *Nature* **505**, 528 (2013), arXiv:1307.5104.

[310]   F. Chiappini *et al.*, *Phys. Rev. B* **92**, 201412 (2015), arXiv:1511.08384.

[311]   J. Schwinger, *Phys. Rev.* **82**, 664 (1951).

[312]   V. P. Gusynin, V. A. Miransky and I. A. Shovkovy, *Phys. Rev. Lett.* **73**, 3499 (1994); arXiv:hep-ph/9405262; Erratum: Ibid., *Phys. Rev. Lett.* **76**, 1005 (1996).

[313]   K. Nomura and A. H. MacDonald, *Phys. Rev. Lett.* **96**, 256602 (2006), arXiv:cond-mat/0604113.

[314]   D. A. Abanin, P. A. Lee and L. S. Levitov, *Phys. Rev. Lett.* **96**, 176803 (2006), arXiv:cond-mat/0602645.

[315]   M. O. Goerbig, R. Moessner and B. Douçot, *Phys. Rev. B* **74**, 161407 (2006), arXiv:cond-mat/0604554.

[316]   J. Alicea and M. P. A. Fisher, *Phys. Rev. B* **74**, 075422 (2006), arXiv:cond-mat/0604601.

[317]   L. Sheng, D. N. Sheng, F. D. M. Haldane and L. Balents, *Phys. Rev. Lett.* **99**, 196802 (2007), arXiv:0706.0371.

[318]   G. W. Semenoff and F. Zhou, *J. High. Energy Phys.* **07**, 037 (2011), arXiv:1104.4714.

[319]   D. V. Khveshchenko, *Phys. Rev. Lett.* **87**, 246802 (2001), arXiv:cond-mat/0101306.

[320]   V. P. Gusynin, V. A. Miransky, S. G. Sharapov and I. A. Shovkovy, *Phys. Rev. B* **74**, 195429 (2006), arXiv:cond-mat/0605348.

[321]   M. Ezawa, *Physica E: Low Dimens. Syst. Nanostruct.* **40**, 269 (2007), arXiv:cond-mat/0609612.

[322]   E. V. Gorbar, V. P. Gusynin, V. A. Miransky and I. A. Shovkovy, *Phys. Rev. B* **78**, 085437 (2008), arXiv:0806.0846.

[323]   I. F. Herbut, *Phys. Rev. Lett.* **97**, 146401 (2006), arXiv:cond-mat/0606195.

[324]   I. F. Herbut, *Phys. Rev. B* **75**, 165411 (2007), arXiv:cond-mat/0610349.

[325]   I. F. Herbut, *Phys. Rev. B* **76**, 085432 (2007), arXiv:0705.4039.

[326] I. F. Herbut, V. Juričić and B. Roy, *Phys. Rev. B* **79**, 085116 (2009), arXiv:0811.0610.

[327] E. V. Gorbar, V. P. Gusynin, V. A. Miransky and I. A. Shovkovy, *Phys. Scripta T* **146**, 014018 (2012), arXiv:1105.1360.

[328] I. A. Shovkovy and L. Xia, *Phys. Rev. B* **93**, 035454 (2016), arXiv:1508.04471.

[329] B. I. Halperin, *Phys. Rev. B* **25**, 2185 (1982).

[330] H. Kleinert, *Gauge Fields in Condensed Matter*. World Scientific (1989).

[331] M. Vozmediano, M. Katsnelson and F. Guinea, *Phys. Rep.* **496**, 109 (2010), arXiv:1003.5179.

[332] H. Suzuura and T. Ando, *Phys. Rev. B* **65**, 235412 (2002).

[333] K.-i. Sasaki, Y. Kawazoe and R. Saito, *Prog. Theor. Phys.* **113**, 463 (2005), arXiv:cond-mat/0401317.

[334] M. Katsnelson and K. Novoselov, *Solid State Commun.* **143**, 3 (2007), arXiv:cond-mat/0703374.

[335] M. Ramezani Masir, D. Moldovan and F. Peeters, *Solid State Commun.* **175–176**, 76 (2013), arXiv:1304.0629.

[336] J. L. Mañes, F. de Juan, M. Sturla and M. A. H. Vozmediano, *Phys. Rev. B* **88**, 155405 (2013), arXiv:1308.1595.

[337] J. L. Mañes, *Phys. Rev. B* **76**, 045430 (2007), arXiv:cond-mat/0702465.

[338] F. Guinea, M. I. Katsnelson and A. K. Geim, *Nat. Phys.* **6**, 30 (2009), arXiv:0909.1787.

[339] F. Guinea, A. K. Geim, M. I. Katsnelson and K. S. Novoselov, *Phys. Rev. B* **81**, 035408 (2010), arXiv:0910.5935.

[340] F. Liu, P. Ming and J. Li, *Phys. Rev. B* **76**, 064120 (2007).

[341] N. Levy *et al.*, *Science* **329**, 544 (2010).

[342] V. M. Pereira and A. H. Castro Neto, *Phys. Rev. Lett.* **103**, 046801 (2009), arXiv:1310.3622.

[343] M. M. Fogler, F. Guinea and M. I. Katsnelson, *Phys. Rev. Lett.* **101**, 226804 (2008), arXiv:0807.3165.

[344] S. V. Morozov *et al.*, *Phys. Rev. Lett.* **97**, 016801 (2006), arXiv:cond-mat/0603826.

[345] M. I. Katsnelson and A. K. Geim, *Philos. Trans. R. Soc. A* **366**, 195 (2008), arXiv:0706.2490.

[346] A. Cortijo and M. A. H. Vozmediano, *Phys. Rev. B* **79**, 184205 (2009), arXiv:0709.2698.

[347] F. Guinea, M. I. Katsnelson and M. A. H. Vozmediano, *Phys. Rev. B* **77**, 075422 (2008), arXiv:0707.0682.

[348] D. J. Thouless, *Topological Quantum Numbers In Nonrelativistic Physics*. World Scientific (1998).

[349] L. Schubnikow and W. de Haas, *Proc. R. Neth. Acad. Arts Sci.* **33**, 130 (1930).

[350] L. Schubnikow and W. de Haas, *Proc. R. Neth. Acad. Arts Sci.* **33**, 363 (1930).

[351] L. Schubnikow and W. de Haas, *Proc. R. Neth. Acad. Arts Sci.* **33**, 418 (1930).

[352] L. Schubnikow and W. de Haas, *Proc. R. Neth. Acad. Arts Sci.* **33**, 433 (1930).

[353] W. de Haas and P. van Alphen, *Proc. Acad. Sci. Amsterdam* **33**, 1106 (1930).

[354] F. Bloch, *Z. Phys.* **52**, 555 (1928).

[355] C. Herring, *Phys. Rev.* **57**, 1169 (1940).

[356] N. W. Ashcroft and N. D. Mermin, *Solid State Physics*. Cengage Learning (1976).

[357] T. Cheng and L. Li, *Gauge Theory of Elementary Particle Physics*. Oxford University Press (2000).

[358] I. Lifshitz, *Soviet Phys. JETP* **11**, 1130 (1960).

[359] T. M. McCormick, I. Kimchi and N. Trivedi, *Phys. Rev. B* **95**, 075133 (2017), arXiv:1604.03096.

[360] D. Hilbert and S. Cohn-Vossen, *Geometry and the Imagination*. AMS Chelsea Pub. (1999).

[361] G. Xu *et al.*, *Phys. Rev. Lett.* **107**, 186806 (2011), arXiv:1106.3125.

[362] S.-M. Huang *et al.*, *Phys. Rev. Lett.* **113**, 1180 (2016), arXiv:1503.05868.

[363] H. Kramers, *Proc. Amsterdam Acad.* **33**, 959 (1930).

[364] S. Murakami, *New J. Phys.* **9**, 356 (2007), arXiv:0710.0930.

[365] S. Murakami, *Physica E: Low Dimens. Syst. Nanostruct.* **43**, 748 (2011), arXiv:1006.1188.

[366] S. M. Young *et al.*, *Phys. Rev. Lett.* **108**, 140405 (2012), arXiv:1111.6483.

[367] B.-J. Yang and N. Nagaosa, *Nat. Commun.* **5**, 4898 (2014), arXiv:1404.0754.

[368] X.-G. Wen, *Quantum Field Theory of Many-Body Systems: From the Origin of Sound to an Origin of Light and Electrons*. Oxford University Press (2007).

[369] E. Hall, *Am. J. Math.* **2**, 287 (1879).

[370] E. Hall, *Phil. Mag.* **12**, 157 (1881).

[371] N. Nagaosa *et al.*, *Rev. Mod. Phys.* **82**, 1539 (2010), arXiv:0904.4154.

[372] R. Karplus and J. M. Luttinger, *Phys. Rev.* **95**, 1154 (1954).

[373] F. Wilczek and A. Zee, *Phys. Rev. Lett.* **52**, 2111 (1984).

[374] R. Shindou and K.-I. Imura, *Nucl. Phys. B* **720**, 399 (2005), arXiv: cond-mat/0411105.

[375] D. Culcer, Y. Yao and Q. Niu, *Phys. Rev. B* **72**, 085110 (2005), arXiv:cond-mat/0411285.

[376] M.-C. Chang and Q. Niu, *J. Phys.: Condens. Matter* **20**, 193202 (2008).

[377] D. Xiao, M.-C. Chang and Q. Niu, *Rev. Mod. Phys.* **82**, 1959 (2010), arXiv:0907.2021.

[378] E. V. Gorbar, V. A. Miransky, I. A. Shovkovy and P. O. Sukhachov, *Phys. Rev. B* **98**, 045203 (2018), arXiv:1805.03222.

[379] M.-C. Chang and Q. Niu, *Phys. Rev. Lett.* **75**, 1348 (1995), arXiv:cond-mat/9511014.

[380] M.-C. Chang and Q. Niu, *Phys. Rev. B* **53**, 7010 (1996), arXiv:cond-mat/9511014.

[381]   D. Xiao, J. Shi and Q. Niu, *Phys. Rev. Lett.* **95**, 137204 (2005), arXiv:cond-mat/0502340.

[382]   P. Dirac, *Proc. R. Soc. Lond. A* **133**, 60 (1931).

[383]   M. Nakahara, *Geometry, Topology and Physics.* CRC Press (2003).

[384]   T. T. Wu and C. N. Yang, *Phys. Rev. D* **12**, 3845 (1975).

[385]   P. Hosur, S. A. Parameswaran and A. Vishwanath, *Phys. Rev. Lett.* **108**, 046602 (2012), arXiv:1109.6330.

[386]   H. Isobe and N. Nagaosa, *Phys. Rev. B* **86**, 165127 (2012), arXiv:1205. 2427.

[387]   H. Isobe and N. Nagaosa, *Phys. Rev. B* **87**, 205138 (2013), arXiv:1303. 2822.

[388]   R. E. Throckmorton, J. Hofmann, E. Barnes and S. Das Sarma, *Phys. Rev. B* **92**, 115101 (2015), arXiv:1505.05154.

[389]   K.-Y. Yang, Y.-M. Lu and Y. Ran, *Phys. Rev. B* **84**, 075129 (2011), arXiv:1105.2353.

[390]   M. M. Vazifeh and M. Franz, *Phys. Rev. Lett.* **111**, 027201 (2013), arXiv:1303.5784.

[391]   A. A. Burkov and L. Balents, *Phys. Rev. Lett.* **107**, 127205 (2011), arXiv:1105.5138.

[392]   C. Kittel, *Quantum Theory of Solids.* Wiley (1987).

[393]   L. Voon and M. Willatzen, *The k · p Method: Electronic Properties of Semiconductors.* Springer-Verlag (2009).

[394]   B.-J. Yang, T. Morimoto and A. Furusaki, *Phys. Rev. B* **92**, 165120 (2015), arXiv:1404.0754.

[395]   E. V. Gorbar, V. A. Miransky, I. A. Shovkovy and P. O. Sukhachov, *Phys. Rev. B* **91**, 121101 (2015), arXiv:1412.5194.

[396]   C. Fang, Y. Chen, H.-Y. Kee and L. Fu, *Phys. Rev. B* **92**, 081201 (2015), arXiv:1506.03449.

[397]   S. Murakami *et al.*, *Phys. Rev. B* **76**, 205304 (2007), arXiv:0705.3696.

[398]   S. Murakami and S.-i. Kuga, *Phys. Rev. B* **78**, 165313 (2008), arXiv:0806. 3309.

[399]   J. A. Steinberg *et al.*, *Phys. Rev. Lett.* **112**, 036403 (2014), arXiv:1309. 5967.

[400]   J. E. Moore and L. Balents, *Phys. Rev. B* **75**, 121306 (2007), arXiv:cond-mat/0607314.

[401]   J.-H. Zhou, H. Jiang, Q. Niu and J.-R. Shi, *Chin. Physics Lett.* **30**, 027101 (2013), arXiv:1211.0772.

[402]   M. Zubkov, *Ann. Phys. (N. Y.)* **360**, 655 (2015), arXiv:1501.04998.

[403]   A. Cortijo, Y. Ferreirós, K. Landsteiner and M. A. H. Vozmediano, *Phys. Rev. Lett.* **115**, 177202 (2015), arXiv:1603.02674.

[404]   H. Shapourian, T. L. Hughes and S. Ryu, *Phys. Rev. B* **92**, 165131 (2015), arXiv:1505.03868.

[405]   A. Cortijo, D. Kharzeev, K. Landsteiner and M. A. H. Vozmediano, *Phys. Rev. B* **94**, 241405 (2016), arXiv:1607.03491.

[406]   A. Zabolotskiy and Y. Lozovik, *J. Magn. Magn. Mater.* **459**, 43 (2018), arXiv:1707.02781.

[407] E. J. Sie *et al.*, *Nature* **565**, 61 (2019).

[408] D. I. Pikulin, A. Chen and M. Franz, *Phys. Rev. X* **6**, 041021 (2016), arXiv:1607.01810.

[409] Z. Song, J. Zhao, Z. Fang and X. Dai, *Phys. Rev. B* **94**, 214306 (2016), arXiv:1609.05442.

[410] P. Rinkel, P. L. S. Lopes and I. Garate, *Phys. Rev. Lett.* **119**, 107401 (2017), arXiv:1610.03073.

[411] P. O. Sukhachov and H. Rostami, *Phys. Rev. Lett.* **124**, 126602 (2020), arXiv:1911.04526.

[412] R. Ilan, A. G. Grushin and D. I. Pikulin, *Nat. Rev. Phys.* (2019), arXiv:1903.11088.

[413] A. G. Grushin, J. W. F. Venderbos, A. Vishwanath and R. Ilan, *Phys. Rev. X* **6**, 041046 (2016), arXiv:1607.04268.

[414] M. N. Chernodub and M. A. Zubkov, *Phys. Rev. B* **95**, 115410 (2017), arXiv:1508.03114.

[415] Z.-M. Huang, J. Zhou and S.-Q. Shen, *Phys. Rev. B* **96**, 085201 (2017), arXiv:1705.04576.

[416] O. Parrikar, T. L. Hughes and R. G. Leigh, *Phys. Rev. D* **90**, 105004 (2014), arXiv:1407.7043.

[417] H. Sumiyoshi and S. Fujimoto, *Phys. Rev. Lett.* **116**, 166601 (2016), arXiv:1509.03981.

[418] Z. V. Khaidukov and M. A. Zubkov, *JETP Lett.* **108**, 670 (2018), arXiv:1812.00970.

[419] T. Liu, D. I. Pikulin and M. Franz, *Phys. Rev. B* **95**, 041201 (2017), arXiv:1608.04678.

[420] E. V. Gorbar, V. A. Miransky, I. A. Shovkovy and P. O. Sukhachov, *Phys. Rev. Lett.* **118**, 127601 (2017), arXiv:1610.07625.

[421] E. V. Gorbar, V. A. Miransky, I. A. Shovkovy and P. O. Sukhachov, *Phys. Rev. B* **95**, 115202 (2017), arXiv:1611.05470.

[422] E. V. Gorbar, V. A. Miransky, I. A. Shovkovy and P. O. Sukhachov, *Phys. Rev. B* **95**, 115422 (2017), arXiv:1612.06397.

[423] E. V. Gorbar, V. A. Miransky, I. A. Shovkovy and P. O. Sukhachov, *Phys. Rev. B* **95**, 205141 (2017), arXiv:1702.02950.

[424] E. V. Gorbar, V. A. Miransky, I. A. Shovkovy and P. O. Sukhachov, *Phys. Rev. B* **95**, 241114 (2017), arXiv:1703.03415.

[425] A. Weststrôm and T. Ojanen, *Phys. Rev. X* **7**, 041026 (2017), arXiv:1703.10408.

[426] R. Soto-Garrido and E. Muñoz, *J. Phys.: Condens. Matter* **30**, 195302 (2018), arXiv:1803.10272.

[427] A. Cortijo and M. Zubkov, *Ann. Phys. (N. Y.)* **366**, 45 (2016), arXiv:1508.04462.

[428] C.-X. Liu, P. Ye and X.-L. Qi, *Phys. Rev. B* **87**, 235306 (2013), arXiv:1204.6551.

[429] Y. Araki and K. Nomura, *Phys. Rev. Appl.* **10**, 014007 (2018), arXiv:1711.03135.

[430] V. Arjona and M. A. H. Vozmediano, *Phys. Rev. B* **97**, 201404 (2018), arXiv:1709.02394.

[431] L. Landau and E. Lifshitz, *Theory of Elasticity*. Butterworth-Heinemann (2012).

[432] E. V. Gorbar, V. A. Miransky, I. A. Shovkovy and P. O. Sukhachov, *Phys. Rev. B* **96**, 085130 (2017a), arXiv:1706.02705.

[433] E. V. Gorbar, V. A. Miransky, I. A. Shovkovy and P. O. Sukhachov, *Phys. Rev. B* **96**, 125123 (2017b), arXiv:1706.09419.

[434] P. O. Sukhachov, E. V. Gorbar, I. A. Shovkovy and V. A. Miransky, *Annalen Phys.* **530**, 1800219 (2018), arXiv:1806.03302.

[435] D. Shao *et al.*, *Phys. Rev. B* **96**, 075112 (2017), arXiv:1708.04094.

[436] J. Mutch *et al.*, *Sci. Adv.* **5** (2019), arXiv:1808.07898.

[437] H.-H. Lai, S. E. Grefe, S. Paschen and Q. Si, *Proc. Natl. Acad. Sci. USA* **115**, 93 (2018), arXiv:1612.03899.

[438] P. E. Ashby and J. P. Carbotte, *Eur. Phys. J. B* **87**, 92 (2014), arXiv:1310.2223.

[439] H. Murakawa *et al.*, *Science* **342**, 1490 (2013).

[440] C. M. Wang, H.-Z. Lu and S.-Q. Shen, *Phys. Rev. Lett.* **117**, 077201 (2016), arXiv:1604.01681.

[441] A. Alexandradinata, C. Wang, W. Duan and L. Glazman, *Phys. Rev. X* **8**, 011027 (2018), arXiv:1707.08586.

[442] D. Cangemi and G. Dunne, *Ann. Phys. (N. Y.)* **249**, 582 (1996), arXiv:hep-th/9601048.

[443] H. Bateman and A. Erdélyi, *Higher Transcendental Functions*, Vol. 1. McGraw-Hill (1953).

[444] I. M. Lifshitz and A. M. Kosevich, *Sov. Phys. JETP* **2**, 636 (1956).

[445] L. P. He *et al.*, *Phys. Rev. Lett.* **113**, 246402 (2014), arXiv:1404.2557.

[446] A. Narayanan *et al.*, *Phys. Rev. Lett.* **114**, 117201 (2015), arXiv:1412.4105.

[447] J. Cao *et al.*, *Nat. Commun.* **6**, 7779 (2015), arXiv:1412.0824.

[448] Y. Zhao *et al.*, *Phys. Rev. X* **5**, 031037 (2015), arXiv:1412.0330.

[449] X. Yang *et al.*, arXiv:1506.02283 (2015).

[450] Y. Luo *et al.*, *Phys. Rev. B* **92**, 205134 (2015), arXiv:1506.01751.

[451] C. Zhang *et al.*, *Phys. Rev. B* **92**, 041203 (2015), arXiv:1507.01298.

[452] J. Hu *et al.*, *Sci. Rep.* **6**, 18674 (2016), arXiv:1507.08346.

[453] P. Sergelius *et al.*, *Sci. Rep.* **6**, 33859 (2016).

[454] F. Arnold *et al.*, *Nat. Commun.* **7**, 11615 (2016), arXiv:1506.06577.

[455] P. J. W. Moll *et al.*, *Nat. Commun.* **7**, 12492 (2016), arXiv:1507.06981.

[456] C.-L. Zhang *et al.*, *Phys. Rev. B* **95**, 085202 (2017), arXiv:1702.01245.

[457] S. Dzsaber *et al.*, *Phys. Rev. Lett.* **118**, 246601 (2017), arXiv:1612.03972.

[458] C. Y. Guo *et al.*, *Nat. Commun.* **9**, 4622 (2018), arXiv:1710.05522.

[459] H. Weng, X. Dai and Z. Fang, *MRS Bull.* **39**, 849 (2014), arXiv:1410.4614.

[460] H. Bonzel and C. Kleint, *Prog. Surf. Sci.* **49**, 107 (1995).

[461] A. Damascelli, Z. Hussain and Z.-X. Shen, *Rev. Mod. Phys.* **75**, 473 (2003), arXiv:cond-mat/0208504.

[462] F. Reinert and S. Hüfner, *New J. Phys.* **7**, 97 (2005).

[463] W. Schattke and M. Van Hove, *Solid-State Photoemission and Related Methods: Theory and Experiment.* Wiley (2008).

[464] S. Kevan, *Angle-Resolved Photoemission: Theory and Current Applications.* Elsevier Science (1992).

[465] A. Einstein, *Ann. Phys. (Berl.)* **322**, 132 (1905).

[466] K. Oura *et al.*, *Surface Science: An Introduction.* Springer-Verlag (2003).

[467] D. Bonnell, *Scanning Probe Microscopy and Spectroscopy: Theory, Techniques, and Applications.* Wiley (2001).

[468] H. J. Zandvliet and A. van Houselt, *Annu. Rev. Anal. Chem.* **2**, 37 (2009).

[469] R. Peierls, *Ann. Phys. (Berl.)* **396**, 121 (1930).

[470] F. Bloch, *Z. Phys.* **52**, 555 (1929).

[471] P. Drude, *Ann. Phys. (Berl.)* **306**, 566 (1900).

[472] P. Drude, *Ann. Phys. (Berl.)* **308**, 369 (1900).

[473] A. Sommerfeld, *Z. Phys.* **47**, 1 (1928).

[474] D. R. Hartree, *Math. Proc. Cambridge Philos. Soc.* **24**, 89 (1928).

[475] V. Fock, *Z. Phys.* **61**, 126 (1930).

[476] J. C. Slater, *Phys. Rev.* **35**, 210 (1930).

[477] P. Hohenberg and W. Kohn, *Phys. Rev.* **136**, B864 (1964).

[478] R. Martin, *Electronic Structure: Basic Theory and Practical Methods.* Cambridge University Press (2008).

[479] D. Sholl and J. Steckel, *Density Functional Theory: A Practical Introduction.* Wiley (2009).

[480] W. Kohn and L. J. Sham, *Phys. Rev.* **140**, A1133 (1965).

[481] M. Marder, *Condensed Matter Physics.* Wiley (2010).

[482] J. C. Phillips and L. Kleinman, *Phys. Rev.* **116**, 287 (1959).

[483] P. E. Blöchl, *Phys. Rev. B* **50**, 17953 (1994).

[484] A. A. Soluyanov and D. Vanderbilt, *Phys. Rev. B* **83**, 035108 (2011), arXiv:1009.1415.

[485] Z. Ringel and Y. E. Kraus, *Phys. Rev. B* **83**, 245115 (2011), arXiv:1010.5357.

[486] R. Yu *et al.*, *Phys. Rev. B* **84**, 075119 (2011), arXiv:1101.2011.

[487] N. Marzari and D. Vanderbilt, *Phys. Rev. B* **56**, 12847 (1997), arXiv:cond-mat/9707145.

[488] I. Souza, N. Marzari and D. Vanderbilt, *Phys. Rev. B* **65**, 035109 (2001), arXiv:cond-mat/0108084.

[489] N. Marzari *et al.*, *Rev. Mod. Phys.* **84**, 1419 (2012), arXiv:1112.5411.

[490] G. H. Wannier, *Phys. Rev.* **52**, 191 (1937).

[491] X. Wang, J. R. Yates, I. Souza and D. Vanderbilt, *Phys. Rev. B* **74**, 195118 (2006), arXiv:cond-mat/0608257.

[492] J. R. Yates, X. Wang, D. Vanderbilt and I. Souza, *Phys. Rev. B* **75**, 195121 (2007), arXiv:cond-mat/0702554.

[493] S. Yip, *Handbook of Materials Modeling.* Springer Netherlands (2005).

[494] W. Harrison, *Electronic Structure and the Properties of Solids: The Physics of the Chemical Bond.* Dover Publications (1989).

[495]   M. P. L. Sancho, J. M. L. Sancho and J. Rubio, *J. Phys. F* **14**, 1205 (1984).

[496]   M. P. L. Sancho, J. M. L. Sancho and J. Rubio, *J. Phys. F* **15**, 851 (1985).

[497]   E. M. Godfrin, *J. Phys.: Condens. Matter* **3**, 7843 (1991).

[498]   G. Kresse and J. Hafner, *Phys. Rev. B* **48**, 13115 (1993).

[499]   G. Kresse and J. Furthmüller, *Comput. Mater. Sci.* **6**, 15 (1996).

[500]   G. Kresse and J. Furthmüller, *Phys. Rev. B* **54**, 11169 (1996).

[501]   G. Kresse *et al.*, Vienna Ab initio Simulation Package, www.vasp.at.

[502]   T. Ozaki *et al.*, OpenMX, www.openmx-square.org.

[503]   P. Blaha, K. Schwarz, P. Sorantin and S. Trickey, *Comput. Phys. Commun.* **59**, 399 (1990).

[504]   K. Schwarz and P. Blaha, *Comput. Mater. Sci.* **28**, 259 (2003).

[505]   P. Giannozzi *et al.*, *J. Phys. Condens. Matter* **21**, 395502 (2009), arXiv: 0906.2569.

[506]   P. Giannozzi *et al.*, *J. Phys. Condens. Matter* **29**, 465901 (2017), arXiv: 1709.10010.

[507]   Quantum ESPRESSO, www.quantum-espresso.org.

[508]   K. Koepernik and H. Eschrig, *Phys. Rev. B* **59**, 1743 (1999).

[509]   K. Koepernik *et al.*, FPLO, www.fplo.de.

[510]   Q.-S. Wu *et al.*, *Comput. Phys. Commun.* **224**, 405 (2018), arXiv:1703. 07789.

[511]   C. W. Groth, M. Wimmer, A. R. Akhmerov and X. Waintal, *New J. Phys.* **16**, 063065 (2014), arXiv:1309.2926.

[512]   Q. D. Gibson *et al.*, *Phys. Rev. B* **91**, 205128 (2015), arXiv:1411.0005.

[513]   J. C. Y. Teo, L. Fu and C. L. Kane, *Phys. Rev. B* **78**, 045426 (2008).

[514]   T. Sato *et al.*, *Nat. Phys.* **7**, 840 (2011), arXiv:1205.3654.

[515]   S.-Y. Xu *et al.*, *Science* **332**, 560 (2011), arXiv:1104.4633.

[516]   S. Souma *et al.*, *Phys. Rev. Lett.* **109**, 186804 (2012), arXiv:1204.5812.

[517]   B. Singh *et al.*, *Phys. Rev. B* **86**, 115208 (2012), arXiv:1209.5896.

[518]   M. Brahlek *et al.*, *Phys. Rev. Lett.* **109**, 186403 (2012), arXiv:1209.2840.

[519]   L. Wu *et al.*, *Nat. Phys.* **9**, 410 (2013), arXiv:1209.3290.

[520]   P. Dziawa *et al.*, *Nat. Mater.* **11**, 1023 (2012), arXiv:1206.1705.

[521]   S.-Y. Xu *et al.*, *Nat. Commun.* **3**, 1192 (2012), arXiv:1206.2088.

[522]   J. A. Steinberg *et al.*, *Phys. Rev. Lett.* **112**, 036403 (2014), arXiv: 1309.5967.

[523]   G. F. Koster, J. D. Dimmock, R. G. Wheeler and H. Statz, *Properties of the Thirty-Two Point Groups*. MIT Press (1963).

[524]   G. Brauer and E. Zintl, *Z. Phys. Chem. B* **37**, 323 (1937).

[525]   J. Heyd, G. E. Scuseria and M. Ernzerhof, *J. Chem. Phys.* **118**, 8207 (2003).

[526]   C.-X. Liu *et al.*, *Phys. Rev. B* **82**, 045122 (2010), arXiv:1005.1682.

[527]   J. Cano *et al.*, *Phys. Rev. B* **95**, 161306 (2017), arXiv:1604.08601.

[528]   W. Zdanowicz and L. Zdanowicz, *Annu. Rev. Mater. Sci.* **5**, 301 (1975).

[529]   H. Huang, S. Zhou and W. Duan, *Phys. Rev. B* **94**, 121117 (2016), arXiv:1607.07965.

[530] T.-R. Chang *et al.*, *Phys. Rev. Lett.* **119**, 026404 (2017), arXiv:1606.07555.

[531] C. Chen *et al.*, *Phys. Rev. Materials* **1**, 044201 (2017).

[532] C. Le *et al.*, *Phys. Rev. B* **96**, 115121 (2017).

[533] P.-J. Guo, H.-C. Yang, K. Liu and Z.-Y. Lu, *Phys. Rev. B* **95**, 155112 (2017), arXiv:1612.07456.

[534] S. Borisenko *et al.*, *Nat. Commun.* **10**, 3424 (2019), arXiv:1507.04847.

[535] D. Chaudhuri *et al.*, *Phys. Rev. B* **96**, 075151 (2017), arXiv:1701.08693.

[536] G. Chang *et al.*, *Sci. Rep.* **6**, 38839 (2016), arXiv:1603.01255.

[537] Z. Wang *et al.*, *Phys. Rev. Lett.* **117**, 236401 (2016), arXiv:1603.00479.

[538] M. Hirschberger *et al.*, *Nat. Mater.* **15**, 1161 (2016), arXiv:1602.07219.

[539] T. Suzuki *et al.*, *Nat. Phys.* **12**, 1119 (2016).

[540] K. Manna *et al.*, *Nat. Rev. Mater.* **3**, 244 (2018), arXiv:1802.03771.

[541] H. Yang *et al.*, *New J. Phys.* **19**, 015008 (2017), arXiv:1608.03404.

[542] Y. Zhang *et al.*, *Phys. Rev. B* **95**, 075128 (2017), arXiv:1610.04034.

[543] E. Liu *et al.*, *Nat. Phys.* **14**, 1125 (2018), arXiv:1712.06722.

[544] Q. Xu *et al.*, *Phys. Rev. B* **97**, 235416 (2018), arXiv:1801.00136.

[545] Q. Wang *et al.*, *Nat. Commun.* **9**, 3681 (2018), arXiv:1712.09947.

[546] Y. Xu *et al.*, arXiv:1908.04561 (2019).

[547] A. A. Burkov, M. D. Hook and L. Balents, *Phys. Rev. B* **84**, 235126 (2011), arXiv:1110.1089.

[548] E. V. Gorbar, V. A. Miransky and I. A. Shovkovy, *Phys. Rev. B* **88**, 165105 (2013), arXiv:1307.6230.

[549] L.-L. Wang *et al.*, *Phys. Rev. B* **99**, 245147 (2019), arXiv:1901.08234.

[550] H. Su *et al.*, arXiv:1903.12532 (2019).

[551] E. V. Gorbar, V. A. Miransky and I. A. Shovkovy, *Phys. Rev. D* **83**, 085003 (2011), arXiv:1101.4954.

[552] L.-L. Wang *et al.*, *Phys. Rev. B* **99**, 245147 (2019), arXiv:1901.08234.

[553] J.-R. Soh *et al.*, *Phys. Rev. B* **100**, 201102 (2019), arXiv:1901.10022.

[554] J.-Z. Ma *et al.*, *Sci. Adv.* **5** (2019), arXiv:1907.05956.

[555] G. B. Halász and L. Balents, *Phys. Rev. B* **85**, 035103 (2012), arXiv:1109.6137.

[556] A. A. Zyuzin, S. Wu and A. A. Burkov, *Phys. Rev. B* **85**, 165110 (2012), arXiv:1201.3624.

[557] J. Liu and D. Vanderbilt, *Phys. Rev. B* **90**, 155316 (2014), arXiv:1409.6399.

[558] M. Hirayama *et al.*, *Phys. Rev. Lett.* **114**, 206401 (2015), arXiv:1409.7517.

[559] S. Souma *et al.*, *Phys. Rev. B* **93**, 161112 (2016), arXiv:1510.01503.

[560] Y. Sun *et al.*, *Phys. Rev. B* **92**, 161107 (2015), arXiv:1508.03501.

[561] Z. Wang *et al.*, *Phys. Rev. Lett.* **117**, 056805 (2016), arXiv:1511.07440.

[562] L. Huang *et al.*, *Nat. Mater.* **15**, 1155 (2016), arXiv:1603.06482.

[563] G. Autès *et al.*, *Phys. Rev. Lett.* **117**, 066402 (2016), arXiv:1603.04624.

[564] T.-R. Chang *et al.*, *Nat. Commun.* **7**, 10639 (2016a), arXiv:1508.06723.

[565] G. Chang *et al.*, *Sci. Adv.* **2**, e1600295 (2016b), arXiv:1512.08781.

[566] K. Koepernik *et al.*, *Phys. Rev. B* **93**, 201101 (2016), arXiv:1603.04323.

[567]   G. Chang *et al.*, *Phys. Rev. B* **97**, 041104 (2018), arXiv:1604.02124.

[568]   A. Damascelli, *Phys. Scr. T* **109**, 61 (2004), arXiv:cond-mat/0307085.

[569]   L. Simon, F. Vonau and D. Aubel, *J. Phys.: Condens. Matter* **19**, 355009 (2007).

[570]   A. Gyenis *et al.*, *New J. Phys.* **18**, 105003 (2016), arXiv:1610.07197.

[571]   S. Kourtis *et al.*, *Phys. Rev. B* **93**, 041109 (2016), arXiv:1512.02646.

[572]   A. K. Mitchell and L. Fritz, *Phys. Rev. B* **93**, 035137 (2016), arXiv:1512.06392.

[573]   S. Jeon *et al.*, *Nat. Mater.* **13**, 851 (2014), arXiv:1403.3446.

[574]   M. Yan *et al.*, *Nat. Commun.* **8**, 257 (2017), arXiv:1607.03643.

[575]   K. Zhang *et al.*, *Phys. Rev. B* **96**, 125102 (2017), arXiv:1703.04242.

[576]   H.-J. Noh *et al.*, *Phys. Rev. Lett.* **119**, 016401 (2017), arXiv:1612.06946.

[577]   H. F. Yang *et al.*, *Nat. Commun.* **10**, 3478 (2019), arXiv:1908.01155.

[578]   N. Morali *et al.*, *Science* **365**, 1286 (2019), arXiv:1903.00509.

[579]   B. Q. Lv *et al.*, *Phys. Rev. Lett.* **115**, 217601 (2015), arXiv:1510.07256.

[580]   G. Chang *et al.*, *Phys. Rev. Lett.* **116**, 066601 (2016).

[581]   R. Batabyal *et al.*, *Sci. Adv.* **2**, e1600709 (2016), arXiv:1603.00283.

[582]   H. Inoue *et al.*, *Science* **351**, 1184 (2016), arXiv:1603.03045.

[583]   H. Zheng *et al.*, *ACS Nano* **10**, 1378 (2016), arXiv:1511.02216.

[584]   K. Deng *et al.*, *Nat. Phys.* **12**, 1105 (2016), arXiv:1603.08508.

[585]   A. Liang *et al.*, arXiv:1604.01706 (2016).

[586]   A. Tamai *et al.*, *Phys. Rev. X* **6**, 031021 (2016), arXiv:1604.08228.

[587]   J. Jiang *et al.*, *Nat. Commun.* **8**, 13973 (2017), arXiv:1604.00139.

[588]   I. Belopolski *et al.*, *Phys. Rev. B* **94**, 085127 (2016), arXiv:1604.07079.

[589]   I. Belopolski *et al.*, *Nat. Commun.* **7**, 13643 (2016), arXiv:1612.05990.

[590]   I. Belopolski *et al.*, *Nat. Commun.* **8**, 942 (2017), arXiv:1610.02013.

[591]   E. Haubold *et al.*, *Phys. Rev. B* **95**, 241108 (2017), arXiv:1609.09549.

[592]   S.-Y. Xu *et al.*, *Sci. Adv.* **3** (2017), arXiv:1603.07318.

[593]   M.-Y. Yao *et al.*, *Phys. Rev. Lett.* **122**, 176402 (2019), arXiv:1904.03523.

[594]   S. G. Davison and M. Stęślicka, *Basic Theory of Surface States.* Clarendon Press (1996).

[595]   I. Tamm, *Phys. Z. Soviet Union* **1**, 733 (1932).

[596]   I. M. Lifshitz and S. I. Pekar, *Usp. Fiz. Nauk* **56**, 531 (1955).

[597]   C. Kittel, *Introduction to Solid State Physics.* Wiley (2004).

[598]   F. Reinert and S. Hufner, *New J. Phys.* **7**, 97 (2005).

[599]   W. Shockley, *Phys. Rev.* **56**, 317 (1939).

[600]   P. Hosur, *Phys. Rev. B* **86**, 195102 (2012), arXiv:1208.0027.

[601]   J. Behrends *et al.*, *Phys. Rev. B* **99**, 140201 (2019), arXiv:1807.06615.

[602]   Y. Aharonov and A. Casher, *Phys. Rev. A* **19**, 2461 (1979).

[603]   J. Behrends, R. Ilan and J. H. Bardarson, *Phys. Rev. Res.* **1**, 032028 (2019), arXiv:1906.08277.

[604]   R. Okugawa and S. Murakami, *Phys. Rev. B* **89**, 235315 (2014), arXiv:1402.7145.

[605]   E. V. Gorbar, V. A. Miransky, I. A. Shovkovy and P. O. Sukhachov, *Phys. Rev. B* **91**, 235138 (2015), arXiv:1503.07913.

[606] S. Li and A. V. Andreev, *Phys. Rev. B* **92**, 201107 (2015), arXiv: 1506.06803.

[607] E. V. Gorbar, V. A. Miransky, I. A. Shovkovy and P. O. Sukhachov, *Phys. Rev. B* **90**, 115131 (2014), arXiv:1407.1323.

[608] P. Baireuther, J. A. Hutasoit, J. Tworzydło and C. W. J. Beenakker, *New J. Phys.* **18**, 045009 (2016), arXiv:1512.02144.

[609] N. Bovenzi *et al.*, *New J. Phys.* **20**, 023023 (2018), arXiv:1707.01038.

[610] E. Benito-Matías and R. A. Molina, *Phys. Rev. B* **99**, 075304 (2019), arXiv:1810.10448.

[611] V. Kaladzhyan and J. H. Bardarson, *Phys. Rev. B* **100**, 085424 (2019), arXiv:1905.11405.

[612] P. O. Sukhachov, M. V. Rakov, O. M. Teslyk and E. V. Gorbar, *Ann. Phys. (Berl.)*, 1900449 (2020), arXiv:1909.10587.

[613] M. Kargarian, M. Randeria and Y.-M. Lu, *Proc. Natl. Acad. Sci. USA* **113**, 8648 (2016), arXiv:1509.02180.

[614] J. González and R. A. Molina, *Phys. Rev. B* **96**, 045437 (2017), arXiv:1702.02521.

[615] R. B. Laughlin, *Phys. Rev. B* **23**, 5632 (1981).

[616] Z. Ezawa, *Quantum Hall Effects: Field Theoretical Approach and Related Topics*. World Scientific (2008).

[617] J. Alicea, *Rep. Prog. Phys.* **75**, 076501 (2012), arXiv:1202.1293.

[618] C. Beenakker, *Annu. Rev. Condens. Matter Phys.* **4**, 113 (2013), arXiv:1112.1950.

[619] T. D. Stanescu and S. Tewari, *J. Phys.: Condens. Matter* **25**, 233201 (2013), arXiv:1302.5433.

[620] A. P. Schnyder and P. M. R. Brydon, *J. Phys.: Condens. Matter* **27**, 243201 (2015), arXiv:1502.03746.

[621] M. Sato and S. Fujimoto, *J. Phys. Soc. Jpn.* **85**, 072001 (2016), arXiv: 1601.02726.

[622] M. Sato and Y. Ando, *Rep. Prog. Phys.* **80**, 076501 (2017), arXiv:1608. 03395.

[623] P. G. Derry, A. K. Mitchell and D. E. Logan, *Phys. Rev. B* **92**, 035126 (2015), arXiv:1503.04712.

[624] E. V. Gorbar, V. A. Miransky, I. A. Shovkovy and P. O. Sukhachov, *Phys. Rev. B* **93**, 235127 (2016), arXiv:1603.06004.

[625] R.-J. Slager, V. Juričić and B. Roy, *Phys. Rev. B* **96**, 201401 (2017), arXiv:1703.09706.

[626] J. H. Wilson *et al.*, *Phys. Rev. B* **97**, 235108 (2018), arXiv:1801.05438.

[627] R. Nandkishore, D. A. Huse and S. L. Sondhi, *Phys. Rev. B* **89**, 245110 (2014), arXiv:1405.2336.

[628] L. Van Hove, *Phys. Rev.* **89**, 1189 (1953).

[629] B. Skinner, *Phys. Rev. B* **90**, 060202 (2014), arXiv:1406.2318.

[630] Y. I. Rodionov and S. V. Syzranov, *Phys. Rev. B* **91**, 195107 (2015), arXiv:1503.02078.

[631] M. Breitkreiz and P. W. Brouwer, *Phys. Rev. Lett.* **123**, 066804 (2019), arXiv:1903.12537.

[632] A. C. Potter, I. Kimchi and A. Vishwanath, *Nat. Commun.* **5**, 5161 (2014), arXiv:1402.6342.

[633] D. Bulmash and X.-L. Qi, *Phys. Rev. B* **93**, 081103 (2016), arXiv:1512.03437.

[634] Y. Zhang *et al.*, *Sci. Rep.* **6**, 23741 (2016), arXiv:1512.06133.

[635] P. E. Ashby and J. P. Carbotte, *Eur. Phys. J. B* **87**, 92 (2014), arXiv:1310.2223.

[636] P. J. W. Moll *et al.*, *Nature* **535**, 266 (2016), arXiv:1505.02817.

[637] Y. Ominato and M. Koshino, *Phys. Rev. B* **93**, 245304 (2016), arXiv:1512.08223.

[638] H. L. Stormer, D. C. Tsui and A. C. Gossard, *Rev. Mod. Phys.* **71**, S298 (1999).

[639] C. M. Wang, H.-P. Sun, H.-Z. Lu and X. C. Xie, *Phys. Rev. Lett.* **119**, 136806 (2017), arXiv:1705.07403.

[640] A. Igarashi and M. Koshino, *Phys. Rev. B* **95**, 195306 (2017), arXiv:1703.03532.

[641] Y. Baum, E. Berg, S. A. Parameswaran and A. Stern, *Phys. Rev. X* **5**, 041046 (2015), arXiv:1508.03047.

[642] S. Treiman, R. Jackiw and D. Gross, *Lectures on Current Algebra and Its Applications.* Princeton University Press (2015).

[643] K. Fujikawa, *Phys. Rev. Lett.* **42**, 1195 (1979).

[644] A. A. Abrikosov, *Phys. Rev. B* **58**, 2788 (1998).

[645] R. D. dos Reis *et al.*, *New J. Phys.* **18**, 085006 (2016), arXiv:1606.03389.

[646] P. Goswami, J. H. Pixley and S. Das Sarma, *Phys. Rev. B* **92**, 075205 (2015), arXiv:1503.02069.

[647] A. V. Andreev and B. Z. Spivak, *Phys. Rev. Lett.* **120**, 026601 (2018), arXiv:1705.01638.

[648] S. A. Parameswaran *et al.*, *Phys. Rev. X* **4**, 031035 (2014), arXiv:1306.1234.

[649] C. Zhang *et al.*, *Nat. Commun.* **8**, 13741 (2017), arXiv:1504.07698.

[650] J. C. de Boer *et al.*, *Phys. Rev. B* **99**, 085124 (2019), arXiv:1901.11469.

[651] M. R. Douglas and N. A. Nekrasov, *Rev. Mod. Phys.* **73**, 977 (2001), arXiv:hep-th/0106048.

[652] E. V. Gorbar and V. A. Miransky, *Phys. Rev. D* **70**, 105007 (2004), arXiv:hep-th/0407219.

[653] E. Gorbar, M. Hashimoto and V. Miransky, *Phys. Lett. B* **611**, 207 (2005a), arXiv:hep-th/0501135.

[654] E. V. Gorbar, S. Homayouni and V. A. Miransky, *Phys. Rev. D* **72**, 065014 (2005b), arXiv:hep-th/0503028.

[655] Y. Ominato and M. Koshino, *Phys. Rev. B* **89**, 054202 (2014a), arXiv:1309.4206.

[656] Y. Ominato and M. Koshino, *Phys. Rev. B* **89**, 054202 (2014b), arXiv:1309.4206.

[657] A. A. Burkov, *Phys. Rev. Lett.* **113**, 187202 (2014), arXiv:1406.3033.

[658] R. R. Biswas and S. Ryu, *Phys. Rev. B* **89**, 014205 (2014), arXiv:1309.3278.

[659] J. Klier, I. V. Gornyi and A. D. Mirlin, *Phys. Rev. B* **92**, 205113 (2015), arXiv:1507.03481.

[660] N. Ramakrishnan, M. Milletari and S. Adam, *Phys. Rev. B* **92**, 245120 (2015), arXiv:1501.03815.

[661] H.-Z. Lu and S.-Q. Shen, *Phys. Rev. B* **92**, 035203 (2015), arXiv: 1411.2686.

[662] A. A. Burkov, M. D. Hook and L. Balents, *Phys. Rev. B* **84**, 235126 (2011), arXiv:1110.1089.

[663] A. G. Grushin, *Phys. Rev. D* **86**, 045001 (2012), arXiv:1205.3722.

[664] A. A. Zyuzin and A. A. Burkov, *Phys. Rev. B* **86**, 115133 (2012), arXiv:1206.1868.

[665] P. Goswami and S. Tewari, *Phys. Rev. B* **88**, 245107 (2013), arXiv:1210. 6352.

[666] B. I. Abelev *et al.*, *Phys. Rev. Lett.* **103**, 251601 (2009), arXiv:0909.1739.

[667] B. I. Abelev *et al.*, *Phys. Rev. C* **81**, 054908 (2010), arXiv:0909.1717.

[668] H. Ke *et al.* (STAR Collaboration), *J. Phys. Conf. Ser.* **389**, 012035 (2012), arXiv:1211.3216.

[669] G. Wang *et al.* (STAR Collaboration), *Nucl. Phys. A* **904–905**, 248c (2013), arXiv:1210.5498.

[670] L. Adamczyk *et al.*, *Phys. Rev. C* **89**, 044908 (2014), arXiv:1303.0901.

[671] L. Adamczyk *et al.*, *Phys. Rev. Lett.* **114**, 252302 (2015), arXiv: 1504.02175.

[672] I. Selyuzhenkov *et al.* (ALICE Collaboration), *Prog. Theor. Phys. Suppl.* **193**, 153 (2012), arXiv:1111.1875.

[673] S. A. Voloshin, *Phys. Rev. C* **70**, 057901 (2004), arXiv:hep-ph/0406311.

[674] D. E. Kharzeev, *Ann. Phys. (N. Y.)* **325**, 205 (2010), arXiv:0911.3715.

[675] K. Fukushima, D. E. Kharzeev and H. J. Warringa, *Nucl. Phys. A* **836**, 311 (2010), arXiv:0912.2961.

[676] V. Khachatryan *et al.*, *Phys. Rev. Lett.* **118**, 122301 (2017), arXiv:1610.00263.

[677] A. M. Sirunyan *et al.*, *Phys. Rev. C* **97**, 044912 (2018), arXiv:1708.01602.

[678] G. Başar, D. E. Kharzeev and H.-U. Yee, *Phys. Rev. B* **89**, 035142 (2014), arXiv:1305.6338.

[679] J. Ma and D. A. Pesin, *Phys. Rev. B* **92**, 235205 (2015), arXiv:1510. 01304.

[680] S. Zhong, J. E. Moore and I. Souza, *Phys. Rev. Lett.* **116**, 077201 (2016), arXiv:1510.02167.

[681] K. Landsteiner, *Acta Phys. Polon. B* **47**, 2617 (2016), arXiv:1610.04413.

[682] M. A. Zubkov, *Phys. Rev. D* **93**, 105036 (2016), arXiv:1605.08724.

[683] E. V. Gorbar, V. A. Miransky, I. A. Shovkovy and P. O. Sukhachov, *Phys. Rev. B* **92**, 245440 (2015), arXiv:1509.06769.

[684] M.-C. Chang and M.-F. Yang, *Phys. Rev. B* **91**, 115203 (2015), arXiv:1508.05187.

[685] B. Lenoir *et al.*, *J. Phys. Chem. Solids* **57**, 89 (1996).

[686] A. R. Akhmerov and C. W. J. Beenakker, *Phys. Rev. Lett.* **98**, 157003 (2007), arXiv:cond-mat/0612698.

[687] P. Recher *et al.*, *Phys. Rev. B* **76**, 235404 (2007), arXiv:0706.2103.

[688] P. N. Bogolioubov, *Ann. Inst. Henri Poincare* **8**, 163 (1967).

[689] A. W. Thomas, Chiral symmetry and the BAG model: a new starting point for nuclear physics, in J. Negele (ed.), *Advances in Nuclear Physics, Vol. 13*, Chap. 1. Springer US, 1 (1984).

[690] Y. A. Sitenko, *Phys. Rev. D* **91**, 085012 (2015), arXiv:1411.2460.

[691] Y. A. Sitenko, *Euro. Phys. Lett.* **114**, 61001 (2016), arXiv:1603.09268.

[692] Y. A. Sitenko, *Phys. Rev. D* **94**, 085014 (2016), arXiv:1606.08241.

[693] A. Akhiezer and V. Berestetskii, *Elements of Quantum Electrodynamics*. Oldbourne Press (1962).

[694] V. G. Bagrov and D. Gitman, *The Dirac Equation and its Solutions*. De Gruyter (2014).

[695] H. Bateman and A. Erdélyi, *Higher Transcendental Functions*, Vol. 2. McGraw-Hill (1953).

[696] P. H. Ginsparg and K. G. Wilson, *Phys. Rev. D* **25**, 2649 (1982).

[697] D. B. Kaplan, *Phys. Lett. B* **288**, 342 (1992).

[698] A. A. Burkov, *Phys. Rev. B* **96**, 041110 (2017), arXiv:1704.05467.

[699] J.-P. Jan, *Galvanomagnetic and Thermomagnetic Effects in Metals*, Vol. 1. Academic Press, (1957).

[700] H. X. Tang, R. K. Kawakami, D. D. Awschalom and M. L. Roukes, *Phys. Rev. Lett.* **90**, 107201 (2003), arXiv:cond-mat/0210118.

[701] S. Nandy, G. Sharma, A. Taraphder and S. Tewari, *Phys. Rev. Lett.* **119**, 176804 (2017), arXiv:1705.09308.

[702] H.-W. Wang, B. Fu and S.-Q. Shen, *Phys. Rev. B* **98**, 081202 (2018), arXiv:1804.00246.

[703] V. A. Zyuzin, *Phys. Rev. B* **98**, 165205 (2018), arXiv:1803.00723.

[704] S.-K. Yip, arXiv:1508.01010 (2015).

[705] V. A. Zyuzin, *Phys. Rev. B* **95**, 245128 (2017), arXiv:1608.01286.

[706] N. Kumar, S. N. Guin, C. Felser and C. Shekhar, *Phys. Rev. B* **98**, 041103 (2018), arXiv:1711.04133.

[707] S. Liang *et al.*, *Phys. Rev. X* **8**, 031002 (2018), arXiv:1802.01544.

[708] H. Li *et al.*, *Phys. Rev. B* **97**, 201110 (2018), arXiv:1711.03671.

[709] M. Wu *et al.*, *Phys. Rev. B* **98**, 161110 (2018), arXiv:1710.01855.

[710] P. Li *et al.*, *Phys. Rev. B* **98**, 121108 (2018), arXiv:1803.01213.

[711] A. A. Burkov, *Phys. Rev. Lett.* **113**, 247203 (2014), arXiv:1409.0013.

[712] G. Sundaram and Q. Niu, *Phys. Rev. B* **59**, 14915 (1999), arXiv:cond-mat/9908003.

[713] D. T. Son and N. Yamamoto, *Phys. Rev. Lett.* **109**, 181602 (2012), arXiv:1203.2697.

[714] D. T. Son and N. Yamamoto, *Phys. Rev. D* **87**, 085016 (2013), arXiv:1210.8158.

[715] D. T. Son and B. Z. Spivak, *Phys. Rev. B* **88**, 104412 (2013), arXiv:1206.1627.

[716] C. Duval *et al.*, *Mod. Phys. Lett. B* **20**, 373 (2006), arXiv:cond-mat/0506051.

[717] J.-W. Chen, S. Pu, Q. Wang and X.-N. Wang, *Phys. Rev. Lett.* **110**, 262301 (2013), arXiv:1210.8312.

[718] Y. Hidaka, S. Pu and D.-L. Yang, *Phys. Rev. D* **95**, 091901 (2017), arXiv:1612.04630.

[719] J.-h. Gao, S. Pu and Q. Wang, *Phys. Rev. D* **96**, 016002 (2017), arXiv:1704.00244.

[720] A. Sekine, D. Culcer and A. H. MacDonald, *Phys. Rev. B* **96**, 235134 (2017), arXiv:1706.01200.

[721] E. V. Gorbar, V. A. Miransky, I. A. Shovkovy and P. O. Sukhachov, *J. High Energy Phys.* **2017**, 103 (2017), arXiv:1707.01105.

[722] S. Carignano, C. Manuel and J. M. Torres-Rincon, *Phys. Rev. D* **98**, 076005 (2018), arXiv:1806.01684.

[723] A. Huang *et al.*, *Phys. Rev. D* **98**, 036010 (2018), arXiv:1801.03640.

[724] E. Wigner, *Phys. Rev.* **40**, 749 (1932).

[725] H.-T. Elze, M. Gyulassy and D. Vasak, *Nucl. Phys. B* **276**, 706 (1986).

[726] D. Vasak, M. Gyulassy and H.-T. Elze, *Ann. Phys. (N. Y.)* **173**, 462 (1987).

[727] H.-T. Elze and U. Heinz, *Phys. Rep.* **183**, 81 (1989).

[728] G. R. Shin, I. Bialynicki-Birula and J. Rafelski, *Phys. Rev. A* **46**, 645 (1992).

[729] C. K. Zachos, D. B. Fairlie and T. L. Curtright, *Quantum Mechanics In Phase Space: An Overview With Selected Papers*. World Scientific (2005).

[730] A. Polkovnikov, *Ann. Phys. (N. Y.)* **325**, 1790 (2010), arXiv:0905.3384.

[731] J.-H. Gao *et al.*, *Phys. Rev. Lett.* **109**, 232301 (2012), arXiv:1203.0725.

[732] X.-l. Sheng, D. H. Rischke, D. Vasak and Q. Wang, *Eur. Phys. J. A* **54**, 21 (2018), arXiv:1707.01388.

[733] G. 't Hooft, *Nucl. Phys. B* **190**, 455 (1981).

[734] G. Panati, H. Spohn and S. Teufel, *Commun. Math. Phys.* **242**, 547 (2003), arXiv:math-ph/0212041.

[735] Y. Gao, S. A. Yang and Q. Niu, *Phys. Rev. Lett.* **112**, 166601 (2014), arXiv:1402.2538.

[736] Y. Gao, S. A. Yang and Q. Niu, *Phys. Rev. B* **91**, 214405 (2015), arXiv:1411.0324.

[737] J.-Y. Chen *et al.*, *Phys. Rev. Lett.* **113**, 182302 (2014), arXiv:1404.5963.

[738] J.-Y. Chen, D. T. Son and M. A. Stephanov, *Phys. Rev. Lett.* **115**, 021601 (2015), arXiv:1502.06966.

[739] E. M. Lifshitz and L. P. Pitaevskii, *Physical Kinetics*. Butterworth-Heinemann (2012).

[740] K. Huang, *Statistical Mechanics*. Wiley (1988).

[741] A. A. Vlasov, *J. Exp. Theor. Phys.* **8**, 291 (1938).

[742] A. A. Vlasov, *Sov. Phys. Usp.* **10**, 721 (1968).

[743] M. Stephanov, H.-U. Yee and Y. Yin, *Phys. Rev. D* **91**, 125014 (2015), arXiv:1501.00222.

[744] A. A. Zyuzin, M. Silaev and V. A. Zyuzin, *Phys. Rev. B* **98**, 205149 (2018), arXiv:1807.01728.

[745] K. Landsteiner, *Phys. Rev. B* **89**, 075124 (2014), arXiv:1306.4932.

[746] W. A. Bardeen, *Phys. Rev.* **184**, 1848 (1969).

[747] W. A. Bardeen and B. Zumino, *Nucl. Phys. B* **244**, 421 (1984).

[748] F. Wilczek, *Phys. Rev. Lett.* **58**, 1799 (1987).

[749] S. M. Carroll, G. B. Field and R. Jackiw, *Phys. Rev. D* **41**, 1231 (1990).

[750] D. Colladay and V. A. Kostelecký, *Phys. Rev. D* **55**, 6760 (1997), arXiv:hep-ph/9703464.

[751] D. Colladay and V. A. Kostelecký, *Phys. Rev. D* **58**, 116002 (1998), arXiv:hep-ph/9809521.

[752] R. Jackiw and V. A. Kostelecký, *Phys. Rev. Lett.* **82**, 3572 (1999), arXiv: hep-ph/9901358.

[753] G. E. Volovik, *J. Exper. Theoret. Phys. Lett.* **70**, 1 (1999), arXiv:hep-th/9905008.

[754] X.-L. Qi, T. L. Hughes and S.-C. Zhang, *Phys. Rev. B* **78**, 195424 (2008), arXiv:0802.3537.

[755] H. Rose, *Geometrical Charged-Particle Optics.* Springer-Verlag (2012).

[756] H. Busch, *Ann. Phys. (Berl.)* **386**, 974 (1926).

[757] M. Knoll and E. Ruska, *Ann. Phys. (Berl.)* **404**, 607 (1932).

[758] M. Knoll and E. Ruska, *Ann. Phys. (Berl.)* **404**, 641 (1932).

[759] M. Knoll and E. Ruska, *Z. Phys.* **78**, 318 (1932).

[760] E. Ruska, *J. Ultrastruct. Mol. Struct. Res.* **95**, 3 (1986).

[761] L. Landau, *The Classical Theory of Fields.* Butterworth-Heinemann (1987).

[762] H. Jeffreys, *Proc. London Math. Soc.* **23**, 428 (1924).

[763] R. Dingle, *Asymptotic Expansions: Their Derivation and Interpretation.* Academic Press (1973).

[764] L. Molenkamp and M. de Jong, *Solid State Electron.* **37**, 551 (1994).

[765] M. J. M. de Jong and L. W. Molenkamp, *Phys. Rev. B* **51**, 13389 (1995).

[766] S. Das Sarma and E. H. Hwang, *Phys. Rev. B* **91**, 195104 (2015), arXiv:1501.05642.

[767] K. B. Efetov, *Adv. Phys.* **32**, 53 (1983).

[768] S. F. Edwards and P. W. Anderson, *J. Phys. F* **5**, 965 (1975).

[769] C. J. Tabert, J. P. Carbotte and E. J. Nicol, *Phys. Rev. B* **93**, 085426 (2016), arXiv:1603.00866.

[770] B. Z. Spivak and A. V. Andreev, *Phys. Rev. B* **93**, 085107 (2016), arXiv:1510.01817.

[771] K.-S. Kim, *Phys. Rev. B* **90**, 121108 (2014), arXiv:1409.0082.

[772] R. Lundgren, P. Laurell and G. A. Fiete, *Phys. Rev. B* **90**, 165115 (2014), arXiv:1407.1435.

[773] G. Sharma, P. Goswami and S. Tewari, *Phys. Rev. B* **93**, 035116 (2016), arXiv:1507.05606.

[774] S. Das Sarma, E. H. Hwang and H. Min, *Phys. Rev. B* **91**, 035201 (2015), arXiv:1408.0518.

[775] D. A. Pesin, E. G. Mishchenko and A. Levchenko, *Phys. Rev. B* **92**, 174202 (2015), arXiv:1507.05349.

[776] S. Doniach and E. H. Sondheimer, *Green's Functions For Solid State Physicists.* World Scientific (1998).

[777] A. V. Balatsky, I. Vekhter and J.-X. Zhu, *Rev. Mod. Phys.* **78**, 373 (2006), arXiv:cond-mat/0411318.

[778] T. O. Wehling, M. I. Katsnelson and A. I. Lichtenstein, *Chemical Phys. Lett.* **476**, 125 (2009).

[779] Z. Huang, T. Das, A. V. Balatsky and D. P. Arovas, *Phys. Rev. B* **87**, 155123 (2013), arXiv:1210.6121.

[780] B. Shklovskii and A. Efros, *Electronic Properties of Doped Semiconductors.* Springer-Verlag (1984).

[781] P. Goswami and S. Chakravarty, *Phys. Rev. Lett.* **107**, 196803 (2011), arXiv:1101.2210.

[782] B. Sbierski, G. Pohl, E. J. Bergholtz and P. W. Brouwer, *Phys. Rev. Lett.* **113**, 026602 (2014), arXiv:1402.6653.

[783] S. V. Syzranov, L. Radzihovsky and V. Gurarie, *Phys. Rev. Lett.* **114**, 166601 (2015), arXiv:1402.3737.

[784] I. Lifshitz, *Adv. Phys.* **13**, 483 (1964).

[785] J. H. Pixley, P. Goswami and S. Das Sarma, *Phys. Rev. Lett.* **115**, 076601 (2015), arXiv:1502.07778.

[786] J. H. Pixley, D. A. Huse and S. Das Sarma, *Phys. Rev. X* **6**, 021042 (2016a), arXiv:1602.02742.

[787] J. H. Pixley, D. A. Huse and S. Das Sarma, *Phys. Rev. B* **94**, 121107 (2016b), arXiv:1606.08860.

[788] M. Buchhold, S. Diehl and A. Altland, *Phys. Rev. Lett.* **121**, 215301 (2018a), arXiv:1805.00018.

[789] M. Buchhold, S. Diehl and A. Altland, *Phys. Rev. B* **98**, 205134 (2018b), arXiv:1809.04615.

[790] I. Herbut, *A Modern Approach to Critical Phenomena.* Cambridge University Press (2010).

[791] S. Sachdev, *Quantum Phase Transitions.* Cambridge University Press (2011).

[792] H. Isobe, *Theoretical Study on Correlation Effects in Topological Matter.* Springer Singapore (2017).

[793] V. Zlatić and R. Monnier, *Modern Theory of Thermoelectricity.* Oxford University Press (2014).

[794] H. Goldsmid, *Introduction to Thermoelectricity.* Springer-Verlag (2010).

[795] E. V. Gorbar, V. A. Miransky, I. A. Shovkovy and P. O. Sukhachov, *Phys. Rev. B* **96**, 155138 (2017), arXiv:1708.04248.

[796] G. Sharma, C. Moore, S. Saha and S. Tewari, *Phys. Rev. B* **96**, 195119 (2017), arXiv:1605.00299.

[797] T. M. McCormick, S. J. Watzman, J. P. Heremans and N. Trivedi, *Phys. Rev. B* **97**, 195152 (2018), arXiv:1703.04606.

[798] B. Skinner and L. Fu, *Sci. Adv.* **4** (2018), arXiv:1706.06117.

[799] M. N. Chernodub, A. Cortijo and M. A. H. Vozmediano, *Phys. Rev. Lett.* **120**, 206601 (2018), arXiv:1712.05386.

[800] E. Muñoz and R. Soto-Garrido, *J. Appl. Phys.* **125**, 082507 (2019), arXiv:1809.10291.

[801]	V. Arjona, M. N. Chernodub and M. A. H. Vozmediano, *Phys. Rev. B* **99**, 235123 (2019), arXiv:1902.02358.

[802]	J. Rammer, *Quantum Transport Theory*. CRC Press (2019).

[803]	J. M. Luttinger, *Phys. Rev.* **135**, A1505 (1964).

[804]	N. R. Cooper, B. I. Halperin and I. M. Ruzin, *Phys. Rev. B* **55**, 2344 (1997), arXiv:cond-mat/9607001.

[805]	D. Xiao, Y. Yao, Z. Fang and Q. Niu, *Phys. Rev. Lett.* **97**, 026603 (2006), arXiv:cond-mat/0604561.

[806]	T. Thonhauser, D. Ceresoli, D. Vanderbilt and R. Resta, *Phys. Rev. Lett.* **95**, 137205 (2005), arXiv:cond-mat/0505518.

[807]	T. Qin, Q. Niu and J. Shi, *Phys. Rev. Lett.* **107**, 236601 (2011), arXiv:1108.3879.

[808]	D. L. Bergman and V. Oganesyan, *Phys. Rev. Lett.* **104**, 066601 (2010), arXiv:0910.2286.

[809]	R. R. Biswas and S. Ryu, *Phys. Rev. B* **89**, 014205 (2014), arXiv:1309.3278.

[810]	A. Lucas, R. A. Davison and S. Sachdev, *Proc. Natl. Acad. Sci. USA* **113**, 9463 (2016), arXiv:1604.08598.

[811]	K. Landsteiner, E. Megías and F. Pena-Benitez, *Phys. Rev. Lett.* **107**, 021601 (2011), arXiv:1103.5006.

[812]	A. Lucas, R. A. Davison and S. Sachdev, *Proc. Natl. Acad. Sci. USA* **113**, 9463 (2016), arXiv:1604.08598.

[813]	J. Gooth *et al.*, *Nature* **547**, 324 (2017), arXiv:1703.10682.

[814]	J. Ziman, *Principles of the Theory of Solids*. Cambridge University Press (1979).

[815]	V. Kozii, B. Skinner and L. Fu, *Phys. Rev. B* **99**, 155123 (2019), arXiv:1902.10123.

[816]	L. Smrčka and P. Středa, *J. Phys. C Solid State Phys.* **10**, 2153 (1977).

[817]	Z. Jia *et al.*, *Nat. Commun.* **7**, 13013 (2016).

[818]	T. Zhou *et al.*, *Inorg. Chem. Front.* **3**, 1637 (2016), arXiv:1611.00132.

[819]	C. Zhang *et al.*, *Chin. Phys. B* **25**, 017202 (2016).

[820]	T. Liang *et al.*, *Phys. Rev. Lett.* **118**, 136601 (2017), arXiv:1610.02459.

[821]	S. J. Watzman *et al.*, *Phys. Rev. B* **97**, 161404 (2018), arXiv:1703.04700.

[822]	U. Stockert *et al.*, *J. Phys. Condens. Matter* **29**, 325701 (2017), arXiv:1704.02241.

[823]	S. J. Watzman *et al.*, *Phys. Rev. B* **97**, 161404 (2018), arXiv:1703.04700.

[824]	F. Caglieris *et al.*, *Phys. Rev. B* **98**, 201107 (2018), arXiv:1805.09286.

[825]	J. P. Heremans, C. M. Thrush and D. T. Morelli, *Phys. Rev. B* **65**, 035209 (2001).

[826]	M. Hirschberger *et al.*, *Nat. Mater.* **15**, 035209 (2016), arXiv:1602.07219.

[827]	W. Zhang *et al.*, *Nat. Commun.* **11**, 1046 (2020), arXiv:1904.02157.

[828]	F. Han *et al.*, arXiv:1904.03179 (2019).

[829]	H. Wang *et al.*, *Sci. Bull.* **63**, 411 (2018), arXiv:1802.07868.

[830]	H. Beyer *et al.*, *Physica E: Low Dimens. Syst. Nanostruct.* **13**, 965 (2002).

[831]	T. Fu *et al.*, *J. Materiomics* **2**, 141 (2016).

[832]	T. Liang *et al.*, *Nat. Commun.* **4**, 2696 (2013), arXiv:1307.4022.

[833] P. E. C. Ashby and J. P. Carbotte, *Phys. Rev. B* **87**, 245131 (2013), arXiv:1305.0275.

[834] B. Rosenstein and M. Lewkowicz, *Phys. Rev. B* **88**, 045108 (2013), arXiv:1304.7506.

[835] P. E. C. Ashby and J. P. Carbotte, *Phys. Rev. B* **89**, 245121 (2014), arXiv:1405.7034.

[836] C. J. Tabert and J. P. Carbotte, *Phys. Rev. B* **93**, 085442 (2016), arXiv:1603.03722.

[837] B. Roy, V. Juričić and S. Das Sarma, *Sci. Rep.* **6**, 32446 (2016), arXiv:1603.00017.

[838] T. Morimoto, S. Zhong, J. Orenstein and J. E. Moore, *Phys. Rev. B* **94**, 245121 (2016), arXiv:1609.05932.

[839] Y. Sun and A.-M. Wang, *Phys. Rev. B* **96**, 085147 (2017), arXiv:1705.02695.

[840] S. Ahn, E. J. Mele and H. Min, *Phys. Rev. B* **95**, 161112 (2017), arXiv:1609.08566.

[841] A. A. Burkov, *Phys. Rev. B* **98**, 165123 (2018), arXiv:1808.05960.

[842] A. V. Pronin and M. Dressel, arXiv:2003.10361 (2020).

[843] P. Hosur and X.-L. Qi, *Phys. Rev. B* **91**, 081106 (2015), arXiv:1401.2762.

[844] A. Cortijo, *Phys. Rev. B* **94**, 235123 (2016), arXiv:1610.06177.

[845] M. Kargarian, M. Randeria and N. Trivedi, *Sci. Rep.* **5**, 12683 (2015), arXiv:1503.00012.

[846] R. Y. Chen *et al.*, *Phys. Rev. B* **92**, 075107 (2015a), arXiv:1505.00307.

[847] R. Y. Chen *et al.*, *Phys. Rev. Lett.* **115**, 176404 (2015b), arXiv:1506.08676.

[848] D. Neubauer *et al.*, *Phys. Rev. B* **93**, 121202 (2016), arXiv:1601.03299.

[849] X. Yuan *et al.*, *NPG Asia Mater.* **8**, e325 (2016), arXiv:1510.00907.

[850] G. S. Jenkins *et al.*, *Phys. Rev. B* **94**, 085121 (2016), arXiv:1605.02145.

[851] W. Lu, J. Ling, F. Xiu and D. Sun, *Phys. Rev. B* **98**, 104310 (2018), arXiv:1807.03049.

[852] I. Crassee *et al.*, *Phys. Rev. B* **97**, 125204 (2018), arXiv:1712.03147.

[853] K. Ueda *et al.*, *Phys. Rev. Lett.* **109**, 136402 (2012).

[854] A. B. Sushkov *et al.*, *Phys. Rev. B* **92**, 241108 (2015), arXiv:1507.01038.

[855] L. Wu *et al.*, *Nat. Phys.* **13**, 350 (2016), arXiv:1609.04894.

[856] B. Xu *et al.*, *Phys. Rev. B* **93**, 121110 (2016), arXiv:1510.00470.

[857] S.-i. Kimura *et al.*, *Phys. Rev. B* **96**, 075119 (2017), arXiv:1705.08774.

[858] D. Neubauer *et al.*, *Phys. Rev. B* **98**, 195203 (2018), arXiv:1803.09708.

[859] Q. Ma *et al.*, *Nat. Phys.* **13**, 842 (2017), arXiv:1705.00590.

[860] K. Sun *et al.*, *Chin. Physics Lett.* **34**, 117203 (2017), arXiv:1612.07005.

[861] G. B. Osterhoudt *et al.*, *Nat. Mater.* **18**, 471 (2019), arXiv:1712.04951.

[862] S. Patankar *et al.*, *Phys. Rev. B* **98**, 165113 (2018), arXiv:1804.06973.

[863] N. Sirica *et al.*, *Phys. Rev. Lett.* **122**, 197401 (2019), arXiv:1811.02723.

[864] Y. Gao *et al.*, *Nat. Commun.* **11**, 720 (2020), arXiv:1901.00986.

[865] Y. Hochberg *et al.*, *Phys. Rev. D* **97**, 015004 (2018), arXiv:1708.08929.

[866] L. D. Landau, E. M. Lifshits and L. P. Pitaevskii, *Electrodynamics of Continuous Media*. Butterworth-Heinemann (1984).

[867] L. Barron, *Molecular Light Scattering and Optical Activity*. Cambridge University Press (2009).

[868] E. Ivchenko, *Optical Spectroscopy of Semiconductor Nanostructures*. Alpha Science (2005).

[869] P. Goswami, G. Sharma and S. Tewari, *Phys. Rev. B* **92**, 161110 (2015), arXiv:1404.2927.

[870] S. Zhong, J. Orenstein and J. E. Moore, *Phys. Rev. Lett.* **115**, 117403 (2015), arXiv:1503.02715.

[871] J. Ma and D. A. Pesin, *Phys. Rev. Lett.* **118**, 107401 (2017), arXiv: 1611.05491.

[872] Q. Chen *et al.*, *Phys. Rev. B* **99**, 075137 (2019), arXiv:1812.06331.

[873] M. S. Ukhtary, A. R. T. Nugraha and R. Saito, *J. Phys. Soc. Jpn.* **86**, 104703 (2017), arXiv:1703.07092.

[874] O. V. Kotov and Y. E. Lozovik, *Phys. Rev. B* **98**, 195446 (2018), arXiv:1808.00342.

[875] A. Cortijo, *Phys. Rev. B* **94**, 241105 (2016), arXiv:1607.01787.

[876] G. L. J. A. Rikken, J. Fölling and P. Wyder, *Phys. Rev. Lett.* **87**, 236602 (2001).

[877] E. Ivchenko and G. Pikus, *Superlattices and Other Heterostructures: Symmetry and Optical Phenomena*. Springer-Verlag (1997).

[878] C.-K. Chan *et al.*, *Phys. Rev. Lett.* **116**, 026805 (2016), arXiv:1509.05400.

[879] I. Sodemann and L. Fu, *Phys. Rev. Lett.* **115**, 216806 (2015), arXiv:1508.00571.

[880] H. Ishizuka, T. Hayata, M. Ueda and N. Nagaosa, *Phys. Rev. Lett.* **117**, 216601 (2016), arXiv:1607.06537.

[881] K. Taguchi, T. Imaeda, M. Sato and Y. Tanaka, *Phys. Rev. B* **93**, 201202 (2016), arXiv:1601.00379.

[882] C.-K. Chan, N. H. Lindner, G. Refael and P. A. Lee, *Phys. Rev. B* **95**, 041104 (2017), arXiv:1607.07839.

[883] F. de Juan, A. G. Grushin, T. Morimoto and J. E. Moore, *Nat. Commun.* **8**, 15995 (2017), arXiv:1611.05887.

[884] L. E. Golub, E. L. Ivchenko and B. Z. Spivak, *JETP Lett.* **105**, 782 (2017), arXiv:1705.04624.

[885] E. J. König, H.-Y. Xie, D. A. Pesin and A. Levchenko, *Phys. Rev. B* **96**, 075123 (2017), arXiv:1705.03903.

[886] L. E. Golub and E. L. Ivchenko, *Phys. Rev. B* **98**, 075305 (2018), arXiv:1803.02850.

[887] D. E. Kharzeev, Y. Kikuchi, R. Meyer and Y. Tanizaki, *Phys. Rev. B* **98**, 014305 (2018), arXiv:1801.10283.

[888] Z. Li *et al.*, *Phys. Rev. B* **97**, 085201 (2018).

[889] H. Rostami and M. Polini, *Phys. Rev. B* **97**, 195151 (2018), arXiv: 1705.09915.

[890] Y. Zhang *et al.*, *Phys. Rev. B* **97**, 241118 (2018), arXiv:1803.00562.

[891] M. M. Jadidi *et al.*, arXiv:1905.02236 (2019).

[892] E. Martino *et al.*, *Phys. Rev. Lett.* **122**, 217402 (2019), arXiv:1905.00280.

[893] A. Akrap *et al.*, *Phys. Rev. Lett.* **117**, 136401 (2016), arXiv:1604.00038.

[894] P. J. W. Moll *et al.*, *Science* **351**, 1061 (2016), arXiv:1509.05691.

[895] L. Landau and E. Lifshitz, *Fluid Mechanics*. Butterworth-Heinemann (2013).

[896] D. T. Son and P. Surówka, *Phys. Rev. Lett.* **103**, 191601 (2009), arXiv: 0906.5044.

[897] M. V. Isachenkov and A. V. Sadofyev, *Phys. Lett. B* **697**, 404 (2011), arXiv:1010.1550.

[898] Y. Neiman and Y. Oz, *J. High Energy Phys.* **2011**, 23 (2011), arXiv: 1011.5107.

[899] Y. Hidaka, S. Pu and D.-L. Yang, *Phys. Rev. D* **97**, 016004 (2018), arXiv: 1710.00278.

[900] K. Landsteiner, Y. Liu and Y.-W. Sun, *J. High Energy Phys.* **2015**, 127 (2015), arXiv:1410.6399.

[901] E. V. Gorbar, V. A. Miransky, I. A. Shovkovy and P. O. Sukhachov, *Phys. Rev. B* **97**, 121105 (2018), arXiv:1712.01289.

[902] P. O. Sukhachov, E. V. Gorbar, I. A. Shovkovy and V. A. Miransky, *J. Phys.: Condens. Matter* **30**, 275601 (2018), arXiv:1802.10110.

[903] E. V. Gorbar, V. A. Miransky, I. A. Shovkovy and P. O. Sukhachov, *Phys. Rev. B* **97**, 205119 (2018), arXiv:1802.07265.

[904] E. V. Gorbar, V. A. Miransky, I. A. Shovkovy and P. O. Sukhachov, *Phys. Rev. B* **98**, 035121 (2018), arXiv:1804.01550.

[905] J. Gooth *et al.*, *Nat. Commun.* **9**, 4093 (2018), arXiv:1706.05925.

[906] K. Hattori and Y. Yin, *Phys. Rev. Lett.* **117**, 152002 (2016), arXiv: 1607.01513.

[907] K. Rajagopal and A. V. Sadofyev, *J. High Energy Phys.* **2015**, 18 (2015), arXiv:1505.07379.

[908] M. A. Stephanov and H.-U. Yee, *Phys. Rev. Lett.* **116**, 122302 (2016), arXiv:1508.02396.

[909] A. V. Sadofyev and Y. Yin, *Phys. Rev. D* **93**, 125026 (2016), arXiv: 1511.08794.

[910] S. A. Hartnoll, P. K. Kovtun, M. Müller and S. Sachdev, *Phys. Rev. B* **76**, 144502 (2007), arXiv:0706.3215.

[911] P. Kovtun and A. Ritz, *Phys. Rev. D* **78**, 066009 (2008), arXiv:0806.0110.

[912] S. A. Hartnoll, *Nat. Phys.* **11**, 54 (2014), arXiv:1405.3651.

[913] R. A. Davison, B. Goutéraux and S. A. Hartnoll, *J. High Energy Phys.* **2015**, 112 (2015), arXiv:1507.07137.

[914] A. Lucas, *New J. Phys.* **17**, 113007 (2015), arXiv:1506.02662.

[915] J. C. W. Song and M. S. Rudner, *Phys. Rev. B* **96**, 205443 (2017), arXiv:1702.00022.

[916] E. V. Gorbar, V. A. Miransky, I. A. Shovkovy and P. O. Sukhachov, *Phys. Rev. B* **99**, 155120 (2019), arXiv:1901.00006.

[917] E. I. Kiselev and J. Schmalian, *Phys. Rev. B* **99**, 035430 (2019), arXiv:1806.03933.

[918] R. Moessner, N. Morales-Durán, P. Surówka and P. Witkowski, *Phys. Rev. B* **100**, 155115 (2019), arXiv:1903.08037.

[919] N. Kumar *et al.*, *Nat. Commun.* **8**, 1642 (2017), arXiv:1703.04527.

[920]   A. A. Vedenov, *Sov. Phys. Usp.* **7**, 809 (1965).

[921]   L. Tisza, *Nature* **141**, 913 (1938).

[922]   L. D. Landau, *J. Phys. (USSR)* **5**, 71 (1941).

[923]   L. Landau, *Phys. Rev.* **60**, 356 (1941).

[924]   D. E. Kharzeev and H.-U. Yee, *Phys. Rev. D* **83**, 085007 (2011), arXiv:1012.6026.

[925]   Z. Song and X. Dai, *Phys. Rev. X* **9**, 021053 (2019), arXiv:1901.09926.

[926]   W. Freyland *et al.*, *Non-Tetrahedrally Bonded Elements and Binary Compounds I.* Springer-Verlag (1998).

[927]   L. Landau, *J. Phys. (USSR)* **10**, 25 (1946).

[928]   X.-G. Huang and J. Liao, *Phys. Rev. Lett.* **110**, 232302 (2013), arXiv:1303.7192.

[929]   I. A. Shovkovy, D. O. Rybalka and E. V. Gorbar, *PoS* **Confinement 2018**, 029 (2018), arXiv:1811.10635.

[930]   Z. Long *et al.*, *Phys. Rev. Lett.* **120**, 037403 (2018), arXiv:1708.05498.

[931]   O. V. Konstantinov and V. I. Perel', *JETP* **11**, 117 (1960).

[932]   P. Aigrain, in *Proc. Int. Conf. Semiconductor Physics*, (1961), p. 224.

[933]   R. Bowers, C. Legendy and F. Rose, *Phys. Rev. Lett.* **7**, 339 (1961).

[934]   J. A. Ratcliffe and L. R. O. Storey, *Philos. Trans. Royal Soc.* **246**, 113 (1953).

[935]   R. A. Helliwell and M. G. Morgan, *Proc. IRE* **47**, 200 (1959).

[936]   É. A. Kaner and V. G. Skobov, *Sov. Phys. Usp.* **9**, 480 (1967).

[937]   B. W. Maxfield, *Am. J. Phys.* **37**, 241 (1969).

[938]   S. Shinohara, *Adv. Phys.: X* **3**, 1420424 (2018).

[939]   F. M. D. Pellegrino, M. I. Katsnelson and M. Polini, *Phys. Rev. B* **92**, 201407 (2015), arXiv:1507.03140.

[940]   N. Krall and A. Trivelpiece, *Principles of Plasma Physics.* San Francisco Press (1986).

[941]   V. T. Petrashov, *Rep. Prog. Phys.* **47**, 47 (1984).

[942]   R. H. Ritchie, *Phys. Rev.* **106**, 874 (1957).

[943]   E. A. Stern and R. A. Ferrell, *Phys. Rev.* **120**, 130 (1960).

[944]   R. H. Ritchie and R. E. Wilems, *Phys. Rev.* **178**, 372 (1969).

[945]   G. Barton, *Rep. Prog. Phys.* **42**, 963 (1979).

[946]   W. L. Barnes, A. Dereux and T. W. Ebbesen, *Phys. Rev. Lett.* **120**, 037403 (2018).

[947]   J. M. Pitarke, V. M. Silkin, E. V. Chulkov and P. M. Echenique, *Rep. Prog. Phys.* **70**, 1 (2006), arXiv:cond-mat/0611257.

[948]   V. M. Agranovich and D. L. Mills, *Surface Polaritons.* North Holland (1982).

[949]   E. Albuquerque and M. Cottam, *Polaritons in Periodic and Quasiperiodic Structures.* Elsevier Science (2004).

[950]   A. D. Boardman, *Electromagnetic Surface Modes.* Wiley (1982).

[951]   K. Tolpygo, *J. Exp. Theor. Phys.* **20**, 497 (1950).

[952]   L. Novotny and S. J. Stranick, *Annu. Rev. Phys. Chem.* **57**, 303 (2006).

[953]   D. N. Basov, M. M. Fogler and F. J. García de Abajo, *Science* **354** (2016).

[954]   Y. Y. Wang, S. C. Cheng, V. P. Dravid and F. C. Zhang, *Ultramicroscopy* **59**, 109 (1995).

[955]   F. J. García de Abajo, *Rev. Mod. Phys.* **82**, 209 (2010), arXiv:0903.1669.

[956]   E. Ozbay, *Science* **311**, 189 (2006).

[957]   A. Politano, L. Viti and M. S. Vitiello, *APL Mater.* **5**, 035504 (2017), arXiv:1904.00624.

[958]   A. A. Zyuzin and V. A. Zyuzin, *Phys. Rev. B* **92**, 115310 (2015), arXiv:1410.2704.

[959]   J. Hofmann and S. Das Sarma, *Phys. Rev. B* **93**, 241402 (2016), arXiv:1601.07524.

[960]   T. Tamaya *et al.*, *J. Phys. Condens. Matter* **31**, 305001 (2019), arXiv:1811.08608.

[961]   G. M. Andolina, F. M. D. Pellegrino, F. H. L. Koppens and M. Polini, *Phys. Rev. B* **97**, 125431 (2018), arXiv:1706.06200.

[962]   Ž. B. Lošić, *J. Phys.: Condens. Matter* **30**, 365003 (2018).

[963]   K. W. Chiu and J. J. Quinn, *Il Nuovo Cimento B (1971–1996)* **10**, 1 (1972).

[964]   R. F. Wallis, J. J. Brion, E. Burstein and A. Hartstein, *Phys. Rev. B* **9**, 3424 (1974).

[965]   M. S. Kushwaha and P. Halevi, *Phys. Rev. B* **36**, 5960 (1987).

[966]   A. Politano *et al.*, *Adv. Funct. Mater.* **28**, 1800511 (2018).

[967]   G. Chiarello *et al.*, *Phys. Rev. B* **99**, 121401 (2019), arXiv:1811.04639.

[968]   W. Meissner and R. Ochsenfeld, *Naturwissenschaften* **21**, 787 (1933).

[969]   F. London and H. London, *Proc. R. Soc. Lond. A* **149**, 71 (1935).

[970]   V. L. Ginzburg and L. D. Landau, *Zh. Eksp. Teor. Fiz.* **20**, 1064 (1950).

[971]   N. N. Bogoljubov, *Il Nuovo Cimento (1955–1965)* **7**, 794 (1958).

[972]   B. Josephson, *Phys. Lett.* **1**, 251 (1962).

[973]   A. F. Andreev, *JETP* **19**, 1228 (1964).

[974]   A. A. Abrikosov, *J. Phys. Chem. Solids* **2**, 199 (1957).

[975]   R. Holm and W. Meissner, *Z. Phys.* **74**, 715 (1932).

[976]   N. Read and D. Green, *Phys. Rev. B* **61**, 10267 (2000), arXiv:cond-mat/9906453.

[977]   A. Y. Kitaev, *Phys.-Uspekhi* **44**, 131 (2001), arXiv:cond-mat/0010440.

[978]   L. Fu and C. L. Kane, *Phys. Rev. Lett.* **100**, 096407 (2008), arXiv:0707.1692.

[979]   L. Fu and C. L. Kane, *Phys. Rev. B* **79**, 161408 (2009), arXiv:0804.4469.

[980]   C. Nayak *et al.*, *Rev. Mod. Phys.* **80**, 1083 (2008), arXiv:0707.1889.

[981]   H. Suhl, B. T. Matthias and L. R. Walker, *Phys. Rev. Lett.* **3**, 552 (1959).

[982]   T. Meng and L. Balents, *Phys. Rev. B* **86**, 054504 (2012); arXiv:1205.5202; Erratum: Ibid., *Phys. Rev. B* **96**, 019901 (2017).

[983]   G. Y. Cho, J. H. Bardarson, Y.-M. Lu and J. E. Moore, *Phys. Rev. B* **86**, 214514 (2012), arXiv:1209.2235.

[984]   H. Wei, S.-P. Chao and V. Aji, *Phys. Rev. B* **89**, 014506 (2014), arXiv:1305.7233.

[985]   P. Hosur, X. Dai, Z. Fang and X.-L. Qi, *Phys. Rev. B* **90**, 045130 (2014), arXiv:1405.4299.

[986] G. Bednik, A. A. Zyuzin and A. A. Burkov, *Phys. Rev. B* **92**, 035153 (2015), arXiv:1506.05109.

[987] S. Kobayashi and M. Sato, *Phys. Rev. Lett.* **115**, 187001 (2015), arXiv:1504.07408.

[988] T. Hashimoto, S. Kobayashi, Y. Tanaka and M. Sato, *Phys. Rev. B* **94**, 014510 (2016), arXiv:1604.05081.

[989] S. Kobayashi and M. Sato, *Phys. Rev. Lett.* **115**, 187001 (2015), arXiv:1504.07408.

[990] Y. Kim, M. J. Park and M. J. Gilbert, *Phys. Rev. B* **93**, 214511 (2016), arXiv:1604.01040.

[991] Y. Li and F. D. M. Haldane, *Phys. Rev. Lett.* **120**, 067003 (2018), arXiv:1510.01730.

[992] P. Fulde and R. A. Ferrell, *Phys. Rev.* **135**, A550 (1964).

[993] A. Larkin and Y. Ovchinnikov, *Sov. Phys. JETP* **20**, 762 (1965).

[994] A. Bianchi *et al.*, *Phys. Rev. Lett.* **91**, 187004 (2003), arXiv:cond-mat/0304420.

[995] H. A. Radovan *et al.*, *Nature* **425**, 51 (2003), arXiv:cond-mat/0304526.

[996] H. Mayaffre *et al.*, *Nature Phys.* **10**, 928 (2014), arXiv:1409.0786.

[997] L. V. Levitin *et al.*, *Phys. Rev. Lett.* **122**, 085301 (2019), arXiv:1805.02053.

[998] R. Casalbuoni and G. Nardulli, *Rev. Mod. Phys.* **76**, 263 (2004), arXiv:hep-ph/0305069.

[999] P. de Gennes, *Superconductivity of Metals and Alloys*. CRC Press (1999).

[1000] J. Schrieffer, *Theory of Superconductivity*. CRC Press (1971).

[1001] B. Lu, K. Yada, M. Sato and Y. Tanaka, *Phys. Rev. Lett.* **114**, 096804 (2015), arXiv:1406.3804.

[1002] M. Rasolt and Z. Tešanović, *Rev. Mod. Phys.* **64**, 709 (1992).

[1003] B. Rosenstein, B. Y. Shapiro, D. Li and I. Shapiro, *Phys. Rev. B* **96**, 224517 (2017), arXiv:1708.04212.

[1004] T. Matsushita, T. Liu, T. Mizushima and S. Fujimoto, *Phys. Rev. B* **97**, 134519 (2018), arXiv:1801.07401.

[1005] P. O. Sukhachov, E. V. Gorbar, I. A. Shovkovy and V. A. Miransky, *J. Phys.: Condens. Matter* **31**, 055602 (2018), arXiv:1809.00019.

[1006] G. Eilenberger, *Z. Phys. A* **214**, 195 (1968).

[1007] A. Houghton and I. Vekhter, *Phys. Rev. B* **57**, 10831 (1998), arXiv:cond-mat/9712272.

[1008] L. Aggarwal *et al.*, *Nat. Mater.* **15**, 32 (2015), arXiv:1410.2072.

[1009] H. Wang *et al.*, *Nat. Mater.* **15**, 38 (2015), arXiv:1501.00418.

[1010] L. He *et al.*, *npj Quantum Mater.* **1**, 16014 (2016), arXiv:1502.02509.

[1011] S. Zhang *et al.*, *Phys. Rev. B* **91**, 165133 (2015), arXiv:1410.3213.

[1012] Y. Li *et al.*, *npj Quantum Mater.* **2**, 66 (2017), arXiv:1611.02548.

[1013] H. Wang *et al.*, *Sci. Bull.* **62**, 425 (2017), arXiv:1607.00513.

[1014] L. Aggarwal *et al.*, *Nat. Commun.* **8**, 13974 (2017), arXiv:1607.05131.

[1015] Y. Qi *et al.*, *Nat. Commun.* **7**, 11038 (2016), arXiv:1703.02696.

[1016] H. Takahashi *et al.*, *Phys. Rev. B* **95**, 100501 (2017), arXiv:1703.02696.

[1017] Z. Guguchia *et al.*, *Nat. Commun.* **8**, 1082 (2017), arXiv:1704.05185.

[1018] C. Heikes *et al.*, *Phys. Rev. Materials* **2**, 074202 (2018), arXiv:1804.09093.

[1019] B. D. Josephson, *Rev. Mod. Phys.* **46**, 251 (1974).

[1020] H. G. Hugdal, J. Linder and S. H. Jacobsen, *Phys. Rev. B* **95**, 235403 (2017), arXiv:1606.01249.

[1021] M. D. Bachmann *et al.*, *Sci. Adv.* **3** (2017), arXiv:1703.08024.

[1022] H. Leng, C. Paulsen, Y. K. Huang and A. de Visser, *Phys. Rev. B* **96**, 220506 (2017), arXiv:1710.03862.

[1023] H. Leng *et al.*, *J. Phys. Condens. Matter* **32**, 025603 (2019), arXiv:1902.01953.

[1024] Q. Li *et al.*, *Nano Lett.* **18**, 7962 (2018), arXiv:1811.06322.

[1025] L. Zhu *et al.*, *Nano Lett.* **18**, 6585 (2018), arXiv:1903.00639.

[1026] W. Chen *et al.*, *EPL* **103**, 27006 (2013), arXiv:1304.6477.

[1027] S. Uchida, T. Habe and Y. Asano, *J. Phys. Soc. Jpn.* **83**, 064711 (2014), arXiv:1403.7896.

[1028] N. Bovenzi *et al.*, *Phys. Rev. B* **96**, 035437 (2017), arXiv:1704.02838.

[1029] F. I. Fedorov, *Dokl. Akad. Nauk SSSR* **105**, 465 (1955).

[1030] C. Imbert, *Phys. Rev. D* **5**, 787 (1972).

[1031] Q.-D. Jiang *et al.*, *Phys. Rev. Lett.* **115**, 156602 (2015), arXiv:1501.06535.

[1032] Y. Liu, Z.-M. Yu and S. A. Yang, *Phys. Rev. B* **96**, 121101 (2017), arXiv:1706.03688.

[1033] G. E. Blonder, M. Tinkham and T. M. Klapwijk, *Phys. Rev. B* **25**, 4515 (1982).

[1034] S.-B. Zhang, F. Dolcini, D. Breunig and B. Trauzettel, *Phys. Rev. B* **97**, 041116 (2018), arXiv:1711.07882.

[1035] J. Fang *et al.*, *Phys. Rev. B* **97**, 165301 (2018).

[1036] D. Breunig, S.-B. Zhang, M. Stehno and B. Trauzettel, *Phys. Rev. B* **99**, 174501 (2019), arXiv:1903.06229.

[1037] Y. Naidyuk and I. Yanson, *Point-Contact Spectroscopy.* Springer New York (2005).

[1038] P. W. Anderson and J. M. Rowell, *Phys. Rev. Lett.* **10**, 230 (1963).

[1039] W. Yu *et al.*, *Phys. Rev. Lett.* **120**, 177704 (2018), arXiv:1801.04365.

[1040] Y. Nazarov and Y. Blanter, *Quantum Transport: Introduction to Nanoscience.* Cambridge University Press (2009).

[1041] K. Yang and D. F. Agterberg, *Phys. Rev. Lett.* **84**, 4970 (2000), arXiv:cond-mat/9912364.

[1042] C. W. J. Beenakker, *Phys. Rev. Lett.* **67**, 3836 (1991).

[1043] C. W. J. Beenakker, Three "universal" mesoscopic Josephson effects, in H. Fukuyama and T. Ando (eds.), *Transport Phenomena in Mesoscopic Systems.* Springer, Berlin (1992), p. 235.

[1044] P. Brouwer and C. Beenakker, *Chaos Soliton Fractals* **8**, 1249 (1997), arXiv:cond-mat/9611162.

[1045] K. A. Madsen, E. J. Bergholtz and P. W. Brouwer, *Phys. Rev. B* **95**, 064511 (2017), arXiv:1612.03734.

[1046]   S.-B. Zhang, J. Erdmenger and B. Trauzettel, *Phys. Rev. Lett.* **121**, 226604 (2018), arXiv:1806.08111.

[1047]   S. Uddin *et al.*, *Phys. Rev. B* **99**, 045426 (2019), arXiv:1901.01520.

[1048]   S. Datta, P. F. Bagwell and M. P. Anantram, *Phys. Low-Dim. Struct.* **3**, 1 (1996).

[1049]   A. Furusaki and M. Tsukada, *Solid State Commun.* **78**, 299 (1991).

[1050]   I. K. Yanson, V. M. Svistunov and I. M. Dmitrenko, *Sov. Phys. JETP* **21**, 650 (1965).

[1051]   S. Shapiro, *Phys. Rev. Lett.* **11**, 80 (1963).

[1052]   S. Shapiro, A. R. Janus and S. Holly, *Rev. Mod. Phys.* **36**, 223 (1964).

[1053]   U. Khanna, D. K. Mukherjee, A. Kundu and S. Rao, *Phys. Rev. B* **93**, 121409 (2016), arXiv:1509.03166.

[1054]   A. Chen, D. I. Pikulin and M. Franz, *Phys. Rev. B* **95**, 174505 (2017), arXiv:1610.08553.

[1055]   C.-Z. Chen *et al.*, *Phys. Rev. B* **98**, 075430 (2018), arXiv:1802.10389.

[1056]   E. Majorana, *Il Nuovo Cimento (1924-1942)* **14**, 171 (1937).

[1057]   H.-J. Kwon, K. Sengupta and V. Yakovenko, *Eur. Phys. J. B* **37**, 349 (2004), arXiv:cond-mat/0210148.

[1058]   R. M. Lutchyn, J. D. Sau and S. Das Sarma, *Phys. Rev. Lett.* **105**, 077001 (2010), arXiv:1002.4033.

[1059]   Y. Oreg, G. Refael and F. von Oppen, *Phys. Rev. Lett.* **105**, 177002 (2010), arXiv:1003.1145.

[1060]   L. Jiang *et al.*, *Phys. Rev. Lett.* **107**, 236401 (2011), arXiv:1107.4102.

[1061]   D. M. Badiane, M. Houzet and J. S. Meyer, *Phys. Rev. Lett.* **107**, 177002 (2011), arXiv:1108.3870.

[1062]   P. San-Jose, E. Prada and R. Aguado, *Phys. Rev. Lett.* **108**, 257001 (2012), arXiv:1112.5983.

[1063]   L. P. Rokhinson, X. Liu and J. K. Furdyna, *Nat. Phys.* **8**, 795 (2012), arXiv:1204.4212.

[1064]   J. Wiedenmann *et al.*, *Nat. Commun.* **7**, 10303 (2016), arXiv:1503.05591.

[1065]   M. T. Deng *et al.*, *Nano Lett.* **12**, 6414 (2012), arXiv:1204.4130.

[1066]   A. Das *et al.*, *Nat. Phys.* **8**, 887 (2012), arXiv:1205.7073.

[1067]   H. O. H. Churchill *et al.*, *Phys. Rev. B* **87**, 241401 (2013), arXiv:1303.2407.

[1068]   M. T. Deng *et al.*, *Sci. Rep.* **4**, 7261 (2014), arXiv:1406.4435.

[1069]   D. Wang *et al.*, *Science* **362**, 333 (2018), arXiv:1706.06074.

[1070]   A. Chen and M. Franz, *Phys. Rev. B* **93**, 201105 (2016), arXiv:1601.01727.

[1071]   P. J. Mohr, D. B. Newell and B. N. Taylor, *Rev. Mod. Phys.* **88**, 035009 (2016), arXiv:1507.07956.

[1072]   T. Liang *et al.*, *Nat. Mater.* **14**, 250 (2014), arXiv:1404.7794.

[1073]   S.-Y. Xu *et al.*, arXiv:1312.7624 (2013).

[1074]   J. Xiong *et al.*, *Euro. Phys. Lett.* **114**, 27002 (2016), arXiv:1502.06266.

[1075]   M. Hakl *et al.*, *Phys. Rev. B* **97**, 115206 (2018), arXiv:1803.05469.

[1076]   Y. Liu *et al.*, *NPG Asia Mater.* **7**, e221 (2015), arXiv:1412.4380.

[1077]   J.-O. Willerström, *J. Less-Common Met.* **99**, 273 (1984).

[1078]   J. Buckeridge, D. Jevdokimovs, C. R. A. Catlow and A. A. Sokol, *Phys. Rev. B* **93**, 125205 (2016).

[1079]   D. Grassano, O. Pulci, A. Mosca Conte and F. Bechstedt, *Sci. Rep.* **8**, 3534 (1965).

Index

Printed in the USA
CPSIA information can be obtained
at www.ICGtesting.com
LVHW021927061023
760242LV00035B/5

9 789811 207341